Symmetry and Structure

Readable Group Theory for Chemists

Symmetry and Structure

Readable Group Theory for Chemists

Third Edition

Sidney F. A. Kettle

Professorial Fellow in the School of Chemical Sciences and Pharmacy, University of East Anglia

John Wiley & Sons, Inc.

Telephone (+44) 1243 779777

Email (for orders and customer service enquiries): cs-books@wiley.co.uk
Visit our Home Page on www.wileyeurope.com or www.wiley.com

Other Wiley Editorial Offices

John Wiley & Sons Inc., 111 River Street, Hoboken, NJ 07030, USA

Jossey-Bass, 989 Market Street, San Francisco, CA 94103-1741, USA

Wiley-VCH Verlag GmbH, Boschstr. 12, D-69469 Weinheim, Germany

John Wiley & Sons Australia Ltd, 33 Park Road, Milton, Queensland 4064, Australia

John Wiley & Sons (Asia) Pte Ltd, 2 Clementi Loop #02-01, Jin Xing Distripark, Singapore 129809

John Wiley & Sons Canada Ltd, 6045 Freemont Blvd, Mississauga, Ontario, Canada, L5R 4J3

Wiley also publishes its books in a variety of electronic formats. Some content that appears in print may not be available in electronic books.

Anniversary Logo Design: Richard J. Pacifico

Library of Congress Cataloging-in-Publication Data

Kettle, S. F. A. (Sidney Francis Alan)
Symmetry and structure : readable group theory for chemists / Sidney F. A. Kettle. – 3rd ed.
p. cm.
ISBN 978-0-470-06039-1
1. Chemical structure. 2. Symmetry (Physics) 3. Group theory.
4. Chemical bonds. 5. Molecules. I. Title.

QD471.K47516 2007
541′.22015122–dc22 2007014664

British Library Cataloguing in Publication Data

A catalogue record for this book is available from the British Library

ISBN 9780470060391 HB
ISBN 9780470060407 PB

Typeset in 10/12pt Times by Aptara, New Delhi, India

This book is printed on acid-free paper responsibly manufactured from sustainable forestry in which at least two trees are planted for each one used for paper production.

Contents

Preface to Third Edition

Although this third edition of 'Symmetry and Structure' has much in common with previous editions, there are major differences too. Most important is a new emphasis on the fact that irreducible representations characterize particular nodal patterns (or vice versa!). It is possible to draw pictures of these nodal patterns and so to give pictorial illustrations of irreducible representations. This is particularly useful for the simpler groups, where much of the group theory may be done pictorially. To obtain the maximum benefit from this approach, Chapters 2–4 contain a basic but reasonably complete overview of the application of group theory to chemistry (or, more accurately, the water molecule!). The major omission, of course, is that of degeneracy. The nodal pattern approach applies not only to the simple groups. Its use has enabled the inclusion of a chapter on electron spin, double groups and spin-orbit coupling. The inclusion of these has been facilitated by the addition of a chapter which includes the spherical group. Hopefully, the treatment of double groups is both readable and accurate. In general, the mathematical content of the book has been reduced, both in the text and in the Appendices. Although clearly there are limits, I have tried to make each chapter as independent as possible. This has led to some duplication of material – which may be no bad thing. By providing cross-references, the student can obtain a, somewhat, different approach to a difficult point, should the need arise. Above all, I have borne in mind the sub-title of the book, that the content should be readable, and with no loss of accuracy. If at some points the reader finds it fun too, that would be a bonus.

I am particularly grateful to Professors K. Gatterer (Graz) and E. Diana (Turin) for providing material which I have used and also for their comments on the text itself. All deficiencies which remain are, of course, my responsibility.

Sidney F.A. Kettle
Tuttington

1 Theories in conflict

1.1 Introduction

As its title says, this book is concerned with the symmetry and structure of molecules. Of these, the latter – both in the sense of the geometric and of the electronic structure of molecules – has long been of concern to chemists. We shall be interested in both these aspects and will adopt the viewpoint that the geometric structure of a molecule tells us something about its electronic structure. The connection between the two will be provided by the molecular symmetry, or rather its expression in what is called group theory. Ultimately, however, this book is concerned with the chemical consequences of molecular symmetry, the application of group theory to molecules, and these extend far beyond the problems of chemical bonding. Rather, the problem of chemical bonding will be used as a particularly convenient – and important – way of introducing the concepts of symmetry. The concepts revealed in this way can then be extended to other areas of chemistry. In an introductory text such as this there will be no attempt to cover all of the uses of symmetry in chemistry – an objective which it would be difficult to achieve in any text. Rather, the more important aspects will be detailed, but sometimes with more than a hint of the advanced. The aim will be to provide a cover of the basics of the subject sufficient to enable the reader to apply them in other areas. Further, this will be done in a readable, almost entirely non-mathematical manner. The take-home message is that the use of symmetry in chemistry is all about phase patterns: that is, about nodal planes akin to those that distinguish different atomic orbitals. But this is to come; in the present chapter we cover material that, hopefully, is familiar to the reader – explanations of why molecules have the shapes that are observed. The examples covered are chosen to be simple and mostly well known. But the final conclusions are surprising and lead us to query the validity of the simple models that we discuss. Rather than exploring these uncertainties, we will find more value in reversing the argument – and this reversal will be a recurrent theme throughout the book. It has already been mentioned. Start with the observed structure and use this to obtain information about the bonding. But first, the more traditional approach.

1.2 The ammonia molecule

The ammonia molecule provides a convenient starting point for our study and it will be used to see the problem of chemical bonding in a rather unusual perspective, one that leads to the approach indicated above – the attempt to infer molecular bonding *from* molecular

Symmetry and Structure: Readable Group Theory for Chemists Sidney F. A. Kettle

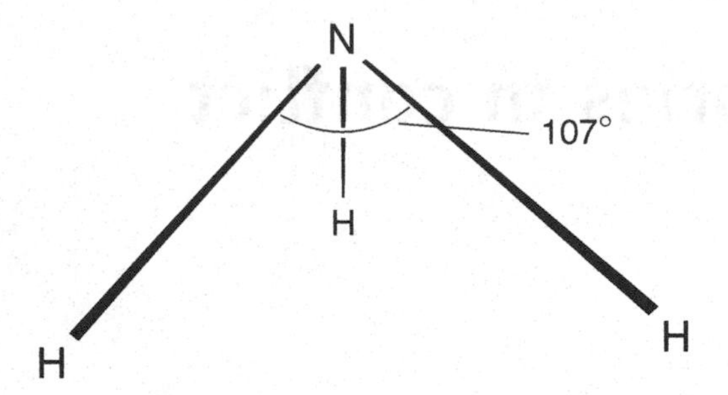

Figure 1.1 The ammonia molecule; the models in the text seek to explain the experimental bond angle

geometry (in contrast to the more common procedure of explaining molecular geometry in terms of chemical bonding). Several approaches to the bonding in the ammonia molecule will first be reviewed, approaches which have been in the chemical literature for many years. The reader may well not be familiar with all of them but he or she should not feel that they have to spend much time trying to master any new ones – our concern is with generalities, not details. However, references are given to enable the reader to explore any of the approaches in more detail, if they so wish.

1.2.1 The atomic orbital model

This model has an historic importance – it is the only description to be found in many pre-1955 texts.[1] Before looking at it, the facts. The ammonia molecule is pyramidal in shape; all three hydrogen atoms are equivalent, the HNH bond angle being 107° (Figure 1.1). Note the restriction that has implicitly been made: we will not attempt to explain bond lengths, only angles. The simplest, and oldest, explanation of the (angular) shape follows from the recognition that the ground state electronic configuration of an isolated nitrogen atom is $(1s)^2 (2s)^2 (2p)^3$, each of the 2p electrons occupying a different p orbital. Each of these 2p electrons may be paired with the electron present in the 1s orbital of a hydrogen atom by placing one hydrogen atom at one end of each 2p orbital so that each nitrogen 2p orbital overlaps with a hydrogen 1s orbital, giving a localized N—H bond. The result is an ammonia molecule which has the correct, pyramidal shape and which has all of three hydrogen atoms equivalently bonded to the nitrogen (Figure 1.2). However, the angle between any pair of 2p orbitals is 90° so that a bond angle of 90° is predicted by this model. Agreement with an experimental value of 107° is obtained by postulating the existence of electrostatic repulsion forces between the hydrogen atoms, each of which, it is assumed, carries a small residual charge. These repulsions cause the H atoms to move further apart – and so the bond angles increase. If, as seems probable, each N—H bond is slightly polar with each hydrogen carrying a small positive charge, this repulsion is nuclear–nuclear in origin. The consequent modification of the original bonding scheme as a result of this distortion of the bond angle from 90° is not usually considered.[2]

[1] See, for example, p. 65 of *Inorganic Chemistry*, by E. de Barry-Barnett and C.L. Wilson, Longman Green, London, 1953.

[2] The reader who wishes to perform this correction should make a note to do it after they have read Chapter 7, when they will be adequately equipped.

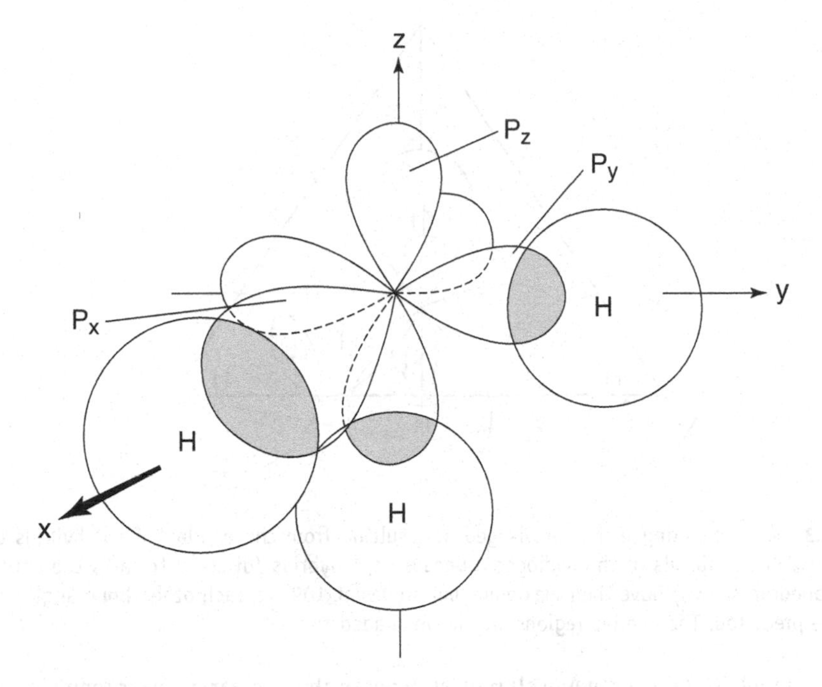

Figure 1.2 N—H bonding in NH_3 envisaged as resulting from the overlap of 2p orbitals of the nitrogen with 1s orbitals of the hydrogens. Because the three nitrogen 2p orbitals have their maximum amplitudes at 90° to each other, bond angles of this value are predicted. The overlap regions are shown shaded

1.2.2 The hybrid orbital model

This is detailed in many post-1955 texts.[3] In this model an alternative description of the bonding in the ammonia molecule is obtained by hybridizing the valence shell orbitals of an isolated nitrogen atom, 2s, $2p_x$, $2p_y$ and $2p_z$, to give four, equivalent, sp^3 hybrid orbitals pointing towards the corners of a regular tetrahedron. Because there are five electrons in the valence shell of the nitrogen atom, three of these hybrid orbitals may be regarded as containing one electron whilst the fourth is occupied by two electrons. As in the previous model, 1s electrons from three hydrogen atoms pair with the unpaired electrons on the nitrogen, now in hybrid orbitals, to give three localized bonds and a pyramidal ammonia molecule (Figure 1.3). Again, the three hydrogen atoms are equivalent but the bond angle is predicted to be 109.5°, the angle between the axes of a pair of sp^3 hybrid orbitals. This value is in closer agreement with experiment than that given by the previous model but again some correction is needed if the experimental value is to be reproduced. This time, the predicted bond angle is too big so a different source has to be found for the correction. It is usually made by invoking the effects of electron–electron repulsion. It is this electron–electron repulsion which forms the basis of a third model for ammonia and so the way that the 'hybrid orbital' model is modified to give agreement with experiment is contained in the description of

[3] See, for example, p. 159 of *Valency and Molecular Structure* by E. Cartmell and G.W.A. Fowles, Butterworth, London, 1956.

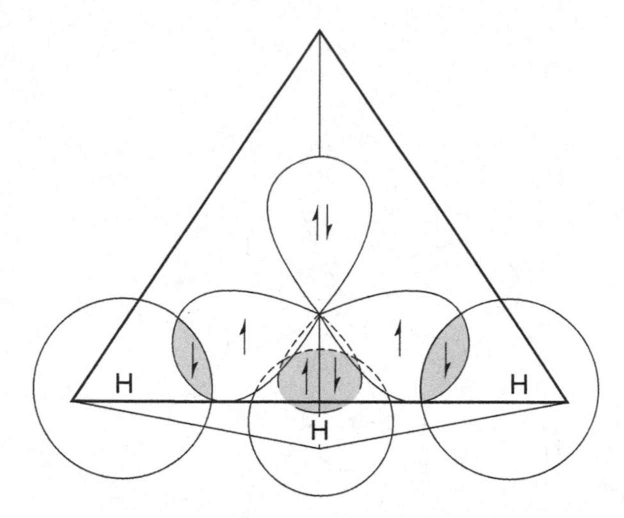

Figure 1.3 N—H bonding in NH_3 envisaged as resulting from the overlap of sp^3 hybrids of the nitrogen with 1s orbitals of the hydrogens. Because sp^3 hybrids (directed towards the corners of the tetrahedron shown) have their maximum amplitudes at 109° to each other, bond angles of this value are predicted. The overlap regions are shown shaded

the next model. The reader may well protest 'what of the nuclear–nuclear repulsion of the previous model and why has it been ignored?'. Indeed this is a good question, but, equally, it is not to be overlooked that in the previous model we ignored electron–electron repulsion. We already detect an element of mutual inconsistency!

1.2.3 The electron-repulsion model

This is the model described in many current texts. The first two models considered above seek to explain the structure of the ammonia molecule in terms of the bonding interactions between the constituent atoms. The atoms adopt that arrangement which makes bonding a maximum. Repulsive forces, of one sort or another, were invoked only to get the predicted angles to agree with experiment.

In contrast, the present and the next model to be discussed explain the structure not in terms of bonding interactions (although these must exist to hold the atoms together) but by electron–electron repulsion. They recognize that electrons repel each other and regard the structure as being determined by the requirement that the inter-electron repulsion energies are minimized. The first of these models is originally due to Sidgwick and Powell, but has been subject to subsequent extensive elaboration and refinement, particularly by Nyholm and Gillespie. Over the years it has been the subject of both debate and further refinement.[4]

In the ammonia molecule there are four electron pairs involving the valence shell of the nitrogen atom. These are the three N—H bonding electron pairs and a non-bonding pair (in the first of the models discussed above these non-bonding electrons were placed in the 2s

[4] For a recent and readable review see 'Models of molecular geometry' by R.J. Gillespie and E.A. Robinson, *Chem. Soc. Rev. (London)* **34** (2005) 396. An idea of the development of the model over a decade can be gained by comparison with an earlier article by the same authors, 'Electron domains and the VSEPR model of molecular geometry', *Angew. Chem. Int. Ed.* **35** (1996) 495.

orbital of the nitrogen; in the second they were placed in an sp^3 hybrid orbital). Because electrons repel each other these four electron pairs would be expected to be as far apart as possible, consistent with still being bound to the nitrogen atom (three of the pairs are also individually bound to a hydrogen atom). It follows that the preferred orientation of these four electron pairs is that in which they point towards the corners of a regular tetrahedron. Remembering that three of the electron pairs are N—H bonding and that their orientation determines the positions of the hydrogen atoms, a HNH bond angle of 109.5° is predicted, the tetrahedral angle, the same as that given by the second model. Figure 1.3 also describes this model; the apical electron pair drawn there is expected to be a bit closer to the nitrogen nucleus than are the other three. It is thought-provoking to recognize that the same bond angle can be predicted either by including bonding interactions (the second model) or by ignoring them (this model)! Equally, by including electron–electron repulsion (this model) or by ignoring it (the second model)!

The refinement of the electron-repulsion model requires the recognition that there are two sorts of electron pairs, the three pairs involved in N—H bonding and a second sort, that which is non-bonding and located on the nitrogen atom. The electron pairs which comprise the N—H bonds are each subject to strong electrostatic attractions from two nuclei, the nitrogen nucleus and that of one of the hydrogen atoms. In contrast, the non-bonding electrons are strongly attracted by one nucleus only, that of nitrogen. It therefore seems reasonable to expect that the centre of gravity of the electron density in the N—H bonds will be located at a distance further away from the nitrogen nucleus than that of the lone pair electron density. The recognition of this difference at once leads to a refinement of the model. The accurate tetrahedral arrangement of four electron pairs resulted from the fact that, at that stage, all the electron pairs were precisely equivalent. In the absence of such equivalence a regular tetrahedral arrangement cannot be expected. It seems reasonable that the repulsive forces occurring between electron densities located in two N—H bonds will be less than the electrostatic repulsions between the non-bonding pair of electrons and a N—H bonding pair, simply because the distance between the centres of gravity of electron density will be greater in the former case. It would be expected that this difference in repulsion will lead the molecule to distort accordingly. The conclusion is that the HNH bond angle will be less than 109.5°. Although no quantitative prediction is possible with this simple model, the qualitative prediction is in accord with experiment – the bond angle is 107°. These same arguments, applied to the 'hybrid orbital' model (Section 1.2.2), also lead to qualitative agreement with experiment.

In more recent years, this, the Valence Shell Electron Repulsion (often abbreviated VSER) model, has become blended with another closely related model – the Ligand Close Packing (often abbreviated LCP) model. The word 'ligand' originates in transition metal chemistry, where it is used as the general name for groups attached to a (central, for simple molecules) metal atom. More loosely, it is used as a name for atoms or groups attached to a central one. The VSER model focuses attention on the central atom and the electron pair arrangement associated with it; almost as a book-keeping exercise, atoms or groups are attached to some of the electron pairs. The LCP model focuses attention on the ligands and explains geometry in terms of repulsions between them (any non-bonding electron pairs on the central atom are treated as if they were ligands). This approach clearly makes sense in those cases where a ligand is really bulky, *tert*-butyl for instance. An increased interest in bulky ligands (they change not only geometries but also reactivities) in recent years has no doubt encouraged interest in the LCP model. One aspect of the LCP model which is perhaps unexpected is that it does not assume a fixed size for any ligand; the size of the ligand depends to some extent on the atom to which it is bonded. Apart from these remarks, we shall not develop the LCP model further; our interest is with the central atom and (ultimately) the symmetry of its surroundings, not repulsion between its components.[5]

1.2.4 The electron-spin-repulsion model

This is a little-used model,[6] although it seems to be undergoing a minor resurgence.[7] It differs from the preceding model principally in its recognition that electrons behave as individuals – and so repel each other as individuals – rather than as pairs. It is therefore more appropriate to consider eight electrons associated with the nitrogen atom, four with spin 'up' and four with spin 'down', than to think of there being four electron pairs (with no explicit mention of spin). In the case of eight individual electrons the preferred orientation (in which the electrons are as well separated spatially as possible) would be expected to be one in which the electrons are located at the corners of a cube. A result of detailed quantum mechanics is the recognition that an additional repulsion exists between electrons of like spin, compared with the repulsion between electrons of unlike spin. So, it would be anticipated that an electron of given spin would have as its nearest neighbours at the corners of the cube electrons of the opposite spin (the reader who would like the relationship between a cube and a tetrahedron described in more detail than given below should take a glance at Chapter 8 and, in particular, Figure 8.1). This means that in the cubic arrangement of electrons there would be four electrons with spin 'up' defining one tetrahedron and four with spin 'down' defining another. If lines are drawn from one corner of a cube across the face diagonals to other corners and this procedure continued, just four corners are reached. These four corners define a regular tetrahedron. Another regular tetrahedron is defined by the four corners which remain – see Figure 1.4. So far in this model all of the electrons have been associated with the nitrogen atom and we have really been thinking of N^{3-}, with eight valence shell electrons. It follows that when the hydrogen atoms are introduced they must be introduced as bare protons – so that an electrically neutral molecule results. These protons attract the eight electrons. The attraction between a proton and an electron does not depend upon whether the electron has its spin 'up' or 'down', although, of course, the extra repulsion between electrons of the same spin persists. The net result is that each proton attracts to its locality just one electron with spin 'up' and one with spin 'down'. This attraction brings the two distinct tetrahedral arrangements of electrons into coincidence to give a single tetrahedral arrangement. The conclusion is that two electrons will be associated with each N—H bond and the remaining two will be non-bonding, just the same as for the previous model. Clearly, this model also predicts a bond angle of 109.5°, the tetrahedral value. It may be corrected in a manner similar to that described above for the electron pair model to give qualitative agreement with experiment. A word of caution. Although in this description there have been phrases such as 'in the cubic arrangement of electrons' this should not be thought of (for the isolated atom) as static; the atom is basically spherical, the electron

[5] Those interested in exploring the LCP model and its relationship with the VSER will find a readable review by R.J. Gillespie and E.A. Robinson, *Chem. Soc. Rev. (London)* **34** (2005) 396.

[6] It is described by J.W. Linnett in *The Electronic Structure of Molecules*, Methuen, London, 1964.

[7] As part of a resurgence of valence bond theory.

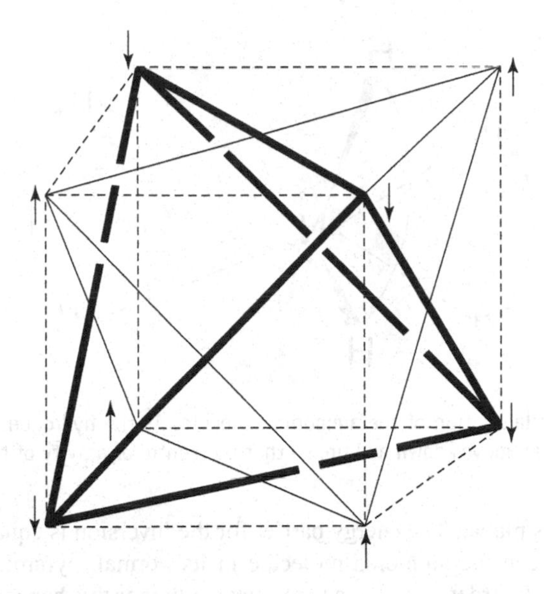

Figure 1.4 The two tetrahedra associated with a cube. Note the association that occurs in Linnett's model between these tetrahedra and the relative spins of the eight electrons placed at the corners of the cube. The thicker tetrahedron has spin 'down' at its corners; the thinner has spin 'up'

distribution smeared out. Rather it should be thought of as a slightly preferred arrangement, with the orientation of the cube totally undefined (and so totally free of constraints).

Although there is considerable overlap between the different models considered above, a survey of them does not lead to any definite conclusion regarding the relationship between the structure of and the bonding in the ammonia molecule. First, they are largely concerned with what is a relatively fine point – bond angles. They say nothing about the more important point (in terms of energy) of bond lengths. Second, all start with the supposition that only valence shell electrons need be considered but then diverge in their explanations. These explanations are not totally distinct but what one model regards as the dominant factor another assumes to be relatively small. The first two models, effectively, say that the geometry is determined by the requirement that bonding interactions be maximized whilst the last two say that it is the consequence of the requirement that non-bonding repulsive forces be minimized. One point that they have in common, however, is the fact that none of them leads to a prediction that the ammonia molecule should be planar.

1.2.5 Accurate calculations

In 1970 Clementi and his co-workers published the results of some very accurate calculations on the ammonia molecule.[8] This is an old paper but the results remain valid – and it has the advantage of presenting the results in a way which provides insights relevant to the present chapter. Clementi and his colleagues were particularly interested in a study of the vibrational motion of the ammonia molecule in which it turns itself inside-out, like an umbrella in a high wind (Figure 1.5). Halfway between the two extremes of this umbrella motion the

[8] A. Rauk, L.C. Allen and E. Clementi, *J. Chem. Phys.* **52** (1970) 4133.

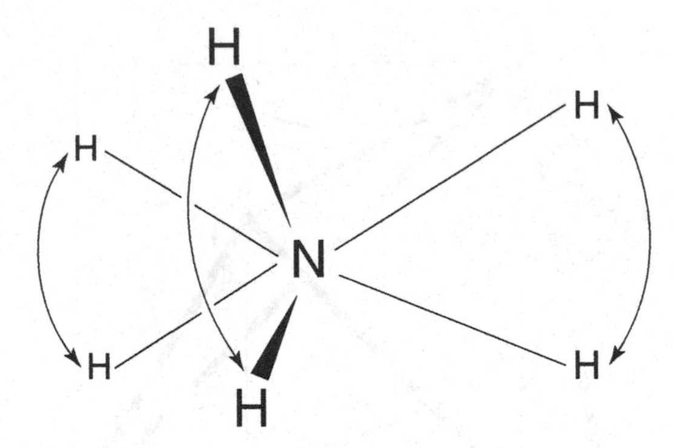

Figure 1.5 The 'umbrella' motion of the ammonia molecule. As the hydrogens move up and down together, so the nitrogen moves down and up, so that the centre of gravity of the molecule remains in the same place

ammonia molecule is planar. The energy barrier for the inversion is equal to the difference in total energy between the ammonia molecule in its normal, pyramidal, shape and the planar configuration. In order to obtain a theoretical value for this barrier, Clementi carried out rather detailed calculations for each geometry. The results obtained by Clementi were very surprising. They showed that the N—H bonding is greater in the planar molecule – there is a loss of N—H bonding energy of approximately 7.0×10^2 kJ mol^{-1} (167 kcal mol^{-1}) in going from the planar to the pyramidal geometry; this loss is accompanied by a slight lengthening of the N—H bond. Bonding favours a planar ammonia molecule. A comparison of the most stable pyramidal and most stable planar geometries shows that the electron–electron and nuclear–nuclear repulsion energies favour the pyramidal molecule over the planar by about 7.2×10^2 kJ mol^{-1} (172 kcal mol^{-1}). Repulsive forces favour a pyramidal molecule. Note the way that the bonding and repulsive energy changes between the two shapes almost exactly cancel each other. It is the slight dominance of the repulsive forces by 20 kJ mol^{-1} (5 kcal mol^{-1}) which leads to the equilibrium geometry of the ammonia molecule in its electronic ground state being pyramidal.

We are left with a most disturbing situation. There is no doubt that the strongest N—H bonding in the ammonia molecule is to be found when it is planar yet two of the simple models considered earlier in this chapter explained its geometry by the assumption that this bonding is a maximum in the pyramidal molecule! Similarly, the models based on electron–electron repulsion ignored both the fact that nuclear–nuclear repulsion is of comparable importance and the fact that their sum is almost exactly cancelled by changes in the bonding energy. This would not matter so much if there were some assurance that repulsive energies would outweigh the bonding in all molecules (molecular geometries could then reliably be explained using a repulsion-based argument). Unfortunately, no such general assurance can be given. This can be seen if the discussion of the ammonia molecule is extended to include some related species.

The molecules NH_3, PH_3, NH_2F, PH_2F NHF_2, PHF_2, NF_3 and PF_3 all have similar, pyramidal, structures and would be treated similarly in all simple models. But calculations by Schmiedekamp and co-workers [9] have shown that the first four owe their pyramidal

[9] A. Schmiedekamp, S. Skaarup, P. Pulay and J.E. Boggs, *J. Chem. Phys.* **66** (1977) 5769.

geometry to the dominance of repulsive forces (bonding is stronger when they are planar) but the last four are pyramidal because the bonding is greatest in this configuration and dominates the repulsive forces (which now favour a planar arrangement)! Although this last sentence is marginally stronger than strictly permitted by the calculations, there is no doubt about the general conclusion. Although these eight compounds all have the same geometrical structure they do not all have it for the same reason, because of the close competition between repulsive and bonding forces. At present there are no rules to enable the prediction of which will win the competition in a particular case. Indeed, a detailed study of the literature leads to this conclusion, or at least forces a person to throw up their hands! So, consider three scandium compounds, all ScX_3 (the electronic structure of Sc is not of immediate relevance). Predict the order of increasing X—Sc—X bond angle for the three compounds: ScH_3, ScF_3, $Sc(CH_3)_3$. The answers (all calculated in a similar way, but reasonably reliable) are: ScF_3, 120°; ScH_3, 140°; $Sc(CH_3)_3$, 60°. The explanation for this sequence is not immediately obvious. [10]

Although simple explanations of molecular shape such as those described earlier in this chapter are very useful to the chemist – and are widely and fruitfully used – they can be considered only as guides because they are not infallible. They are more *aides-mémoire* than correct explanations. It is for this reason, and because it happens to be particularly convenient for our purpose, that in this book the opposite strategy of using the experimentally determined shape of a molecule to infer details of the electronic structure of the molecule *in that shape* will be adopted. Few attempts will be made to explain why a molecule has a particular shape, although there will be many points at which the *consequences* of a particular geometry and its changes will become the focus of attention.

Problem 1.1 Consider each of the models for the structure of the bonding in the ammonia molecule detailed above and for each indicate the importance (if any) that it places on (a) electron–nuclear bonding forces, (b) electron–electron repulsion forces and (c) nuclear–nuclear repulsion forces.

Problem 1.2 Factors which might influence molecular geometry but which have not been included in the present chapter are atom size and bond polarity, although some relevant data have been included. Comment on the possible importance of these additional factors in the light of these data and those in the table below:

Table 1.1

Molecule	N/P—C bond length	C—N/P—C bond angle
$N(CH_3)_3$	1.47	109
$N(CF_3)_3$	1.43	114
$N(CH_3)_3$	1.84	99
$P(CF_3)_3$	1.94	100

[10] For an explanation see the original paper: R.J. Gillespie, S. Noury, J. Pilmé and S. Silvi, *Inorg. Chem.* **43** (2004) 3248. The reader who does not wish to return to the original may note that there is also an unexpected pattern of calculated bond lengths: ScF_3, 1.84; ScH_3, 2.02; $Sc(CH_3)_3$, 1.52 Å.

Problem 1.3 Show that each of the models described in Sections 1.2.1 and 1.2.4 predicts that the water molecule is non-linear (the bond angle is actually 104.5°). Extend the discussion to include the species (bond angles in brackets) $Hg(CH_3)_2$ (180), $O(CH_3)_2$ (111) and $S(CH_3)_3$ (105).

Problem 1.4 Hazard a guess at whether it is bonding or non-bonding forces which lead to NCl_3 having a pyramidal shape.[11]

[11] An answer will be found in a paper by K. Faegri and W. Kosmus, *J. Chem. Soc.* **73** (1977) 1602 (be prepared for a surprise).

2 The symmetry of the water molecule

This chapter begins the work towards the objective of the book, a study into the consequences of molecular symmetry. It, and the following two chapters, will be concerned solely with the water molecule. Although simple, the water molecule – and its symmetry – enables almost all of the important aspects of the subject to be introduced. The way that symmetry simplifies discussions of the chemical bond and of many forms of spectroscopy are two topics that will be covered – but no less important is the introduction to the way that symmetry itself is handled. However, the approach is not without disadvantages. In particular, it is not until Chapter 9 (Section 9.1) that there is a general discussion of the symmetry of molecules, of the allocation of the correct point group to a molecule. This is because it is not until then that the reader will have met enough examples to make the task an understandable one. Hopefully, in this book no new concepts or ideas are introduced without using them – and this includes specific types of symmetry operations. It is not until Chapter 8 that all of these types will have been both introduced and used.

2.1 Symmetry operations and symmetry elements

A good starting point for our discussion is that of the meaning of the word 'symmetry', as applied to molecules. When we say that a molecule has high symmetry we usually mean that within the molecule there are several atoms which have equivalent positions in space. Thus, the tetrahedral symmetry of the methane molecule is evident in the fact that the four hydrogen atoms are equivalent (Figure 2.1). Suppose that you have in front of you a model of the methane molecule so well constructed that no minor blemishes serve to distinguish one hydrogen atom from another. If you were to briefly close and then open your eyes you would have no means of telling whether someone had rotated the model so that, although each hydrogen atom had been moved, the final position of the model was indistinguishable from its starting position. Asking such questions about indistinguishability provides a convenient approach to symmetry and is the one which will be followed in this book. The symmetry of a molecule is characterized by the fact that it is possible, hypothetically at least, to carry out *operations* which, whilst interchanging the positions of some (or all) of the atoms in the molecule, lead to arrangements of atoms which are indistinguishable from the initial arrangement. Note the phrase 'hypothetically

Symmetry and Structure: Readable Group Theory for Chemists Sidney F. A. Kettle

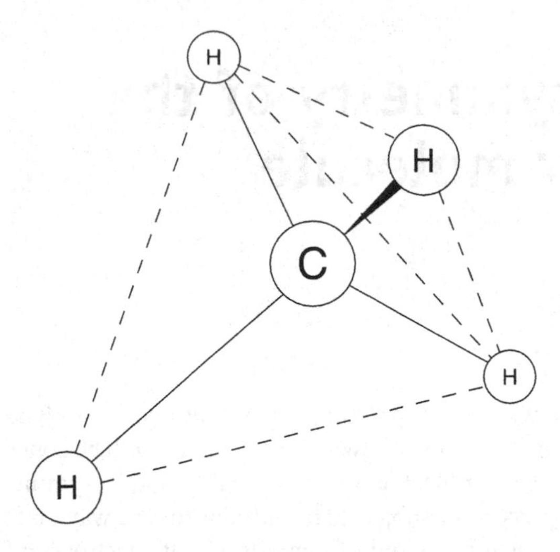

Figure 2.1 The methane molecule, shown in perspective. The important point is that all the hydrogens are equivalent

at least'. There is no requirement that all symmetry operations be physically possible in the way that a rotation is. As will be seen shortly, reflection in a mirror plane is an important symmetry operation. But if an atom lies in such a mirror plane then, presumably, a physical reflection in the mirror plane would mean one side of the atomic nucleus interchanging with the other and this is scarcely physically possible. [1] Of the operations that will be met in the following chapters only the rotation operations are physically possible. The others – such as the operation of reflection in a mirror plane or inversion in a centre of symmetry – are not. Those operations which cannot physically be carried out are called 'improper rotations' in contrast to the 'proper rotations' which are physically possible. A more precise definition of improper rotations will be given at several points in this book; the most detailed discussion is towards the end of Section 8.1. Although the distinction between physically possible and impossible operations is key in such topics as optical activity, the distinction is not one of great fundamental importance. This is because hidden behind everything that we cover in this book there is a mathematics, the mathematics of group theory. What is permissable in this mathematics is what is important; what is possible with ball-and-stick models is of much less significance.

It is helpful to consider a particular example and, as we have said, in this chapter, the symmetry of the water molecule will be the subject of study. This molecule and symmetry have the advantage of simplicity, both geometrically and mathematically. But first a word of caution. We will be working with two closely related concepts. Symmetry elements and symmetry operations. Examples of symmetry elements are rotation axes, mirror planes and centres of symmetry. Examples of symmetry operations are those of rotation about an axis, reflection in a mirror plane and inversion in a centre of symmetry. Care has to be taken

---

[1] The operation of time reversal – scarcely physically possible – is important in some aspects of theoretical chemistry, although it is not one which will be considered in this book. It has the effect (mathematically!) of converting an electron with α spin into one with β.

because in the next few pages, and elsewhere in other chapters, the discussion will flow forth between symmetry elements and symmetry operations. Symmetry elements perhaps seem the more real, in that one can (and we will) draw a line to represent a rotation axis, draw a surface to represent a mirror plane and indicate a centre of symmetry with a point. But this is deceptive. All points in a molecule can be rotated about a symmetry axis, and this means that the axis itself must be of zero thickness. Similarly, a mirror plane must be infinitely thin and a centre of symmetry infinitesimally small. The diagrams used in this and other books must not be taken for reality! In this book it is the symmetry operations which are important – something which we have already signalled, in that we talked about them earlier in this chapter. Our first task, then, is that of obtaining a list of those symmetry operations which turn the water molecule into a configuration indistinguishable from the initial one.

The most evident symmetry operation which turns the water molecule into itself is the act of rotation by 180° about an axis which bisects the HOH angle and lies in the molecular plane. Figure 2.2 shows the water molecule before, in the middle of, and after completion of this operation. Apart from the arrows, which have been added for clarity, the first and third diagrams are indistinguishable. The effect of the operation is to interchange the two hydrogen atoms. We say that 'the two hydrogen atoms are symmetry-related' or 'they are symmetrically equivalent'. A rotation operation such as this is denoted by the letter C (which may be conveniently thought of as derived from the symbol ↻). Because it takes two successive rotations to return each atom to its original position, the rotation operation is called a twofold rotation operation and is denoted C_2, pronounced 'see two'. The same symbol, C_2, is used to denote both the rotation operation and the axis about which the rotation occurs – although, as we have seen, the distinction between the two is very important. Some authors distinguish between an axis and the corresponding operation by writing the latter in bold type, thus $\mathbf{C}_2$. The use of bold type is very helpful if one is developing the mathematics of symmetry theory, group theory. In the present book, however, it will always be clear from the context whether an axis or operation is being discussed and bold type will not be used (because its use tends to make the subject look more daunting than need be the case). Twofold rotation operations are not the only ones which can exist, threefold (C_3), fourfold (C_4), fivefold (C_5) and sixfold (C_6) rotation operations are quite common in chemistry; there are examples later in this book. Such operations have a complication, however, in that there is not a 1:1 correspondence between symmetry axis and symmetry operation. So, for a threefold axis the acts of rotating in clockwise and anticlockwise directions are different; they lead to different atomic interchanges. The absence of such complications for the water molecule is one of the reasons that it is chosen as the first to study. Some molecules have a symmetry such that there is more than one sort of rotation axis (methane, with both twofold and threefold axes, is one example). If a general rotation axis is denoted C_n (where $n = 2$, $3, 4 \ldots$) then the axis of highest symmetry in a molecule is that with the largest value of n. Whenever possible (and methane is a counter-example, one to which we will return later in the book, in Chapter 10), the axis of highest symmetry is chosen as the z axis.

The twofold rotation operation is not the only manifestation of symmetry in the water molecule. If the plane defining the water molecule were to be replaced by an infinitely thin mirror, as shown in Figure 2.3, then reflection in this mirror plane would have the effect of turning the water molecule into a configuration indistinguishable from the original one. This operation has the effect of turning the 'front' of the two hydrogen atoms and of the oxygen atom into the 'back' and vice versa. Mirror planes and the operation of reflection

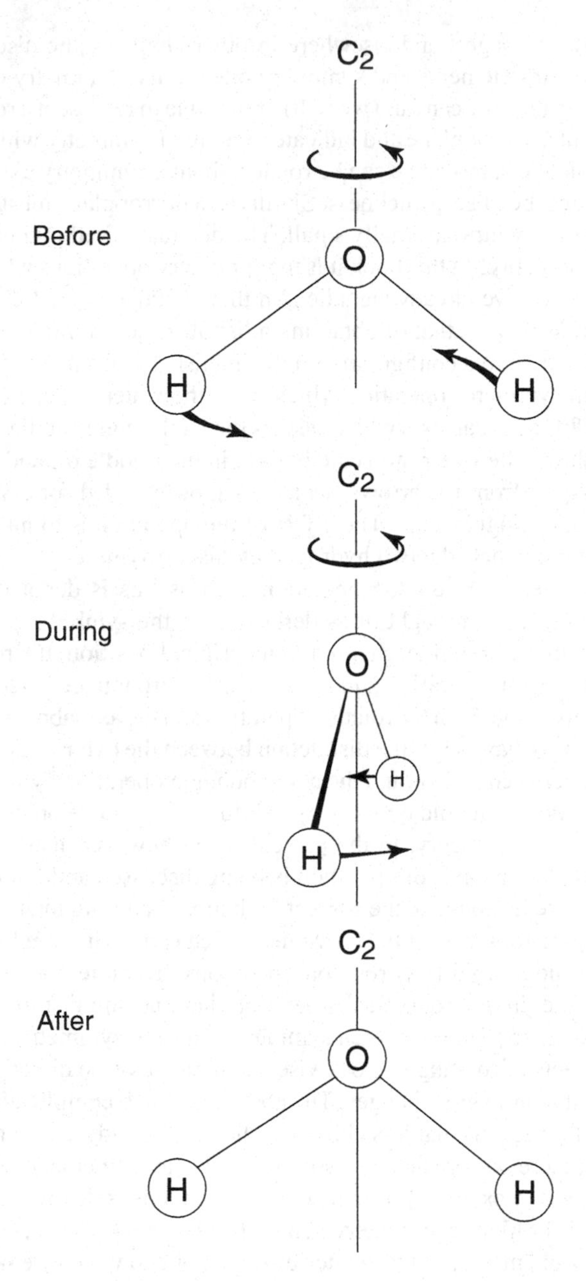

Figure 2.2 The conversion of the H_2O molecule into an arrangement which is indistinguishable from the original by a rotation of 180° ($360/2 \equiv C_2$). In general, it is not possible to give symmetry operations the sort of physical reality which is attempted here

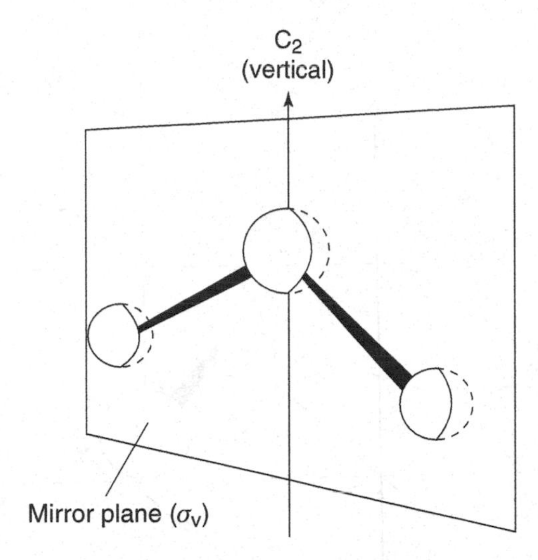

Figure 2.3 The mirror plane of symmetry in the molecular plane of the H_2O molecule. The plane should be thought of as infinitely thin and serving to reflect one 'side' of the molecule into the other. It is not possible to attempt to picture the operation in a way akin to that in Figure 2.2. This points to a fundamental difference in the operations, which will be explored in later chapters

in them are both denoted by sigma – σ – (the operation sometimes being distinguished by bold type) and, just as for rotation axes, various subscripts are used. In the present case the subscript v is added, to give the symbol σ_v. This subscript arises because when, as is the convention, the axis of highest symmetry (C_2 in the present case) is arranged so as to be vertical, as in Figure 2.3, then the mirror plane is also vertical. The subscript v on σ_v is the initial letter of vertical. Thus, more jargon: a σ_v mirror plane is vertical with respect to the axis of highest symmetry (this axis always lies in the σ_v mirror plane). Other subscripts on σ which will be met are h (for horizontal) and d (for dihedral). They will be discussed in detail later in the book (Chapter 8). The symbol σ is derived from the initial letter of the German for mirror, 'spiegel', translated into Greek. Strange, but it seems to have been so since the beginning of the subject.

The C_2 and σ_v symmetry operations do not exhaust the symmetry possessed by the water molecule. Another feature of this symmetry is the existence of a second mirror plane. This mirror plane, which lies perpendicular to the molecular plane, is shown in Figure 2.4. Like the first, the second mirror plane contains the twofold axis (indeed, the line of intersection between the two mirror planes defines the twofold axis). It follows that, like the first mirror plane, the second is denoted σ_v. However, its effect on the molecule is quite different to that of the first – reflection in it has the effect of interchanging the two hydrogen atoms, for instance – and so it is necessary to distinguish between them. This is done by adding a prime to the symbol for the second mirror plane, thus: σ_v'. Had the water molecule possessed a third type of vertical mirror plane (which it does not!) then this would have been denoted σ_v'' and so on. Note that a σ_v is an *improper* symmetry operation – although its effect can be seen, it cannot physically be carried out.

As can be shown by an abortive search, no other rotation axes or mirror planes exist in the water molecule, so it would seem that the three symmetry operations which we have

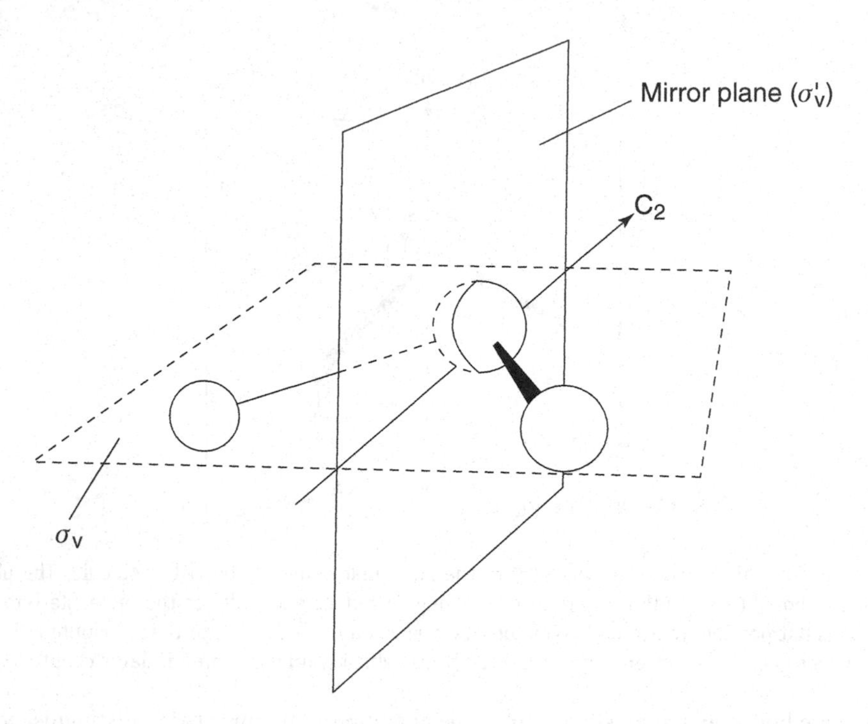

Figure 2.4 A second mirror plane of symmetry, perpendicular to the first, in the H_2O molecule. The line of intersection of the two mirror planes is the twofold rotation axis. The molecule is shown from a different viewpoint

recognized serve to define the symmetry of the water molecule. This, however, is not so. We have already seen that the application of the C_2 operation twice over regenerates the original molecule, with each atom restored to precisely its original position. It is easy to see that the same is true of the σ_v and σ_v' operations. That is, the end result of carrying out any of these symmetry operations twice is the same as that of leaving the molecule alone. The implication of this is that we should formally recognize the possibility that one way of turning a molecule into a configuration indistinguishable from the original is simply to leave the molecule alone. This, so called, *identity operation* will be denoted by the letter E (some books use I). No matter how much or how little symmetry a molecule possesses the identity operation always exists for it. [2] It is the set of four symmetry operations E, C_2, σ_v and σ_v' which completely defines the symmetry of the water molecule. So, one way of talking about the symmetry of the water molecule would be to give this list. However, rather than give a complete listing of symmetry operations (which for some high symmetry molecules could be rather tedious) this information is compressed into a shorthand symbol which for the set of operations of the water molecule is C_{2v} (pronounced 'see two vee'). One talks of the water molecule as 'having C_{2v} symmetry' or we talk of 'the symmetry operations of the C_{2v} point group' (by which is meant E, C_2, σ_v and σ_v'). As for C_{2v}, most, but not all, of

[2] It is sometimes convenient to regard the identity operation as a C_1 rotation, i.e. a rotation of 360°. This interpretation highlights the fact that although there is an infinite number of choices of C_1 axis there is only one distinct $C_1(E)$ *operation*.

the labels conventionally used for groups are a sort of summary of some of their operations. Problems arise when there are lots of operations; again, methane is an example – its group of twenty-four operations is denoted T_d, where the T is the initial letter of tetrahedral.

The last paragraph contained two new words, 'point' and 'group'. The word 'group' arises from the fact that the set of operations satisfy all of the requirements of mathematical group theory. It is helpful to see an example of these requirements. Apply any two of the symmetry operations to the water molecule one after the other. The result is always equivalent to the effect of applying just *one* of the operations of the group (which may be different from the two that were used). Thus, and this is something that we will explore in detail later, for the case of the water molecule, following the σ_v operation by C_2 gives the same result as the application of the σ_v'. Indeed, this combination method is sometimes a useful method of making sure that all of the symmetry operations of a particular molecule have been found – the 'look and see' method that has just been used for the water molecule becomes increasingly fallible as the number of symmetry operations increases. [3] There is a limit to the process, however. Eventually all of the symmetry operations that turn a particular molecule into itself will have been obtained. [4] The successive application of two members of the set of operations will always produce a result which is equivalent to the application of a member of the same set. Sets which are closed in this fashion are called *groups*. Our interest is in *groups* of symmetry operations (although there are many other types of group). For any group there has to be a specified method of combining the group elements – for symmetry operations it is that of applying them one after another. Other types of groups may have very different methods of combination. The complete, formal, definition of a 'group' requires some mathematics but it may help to give two more examples.

Consider the three numbers 1, 0, −1. Do these form a group under the operations of addition and subtraction of the number 1? Whilst it is clear that all three numbers can be interrelated by these operations, it is equally clear that when the operation (+1) is applied to the number 1 the number 2 is generated. Similarly, (−1) applied to the number −1 gives the number −2. Clearly, 1, 0, −1 do not comprise the entire group because 2 and −2 have to be included. In similar fashion it can be seen that 3, −3 and, indeed, all integers between ∞ and $-\infty$ (plus and minus infinity) have to be included. This is an example of an infinite group, the group of all integers. Groups such as this are of importance in the description of the translational symmetry found in crystal lattices, a topic which will be dealt with in Chapter 13, although the detailed group theory is not included in that chapter. As a second example consider the rectangular table shown in Figure 2.5. The top of the table has been divided into quarters and two of these are coloured black and two white. Were there no such colouration, the table would have the same symmetry as the water molecule, as shown in Figure 2.6. However, the presence of the coloured sections means that the σ_v and σ_v' operations are no longer symmetry operations unless they are each combined with quite a new type of symmetry operation, that of changing colour, black into white and white into

[3] In practice, the combination method usually serves to highlight two points in space connected by an overlooked operation without giving the operation explicitly. But, even so, this is a great help in finding the single operation connecting the two points.

[4] This statement is not quite true. There are infinite groups. So, rotation by any angle about the molecular axis of the linear molecule CO_2 is a symmetry operation. The choice of rotation angle is infinite and so too, therefore, is the number of symmetry operations. Such cases are best discussed individually, although commonly one can work with some sort of basic, primitive, operation. So, for CO_2, rotation by an infinitesimally small angle is important. This topic is covered in Chapter 10. The following text gives another example.

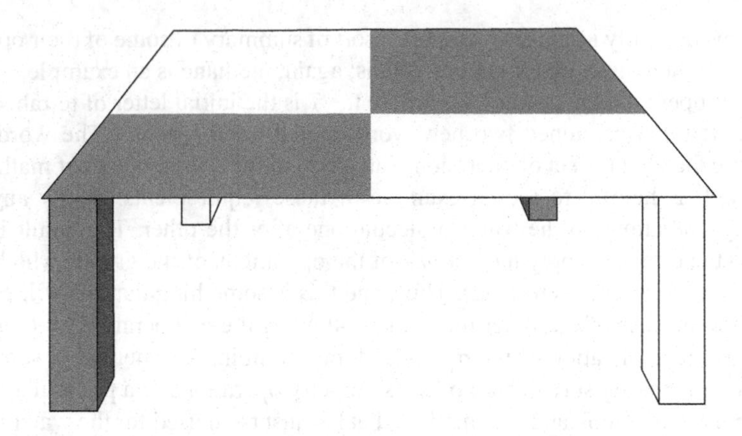

Figure 2.5 A table showing back and white colour-change symmetry. The legs of the table reduce the symmetry so that it is not necessary to compare the top surface of the table with the bottom

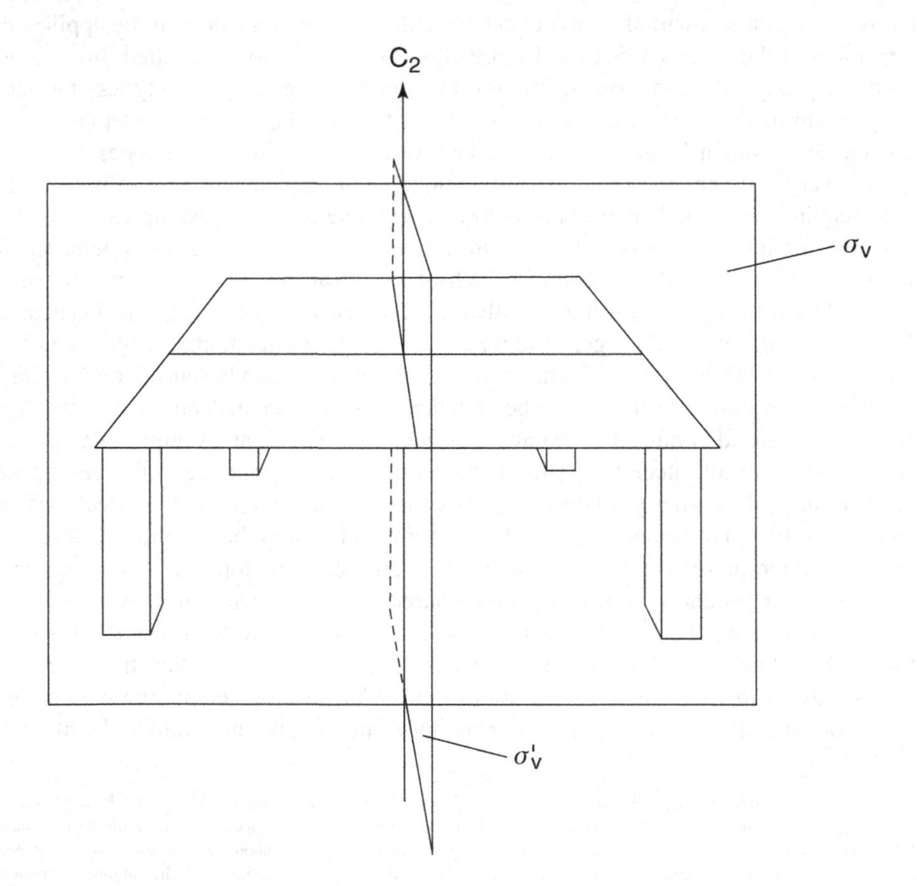

Figure 2.6 The C_{2v} symmetry of the table of Figure 2.5 when uncoloured

black. If these (reflection and colour change) operations are labelled (σ_v) and (σ_v'), then the operations E, C_2, (σ_v) and (σ_v') form a group. [5] There is a more detailed, and general, discussion of groups in Appendix 1.

As has been mentioned, a geometrical feature corresponding to a symmetry operation is called a symmetry *element*. Thus, corresponding to a rotation operation is a rotation axis; corresponding to a reflection operation is a mirror plane. Rotation axes and mirror planes (and also other similar things, such as a centre of symmetry) are examples of symmetry elements. For all molecules it is true that all the symmetry elements which they possess pass through a common point in the molecule (in the case of the C_{2v} point group, perhaps confusingly, they pass through an infinite number of common points along the C_2 axis). This is the reason that all such groups (of operations) are called *point* groups. Put another way, there is always at least one point which is left invariant (unchanged) by all of the operations of a point group. This point does not have to be a point at which an atom is located, although it may well be.

Problem 2.1 Give one-sentence definitions of each of the following in the context of the present chapter (for some there may be more than one acceptable answer): σ, point group; C_2, E, σ_v.

Problem 2.2 How do the terms 'symmetry element' and 'symmetry operation' differ when applied to C_3, C_4 and C_5 rotations?

Problem 2.3 Explain to someone new to the subject the meaning of the phrase 'the symmetry operations of the C_{2v} group'.

2.2 Multipliers associated with symmetry operations

From the way that they have been defined above, it is evident that the effect of each of the symmetry operations of the C_{2v} point group when applied to the water molecule, considered as a whole, is to turn the molecule into itself. An alternative way of putting this is to say that the effect of each of the symmetry operations on the molecule is equivalent to multiplication by the number 1. That is, the effect of each of the operations can be represented as shown in the following table:

Symmetry operation	Effect of the operation on the water molecule (considered as a whole)
E	1
C_2	1
σ_v	1
σ_v'	1

[5] This may seem a contrived example and perhaps it is. However, it is not too far from real applications in chemistry. Suppose we have a molecule which contains an unpaired electron. What is the symmetry relationship between a molecule with spin 'up' and one with spin 'down'? One has to introduce an operation of 'spin change', analogous to that of 'colour change'.

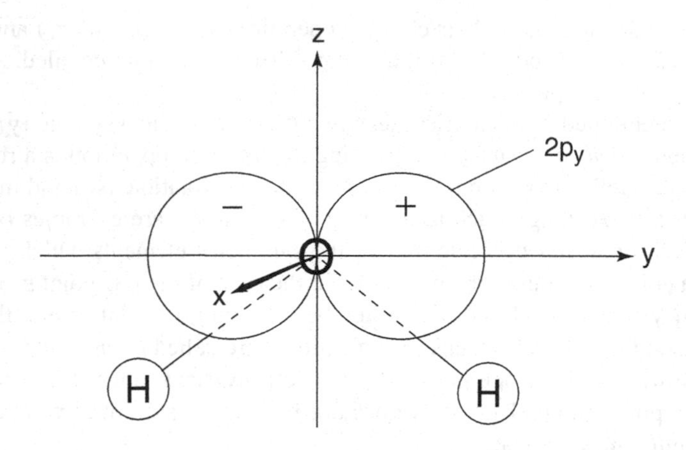

Figure 2.7 The $2p_y$ orbital of the oxygen atom in H_2O. By convention, the *y* axis is taken to lie in the plane of a planar molecule. Notice that the positive lobe of the orbital coincides with the positive *y* direction

The apparently pointless exercise of representing by the number 1 the effects of the behaviour of the water molecule under the symmetry operations begins to acquire some significance when we ask whether all quantities associated with the water molecule are, like the water molecule itself, turned into themselves by the operations of the C_{2v} point group? It will be seen that they are not. Consider, for example, the oxygen $2p_y$ orbital shown in Figure 2.7. In Figure 2.8 are pictured the effects of the symmetry operations of the C_{2v} point group on this orbital. The identity operation, E, of course, leaves the orbital unchanged, giving a multiplier of 1. The σ_v operation also leaves the phases unchanged (although the 'front' and 'back' of each lobe are interchanged), again a multiplier of 1. In contrast, the C_2 and σ_v' operations have the effect of reversing the phases of the lobes, although they do so in different ways, each giving a multiplier of -1. In summary, the association between symmetry operations and multiplicative factors is:

Symmetry operation	Effect on the oxygen $2p_y$ orbital
E	1
C_2	−1
σ_v	1
σ_v'	−1

Having obtained two such sets of numbers the question at once arises of how many such sets can be found? Would any combination of 1 and -1 be acceptable? The answer is 'no'. To explore this further, consider the effect of the symmetry operations of the C_{2v} point group on the $2p_x$ and $2p_z$ orbitals of the oxygen. The $2p_x$ orbital is shown in Figure 2.9,

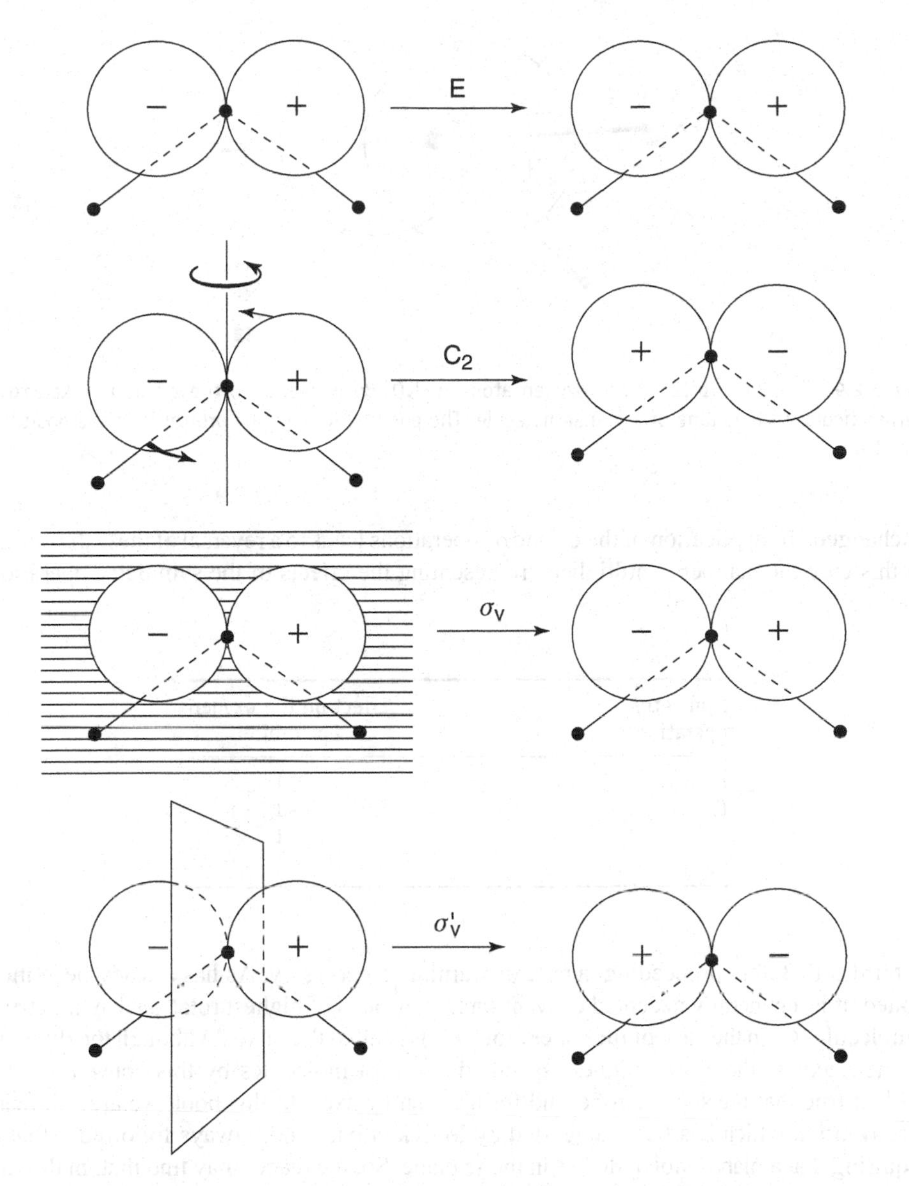

Figure 2.8 The effects of the symmetry operations of the C_{2v} point group on the oxygen $2p_y$ orbital in the water molecule. The point of importance is the relative phases of the orbital 'before' (left) and 'after' (right). On the left, each symmetry operation is diagrammatically represented

where the fact that the positive lobe is located above the plane of the page and the negative lobe beneath this plane is indicated by the perspective of the diagram. In order to avoid completely obscuring the negative lobe behind the positive, the water molecule is viewed from a slightly skew position. In Figure 2.10 are shown the effects of the four symmetry operations of the C_{2v} point group on the oxygen $2p_x$ orbital. It is evident that, whilst the application of the E and σ_v operations result in the phases of the lobes of the orbital being

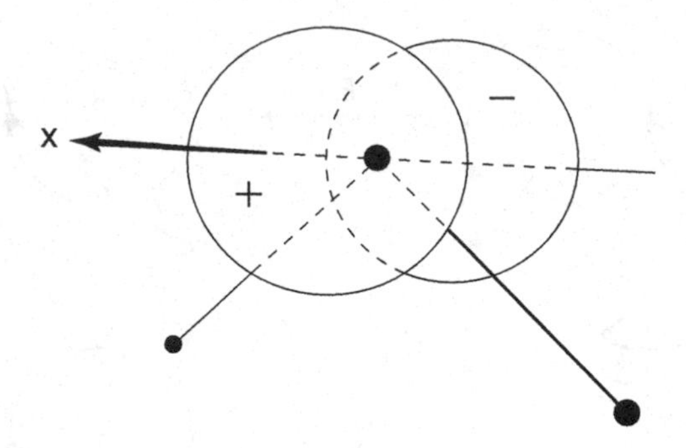

Figure 2.9 The $2p_x$ orbital of the oxygen atom in H_2O. By convention, the x axis is taken to be perpendicular to the plane of a planar molecule. The positive lobe of the orbital is in the positive x direction

unchanged, the application of the C_2 and σ_v' operations leads to a reversal of these the phases. In this case the numbers, multipliers, representing the effects of the symmetry operations are:

Symmetry operation	Effect on the oxygen $2p_x$ orbital
E	1
C_2	−1
σ_v	−1
σ_v'	1

A third set! Before proceeding, a note of warning is necessary. As has already been mentioned, it is a generally accepted convention that the axis of highest rotational symmetry in a molecule (C_2 in the case of the water molecule) is called the z axis. Although the direction of the z axis is therefore uniquely specified for most molecules by this convention it is seldom true that the same can be said for the x and y axes. In this book we are following a convention which has been suggested by Mulliken but is not always followed – that of requiring that a planar molecule lies in the yz plane. So, the reader may find that, in the case of the water molecule, what we have called the x axis some authors will call the y (so that the zx plane, rather than the yz, is the molecular plane). Had the x and y axes been interchanged then, of course, the sets of numbers to which they give rise in the above discussion would also be interchanged. This would lead to similar interchanges throughout the remainder of this and the next two chapters.

We now return to the problem of the symmetry properties of the orbitals of the oxygen atom and consider the $2p_z$ orbital. This orbital is shown in Figure 2.11 and its behaviour under the symmetry operations of the group in Figure 2.12. It is evident from this latter figure that, although the symmetry operations may have the effect of turning one side of the orbital into the other, this change is always accompanied by the retention of the *phase*

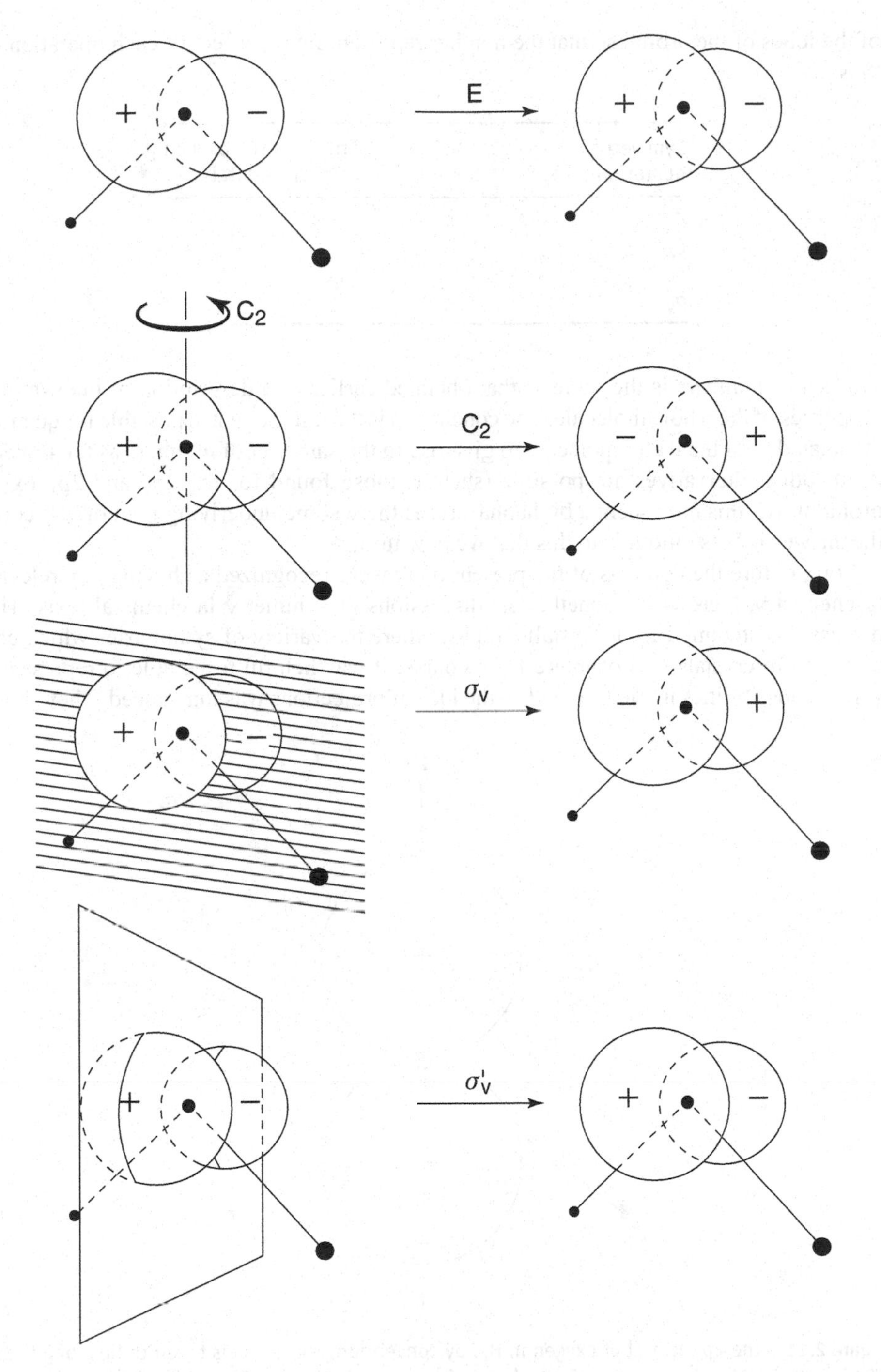

Figure 2.10 The effects of the symmetry operations of the C_{2v} point group on the oxygen $2p_x$ orbital in the water molecule. The point of importance is the relative phases of the orbital 'before' (left) and 'after' (right)

of the lobes of the orbital so that the number representing the effect of each operation is 1, thus:

Symmetry operation	Effect on the oxygen $2p_z$ orbital
E	1
C_2	1
σ_v	1
σ_v'	1

This set of numbers is the same as that obtained earlier as a description of the symmetry properties of the whole molecule. The conclusion is that although it is possible for quantities associated with the water molecule to give rise to the same set of numbers as the molecule itself, other alternatives are possible (such as those found for the $2p_y$ and $2p_x$ oxygen orbitals). All this may seem a bit haphazard; is there some underlying pattern? Of course, the answer is 'yes' and it is to this that we now turn.

Long before the contents of the present book were recognized as having any relevance to chemistry, there were, nonetheless, discussions of symmetry in chemical texts. These discussions stemmed from crystallography, where the variety of symmetries which could be found in crystals was of interest. Of course, it was helpful to be able to represent the various possibilities in pictures and so the idea of projections was introduced. They showed

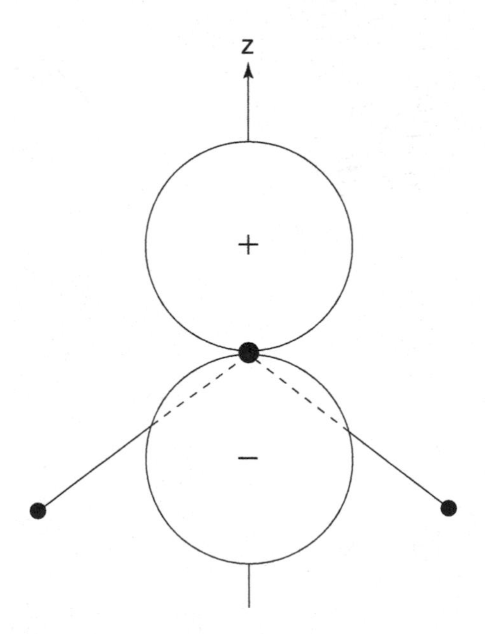

Figure 2.11 The $2p_z$ orbital of oxygen in H_2. By convention, the z axis is taken to lie along the axis of highest rotational symmetry of a molecule (there are departures from this rule for molecules of very high symmetry, discussed in Chapter 10). The positive lobe of the orbital lies in the positive z direction

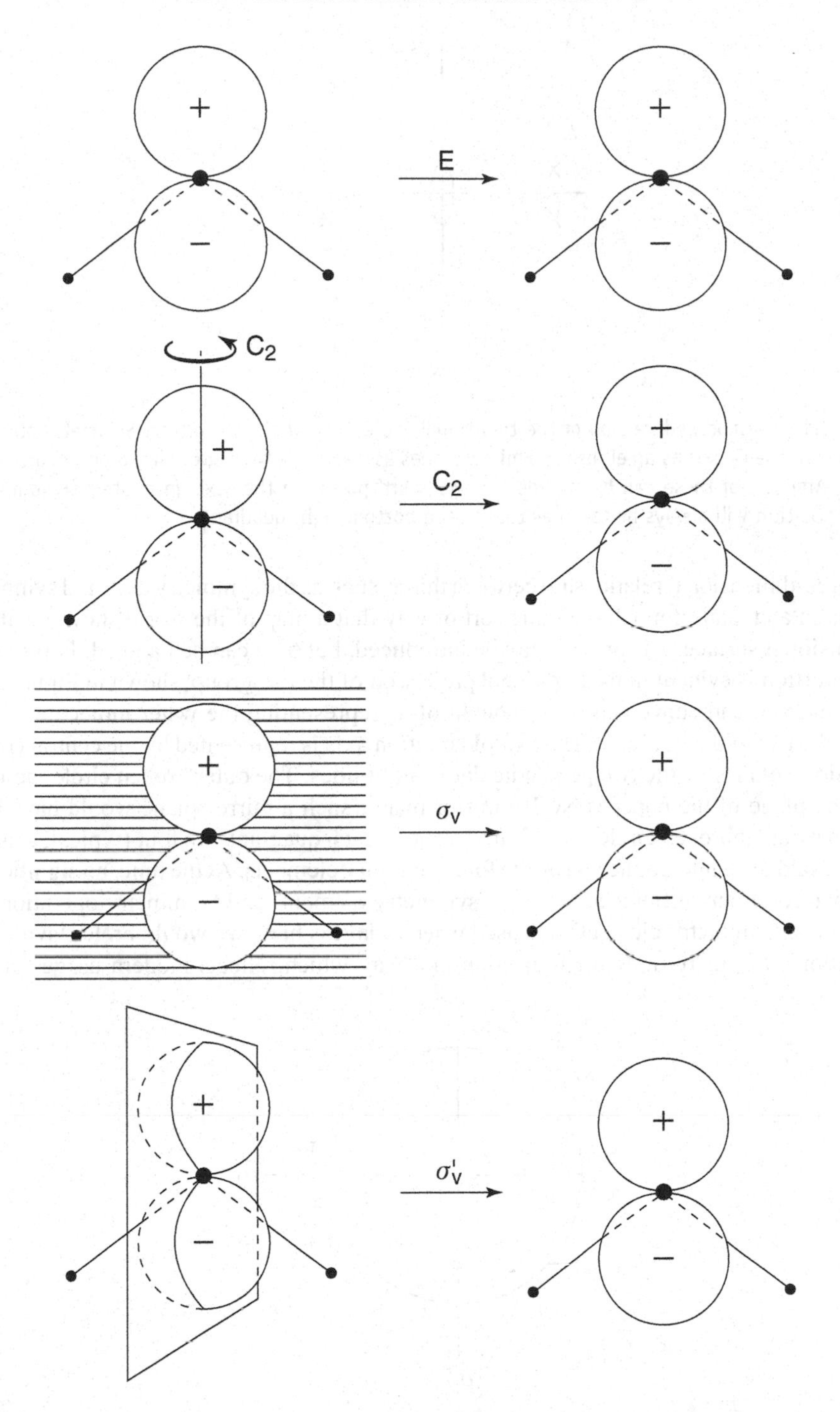

Figure 2.12 The effects of the symmetry operations of the C_{2v} point group on the oxygen $2p_z$ orbital in the water molecule. The point of importance is the relative phases of the orbital 'before' (left) and 'after' (right)

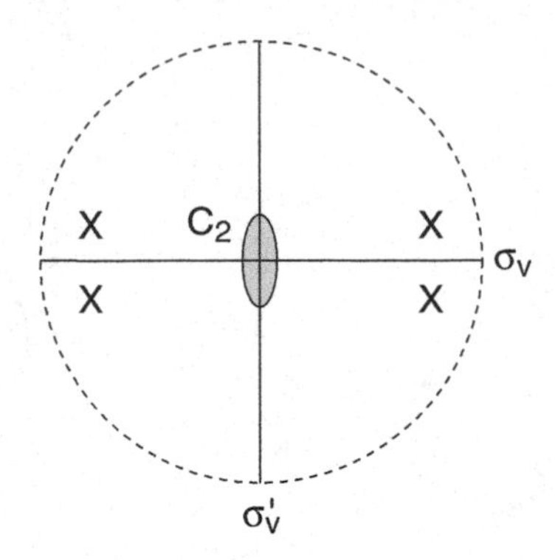

Figure 2.13 A modified version of the traditional projection of the C_{2v} group. Symmetry elements are shown, the C_2 axis as an ellipse and mirror planes as lines. The four interrelated points are shown as x's. Any one of these can be regarded as the 'start' point. In the text, in related diagrams, the 'start' position will always be taken as that in the bottom right quadrant

the three-dimensional relationship between things such as the symmetry axes and symmetry planes in a crystal in much the same sort of way that a map of the world shows its three-dimensional surface. Distortions may be introduced, but they can be handled. Fortunately, no distortion is evident in the traditional projection of the C_{2v} group, shown in Figure 2.13. This diagram can conveniently be thought of as representing the water molecule viewed down the twofold, C_2, z, axis. The twofold rotation axis is represented by the central ()2 and the mirror planes by the two perpendicular straight lines. The outer broken circle indicates that the plane of the paper is NOT a mirror plane (such a mirror plane would have been shown by an unbroken circle). The four x's, one in each quadrant, are four (typical, general) points, said to be interconnected by the four symmetry elements. At the time, no significance was placed on any difference between 'symmetry element' and 'symmetry operation', so the phrase 'symmetry element' was used when today, as here, we would prefer 'symmetry operation'. Let us re-draw the projection in a way which reflects modern usage. This is

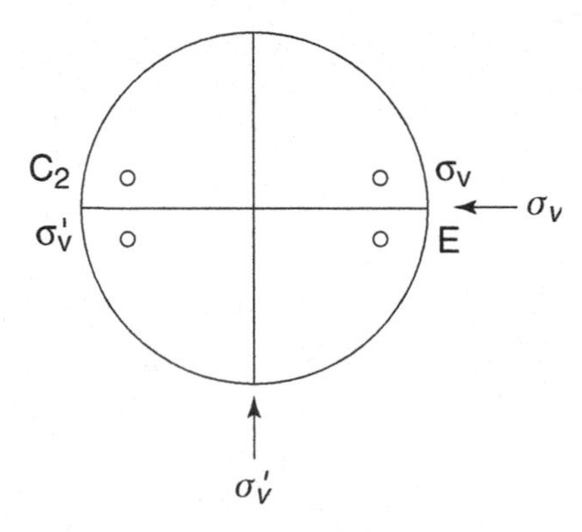

Figure 2.14 An updated version of Figure 2.13. The effects of symmetry operations are shown and symmetry elements not indicated (although here the mirror planes are drawn as an aid). Each point is labelled by the operation which generates it, starting from the E point

done in Figure 2.14, in which we are unambiguously talking of symmetry operations. We have to nominate a point to represent a starting point, but, like the x's before it, placed in a general position. This is the small circle labelled E, indicating that when operated on by the identity, leave alone, operation, this point remains unmoved. Diametrically opposite in the diagram is the point which is reached by the C_2 operation acting on the E point. Because this point is labelled, there is no need to include a ()3 at the centre of the diagram. Similarly, there is no need to represent the σ_v and σ_v' mirror planes; we are concerned with operations not axes – and the points resulting from the application of each of these operations on E are marked. Nonetheless, we show these mirror planes because they acquire a new significance as phase boundaries, as we shall soon see (and the labels on them will also be dropped). Finally, it makes for a tidier diagram if it is enclosed by a complete circle. Such a circle has no symmetry-element implication; if there were a mirror plane in the paper, we would demonstrate it by showing the points reached, making representation of the plane itself redundant (although we might want to use it to show phase relationships). Such a mirror plane will be met in Chapter 5.

The next step is to take the three oxygen 2p orbitals which we have studied earlier and to superimpose each of them on the diagram of Figure 2.14. This is done in Figure 2.15, where it must be remembered that we are viewing everything down the z axis (so that we lose sight of the nodal plane of thc $2p_z$ orbital). Because they have been given in Figure 2.14, there is no need to label the effects of the symmetry operations on the E point. Even so, these labels are given in the hope that they will clarify the discussion. They will be retained for a few more diagrams but soon they will be omitted. Of course, for present purposes, the p orbitals are no more than convenient vehicles. We have seen that the symmetry properties

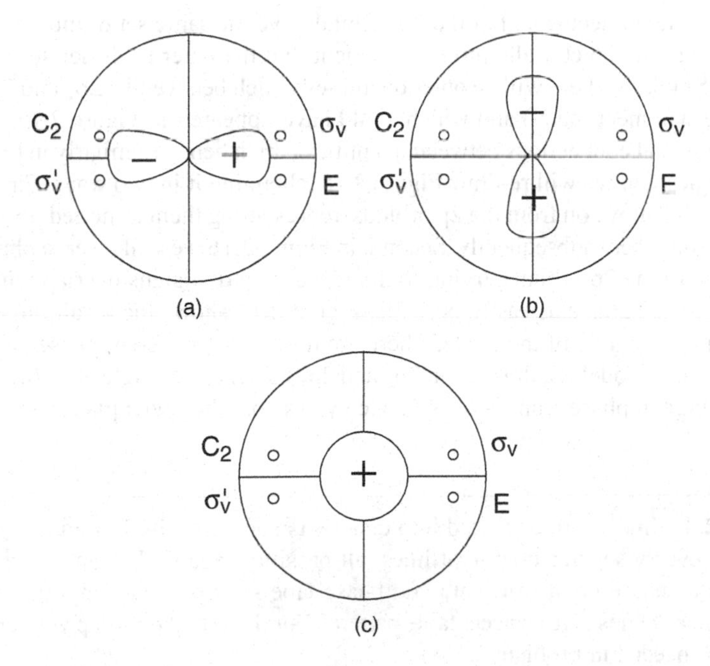

Figure 2.15 The oxygen $2p_y$ (a), $2p_x$ (b) and $2p_z$ (c) orbitals superimposed on Figure 2.14. Each lobe of the orbital will determine the phase of the quadrants with which it overlaps

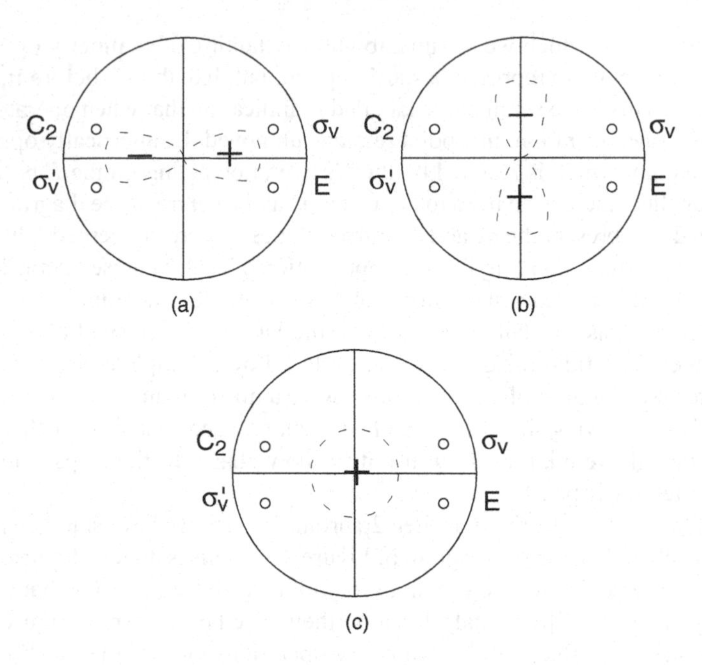

Figure 2.16 Figures 2.15a–c repeated, but with the phases becoming more important than the orbital which led to them

of the entire water molecule and of the $2p_z$ orbital gave the same set of multipliers and so it is only because it is pictorially more convenient that the latter is chosen to be shown in Figure 2.15. Similarly, there will be other quantities which behave like $2p_x$ and $2p_y$ (indeed there are, we will meet many) and which could have appeared in Figure 2.15. What then are the fundamental connections between quantities which behave similarly in this way? To answer this question, we will re-draw Figure 2.15, changing it in two ways. First, as befits vehicles, we shall move on from the 2p orbitals, representing them as dotted lines in Figure 2.16 and omitting them subsequently. Second, in Figure 2.16 we will insert a phase symbol into each lobe of the 2p orbitals, trying to make the + and − signs overlap with as many quadrants in the diagrams as easily possible. Figure 2.17 shows the result; no 2p orbitals but a phase in every lobe of the circle. There are three quite different phase patterns: that derived from $2p_z$ is nodeless; those from $2p_x$ and $2p_y$ each have a single node (across a node there is a change in phase from + to − or vice versa) but the nodal planes are differently orientated.

Problem 2.4 Into a circle divided into quadrants such as those in Figure 2.17 add a number of evenly separated straight lines, all passing through the centre point. Treat these lines as phase boundaries and add phase labels (+ or −) as appropriate. Does every number of lines give an acceptable pattern? For those that do, keep your answers – they will be needed in Problem 2.5.

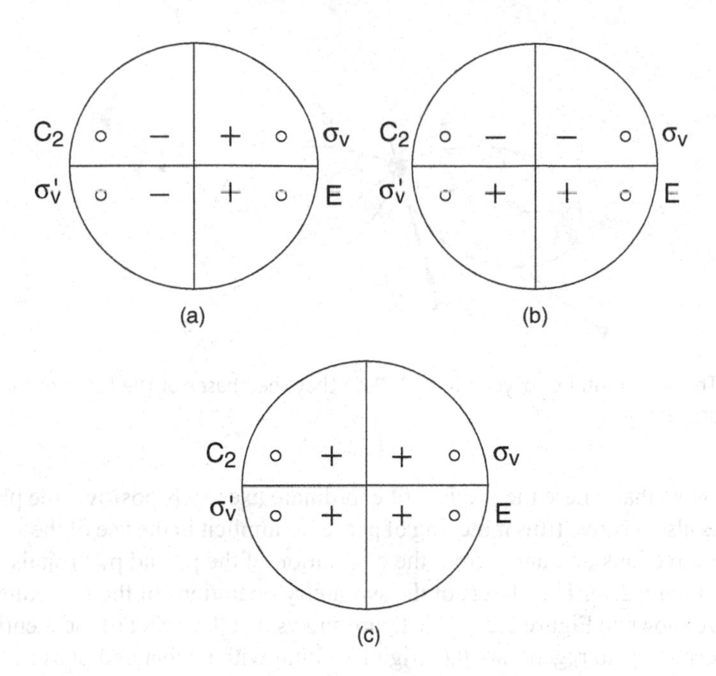

Figure 2.17 Figures 2.16a–c repeated, but with only the phases shown

Does Figure 2.17 exhaust the list of possible nodal patterns? Probably not. What about that shown in Figure 2.18? Can we find an orbital of the oxygen atom that leads to this pattern when it is transformed under the operations of the C_{2v} point group? Yes! Consider the symmetry properties of the $3d_{xy}$ orbital of the oxygen atom. Although this orbital is not commonly included in elementary discussions of the electronic structure of oxygen-containing compounds (because it is not a valence shell orbital) it does nonetheless exist and would be included in most sophisticated electronic structure calculations. It is shown in

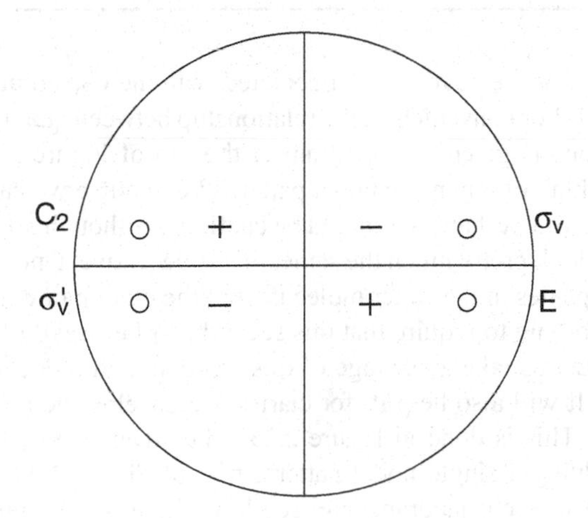

Figure 2.18 A possible nodal pattern, but not included in Figure 2.17

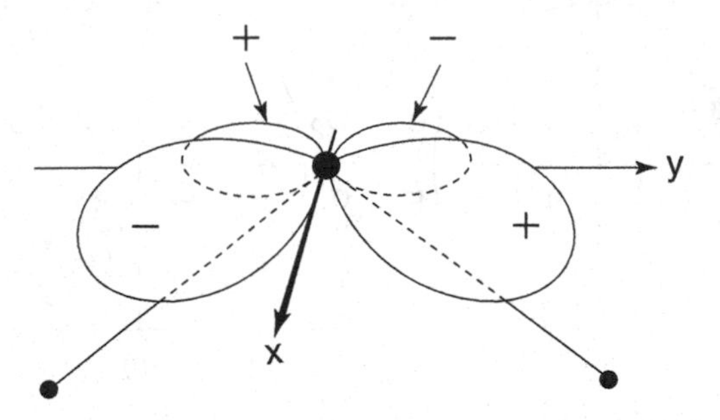

Figure 2.19 The $3d_{xy}$ orbital of oxygen in H_2O. Note that the phases of the lobes of the orbital are those of the product *xy*

Figure 2.19. Note that where the product of coordinate axes xy is positive, the phase of the $3d_{xy}$ orbital is also positive (this matching of phases is implicit in the use of the xy subscript; we know the directions of x and y from the orientation of the p_x and p_y orbitals – and they are shown in Figure 2.7). The effects of the symmetry operations of the C_{2v} point group on this orbital are shown in Figure 2.20. This figure shows that the effect of the identity (E) and of the C_2 operation is to regenerate the original orbital with unchanged phases. In the case of the σ_v and σ_v' operations, however, the phase of each lobe of the orbital is reversed. For the record, the appropriate multiplicative factors representing the effects of the operations are therefore:

Symmetry operation	Effect on the oxygen $3d_{xy}$ orbital
E	1
C_2	1
σ_v	−1
σ_v'	−1

The four nodal patterns we have found associated with the C_{2v} point group are brought together in Figure 2.21. For convenience, the relationship between quadrants and the effects of symmetry operations is given in a diagram at the top of Figure 2.21. Is Figure 2.21 exhaustive? Can we think of any more nodal patterns? Of course we can. What about that shown in Figure 2.22? But perhaps we should be cautious. Although all of the nodal planes shown in Figure 2.22 look pretty much the same, this is deceptive. One pair of nodal planes coincide with mirror planes in the water molecule, but the other pair coincide with nothing similar. So, there is nothing to require that this second pair be straight lines: they could be curved for instance. Let us take advantage of this freedom and reduce the second pair of lines to short pieces. It will also help if, for clarity, we envelop the phase patterns within rabbit-ear-like shells. This is done in Figure 2.23. The advantage of Figure 2.23 is that, although we are studying a single nodal pattern, that of Figure 2.22, we can break this pattern into bits (one in each quadrant) and see how these behave under the operations of the C_{2v} point group. This will lead us to recognize how the whole pattern behaves.

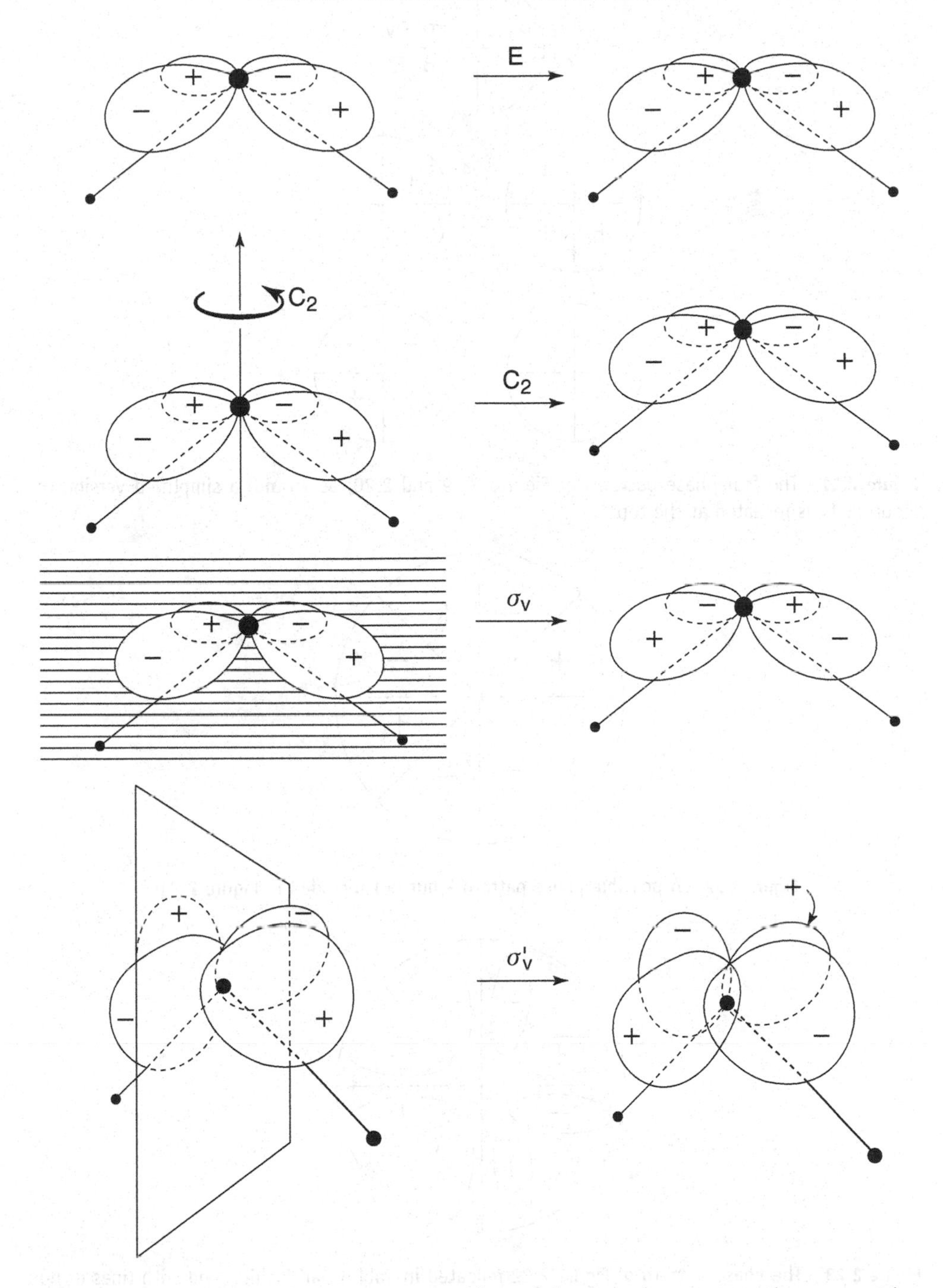

Figure 2.20 The effects of the symmetry operations of the C_{2v} point group on the oxygen $3d_{xy}$ orbital in the water molecule. The point of importance is the relative phases of the orbital 'before' (left) and 'after' (right)

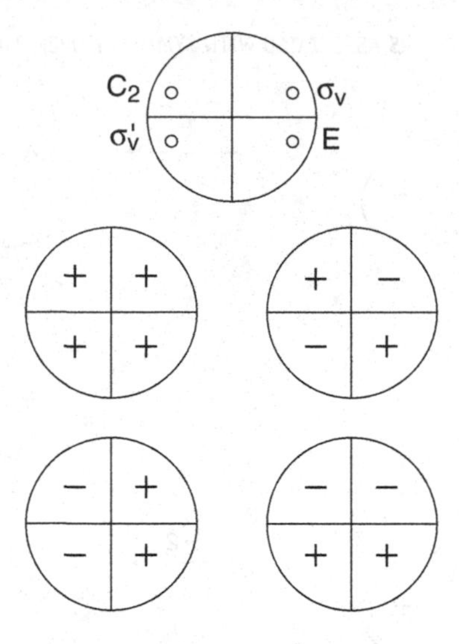

Figure 2.21 The four phase patterns of Figures 2.19 and 2.20; as an aid, a simplified version of Figure 2.14 is repeated at the top

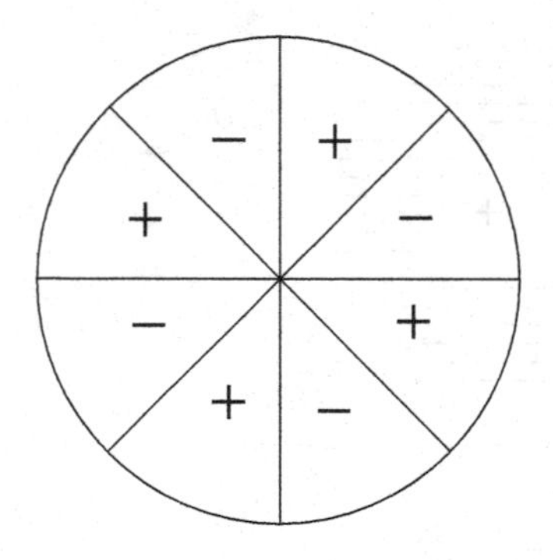

Figure 2.22 A possible phase pattern – but not included in Figure 2.21

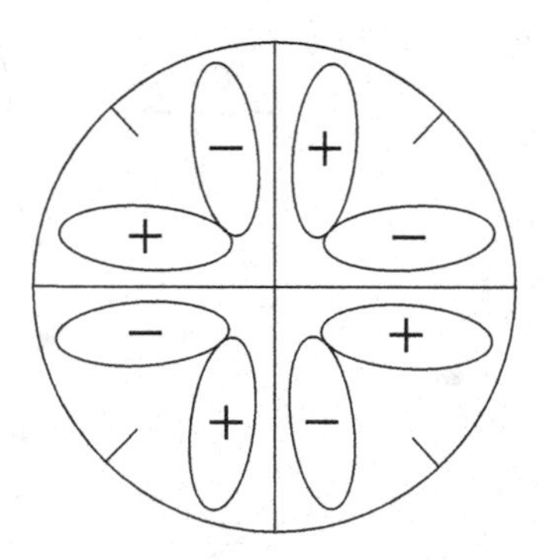

Figure 2.23 The phase pattern of Figure 2.22 repeated in 'rabbit ear' fashion and with lines of no real significance reduced in size. If each rabbit ear pattern is regarded as a single entity, then it can be seen that a C_2 rotation gives the same pattern but the mirror plane reflections gives the negative of the original pattern. An alternative approach is to regard the set of four rabbit ears as a single entity and to transform this set under the symmetry operations. The same result is obtained

Figure 2.23 shows that the rabbit ear bits interrelated by σ_v and σ_v' reflections have opposite phases but the bits related by the C_2 rotation have the same phase. That is, we get the multipliers:

Symmetry operation	Effect on the nodal pattern of Figure 2.22
E	1
C_2	1
σ_v	−1
σ_v'	−1

This is NOT new. It is the same result as we obtained for the oxygen d_{xy} orbital – it is the fourth nodal pattern of Figure 2.21. It is not difficult to see what has happened. We obtained Figure 2.21 by considering orbitals chosen so that in each quadrant of Figure 2.14 there was a unique phase, either + or −. Essentially, we looked at four points and considered their possible relative phases. But in Figure 2.22 we moved away from considering points; we replaced them with (local) nodal (rabbit ear) patterns – and there is no limit to the number of choices that we can make for such multi-nodal patterns. Figures 2.22 and 2.23 suggest that perhaps the diagrams of Figure 2.21 are the only independent ones that exist for the C_{2v} group, that all others we might invent are all built on these four. Indeed, this is so – and this simplicity indicates the value of symmetry concepts. But it would be helpful to have some independent confirmation that just four, and only four, basic phase pattern diagrams exist. The next two sections each provide such a confirmation.

Problem 2.5 Take the patterns left after Problem 2.4 and assess each in the light of the contention that all must be, or be based on, one of the four acceptable patterns.

The first confirmation comes from something of a buckshot approach; the abortive study of the symmetry properties of quantities, orbitals, which might be expected to lead to new symmetry patterns if any exist. The twin facts that the number of orbitals to be studied is limited and that the search will prove abortive might well indicate that the task is pointless. This is not so; the search will also provide the reader with invaluable experience in the transformation of complicated-looking (but actually simple) objects. Specifically, we will consider the transformation of the remaining four 3d orbitals of the oxygen ($3d_{z^2}$, $3d_{x^2-y^2}$, $3d_{zx}$ and $3d_{yz}$). The case of the $3d_{xy}$ orbital was considered above.

There are two ways of tackling this problem. The first, more tedious but more valuable, is to follow the pattern adopted for the oxygen 2p orbitals (Figures 2.8, 2.10 and 2.12) and the $3d_{xy}$ orbital (Figure 2.20). To encourage the reader to undertake this task the hard way, in Figure 2.24 are shown each of the above orbitals in the coordinate system of Figure 2.7. For those looking for an easier life or a different insight, in Figure 2.25 are shown the same orbitals in projections similar to those of Figure 2.16. The phase patterns follow. The $3d_{x^2-y^2}$ case in Figure 2.25b may pose a problem, one which is interesting to resolve. It might seem to contain a new pattern of nodes. We can emphasize the new nodes by

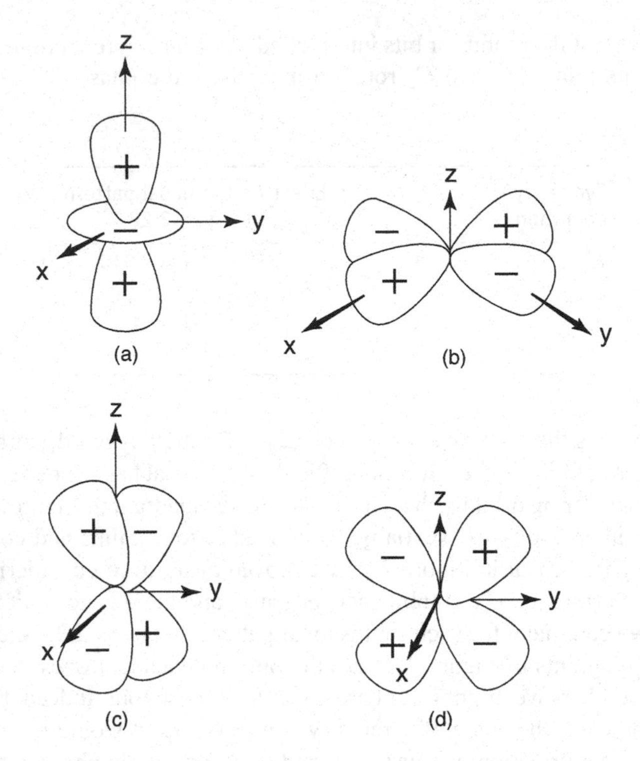

Figure 2.24 The oxygen $3d_z2$ (a), $3d_{x^2-y^2}$ (b), $3d_{zx}$ (c) and $3d_{yz}$ (d) orbitals

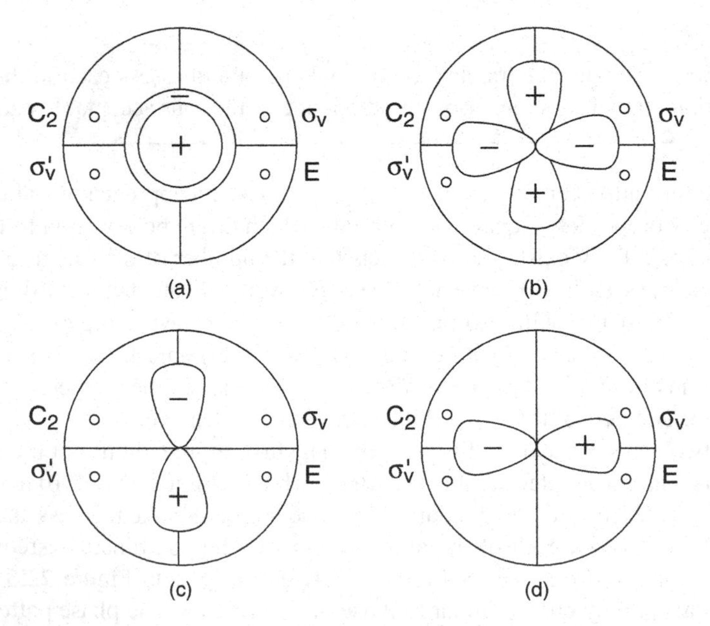

Figure 2.25 The oxygen $3d_z2$ (a), $3d_{x^2-y^2}$ (b), $3d_{zx}$ (c) and $3d_{yx}$ (d) orbitals shown superimposed on Figure 2.14 (simplified). Only the 'top' lobes are shown for (c) and (d). The 'bottom' lobes would give the same pattern, but as its negative

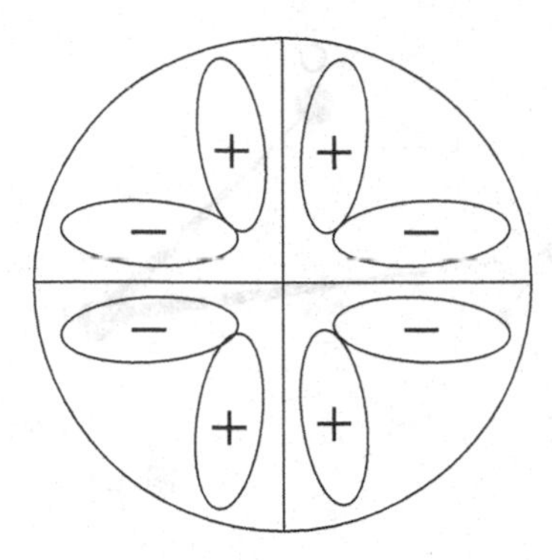

Figure 2.26 The oxygen $3d_{x^2-y^2}$ orbital of Figure 2.25b drawn as a 'rabbit ear' pattern. All of the symmetry operations of the group regenerate the pattern unchanged

redrawing the pattern in a manner akin to that of Figure 2.23. This is done in Figure 2.26. We know from the discussion connected with Figure 2.23 that we have to consider the rabbit ears as single objects. When we do this, perhaps to our surprise, we find that the rabbit ears are turned into themselves by all of the operations of the group. So, the fundamental phase pattern of Figure 2.25b is just that of Figure 2.17c; the apparent simplicity of the latter is a bit deceptive! Both of the approaches of Figures 2.24 and 2.25 (should!) lead to the same set of answers. These are given below in the form of multipliers:

Symmetry Operation	Effect on the oxygen $3d_{z^2}$ orbital	Effect on the oxygen $3d_{x^2-y^2}$ orbital	Effect on the oxygen $3d_{zx}$ orbital	Effect on the oxygen $3d_{yz}$ orbital
E	1	1	1	1
C_2	1	1	−1	−1
σ_v	1	1	−1	1
σ_v'	1	1	1	−1

Of the two approaches to these problems, that which the reader was urged to follow will have shown that not all of the nodal patterns inherent in an individual orbital may be relevant to the solution of problems associated with it. In the projection approach, these irrelevant phases were sometimes discarded without comment (usually those in the *xy* plane). Of course, the fact that they were irrelevant here does not mean that they will always be irrelevant (just as our discussion of Figures 2.22 and 2.23 – and 2.26 – depended on the non-existence of two mirror planes; had these mirror planes existed then we could not have treated the phase patterns in the way that we did).

Problem 2.6 If it has not already been done, generate the multipliers obtained from the transformations of the oxygen $3d_{z^2}$, $3d_{x^2-y^2}$, $3d_{zx}$ and $3d_{yz}$ orbitals under the operations of the C_{2v} point group.

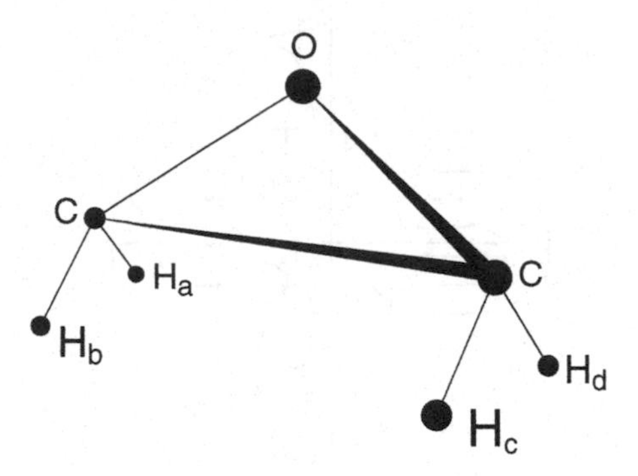

Figure 2.27 The ethylene oxide molecule C_2H_4O. The objects of interest are the labels on the H atoms

2.3 Group multiplication tables

The second way of showing that the set of diagrams in Figure 2.21 is complete, that none can be added, is sufficiently important to be given a section of its own. Earlier in this chapter it was asserted that the effect of the successive application of symmetry operations of a group was always equivalent to the effect of some single operation of the group. This will now be investigated in detail for the C_{2v} point group by considering, in turn, each operation and the effect of following it with all four symmetry operations of the group, each considered in turn. It will be helpful to focus attention on a particular molecule. The water molecule, on its own, is inconvenient for this particular purpose (because of the apparent equivalence of the effects of applying different symmetry operations, a phenomenon encountered several times already in this chapter). To overcome this problem we will consider, instead, four points in space, symmetrically related to the water molecule. This pattern is shown in Figure 2.27, where, to make the problem more understandable, the four points are each associated with one of the four hydrogen atoms of the ethylene oxide molecule. In this figure the four points in space (hydrogen atoms) have been labelled with the suffixes a, b, c or d – so that in order to study the effects of the symmetry operations all that has to be done is to see how these labels are rearranged. The effects of the operations of the C_{2v} point group on these labels are shown in Figure 2.28, a figure that should be studied carefully until the reader is fully conversant with it, feels comfortable with it. It will be noted that each symmetry operation gives rise to a different final arrangement of labels, a feature which would not have been found with the simple H_2O molecule and the reason for the present choice of example.

Because the identity operation does not change the distribution of the labels at all, it is evident that any operation preceded or followed by the identity operation gives rise to the same final arrangement as that operation on its own. It can immediately be concluded that

$$E \text{ followed by } E \equiv E$$
$$E \text{ followed by } C_2 \equiv C_2$$
$$E \text{ followed by } \sigma_v \equiv \sigma_v$$
$$E \text{ followed by } \sigma_v' \equiv \sigma_v'$$
$$C_2 \text{ followed by } E \equiv C_2$$

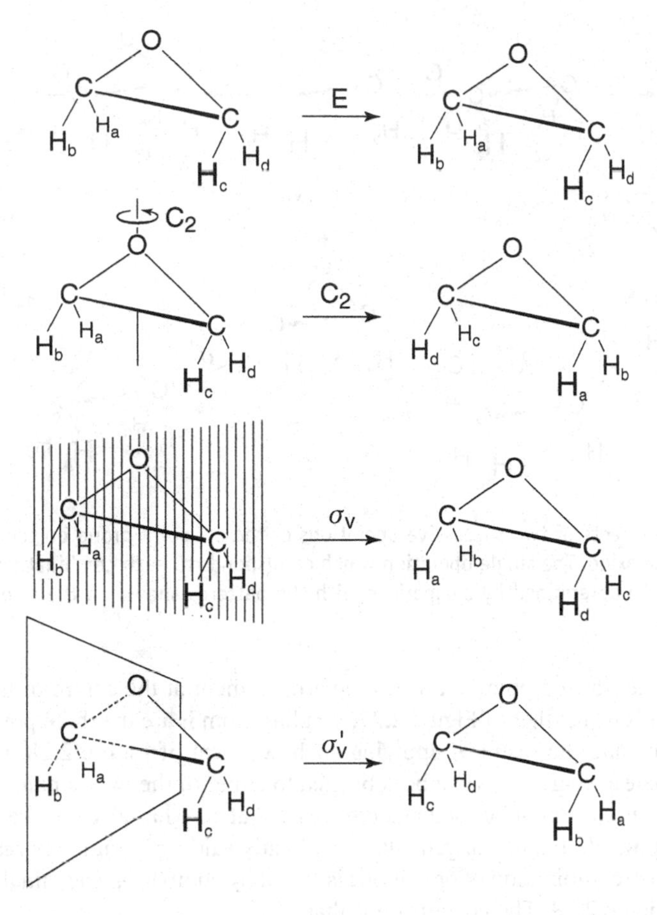

Figure 2.28 The effects of the symmetry operations of the C_{2v} point group on the four hydrogen atoms of ethylene oxide. The point of importance is the relative pattern of the hydrogen atoms 'before' (left) and 'after' (right)

$$\sigma_v \text{ followed by } E \equiv \sigma_v$$
$$\sigma_v' \text{ followed by } E \equiv \sigma_v'$$

Slightly less trivial is the result of the successive application of pairs of operations from the set C_2, σ_v and σ_v'. It has been pointed out earlier in this chapter that any one of these operations followed by itself gives rise to the initial arrangement and so the sequence is equivalent to the identity operation. That is,

$$C_2 \text{ followed by } C_2 \equiv E$$
$$\sigma_v \text{ followed by } \sigma_v \equiv E$$
$$\sigma_v' \text{ followed by } \sigma_v' \equiv E$$

In Figure 2.29 are illustrated the remaining combinations of operations and the reader may, by comparison with Figure 2.28, determine which single operation is equivalent to each

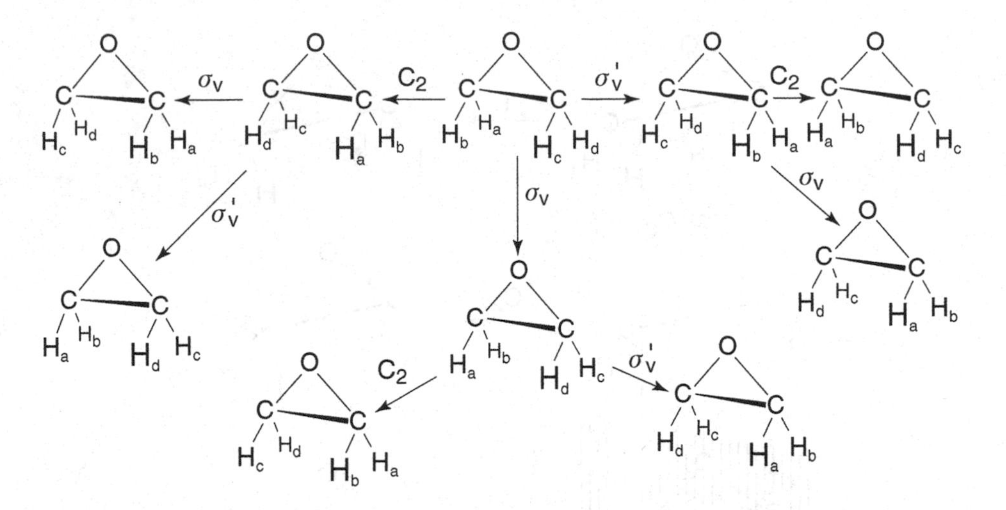

Figure 2.29 The effects of two successive operations of the C_{2v} point group on the four hydrogen atoms of ethylene oxide. The single operation which corresponds to each combination of operations shown here may be determined by comparison with the patterns shown on the right-hand side of Figure 2.28

combination. The starting point is always the arrangement at the centre of the top line in the diagram – this is just that of Figure 2.27. Leading from it are the three possible choices of first operation and the result of applying each: a repeat of Figure 2.28. Each of these three intermediate arrangements can be subjected to either of the two remaining operations (the one already used cannot be used again because that case has already been considered) leading to six possible final arrangements. As already said, the single operation which is equivalent to each combination of operations is found by comparing each final arrangement with those in Figure 2.28. The conclusion is that

$$C_2 \text{ followed by } \sigma_v \equiv \sigma_v'$$
$$C_2 \text{ followed by } \sigma_v' \equiv \sigma_v$$
$$\sigma_v \text{ followed by } C_2 \equiv \sigma_v'$$
$$\sigma_v \text{ followed by } \sigma_v' \equiv C_2$$
$$\sigma_v' \text{ followed by } C_2 \equiv \sigma_v$$
$$\sigma_v' \text{ followed by } \sigma_v \equiv C_2$$

These results are collected together in Table 2.1.

Problem 2.7 Use Figures 2.28 and 2.29 to check that Table 2.1 is correct.

It is usual in the mathematical theory of groups to refer to the law of combination of group elements as 'multiplication' (although only rarely does the operation have anything to do

Table 2.1

C_{2v}		First operation		
	E	C_2	σ_v	σ_v'
Second operation				
E	E	C_2	σ_v	σ_v'
C_2	C_2	E	σ_v'	σ_v
σ_v	σ_v	σ_v'	E	C_2
σ_v'	σ_v'	σ_v	C_2	E

with ordinary arithmetical or algebraic multiplication). So, in the present case, where two symmetry operations combine by being applied in succession, they are said to 'multiply'. Thus we say 'C_2 multiplied by σ_v is equal to σ_v''. Table 2.1 is therefore referred to as the *multiplication table* for the operations of the C_{2v} point group. [6]

This is another point at which the reader is offered a choice of ways forward. The first involves a game, which she or he is encouraged to play. The game consists of taking Table 2.1 above and making substitutions in it. Everywhere that a chosen symbol (and there are only four to chose from) appears in the table it is replaced by a number, to be chosen by the reader. When all four symbols have been replaced the table will contain nothing but numbers. The problem is to find a set of substitutions which lead to a final table which is arithmetically correct. So, if the following substitutions were made:

E	C_2	σ_v	σ_v'
50	−0.3	4.3	−7.8

the final table would contain not a single correct entry (and if the author can be this bad, the reader must surely be able to do better!).

Problem 2.8 Play the game described in the text; systematically substitute numbers for the entries in Table 2.1 and attempt to find a set that give arithmetically correct tables. Four such sets exist.

This sentence is separated as a paragraph to inhibit the reader from reading it by accident. It contains a hint to help with the above problem and is designed to help narrow down the search. The fact that E followed by E equals E – to be replaced by $(E) \times (E) = (E)$, where the symbol (E) stands for the number replacing E – shows that the only number that can be used to substitute for E is 1 (divide each side of the equation by (E)). Rather similar arguments can be used for some of the other entries in the table.

The alternative way forward is similar to that of playing a game, but without the effort. It is convenient to collect together all of the sets of multipliers that have been generated in this chapter. They are listed in Table 2.2, although the order in which they are presented is not that in which they were obtained. Against each set of multipliers is/are shown the oxygen

[6] In Table 2.1 the results of the multiplication are independent of the order in which the operations are applied – the table is symmetric about its leading diagonal. This is not a general property of multiplication tables but it is one that makes that in Table 2.1 a particularly simple case to start with.

Table 2.2

E	C_2	σ_v	σ_v'	
1	1	1	1	$2p_z$, $3d_{z^2}$, $3d_{x^2-y^2}$
1	1	−1	−1	$3d_{xy}$
1	−1	1	−1	$2p_y$, $3d_{yz}$
1	−1	−1	1	$2p_x$, $3d_{zx}$

atom orbital(s) which lead to its generation. The next step is to play the game above, with knowledge of the answers. Choose any row of Table 2.2, say, the second. Abstracting this row from the table, the association between symmetry operations and multipliers is that shown below

$$\begin{array}{cccc} E & C_2 & \sigma_v & \sigma_v' \\ 1 & 1 & -1 & -1 \end{array}$$

Turning now to Table 2.1, everywhere in this table that the operation E is listed, replace it by the number with which it is associated in the chosen row of Table 2.2. That is, in our case it is replaced by the number 1. Similarly, wherever C_2 appears in Table 2.1 it is replaced by 1, whilst both σ_v and σ_v' are replaced by −1. When these replacements have been made, Table 2.3 is obtained. The interesting – and important – thing about this table is that, if it is looked upon simply as a table in which numbers multiply each other, arithmetically, then the products are all correct. It is left to the reader to demonstrate that this statement is true no matter which row of numbers is selected from Table 2.2. Only those sets of numbers contained in Table 2.2 will be found to substitute correctly (hopefully, the reader will have discovered this by playing the game, Problem 2.8). This is a result that could not have been anticipated from the way that the sets of multipliers were obtained. This is the first hint of the fundamental nature of the set of numbers of Table 2.2; more will be met in the next chapter. [7] But this is an unexpected result. It happened very much earlier in the chapter, but the reader may recall that the above discussion was provoked by a desire to show that the set of four diagrams in Figure 2.21 was complete; that no more can be added. Instead, we have found four sets of numbers to which no more sets can be added. Of course, the answer is simple – the diagrams and sets of numbers are equivalent. Replace the + and − phases in Figure 2.21 by the numbers 1 and −1, respectively. Next, pair each number with the symmetry operation associated with the quadrant in which the number falls. One obtains the four sets of Table 2.2 (but without the association with oxygen orbitals!).

2.4 Character tables

We now come to a key point in the argument which is being developed. This is that the differing symmetry properties of, for example, the $2p_y$, $2p_x$ and $2p_z$ orbitals of the oxygen atom in the water molecule (i.e. the fact that their behaviour under the various symmetry operations differs) may be *represented* by the different sets of multipliers which were obtained

[7] This discussion explains why only the numbers 1 and −1 appear in Table 2.2. Had the number 2 been associated with the E operation, for instance, then the product of multiplying E with E, in the group theoretical sense, to give the answer E – and so 2 by substitution – would not be arithmetically correct (arithmetic would call for the number 4).

Table 2.3

	1	1	−1	−1
1	1	1	−1	−1
1	1	1	−1	−1
−1	−1	−1	1	1
−1	−1	−1	1	1

for them. Quantities which have different symmetry properties give rise to different sets of numbers. Because of the close relationship between the multiplication of the operations of the C_{2v} point group (given in Table 2.1) and the multiplication of the numbers in the rows of Table 2.2, each set of numbers may be regarded as representing (i.e. behaving in an analogous way to) the set of symmetry operations. [8] We shall speak of each row of Table 2.2 as being a *representation* of the symmetry operations. Further, we shall call them '*irreducible representations*' (the significance of the word 'irreducible' will not become evident until the next chapter, when the concept of a reducible representation will be introduced). In the discussion that follows it will often be necessary to refer to the individual rows in Table 2.2 and it is convenient to circumvent the need to write each one out in full by giving each a label. The labels commonly used are those shown in Table 2.4.

Thus, the set of numbers given at the beginning of this section (1 1 −1 −1) would be referred to as 'the A_2 irreducible representation of the C_{2v} point group'. This sounds a bit of a mouthful when first encountered but it is the sort of phrase which occurs over and over again in the subject. Because the association between the symmetry operations and irreducible representations given in Table 2.4 is unique to the C_{2v} point group, this is indicated by including the group label in the top left-hand corner of the table.

There is some system about the choice of the labels A_1, A_2, B_1 and B_2 in Table 2.4. The A's are distinguished from the B's by the fact that they have numbers of +1 for the C_2 operation whereas the B's have −1 (in the general case, A's have numbers of +1 for rotation about the axis of highest symmetry whilst B's have a number of −1). A_1, by convention, is the so-called 'totally symmetric' irreducible representation and has +1 for all of its numbers. It is called totally symmetric because *all* the operations of the group turn something of A_1 symmetry into itself. Every group has a totally symmetric irreducible representation. Although it may not be labelled A_1 it is always the first A listed (it could be something like A_g or A' for instance).

The system distinguishes A's from B's and A_1 from A_2. This is really the end, although the distinction can be extended to the B's by noting that irreducible representations with the suffix 1 are symmetric (numbers +1) under the σ_v operation whereas those with suffix 2 are antisymmetric (numbers −1). However, in the case of the B's this distinction is marred by the fact that the distinction between σ_v and σ_v' is somewhat arbitrary – interchange the use of these labels and the labels B_1 and B_2 would have to change too. In practice this means that it is advisable to check the notation used by each author – one worker's notation may not be the same as the next. As long as one is consistent in the notation used for a particular problem there is no ambiguity about the final answer obtained. In fact, this is just the same

[8] Note that the word 'multiplication' in this sentence does not have quite the same meaning when applied to symmetry operations as it does when it refers to numbers.

Table 2.4

C_{2v}	E	C_2	σ_v	σ_v'	
A_1	1	1	1	1	$2p_z(0)$
A_2	1	1	−1	−1	$3d_{xy}(0)$
B_1	1	−1	1	−1	$2p_y(0)$
B_2	1	−1	−1	1	$2p_x(0)$

problem as one discussed above, that of the choice of x and y axes for the water molecule; interchange the choice of x and y and B_1 and B_2 also interchange.

Just as the set of operations (E, C_2, σ_v, σ_v') may be represented by any of the irreducible representations A_1, A_2, B_1, B_2, so, too, individual symmetry operations, such as C_2, are characterized, in each irreducible representation, by a particular number (which, in general, varies from one irreducible representation to the next). These individual numbers are termed *characters* and tables such as Table 2.4 are called *character tables*. As has already been indicated, character tables are of prime importance for the topics discussed in this book. The unexpected properties of the sets of numbers in Table 2.2 become the unexpected properties of character tables. It is these 'unexpected properties' (which are actually fundamental and far from accidental) which are at the heart of their value in chemistry. In the next chapter, when the bonding in the water molecule is discussed, the existence, and to some extent the origin, of these properties will become clear. Because of their importance, this chapter concludes with some further comments on character tables in general and that of the C_{2v} point group in particular.

On the right-hand side of Table 2.4 is indicated the oxygen orbital which was used to generate a particular irreducible representation. Functions which have the property of generating an irreducible representation are commonly listed alongside character tables in this way. Such functions are called *basis functions*. It has been seen that the transformations of the oxygen $2p_y$ orbital under the operations of the C_{2v} point group lead to the B_1 set of characters – the oxygen $2p_y$ orbital is a basis function for the generation of the B_1 characters. This would normally be said a bit more formally:- 'the oxygen $2p_y$ orbital is a basis for the B_1 irreducible representation of the C_{2v} point group'. Alternatively, and more simply, 'the oxygen $2p_y$ orbital has B_1 symmetry in the C_{2v} point group'. [9]

Finally, back to the diagrams of Figure 2.21. We can now see them for what they are. Essentially, they are pictures of the four irreducible representations of the C_{2v} point group. Put another way, the application of group theory to chemical problems is a way of exploiting the differing nodal patterns associated with different phenomena (if this sounds a bit obtuse, there will be plenty of examples in the following chapters). It would be nice if all of this could be done using pictures akin to, or derived from, those of Figure 2.21 – and, indeed, we will do this as far as possible. However, the road may not always be easy. Consider a molecule as simple as methane. Ten diagrams would be needed. That may not be too bad – but they could not have the simple two-dimensional form of those we have met in this chapter and so some form of distortion would have to be introduced to show what are three-dimensional patterns. It gets worse. SF_6 (an octahedral molecule) would need twenty

[9] There is a subtle point here. When the *symmetry* is the focus of attention, an upper case (capital) letter is used. However, when the *orbital* is the focus of attention it is denoted by a lower case symbol: thus, 'the b_1 orbital'. In this book the use of lower case symbols will largely be confined to diagrams.

three-dimensional pictures. C_{60} (an icosahedral molecule) would need thirty-two. It is not surprising that character tables, which have no similar problems, dominate the subject. Further, there are many applications which demand the use of character tables; diagrams will not do – although in such cases it may well be that diagrams may help in understanding the outcome. In this book, diagrams will be used wherever possible and helpful.

One final word, one which is not important at a first reading but which is included to help the reader understand the logic behind the sequence of the chapters in this book and to explain a word that will be used from time to time. The C_{2v} point group is an *Abelian* point group. Abelian groups have multiplication tables which are symmetric about their leading diagonal (top left to bottom right) – inspection of Table 2.1 shows that this is true for the C_{2v} group. That is, the result of multiplying two operations is independent of the order in which they are multiplied – of which operation comes first and which comes second. It is this fact, together with the fact that each operation multiplied by itself gives the identity, that makes the C_{2v} group a particularly simple one to work with. An alternative (but equivalent) definition of an Abelian point group is to regard such point groups as those for which the character tables contain only numbers like 1 and -1.[10] The character tables of Abelian groups never contain numbers such as 3, -3, 2, -2 and 0. The reason why at the beginning of this chapter consideration of the ammonia molecule was deferred is that the character table of its point group contains the numbers 2 and 0 as well as 1 and -1. It is shown in Appendix 1 (Table A1.1) that this originates in the fact that the result of multiplying some of the elements of its group *does* depend on the order in which they are taken; its group is non-Abelian.

2.5 Summary[11]

In a molecule the axis of highest symmetry is conventionally chosen to be the z axis; recommendations for the choice of x and y exist (p. 000). The concern of this book is with point group symmetry operations (p. 000), which are named according to a conventional nomenclature (p. 000). These operations form a group (p. 000). In the present (and the next two) chapters the discussion is restricted to Abelian point groups (p. 000). In such groups, individual quantities – such as atomic orbitals on a central atom (p. 000) – that are transformed into themselves under the operations of the point group may have these transformations described by characters (p. 000). (An example of an Abelian group which shows a more complicated behaviour will be met in Chapter 11). A complete collection of characters is called a character table (p. 000). Each row of characters is called an irreducible representation (p. 000); each of the individual quantities used to generate them are said to be the *basis* for the irreducible representation that it generates (p. 000). Characters multiply together in a way that is isomorphous (p. 000) to the way that the operations of the point group multiply (p. 000). Irreducible representations are given labels in a systematic, but not always unambiguous, way (p. 000). Diagrams may be drawn of irreducible representations and help to demonstrate that irreducible representations differ in their nodal characteristics (p. 000). Each irreducible representation is associated with a unique nodal pattern.

---

[10] In Chapter 11 it will be seen that they can also contain complex numbers, such as i and $-i$, which are such that some power of them equals 1 (thus $i^4 = 1$, for instance).

[11] Page numbers refer to the page in the chapter on which a full discussion commences. Sometimes in the summaries words are used in a way that should be evident from the context but which will be discussed in detail in later chapters, e.g. 'isomorphism'.

2.5 Summary

3 The electronic structure of the water molecule

In Chapter 2 we found that it is possible to obtain sets of four numbers (characters) from the way that the atomic orbitals of the oxygen atom of the water molecule behaved under the symmetry operations of the water molecule. Each set of numbers is called an 'irreducible representation' and one says that 'the atomic orbitals served as bases for the generation of the irreducible representations'. Fortunately, this is not as difficult as it may seem; we were able to draw pictures of each irreducible representation and these showed that the irreducible representations are nothing more than a listing of different nodal patterns. The nodal patterns are a complete set of those which are symmetry-distinct in the point group. Atomic orbitals are not the only things which may serve as bases. Almost anything can; one's imagination is the limiting factor. So, in the following chapters a variety of bases will be met; for instance, when studying the vibrations of a molecule the small displacements of individual atoms will be used as bases. Sometimes, the set of numbers obtained – the representation generated by the transformation properties of a basis set – appear in the character table. This is when an irreducible representation is generated. More commonly, however, the representation generated does not appear in the character table. For the water molecule we can guarantee this when we look at more than one thing simultaneously (for example, the two O—H bond stretch vibrations); all of the examples met so far considered single things. In cases where many things are considered, the representation obtained is often a *reducible* one. One of the representations encountered in the present chapter is a *reducible representation*; by studying it, a method of breaking up a reducible representation into a sum of irreducible representations will be obtained. However, to be able to do this it is necessary to recognize more of the special properties of the irreducible representations than those met in Chapter 2. Again, these will be developed with reference to the character table of the C_{2v} point group.

3.1 The orthonormal properties of irreducible representations

As indicated in Chapter 2, the sets of characters in the C_{2v} character table have properties beyond those which might reasonably be expected from the way that they were derived. There we found that they could be substituted into the group multiplication table to give an arithmetically correct outcome. But there is more, much more. One set of the properties

Table 3.1

C_{2v}	E	C_2	σ_v	$\sigma_v{}'$	
A_1	1	1	1	1	2s(0), $2p_z(0)$, $\psi(A_1)$
A_2	1	1	−1	−1	
B_1	1	−1	1	−1	$2p_y(0)$, $\psi(B_1)$
B_2	1	−1	−1	1	$2p_x(0)$

of irreducible representations proves to be of great importance. Consider any irreducible representation of the C_{2v} point group (Table 3.1) and multiply its individual characters by the corresponding characters of any other irreducible representation. Then sum the products of characters which have been obtained. So, consider as an example the A_2 and B_1 irreducible representations. Giving at the head the operation associated with the characters, the sum of the products of characters is easily obtained:

$$\begin{array}{cccc} E & C_2 & \sigma_v & \sigma_v' \\ (1\times 1) + & (1\times -1) + & (-1\times 1) + & (-1\times -1) = 0 \end{array}$$

In this case, and for *all* others in which the characters of two *different* irreducible representations of the C_{2v} point group are multiplied together, the sum is zero. If, however, instead of multiplying the characters of two different irreducible representations, the characters of an irreducible representation were squared and the answers summed, then a different result is obtained. For the B_2 irreducible representation:

$$\begin{array}{cccc} E & C_2 & \sigma_v & \sigma_v' \\ (1\times 1) + & (-1\ \times -1) + & (-1\ \times -1) + & (1\times 1) = 4 \end{array}$$

The sum of products is equal to four. The same answer would have been obtained no matter which of the irreducible reducible representations had been chosen. Four is also the number of operations in the C_{2v} point group. This is no accidental coincidence. So, as we mentioned in Chapter 2, every character table contains as its first row a series of 1's, the characters of the so-called 'totally symmetric irreducible representation'. It is obvious that for the C_{2v} group if these 1's are squared and added then the result simply counts the number of operations in the group.[1] Because the number of operations in a group turns out to be an important quantity, it is given a name – it is called the *order* of the group. Thus, 'the C_{2v} point group is of order four'.

If, instead of choosing a row of the character table for the calculations of the above paragraph the columns had been selected, a similar result would have been obtained. The sum of the products of the characters in two columns is equal to zero when the characters come from two different columns. If the *same* column is chosen, i.e. the characters squared, then the sum of squares is equal to the order of the group.

These patterns of multiplication between rows and between columns of the character table are known as the character table *orthonormality* relationships; a more general form of them will be discussed in Chapter 6, where it will be shown that they may be used to derive

[1] No matter what the group, as long as it is finite, if the character of each individual operation of the totally symmetric irreducible representation (always 1) is squared and the results added, you just count the number of operations in the group.

character tables as an alternative to the procedure used in Chapter 2. It is in large measure the existence of these relationships which enables symmetry considerations to simplify many problems in the physical sciences. They will be used frequently in this book. The word 'orthonormal' is a composite of the words 'orthogonal' and 'normal' and embodies both. 'Orthogonal' here means 'independent'. When two things are orthogonal it means that one behaves – and can be discussed – without automatically requiring a change to the other. Thus, all the wavefunctions associated with an atom are orthogonal to each other. In the present case, we can talk of different irreducible representations quite independently of each other. If we were talking of atomic orbitals and found that they had an overlap of zero we would say that they are independent, orthogonal. Here, we get the same number, 0, in a different way – but the outcome is the same. 'Normal' or 'normalized' means 'weighted equally' – and equal weighting usually means being given unit weight. This concept is most easily seen for two one-electron wavefunctions of an atom. Each wavefunction is normalized if, when we (mathematically) ask the question 'How many electrons does each wavefunction describe?' we obtain (mathematically) the answer '1'. If we obtained the answer '1' for the first wavefunction but some different answer, say '1.83', for the second we would say that the second was not normalized and we would have to modify it with a multiplicative scale factor so that we did, indeed, get the answer '1' to our question. This scaled wavefunction would also then be said to be normalized. Later in this chapter we shall be effectively normalizing irreducible representations when we divide the number 4 (obtained by simple arithmetic, as shown above) by the order of the C_{2v} point group (the total number of operations in the group), which is also 4, to give the number 1.

Problem 3.1 Check that each of the irreducible representations of Table 3.1 is orthonormal.

3.2 The transformation properties of atomic orbitals in the water molecule

In the present chapter it will be shown that the C_{2v} character table may be used to greatly simplify a discussion of the bonding in the water molecule. As usual, this bonding will be treated as arising from the interaction of orbitals located on the oxygen atom with those on the two hydrogen atoms. For simplicity, the discussion will largely be confined to the valence shell atomic orbitals of these atoms. That is, we shall consider the oxygen 2s, $2p_z$, $2p_x$ and $2p_y$ orbitals together with the two hydrogen 1s orbitals. The transformation properties of the oxygen orbitals have already been discussed (Section 2.2) and symmetry labels placed on them. The results are summarized on the right-hand side of Table 3.1; the hydrogen 1s orbitals will be discussed shortly. It will prove convenient to use phrases like 'orbitals of A_1 symmetry', by which, in the present example, is meant the 2s and $2p_z$ orbitals of the oxygen together with any orbitals of this symmetry which may subsequently be discovered (one arises from the hydrogen 1s orbitals). In a similar way, the $2p_y$ orbital of the oxygen as will be referred to as 'an orbital of B_1 symmetry', by which is meant that the characters of the B_1 irreducible representation describe its transformations under the operations of the C_{2v} point group. All of this sounds more difficult than it really is; remember that labels

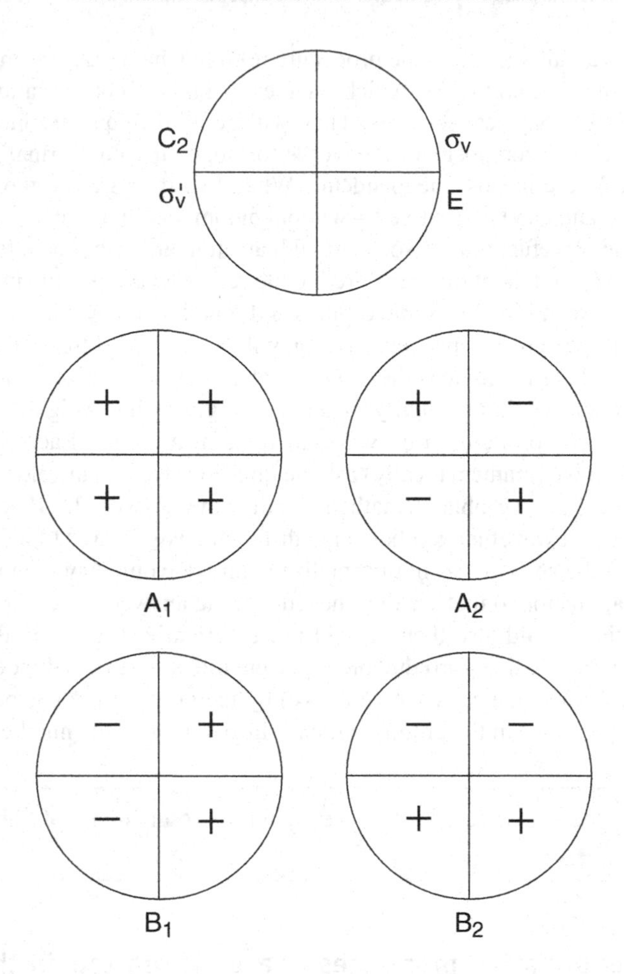

Figure 3.1 Figure 2.21 repeated, but with the addition of labels taken from Table 3.1 (or Table 2.4) added to the phase patterns

like A_1 and B_1 can be regarded as shorthands for the nodal patterns shown in Figure 2.21 as well as for the sets of characters in Table 3.1. To emphasize this point, in Figure 3.1 the irreducible representation labels of Table 3.1 are combined with the nodal patterns of Figure 2.21. In the top line of Figure 3.1 are the definitions of the symmetry relationships of the quadrants (we have just seen how important it is to be aware of these). But these symmetry relationships are not drawn in quite the same way as in Figure 2.21. The marked points are gone. This is because we can chose *any* point in the E quadrant and it will have its symmetry-related equivalents in the other three quadrants. Marked points are helpful when first talking about these diagrams, but if they are kept they tend to obscure the generality of the diagrams.

So far in this book only the transformational properties of individual orbitals have been considered. However, it could happen that it becomes necessary to consider all three 2p orbitals of the oxygen atom together, as a set. What if the z axis had been chosen to lie in some arbitrary direction in the water molecule rather than along the twofold axis (and, similarly, no symmetry constraints placed on the x and y axes – other than that all axes be mutually perpendicular)? It is only convention (and that is motivated by a desire for simple mathematics, as immediately becomes evident) that leads to the choice of C_2 as z axis. A $2p_z$ orbital pointing in an arbitrary direction would not be turned neatly into itself by all of the symmetry operations of the group. The behaviour of the $2p_x$ and $2p_y$ would similarly be complicated. In fact, any one of them, after being rotated or reflected, would have to be described as a linear combination of all three of the starting orbitals. We would have had to treat the orbitals as a set. Evidently, a careful choice of direction for coordinate axes can simplify symmetry discussions![2] Had we persisted in choosing arbitrary (but, of course, mutually perpendicular) directions for our axes the final result would have been the same – we would have ended up with 2p orbitals transforming as A_1, B_1 and B_2. However, the work involved would have been more difficult, although it could be simplified a bit by the use of matrix algebra (there is more on this in Appendix 2).

When we turn to the two hydrogen 1s orbitals in the water molecule and attempt to place symmetry labels on them we are confronted with a similar problem. Should they be considered as individuals or as a pair? The answer to this question is simple (and covers the case of oxygen 2p orbitals oriented along arbitrary axes). Whenever one (or more) operation of the point group has the effect of interchanging or mixing orbitals (or, as sometimes happens, a bit of both) then all of the orbitals which are scrambled must be considered together, as a set. This statement applies not just to atomic orbitals; it also holds for other quantities. For instance, in Chapter 4 it will be seen that the small atomic displacements used in the study of molecular vibrations are often scrambled by symmetry operations and these have to be treated as a set. We return now to the specific problem of the transformation of the two hydrogen 1s orbitals in the water molecule. In Figure 3.2 the behaviour of these two orbitals (which will be denoted h_1 and h_2) under the symmetry operations of the C_{2v} point group is shown. For the C_2 and σ_v' operations the two hydrogen 1s orbitals interchange, but under E and σ_v each remains itself. Something which remains unchanged under an operation gives rise to a character of 1 (the numbers which were introduced as multiplicative factors in Chapter 2 will now be referred to as 'characters', a name first introduced in that chapter). So, when two things remain unchanged it is both reasonable and correct to conclude that each makes a contribution of 1 to the character. An aggregate character of 2 is obtained. This can be generalized:

> When the transformation of several things is being considered together the character which they together generate under a symmetry operation is the sum of the characters which they generate as individuals.

It follows that for the σ_v symmetry operation, like the E, the character is 2, because under both operations the two orbitals remain themselves. However, for the C_2 and $\sigma_v{}'$ operations a situation is encountered which has not previously been met, because the orbitals h_1 and

[2] Throughout this book the author will be making educated choices of coordinate axes which simplify the subsequent discussion. The reader may find it amusing to try to catch him making these simplifications. It is also a helpful exercise.

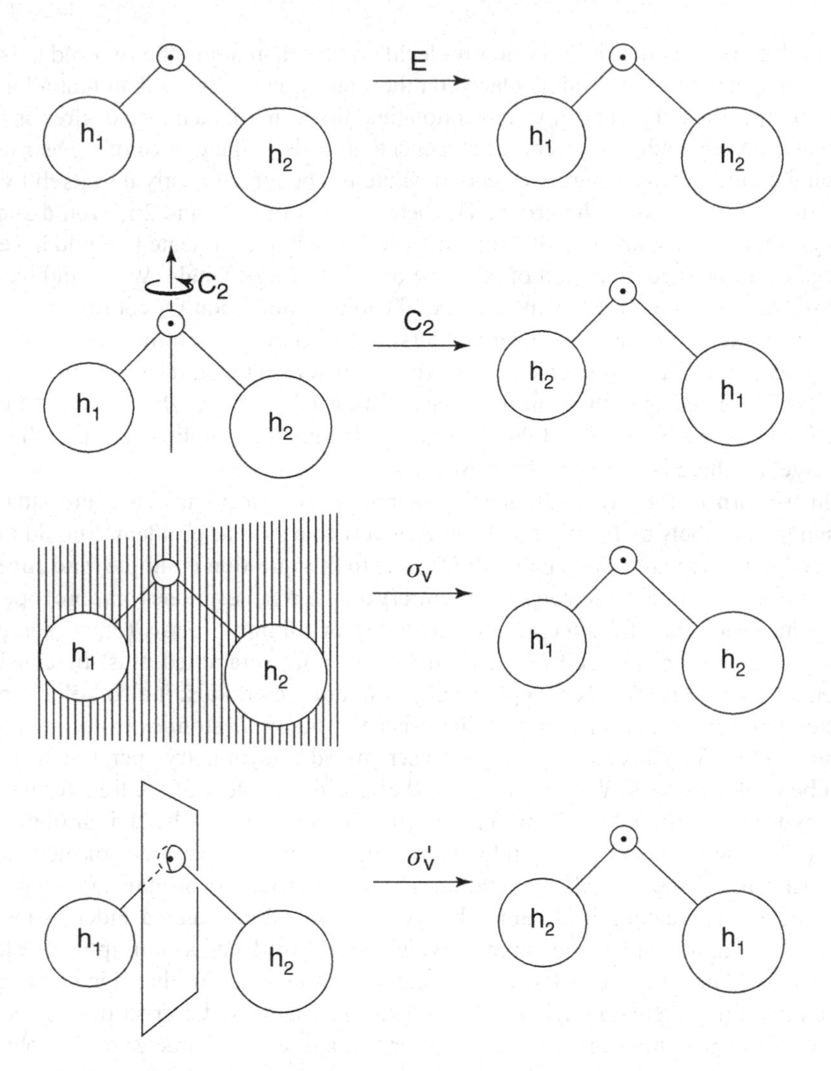

Figure 3.2 The behaviour of the two hydrogen 1s orbitals of H_2O under the symmetry operations of the C_{2v} point group. The point of interest is the permutation of the labels h_1 and h_2

h_2 interchange under these operations. The fact that h_1, for instance, *disappears* from its original position has to be described. The only evident way of doing this is by using a multiplicative factor (character) of zero. The same is true for h_2. We can generalize this result:

> Symmetry operations which lead to all of the members of a set interchanging with each other give rise to a resultant character of zero. If a symmetry operation results in some members interchanging whilst others remain in the same position, it is only the latter which make non-zero contributions to the character.

This discussion contains no explicit recognition of the fact that h_1 and h_2 *interchange* under the C_2 and σ'_v operations – only of their disappearance from their original positions. The correct, and full, way of describing the situation is by matrix algebra. The present discussion is one that avoids the use of matrix algebra and so it is not able to provide a description of the fact that h_1 and h_2 interchange, only that each moves away somewhere. A more detailed treatment, using matrix algebra, is given in Appendix 2; it contains a proof of the two important rules given in boxes above and shows that the transformation of h_1 and h_2 under C_2 and σ_v leads to each contributing zero to the aggregate character.[3]

The set of characters which has just been obtained is:

E	C_2	σ_v	σ'_v
2	0	2	0

This set has properties which are rather different to those of corresponding sets which we obtained in Chapter 2 (and which we called irreducible representations). For instance, when these characters are substituted for the corresponding symmetry operations in the group multiplication table (Table 2.1), the multiplication table obtained is not arithmetically correct.[4]

Problem 3.2 Substitute characters for the corresponding operations in Table 2.1 using the correspondence

E	C_2	σ_v	σ'_v
2	0	2	0

and check that the table obtained is not arithmetically correct.

Problem 3.3 Show that the 1s orbitals of the four hydrogen atoms of the ethylene oxide molecule discussed in Section 2.3 and, in particular, Figures 2.26, 2.27 and 2.28 form a basis for the following representation of the C_{2v} point group

E	C_2	σ_v	σ'_v
4	0	0	0

3.3 A reducible representation

Although the set of characters generated at the end of the previous section is not identical to any of the irreducible representations of the C_{2v} point group, it *is* equal to a sum of two of them. If the corresponding characters of the A_1 and B_1 irreducible representations are added

[3] The zeros encountered in the text are shown to be the diagonal elements of the appropriate transformation matrix.

[4] It is very significant that when the transformation matrices of Appendix 2 are substituted, rather than characters, then the multiplication table obtained *is* correct provided that the rules of matrix multiplication are applied to the matrices.

together, there the same set of characters is generated as those obtained using h_1 and h_2 as a basis:

	E	C_2	σ_v	σ_v'
A_1	1	1	1	1
B_1	1	−1	1	−1
$A_1 + B_1$	2	0	2	0

That is, the representation which was generated by the transformations of h_1 and h_2 can be decomposed into a sum of irreducible representations. A representation which can be reduced to a sum of other representations is, reasonably enough, called a *reducible representation*. The use of the name 'irreducible representation' for the representations appearing in the character table should now be clear. These representations cannot be reduced further, they are irreducible. There are similarities between reducible representations and irreducible representations, but there are also important differences.[5] A most important connection between reducible and irreducible representations is found in the orthonormality relationships (Section 3.1). These relationships provide a systematic way of reducing a reducible representation into its irreducible components. The relationships were introduced by multiplying the characters of two different irreducible representations together. What if, instead, one of the selected representations was reducible and only one was irreducible? That is, if one was a *selected* irreducible representation and the other was a reducible representation? We will see that this provides us with a way of finding out whether the selected irreducible representation is contained within the reducible. By selecting each and every irreducible representation in turn we can discover the breakdown of the reducible representation. This is important. Let us do it carefully for the example of the two hydrogen 1s orbitals in water.

Multiply the individual characters of the reducible representation generated by the transformation of h_1 and h_2 by the corresponding characters of one of the irreducible representations of the C_{2v} character table and sum the products. Choose, for example, the A_2 irreducible representation:

	E	C_2	σ_v	σ_v'	
	2	0	2	0	
A_2	1	1	−1	−1	
multiply:	2	0	−2	0	add; the sum = 0

We get an answer of zero, which tells us that the A_2 irreducible representation is not contained within our reducible representation (something that we already know, because we have already seen that it is the sum of A_1 and B_2).

For the B_2 irreducible representations the answer zero would also have been obtained; no surprise here! For the A_1 and B_1 irreducible representations, however, non-zero answers

[5] One of these has already been seen – for the C_{2v} group the characters of a reducible representation do not multiply arithmetically to give a multiplication table in which there is a consistent correspondence between the numbers it contains and the operations in the corresponding group multiplication table.

result. For example, for the B_1:

	E	C_2	σ_v	σ_v'	
	2	0	2	0	
B_1	1	−1	1	−1	
multiply:	2	0	2	0	add; the sum = 4

It is not difficult to understand in more detail why A_2 and B_1 give different results. We know that the reducible representation is a sum of the A_1 and B_1 irreducible representations. So, in the above procedure we were forming products of A_2 (and, as the second worked example, of B_1) with $(A_1 + B_1)$. That is, in the latter, the B_1 case, we were forming products between B_1 and A_1 and between B_1 and B_1 simultaneously. But the first of these gives a sum which is equal to zero, whilst the second gives a sum of 4 – as was seen in Section 3.1. That is, non-zero answers are obtained by the above procedure when from the character table there is selected an irreducible representation which is contained in the reducible. It is not surprising that an answer of zero was obtained in the A_2 case, because A_2 is not contained in the reducible representation. This recognition leads to a general method for reducing reducible representations into their irreducible components. This method is that which has just been used for the B_1 reducible representation but, because it is so important, it is worthwhile repeating it in detail, applied to the A_1 case:

The steps involved are:

Write down the reducible representation

E	C_2	σ_v	σ_v'
2	0	2	0

Write down the characters of the selected (here, A_1) irreducible representation

1	1	1	1

Multiply the characters in the same column

2	0	2	0

Add these products together and then divide the sum by the order of the group

$$4/4 = 1$$

We conclude that our reducible representation contains the A_1 irreducible representation once. Had it contained A_1 twice (a possible situation) the final answer would have been 2 – and so on.

Problem 3.4 Use the method described above to reduce the following reducible representations of the C_{2v} point group.

	E	C_2	σ_v	σ_v'
(a)	2	2	0	0
(b)	2	−2	0	0
(c)	2	0	0	−2
(d)	3	1	1	−1
(e)	3	−1	1	1

Problem 3.5 Reduce the following reducible representations of the C_{2v} point group and for each check your answer by adding together the characters of the irreducible representations to regenerate those given below (there is an aspect of the irreducible representations in this problem which distinguishes them from those in Problem 3.4 and which makes this check worthwhile).

	E	C_2	σ_v	σ_v'
(a)	3	−1	−3	1
(b)	4	−4	0	0
(c)	4	0	−2	−2
(d)	6	−4	−2	0
(e)	9	1	1	1
(f)	10	−4	−2	0

3.4 Symmetry-adapted combinations

What is the significance of the fact that the reducible representation generated by the transformation of the two hydrogen 1s orbitals h_1 and h_2 may be reduced into a sum of A_1 and B_1 irreducible representations? As has been seen, an irreducible representation such as A_1 describes the transformation properties of a *single* orbital, as too does the B_1 irreducible representation.[6] The significance therefore has to be that it is possible to derive from the orbitals h_1 and h_2 *one* orbital, the transformations of which are described by the A_1 irreducible representation and a second orbital which transforms as B_1. Evidently, the next step is to investigate the form of these orbitals – to find out what they look like. There is a systematic method of carrying out this task but it will not be introduced until Chapter 5 (when it can be given a wider applicability than is possible here). For the present example two rather simpler arguments will suffice.

The reader will recall that in a discussion of the electronic structure of the hydrogen molecule, H_2, two hydrogen 1s orbitals combine to give bonding and antibonding combinations. If, hypothetically, the oxygen atom is removed from a water molecule a hydrogen molecule is left, albeit with a rather stretched H—H bond. It would be reasonable to expect that the combinations of hydrogen 1s orbitals in this stretched H_2 molecule would be related to the correct combinations of hydrogen 1s orbitals in the water molecule. With neglect of overlap between the two atomic orbitals, the bonding and antibonding combinations of hydrogen 1s orbitals in the H_2 molecule have the form

$$\psi(\text{bonding}) = \frac{1}{\sqrt{2}}(h_1 + h_2)$$

$$\psi(\text{antibonding}) = \frac{1}{\sqrt{2}}(h_1 - h_2)$$

[6] This is perhaps best seen in the context of the contents of the first box in Section 3.2, applied to the identity operation, E. The character under E simply counts the number of objects under consideration.

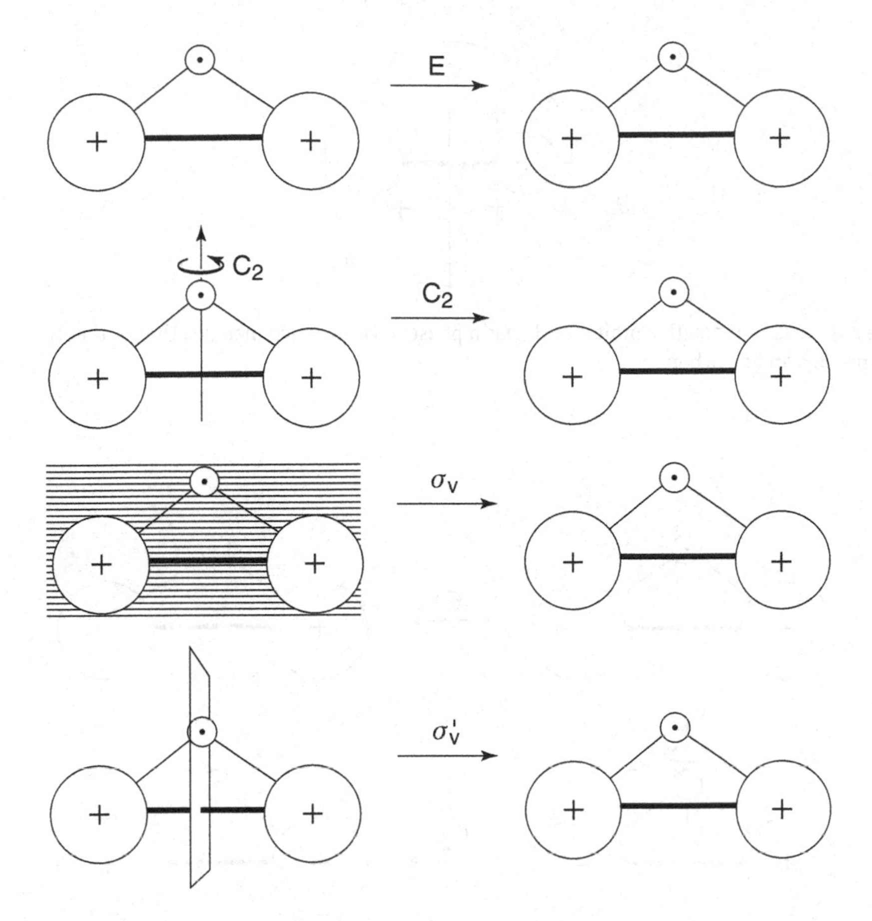

Figure 3.3 The transformations of the H—H bonding orbital of H_2 under the symmetry operations of the C_{2v} point group

where the same labels for the hydrogen atomic orbitals have been used as in the water molecule ($1/\sqrt{2}$ is a normalizing factor – recall the discussion of normality earlier in the chapter; here, without the normalizing factor, we would have squared to get the answer 2). Consider the transformation of these bonding and antibonding combinations under the operations of the C_{2v} point group. The transformations of the bonding combination are shown in Figure 3.3 where, to emphasize the fact that it is a single orbital which is drawn, the component hydrogen 1s orbitals have been joined together. Clearly, under all of the operations of the C_{2v} point group this orbital is transformed into itself. That is, the combination $(1/\sqrt{2})(h_1 + h_2)$ is of A_1 symmetry in the C_{2v} point group. An even simpler way of showing this is to place the picture of the hydrogen 1s bonding combination into the nodal patterns of Figure 3.1. Clearly, and evident in Figure 3.4, the orbital in Figure 3.3 has the nodality of the A_1 irreducible representation.

In Figure 3.5 are shown the transformation properties of the antibonding combination of hydrogen 1s orbitals in the hydrogen molecule. In this figure, again to emphasize the fact that it shows the transformations of a single orbital, the two parts of the orbital are linked.

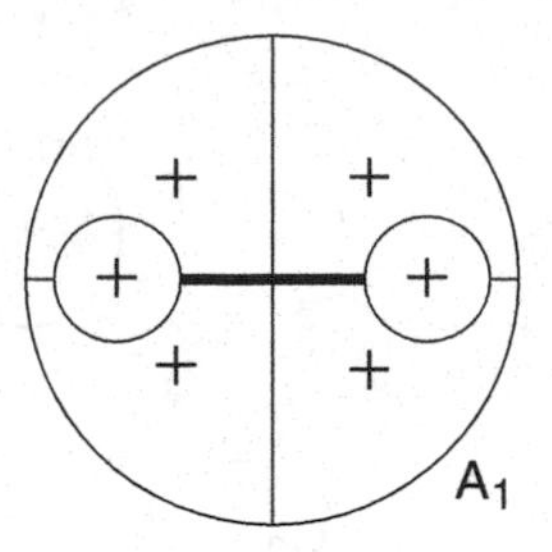

Figure 3.4 The H–H bonding orbital of H_2 has a phase pattern which matches that of the A_1 pattern of Figure 3.1 and no other

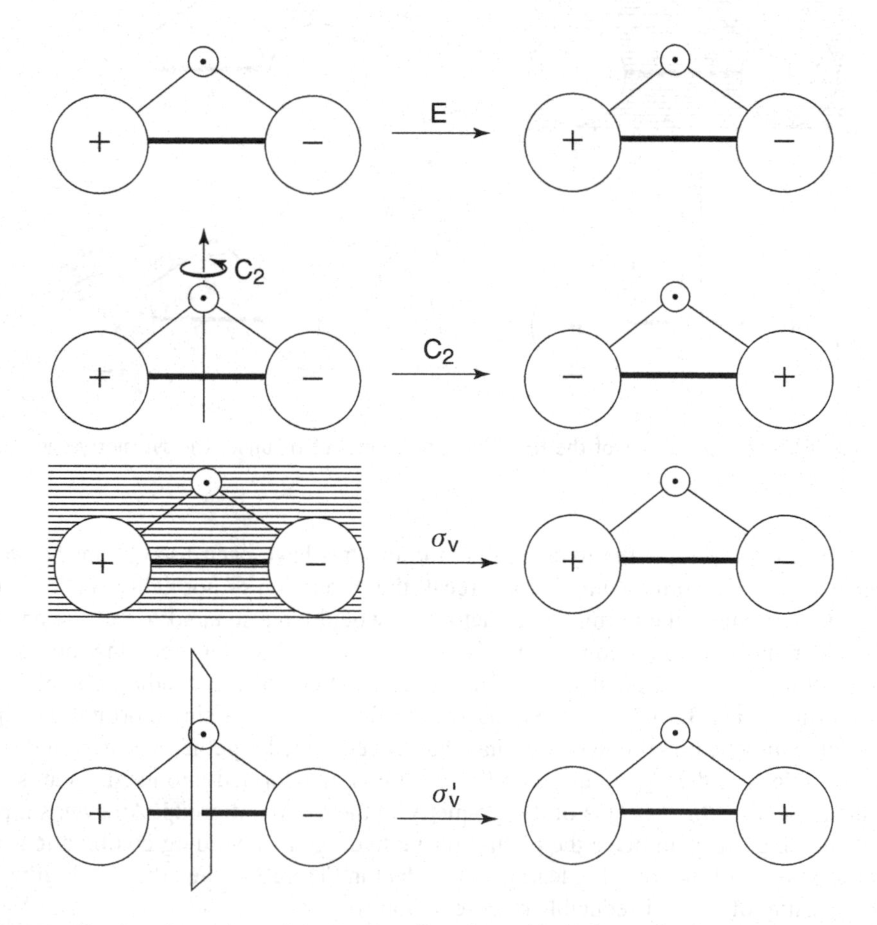

Figure 3.5 The transformations of the H—H antibonding orbital of H_2 under the symmetry operations of the C_{2v} point group. The point of interest is a comparison of the phases of this orbital 'before' (left) and 'after'

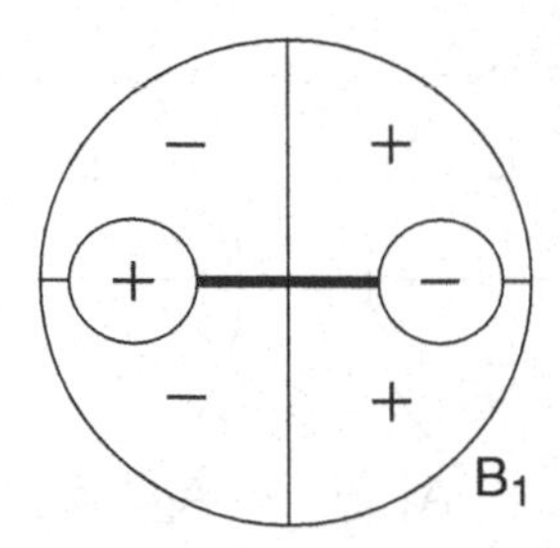

Figure 3.6 The H—H antibonding orbital of H_2 has a phase pattern which matches that of the B_1 pattern of Figure 3.1 and no other. The phase pattern of the orbital is the negative of that of Figure 3.1 but this in no way invalidates the match

The characters generated by the action of the operations of the C_{2v} point group on this orbital are:

E	C_2	σ_v	σ_v'
1	−1	1	−1

so that, as expected, it is of B_1 symmetry. Simpler, of course, is to offer the orbital of Figure 3.5 up into the patterns of Figure 3.1. Provided we do not change our labels in any way, it is for the B_1 pattern that a match is obtained, albeit with the phases all changed (Figure 3.6). Had we been sloppy and placed the nodal plane of Figure 3.5 in a horizontal (in the page) position, instead we would have obtained the B_2. This wrong result appears because we have ignored the underlined instruction in the previous sentence. Effectively, we have changed our definitions of σ_v and σ_v', making the latter, not the former, the molecular plane. Not for the first time, a warning: take care to work within a single axis system and not to change it inadvertently.

We have found that the bonding and antibonding 1s molecular orbitals of the hydrogen molecule have the symmetries A_1 and B_1, respectively, in the C_{2v} point group. From this we conclude that the two hydrogen 1s orbitals in H_2O also combine to give A_1 and B_1 combinations. This argument depended on the exploitation of the relationship between H_2O and H_2. Is there another way? Fortunately, the answer is 'yes', and we have half used it above. The alternative approach is to use a nodal pattern method. We have already used Figure 3.1, which repeats the nodal patterns drawn in Figure 2.21 and which are equivalent to the irreducible representations of the C_{2v} group. Now we use the entire figure, not bits. It is repeated in Figure 3.7 but in the bottom row of this figure are drawn the two hydrogen 1s orbitals, h_1 and h_2, in the same projection as the rest of the figure. We now ask the simple question: which of the diagrams in the upper two rows are compatible with that in the lower? The two hydrogen is orbitals in the bottom row must have the same phase across the σ_v mirror plane (although the diagram in the bottom row does not detail phases, it is clear that the unspecified phases must have this property). The only diagrams in the upper part of Figure 3.4 which share this property are the A_1 and B_1. We conclude that h_1 and h_2, together, somehow participate in combinations of A_1 and B_1 symmetry. It follows that we have also answered the question of what these combinations look like; one is a zero node combination and the other has a single node. We have not generated a normalizing factor, but in truth we did not when working with H_2; we imported it. But clearly, we have

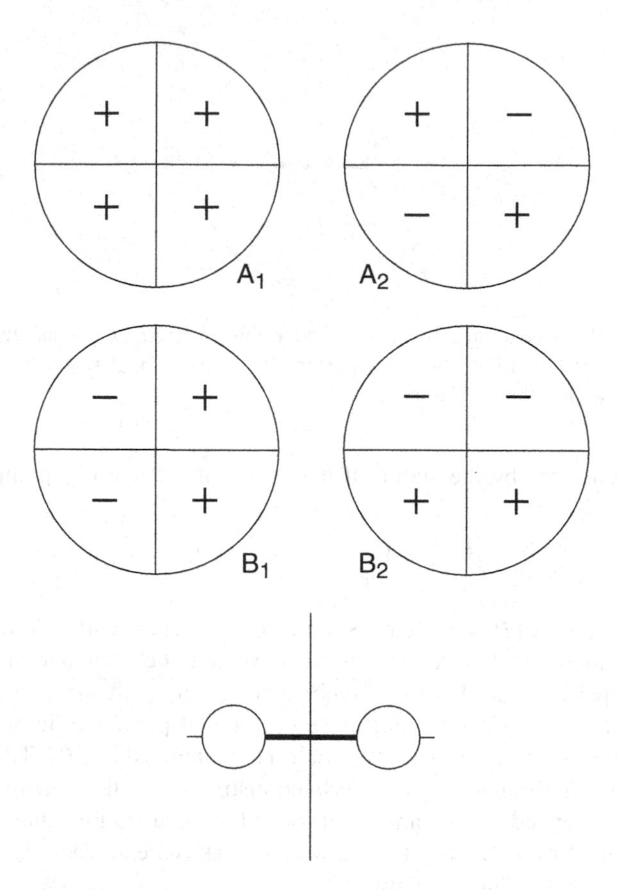

Figure 3.7 Part of Figure 3.1 repeated, along with the two hydrogen 1s orbitals of H_2 drawn superimposed on the phase lines of the diagrams above. The bottom diagram has to be offered up to each of the upper four in turn. There is no specification of phase in the H orbitals; one has to ask whether it is possible to add then in such a way that they match those of the upper pattern being tested

moved on from the point where, at the beginning of this section, it was shown that there are A_1 and B_1 combinations of hydrogen 1s orbitals in the water molecule but it was not possible to say what they looked like. Evidently, in the water molecule, rather than treating the two hydrogen 1s orbitals separately, it is both necessary and possible to work with two combinations, one of A_1 symmetry and one of B_1. The A_1 combination is:

$$\psi(A_1) = \frac{1}{\sqrt{2}}(h_1 + h_2)$$

and the B_1 combination is:

$$\psi(B_1) = \frac{1}{\sqrt{2}}(h_1 - h_2)$$

The arguments used to obtain the mathematical forms of $\psi(A_1)$ and $\psi(B_1)$ were based as much on plausibility as mathematics. As indicated above, a more rigorous method will be developed in Chapter 5.

3.5 The bonding interactions in H_2O and their angular dependence

The two linear combinations of hydrogen 1s orbitals in the water molecule which transform as the A_1 and B_1 irreducible representations have now been obtained. Although the mathematical form of these orbitals is one which neglects overlap between h_1 and h_2, this neglect in no way affects their symmetry species.

We now come to a vital point in our argument. It involves as the key step an assertion which, for the moment, the reader is asked to take to some extent on trust. A proof will be given in Chapter 4 although a partial justification is included here. The assertion is that:

> Interactions between orbitals transforming as different irreducible representations are always zero.[7]

That is, in a discussion of the bonding in a molecule the argument can be broken up into smaller, separate, discussions, one for each irreducible representation. This is an enormous simplification; the more the molecular symmetry, the greater its value – the more it breaks the discussion up into bits, bits which of course get smaller and simpler the more there are of them. In the case of the water molecule, for example, the only orbital of B_2 symmetry is the $2p_x$ orbital of the oxygen. There is no hydrogen 1s combination of this symmetry and so the assertion in the box above leads us to conclude that the oxygen $2p_x$ orbital does not interact with any other orbital in the molecule. That is, it is a non-bonding orbital located on the oxygen atom. This conclusion required virtually no work to obtain, yet it gives us chemically useful information on one of the orbitals of the water molecule.[8] Symmetry arguments are useful! Incidentally, as a little thought about their transformation properties should confirm, when two (or more) orbitals of the same symmetry species interact, the final molecular orbitals are all of the same symmetry species as the initial orbitals. For the water molecule, for instance, they must have in common one of the nodal patterns in Figure 3.1 (they can have additional nodal planes too, but these cannot include those in Figure 3.1).

Figure 3.8 provides some justification for the assertion that consideration of bonding interactions can be confined to those between orbitals of the same symmetry species. It shows the overlap between the 2s orbital of the oxygen – of A_1 symmetry – and the B_1 combination of hydrogen 1s orbitals, $\psi(B_1)$. It is evident that, although these orbitals overlap with one another, the overlap *integral* is zero since the regions of positive overlap are exactly cancelled by the regions of negative overlap. The zero overlap integral between $2p_x$ and the A_1 or B_1 combinations of hydrogen 1s orbitals $\psi(A_1)$ and $\psi(B_1)$ – very relevant to the conclusion that $2p_x$ is a non-bonding orbital – can similarly be demonstrated. Of course, all this is making a simple job look difficult. The results can all be seen as stemming from Figure 3.1. The overlap between any two different diagrams (of the lower four) is zero because of their different nodalities. And this we have met before when, earlier in this chapter, we

[7] This statement concerns one-electron terms in the Hamiltonian. An analogous statement may be made which covers the two-electron terms. The general form of such statements will become evident in Chapter 4.

[8] We have also found an orbital of A_2 symmetry – the (empty) d_{xy} orbital of Figure 2.20. It could be included in high quality calculations on the water molecule and, almost certainly, in calculations on analogous molecules such as H_2Se and H_2Te.

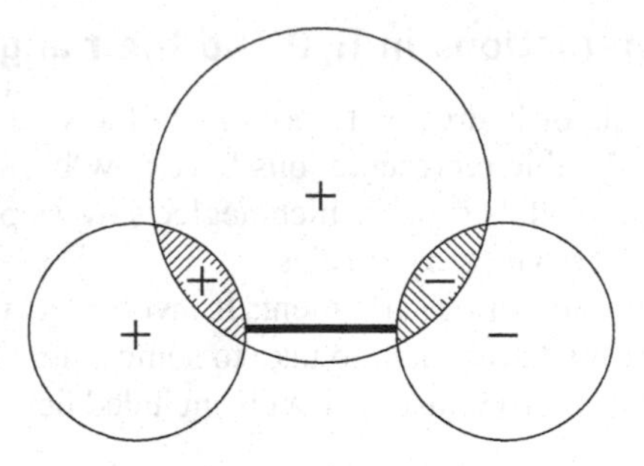

Figure 3.8 The zero overlap *integral* between an orbital of A_1 symmetry (oxygen 2s) and an orbital of B_1 symmetry (a linear combination of hydrogen 1s orbitals)

talked of the orthonormality of the irreducible representations of the C_{2v} point group. And lest this repetition of the same thing from different angles is beginning to become wearing to the reader, the author's reason (or excuse, if the reader prefers) is that it is a topic of vital importance; it underpins all that is to follow and the reader had better feel comfortable with it.

We are now in a position to delve deeper into the bonding in the water molecule. The only interactions which can be involved in the bonding are of A_1 and B_1 symmetries (since these are the only symmetries associated with the two hydrogen 1s orbitals). The basis functions associated with the C_{2v} character table – Table 3.1 – provide a list of all the orbitals which interact with one another. The orbitals of A_1 symmetry which must be discussed are the 2s and $2p_z$ orbitals on the oxygen, each of which interacts with the hydrogen 1s combination $\psi(A_1)$. There are no orbitals of A_2 symmetry and only two of B_1 symmetry – the $2p_y$ orbital of oxygen and a hydrogen 1s combination $\psi(B_1)$. The oxygen $2p_x$, B_2, orbital is non-bonding, of course. We shall first consider the B_1 interactions qualitatively but in some detail.

The interaction between the $2p_y$ orbital of the oxygen and the B_1 combination of hydrogen 1s orbitals will lead to bonding and antibonding molecular orbitals. A schematic representation of the overlap between $2p_y$ and $\psi(B_1)$, together with the form of the resultant bonding and antibonding molecular orbitals, is shown in Figure 3.9. The bonding molecular orbital is an out-of-phase combination of $2p_y$ and $\psi(B_1)$ whilst the antibonding molecular orbital is an in-phase combination (if this pattern seems strange and the reader automatically expects in-phase to be bonding, compare the relative phases of $2p_y$ and $\psi(B_1)$ with Figure 3.6).

The B_1 bonding orbital will surely be occupied by two electrons in the water molecule and so contribute to the molecular stability. There is an important point which must be made concerning this bonding molecular orbital. Consider, qualitatively, the dependence of the molecular stabilization derived from this orbital upon the HOH bond angle, θ. For the (hypothetical!) case of very small θ, shown in Figure 3.10, the lobes of $\psi(B_1)$ overlap with $2p_y$ in a way that leads to a relatively small value for the overlap integral between them; the overlap integral decreases as the bond angle decreases. When θ is very small, then, the interaction between the two orbitals of B_1 symmetry, which varies roughly as the overlap integral, will be small and the B_1 bonding molecular orbital will make little contribution

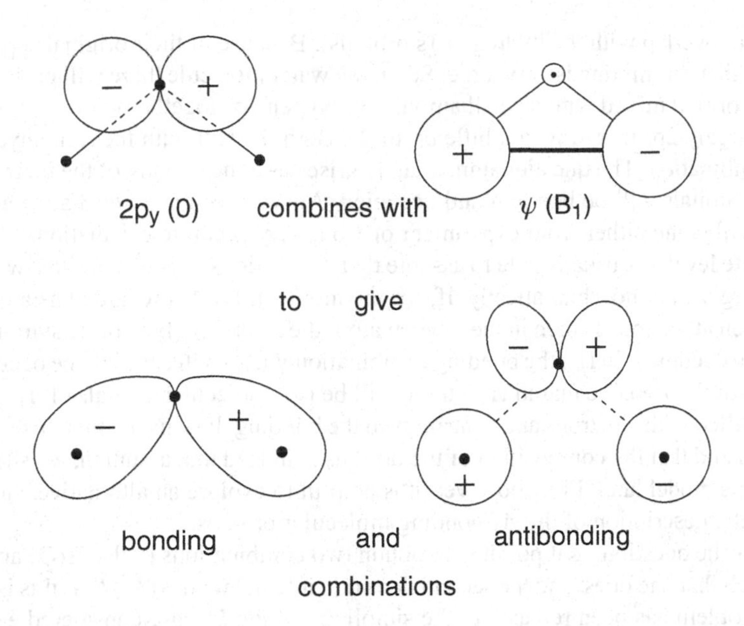

Figure 3.9 Interaction between orbitals of B_1 symmetry, leading to bonding and antibonding combinations

to the molecular stability. As is qualitatively evident from Figure 3.10 (and is confirmed by more detailed calculations), as θ increases (keeping the O—H bond length constant) the interaction between $2p_y(O)$ and $\psi(B_1)$ smoothly increases with θ and reaches a maximum for a bond angle of 180°. It is legitimate to conclude that, were this interaction the only thing determining the geometry of the water molecule, then H_2O would be a linear molecule. But it is not! Clearly, there is more to come!

The interaction between the three orbitals of A_1 symmetry is a more difficult problem simply because there are three orbitals to consider, not two. In the case of the B_1 interaction, the final molecular orbitals were mixtures of the two starting orbitals. Similarly, we would expect the final A_1 molecular orbitals to be mixtures of 2s(O), $2p_z(O)$ and $\psi(A_1)$. Although 2s(O) and $2p_z(O)$ were introduced as separate functions they will be mixed (i.e. contribute to the same molecular orbitals) by virtue of their mutual interaction with $\psi(A_1)$ (induced by

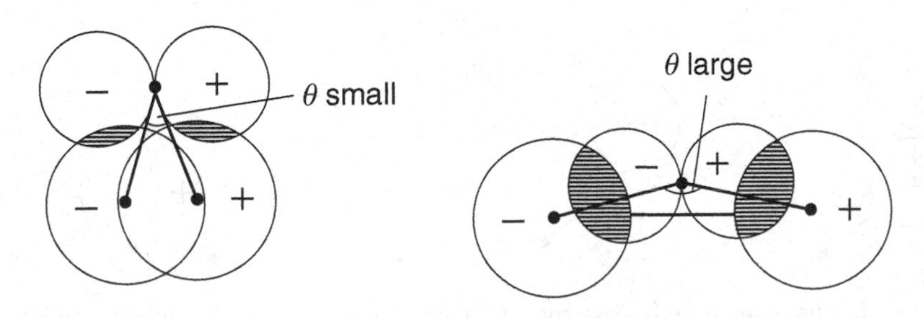

Figure 3.10 Variation of overlap integral of the B_1 orbitals with variation of the bond angle in H_2O

their mutual overlap with the hydrogen 1s orbitals). Because of their original separation, it is unlikely that this mixing is very large. So, in the water molecule there will probably be an oxygen 2s orbital mixed with a small amount of oxygen $2p_z$ together with a second orbital which is largely $2p_z$ mixed with a little bit of 2s. Both interact with the same hydrogen 1s orbital combination. The question immediately arises as to how many of the three resulting molecular orbitals will be bonding and occupied. Answers of guaranteed accuracy to this question will come either from experiment or from very accurate calculations. At a more approximate level it is usually safe to assume that interactions between orbitals will change orbital energies, but not dramatically. If, for the moment, H_2O is regarded as a composite of an oxygen atom and H_2 then in the oxygen atom the 2s orbital, here of A_1 symmetry, will certainly be occupied. In H_2, the bonding combination $\psi(A_1)$ will certainly be occupied. So, it seems entirely probable that in H_2O there will be two molecular orbitals of A_1 symmetry which are filled with electrons and contribute to the bonding. It turns out that this is a correct description and that the composition of the orbitals is indeed that assumed; we shall briefly return to this model later. First, however, it is helpful to explore an alternative, simpler but less accurate, description of the A_1 bonding molecular orbitals.

Consider the question 'is it possible to obtain two combinations of the 2s(O) and $2p_z$(O) orbitals such that one does, and the second does not, interact with $\psi(A_1)$?'. If this is possible then the problem has been reduced to the simplicity of the B_1 case considered earlier; the interaction between two orbitals, not three. The simplification is possible, and the general way that it may be achieved is indicated in Figure 3.11. Figure 3.11 shows, schematically, that if in-phase and out-of-phase combinations of 2s(O) and $2p_z$(O) are taken then one of the resulting mixed (hybrid) orbitals is directed towards the hydrogen atoms whilst the second combination is largely located in a region remote from them. This second combination would be essentially non-bonding and we may, as a first approximation, ignore its interaction with $\psi(A_1)$. It is convenient to choose 2s(O) and $2p_z$(O) combinations which simplify the pictorial representation of the problem and, therefore, to assume that they are sp hybrids of the form:

$$\frac{1}{\sqrt{2}}[2s(\mathrm{O}) + 2p_z(\mathrm{O})] - \text{non-bonding}$$

$$\frac{1}{\sqrt{2}}[2s(\mathrm{O}) - 2p_z(\mathrm{O})] - \text{involved in bonding}$$

and it is these which are shown, qualitatively, in Figure 3.11.

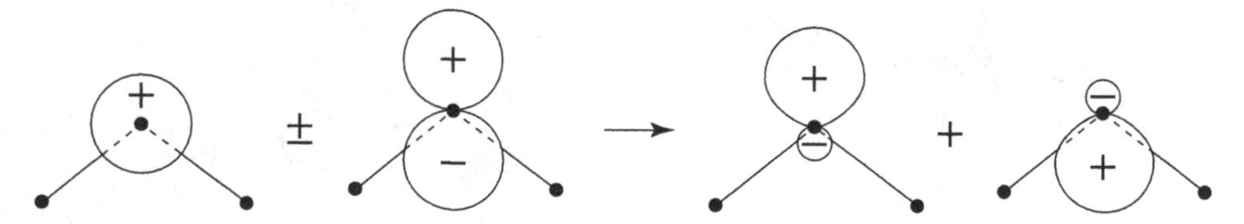

Figure 3.11 In-phase and out-of-phase combinations of oxygen 2s and $2p_z$ orbitals give two (sp) hybrid orbitals

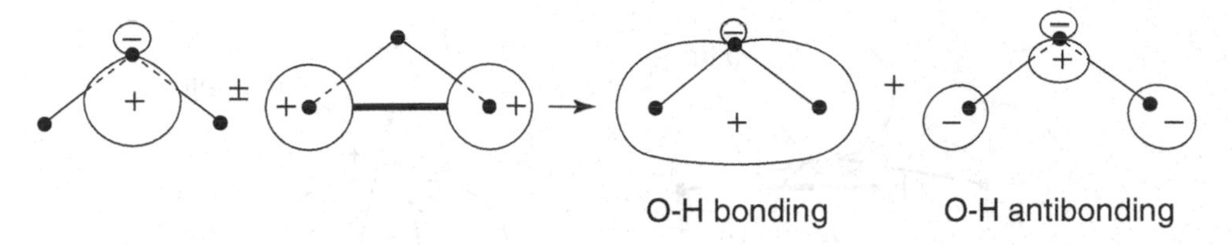

Figure 3.12 Interaction between orbitals of A_1 symmetry, leading to bonding and antibonding combinations

The problem has now been simplified so that it is analogous to that discussed earlier for the case of the B_1 interactions; we have only the interaction between two orbitals, $\psi(A_1)$ and the second given above, to consider. These two orbitals will combine to give in-phase and out-of-phase combinations which are, respectively, bonding and antibonding molecular orbitals. These orbitals are shown schematically in Figure 3.12. Although excluded from participation in the bonding, the non-bonding orbital given above will surely be occupied. It may be identified with a lone-pair orbital in the water molecule, the second lone pair orbital being $2p_x$, which, as has been seen, is of B_2 symmetry. Note two things about the way that this argument was developed. First, it was only allowable for us to mix 2s and $2p_z$ because they are of the same symmetry. Secondly, although they are of the same symmetry, we were not *required* to mix them. We only did so because of the pattern which emerged – one that provided a basis for the idea that the water molecule contains two sterically active lone pairs. In reality, any mixing would have to be caused by the presence of the two hydrogen atoms and not separate from them (although in the last model discussed we mixed the oxygen orbitals with the hydrogens appearing as little more than spectators!).

To complete the picture of the bonding in the water molecule, consider the relationship between the stabilization resulting from the interactions between the various orbitals of A_1 symmetry and the value of the HOH bond angle. For this discussion it proves convenient to consider the interactions involving the 2s(O) and $2p_z$(O) orbitals separately. It is evident, from Figure 3.13, that, if the O—H distance is kept constant, the magnitude of the interaction between 2s(O) and the hydrogen 1s combination $\psi(A_1)$ does not depend upon the HOH bond angle. Because the oxygen orbital is spherically symmetrical the overlap integral is

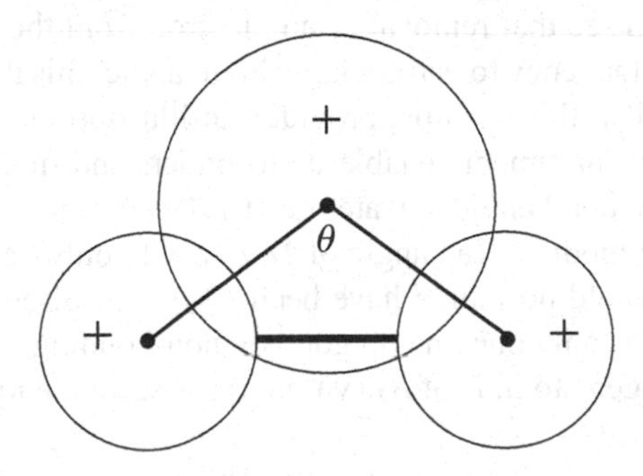

Figure 3.13 An (A_1) overlap integral which does not depend on bond angle because all orbitals concerned are spherical

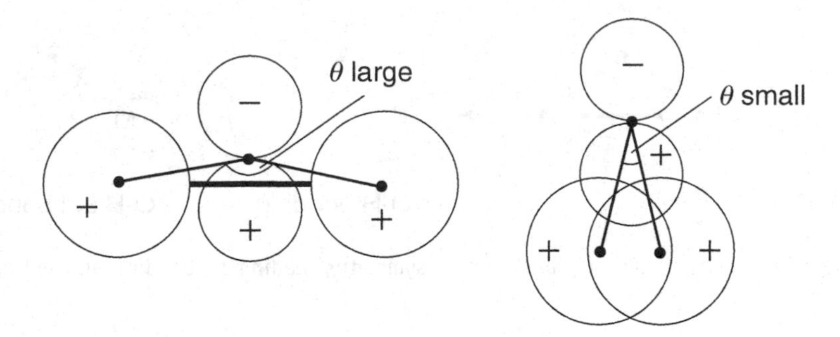

Figure 3.14 An (A_1) overlap integral which varies with bond angle (cf. Figure 3.13) because one of the orbitals involved is non-spherical. When θ is 180° the hydrogen orbitals overlap equally with the two lobes of the p orbital, giving a zero overlap integral

independent of the bond angle. So, this interaction does not vary with bond angle – and this is why it is convenient to consider the 2s(O) and $2p_z$(O) orbitals separately in this and the next paragraph.

The $2p_z$(O)–$\psi(A_1)$ interaction is shown schematically in Figure 3.14. It is evident from Figure 3.14 that when $\theta = 180°$ there is a zero overlap integral and so no interaction between $2p_z$ and $\psi(A_1)$. This result has its origin in molecular symmetry. The symmetry of a linear water molecule is no longer C_{2v} but that of a different point group (called $D_{\infty h}$). In this latter group the $2p_z$(O) orbital and $\psi(A_1)$ are of different symmetries; it follows that they do not interact.[9] Evidently, the interaction between $2p_z$(O) and $\psi(A_1)$ increases smoothly as the H—O—H bond angle decreases from a value of 180° and reaches a maximum at the physically unrealistic value of $\theta = 0°$. Because the interaction of $\psi(A_1)$ with 2s(O) is independent of bond angle it is the interaction of $\psi(A_1)$ with $2p_z$(O) which determines the angular variation of the interaction of $\psi(A_1)$ with mixtures of 2s(O) and $2p_z$(O) orbitals; this interaction will tend towards a maximum at $\theta = 0°$. If the A_1 interactions were solely responsible for the bonding of the water molecule then the bond angle would be very small – eventually, of course, repulsive interactions would prevent a total collapse of the bond angle.

In summary, of the bonding interactions in the water molecule, those of A_1 symmetry favour a bond angle $\theta \to 0°$ and that of B_1 symmetry leads to a stabilization which maximizes as $\theta \to 180°$. The two interactions are opposed and the observed bond angle represents a compromise. It is easy to see that removal of an electron from the B_1 bonding molecular orbital would reduce the tendency towards a large bond angle; this theme is developed later in this chapter. Incidentally, this example provides an illustration of an assertion made in Chapter 1: that symmetry arguments enable us to understand molecular structure rather than to predict it. Had the bond angle in water been 170° – or more – the discussion above could have been suitably modified (an angle of 180° would only have presented problems because the symmetry would no longer have been C_{2v}). Our discussion has also enabled us to conclude that there is only one unambiguously non-bonding orbital. This, a pure $2p_x$ atomic orbital of the oxygen atom, is of B_2 symmetry. A second entirely non-bonding lone

[9] This is an example of a much-exploited trick in group theory. If a molecule *almost* has a higher symmetry (more symmetry operations), then the symmetry-imposed requirements of this higher group *almost* hold. Many a weak spectral band (forbidden in the higher symmetry) has been explained in this way.

pair does not exist in the isolated water molecule (although most simple descriptions of the bonding in the molecule include one). However, in a rather less accurate model, a second non-bonding orbital of A_1 symmetry can be introduced. The physical evidence for two lone pairs, and it is this evidence which provides the motivation for the simple pictures of bonding in the water molecule, comes largely from structural data. Thus, in ice each oxygen is roughly tetrahedrally surrounded by four hydrogens, two close and two distant. It seems reasonable that each of the distant hydrogens should be associated, by hydrogen bonding, with a lone pair. However, it must not be forgotten that attaching two more protons to the water molecule, even loosely, will modify its electronic structure. So, for instance, each of the A_1 molecular orbitals of an isolated water molecule will also be involved in the bonding of these additional protons (just as is the case in methane). Further, the reader should be able to show that the 1s orbitals of the distant hydrogens give rise to a combination of B_2 symmetry, so that even the – genuinely on a symmetry model – non-bonding p_x orbital of H_2O may become weakly involved in bonding in ice.

Problem 3.6 The individual oxygen atoms in ice are surrounded by a distorted tetrahedron of hydrogen atoms. That is, they resemble the carbon atom of Figure 2.1 but two of the oxygen–hydrogen bonds are longer than the other two. The closely bonded pair are those discussed in Section 3.3. Show that the transformation of the 1s orbitals of the more distant hydrogen atoms gives rise to a reducible representation with $A_1 + B_2$ components.

Problem 3.7 In a discussion of the bonding in H_2S (bond angle 93°) the valence shell orbitals on the sulphur are 3s, 3p and 3d. Inclusion of the 3d orbitals would increase the number of possible interactions with $\psi(A_1)$ and $\psi(B_1)$ which would have to be considered. List all of the sulphur valence shell orbitals which could interact with each of them (use Table 2.4 and the results of Problem 2.3).

3.6 The molecular orbital energy level diagram for H_2O

The discussion has now reached the point at which it is possible to obtain a schematic molecular orbital energy level diagram for the water molecule. Rather than work with the first model presented above, the more accurate, we shall consider the approximate. This is because it is the model that is the more compatible with relatively simple ideas about the bonding in H_2O – it is the one most likely to be produced by simply following chemical intuition. Notwithstanding its approximate nature, it gives a good prediction of the *relative* ordering of molecular orbital energies and so gives hope that similar approximate models will be of value for other molecules also. We proceed by presenting a schematic energy level scheme in Figure 3.15 and then detail the arguments used in its derivation. Before doing this note that, in contrast to the discussion in the text, lower case symbols have been used in Figure 3.15. Strictly, as has been briefly mentioned earlier in a footnote, *lower case symbols are used to describe wavefunctions*. Any wavefunctions – they could be vibrational wavefunctions, for instance – and this usage will be met in Chapter 4. Thus, a one-electron wavefunction, reasonably enough, characterizes a single electron. However, an orbital can be occupied by two electrons, each with the same (three-dimensional, spacial) wavefunction.

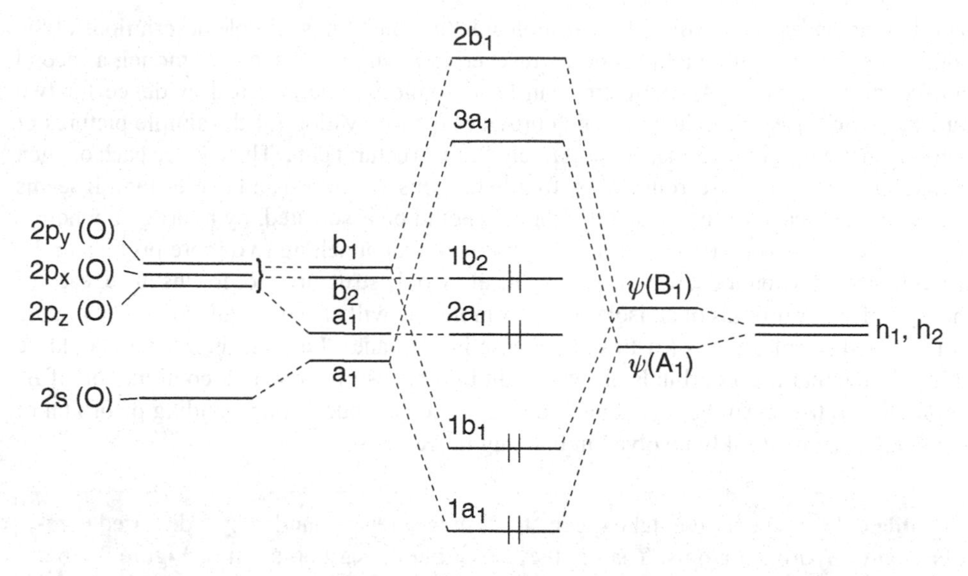

Figure 3.15 A schematic molecular orbital energy level diagram for H_2O

The distinction in usage is a rather fine one and many, but not all, authors use lower case symbols for both orbitals and wavefunctions. The convention seems to be that in diagrams such as Figure 3.15 lower case symbols are used but that in the associated text one is as likely to meet either 'the orbital of A_1 symmetry' as 'the a_1 orbital'. In this book, upper case symbols will largely be used in the text and lower case symbols largely confined to diagrams (the reader is invited to keep an eye open for the points in the coming chapters where the author has deviated – and try to work out why). One important context in which lower case symbols are (almost!) invariably used in the literature is when orbital occupancies are specified. So, Figure 3.15 shows an orbital occupancy, starting with the lowest energy first and labelling orbitals of the same symmetry sequentially ($1a_1, 2a_1, 3a_1$ ..), which is written as

$$1a_1^2 1b_1^2 2a_1^2 1b_2^2$$

Here the superscripts indicate the number of electrons in each orbital, two in each case.

We now return to the details of Figure 3.15 and explain the arguments leading to the orbital energy sequence shown there. Firstly, we have used a nodal plane criterion, which experience shows to be reliable.[10] In the simple model of the water molecule there are just two bonding molecular orbitals, one of A_1 and one of B_1 symmetry. Figures 3.12 and 3.9 reveal that although for both A_1 and B_1 bonding molecular orbitals the oxygen–hydrogen interactions are entirely bonding, in the B_1 case the hydrogen–hydrogen interactions are antibonding – $\psi(B_1)$ corresponds to the antibonding combination of 1s atomic orbitals in the hydrogen molecule. The corresponding component in the A_1 bonding molecular orbital is bonding (Figure 3.12) and so it is reasonable to anticipate that this orbital will be more stable than the B_1. That is, the orbital with the smallest number of nodal planes is usually expected to be the most stable. Secondly, to obtain the most probable order of the two non-bonding

[10] E.B. Wilson, *J. Chem. Phys.* **63** (1975) 4870.

molecular orbitals (of A_1 and B_2 symmetries) note that the 2s orbital of the isolated oxygen atom is of a lower energy than the 2p's. It seems reasonable, therefore, that $\psi(A_1)$, which contains a 2s component, should be of lower energy than the B_2 non-bonding orbital which is pure $2p_x$ (the same argument is also relevant to the A_1 and B_1 bonding molecular orbitals and reinforces the previous conclusions about their relative order). There is another point which must be made in connection with Figure 3.15. In this figure the interaction between the hydrogen 1s orbitals h_1 and h_2 is shown as removing the degeneracy of these two orbitals. This splitting corresponds to the separation between the bonding and antibonding molecular orbitals of H_2 (much reduced in the present case because of the large separation between the two hydrogen atoms). On the other hand, the mixing of the 2s and 2p orbitals of the oxygen combination $\psi(A_1)$ is shown as bringing these two closer together in energy. This is because if the combinations have the idealized sp hybrid forms which they were given earlier then they would be precisely equivalent (although differently orientated in space). It follows that if we were to work out energies associated with these two hybrids then we would expect to obtain the same result for each. The conclusion is, therefore, that when orbitals on the same atom are mixed to give general – not idealized, not like sp – hybrid orbitals these hybrids should be regarded as having energies intermediate between those of their components.

In the water molecule there are eight valence electrons available to be allocated to the four lowest molecular orbitals shown in Figure 3.15 (the electron configuration of the oxygen atom is $1s^22s^22p^4$ and contributes six valence electrons; each hydrogen is $1s^1$ and contributes one). It follows that the lowest four orbitals, two non-bonding and two bonding molecular orbitals, are occupied.

3.7 Comparison with experiment

Is there any experimental test of the model which has just been developed? The most pertinent test would be the observation of individual orbital energy levels. Such data are provided by photoelectron spectroscopic measurements, in which electrons are ejected from individual molecules by high energy monochromatic radiation in a high vacuum. The difference between the (measured) kinetic energy of an ejected electron and the energy of the incident photons is the energy required to remove the electron from the molecule. A variety of electron energies results, corresponding to a variety of molecular ionization energies. These ionization energies correspond very closely to the usual definition of orbital energy. An orbital energy is defined as the energy required to remove an electron from a molecule subject to the restriction that the orbitals of the other electrons in the molecule are unchanged. Evidently, this is a theoretical, rather than practical, definition – some readjustment of the orbitals of the residual electrons would be expected. Fortunately, however, the effect of these readjustments is usually rather small. It is therefore possible to use the ionization energies given by photoelectron spectroscopy to test our model. For larger molecules the test is not always so easy and sometimes there is a problem in that, strange as it seems, the highest energy electron is not necessarily the easiest to remove.[11]

[11] If there is a large electron distribution rearrangement when an electron is removed then this is an additional energy term which has to be included in the balance sheet. For large molecules where some orbitals are very diffuse and others very localized, it is the latter that have large rearrangement energies.

The photoelectron spectrum of water shows four peaks (of energies 12.62 eV, 13.78 eV, 17.02 eV and 32.2 eV). Qualitatively, then, the photoelectron measurements support the energy scheme given in Figure 3.15. There are four different ionization energies arising from valence-shell electrons. A detailed analysis of the photoelectron spectrum can also give some idea of the symmetry species of the molecular orbital from which a particular electron is photo-ejected. In the case of the water molecule the 12.62 eV peak is probably associated with ionization from a B_2 orbital, the 13.78 and 32.2 eV peaks with ionization from A_1 orbitals and the 17.02 eV peak with ionization from a B_1 orbital, all in agreement with the qualitative predictions of Figure 3.15. Agreement with the more accurate model is even better. The ionization from the most stable orbital (that at 32.2 eV) is from a largely 2s(O) orbital; ionizations from the other orbitals are from orbitals which have considerable 2p(O) contributions and so are relatively close together in energy.

With the present-day computers and programs it is possible to do accurate calculations on rather large (perhaps with 100 electrons) molecules. The water molecule with a total of only ten electrons should, therefore, be amenable to quite precise theoretical investigation. All of the accurate calculations which have been performed on this molecule lead to roughly the same orbital energies and demonstrate the presence of molecular orbitals of A_1 symmetry at ca. 14 and 30 eV, a B_1 at ca. 17 eV and one of B_2 symmetry at ca. 12.5 eV. The agreement between these data and the photoelectron results is very good. That with Figure 3.15 is as good as could be hoped for.

It is particularly encouraging to find that the qualitative symmetry-based arguments which have been used to discuss the electronic structure of the water molecule should give results which are in excellent qualitative agreement both with those obtained by experiment and those obtained by detailed calculations. Hopefully, the same techniques may be applied to other molecules and similar qualitatively accurate results obtained. It is obvious that as molecular complexity increases, the difficulty in arriving at an unambiguous energy level scheme will also increase. However, the symmetry of the water molecule is not particularly high. It is not unreasonable to hope that for larger, but higher symmetry, molecules the increase in molecular complexity will be compensated for by the increase in molecular symmetry and so the methodology will remain applicable.

3.8 The Walsh diagram for triatomic dihydrides

We are now in a position to reconsider in more detail a problem which we first encountered in Chapter 1 – that of the significance which can be placed upon the observation that the bond angle in the electronic ground state of the water molecule is 104.5°. As shown in Section 3.5, the bonding interactions responsible for the stability of the water molecule are maximized at quite different bond angles. The stabilization resulting from the B_1 interaction maximizes at a bond angle of 180° whereas that from the A_1 interaction involving the oxygen 2s orbital did not vary with angle. On the other hand, the $2p_z$ A_1 interaction maximizes at a small bond angle. The observed bond angle represents a compromise, showing that interactions involving both $2p_y$(O) and $2p_z$(O) are of importance. However, the total bonding is unlikely to show a strong dependence on bond angle because although a change in θ will reduce the stabilization resulting from interactions involving orbitals of one symmetry species it will increase the stabilization accruing from the other. Only if water had been found to be

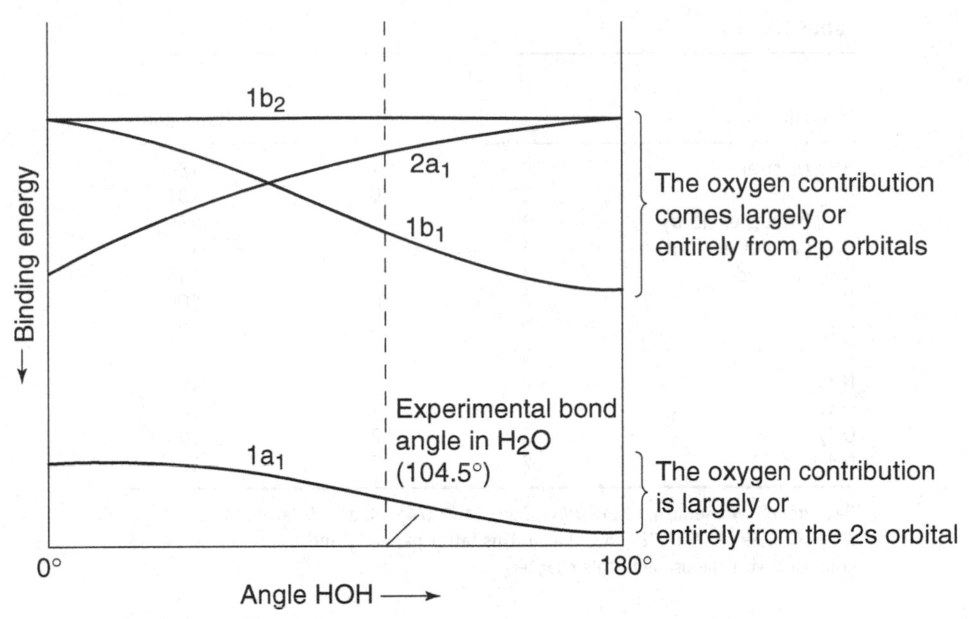

Figure 3.16 A Walsh diagram for H_2O

linear could it have been concluded that the major contribution to the molecular bonding resulted from interactions between those orbitals which we have identified as being of B_1 symmetry. Conversely, only if the bond angle were very small, say 60°, would it then have been valid to conclude that most of the stabilization resulted from the A_1 interaction. A diagram showing this behaviour schematically is given in Figure 3.16 where, again, orbitals have been denoted by lower case symbols. In this diagram there is also recognition of the fact that at the $\theta = 180°$ limit the orbital which has, qualitatively, been called the A_1 non-bonding orbital loses its 2s component and becomes a pure 2p non-bonding orbital. This latter orbital is therefore degenerate with (i.e. has the same energy as) the non-bonding orbital which we have labelled B_2.[12] Conversely, the lower, bonding, A_1 orbital loses its 2p component and becomes a pure 2s orbital at 180° – and so is of lower energy in this limit. The non-bonding B_2 orbital remains unchanged in energy as the bond angle changes (actual calculations show that it increases slightly in energy at the 180° limit but as this is caused by the effects of electron repulsion we have not included it in Figure 3.16). Figure 3.16 is specifically drawn for H_2O – but its general form is applicable to all MH_2 molecules for M

[12] At this point, as has already been noted, the symmetry is different, that of a group called $D_{\infty h}$; in this group the oxygen $2p_x$ and $2p_y$ orbitals are symmetry-required to have the same energy. The $D_{\infty h}$ group is studied in Chapter 10.

Table 3.2

Molecule	Orbital occupancy		Bond angle
	A_1	B_2	
BH_2 (excited)	0	1	180°
BH_2	1	0	131°
CH_2 (triplet excited)	1	1	136°
CH_2 (singlet excited)	1	1	140°
NH_2 (excited)	1	1	144°
BH_2^-	2	0	100°
CH_2	2	0	102°
CH_2^-	2	1	99°
NH_2	2	1	103°
OH_2^+	2	1	107°
OH_2	2	2	105°
NH_2^-	2	2	104°

Data from E. Wasserman, *Chem. Phys. Lett.* **24** (1974) 18, and Y. Takahata, *Chem. Phys. Lett.* **59** (1978) 472 (note that in this latter paper B_1 and B_2 are interchanged compared with the usage in this chapter)

atoms which have similar valence shell orbitals to oxygen. Diagrams of this type were first introduced by Walsh[13] and are therefore commonly known as Walsh diagrams.

It is possible to directly relate the observed geometries of first row MH_2 molecules (in electronic ground and excited states) to the occupancy of the highest B_2 and A_1 orbitals in Figures 3.15 and 3.16. Occupancy of the former, which is non-bonding, would be expected to have little effect on bond angle but the lower the occupancy of the latter the larger we expect the bond angle to be (and vice versa). Table 3.2 details relevant data and shows that when there are two electrons in the highest A_1 orbital an angle of ca. 103° results; one electron in this orbital leads to a bond angle of ca. 140°. When this orbital is empty a linear molecule results.

Problem 3.8 The following species have been the subject of theoretical investigations but their bond angles have yet to be determined experimentally. Predict approximate values for their bond angles:

$$FH_2^{3+},\quad BeH_2^{3-},\quad CH_2^{2-},\quad BH_2^{3-},\quad NH_2^{+}$$

3.9 Simple models for the bonding in H_2O

This chapter is concluded by investigating the relationship between the picture of the electronic structure of the water molecule developed in the chapter and that given by models such as those discussed for ammonia in Chapter 1. The first model is that in which the oxygen atom in the water molecule is regarded as being tetrahedrally surrounded by

[13] A.D. Walsh, *J. Chem. Soc. (London)* (1963) 2250.

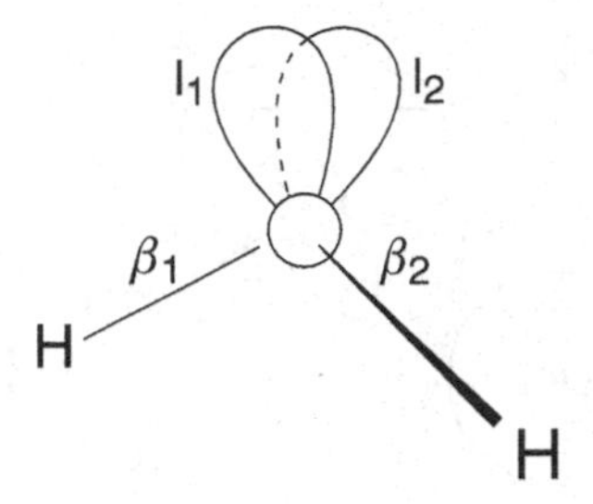

Figure 3.17 The tetrahedral arrangement of bonding electron pairs (β) and lone pairs (l) in H_2O

electron pairs, two bonding and two non-bonding. The concordance between this model and the general pattern of energy levels shown in Figure 3.12 is easy to show. In Figure 3.17 are shown, schematically, the bonding electron pairs (which have been called β_1 and β_2) and the lone pair electrons (labelled l_1 and l_2). It is easy to show that the transformations of the two bonding orbitals β_1 and β_2, considered as a pair, generate the reducible representation:

E	C_2	σ_v	σ_v'
2	0	2	0

which has A_1 and B_1 components. These, of course, are precisely the same symmetries as possessed by the bonding molecular orbitals shown in Figure 3.15. It is also easy to show that the transformations of the lone pair orbitals, l_1 and l_2, generate the reducible representation:

E	C_2	σ_v	σ_v'
2	0	0	2

which is also easily shown to be the sum of the A_1 and B_2 irreducible representations. These, again, are the symmetries of what in Figure 3.15 have been identified as the lone pair orbitals.

Problem 3.9 Generate the two reducible representations discussed above. Use Figure 3.17.

It is interesting to consider the 'tetrahedral oxygen' model for H_2O in more detail. Let us start with the lone pair orbitals, which as we have just seen transform as $A_1 + B_2$. Since they are localized on the oxygen atom we conclude that they must be derived from the valence shell atomic orbitals of the oxygen atom. But the only B_2 valence shell oxygen atomic orbital is $2p_x$(O), so the orbital of this symmetry which is a combination of lone pair orbitals must be identical to that obtained from our symmetry-based discussion. Similarly, the A_1 combination must be some mixture of 2s(O) and $2p_z$(O), again in qualitative agreement with our earlier result. Study of the O—H bonding orbitals leads to a similar agreement with the model we have developed in this chapter. Because β_1 and β_2 are bases

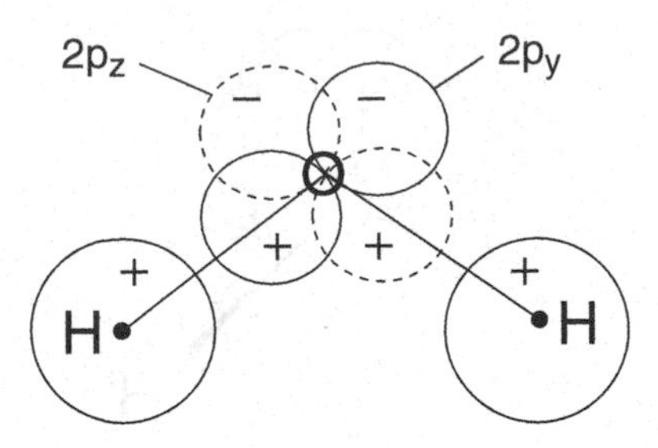

Figure 3.18 Bonding in H_2O with 2p as the only oxygen orbitals involved. Note that the axis labels do not follow the convention adopted elsewhere in this book, although they remain mutually perpendicular

for A_1 and B_1 irreducible representations we conclude that 2s(O), $2p_z$(O) and $\psi(A_1)$ may contribute to the A_1 combination. Similarly, $2p_y$(O) and $\psi(B_1)$ contribute to the B_1 combination – conclusions identical to those reached earlier. Further, the statement that β_1 and β_2 are *bonding* means that the A_1 and B_2 combinations must be in-phase, bonding, combinations of oxygen and hydrogen orbitals. We conclude that the simple two bonding, two non-bonding orbital picture of the water molecule may readily be reinterpreted in the language used in this chapter. Further, by using the energy-level criteria based on the number of nodal planes and the relative energies of atomic orbitals, discussed in the present chapter, the qualitative energy level diagram shown in Figure 3.15 can again be derived.

The second simple model of the water molecule to be considered is one that is sometimes quickly discarded, although of historic importance. This is the model in which only the 2p orbitals of the oxygen atom are considered as being involved in O—H bonding, the 2s orbital being, implicitly, regarded as non-bonding. In this model, the oxygen 2p orbitals which lie in the plane of the water molecule overlap with the hydrogen 1s orbitals. The relevant oxygen 2p orbitals, which because it is a choice in accord with that made in setting up the group theoretical model will be taken to be $2p_z$(O) and $2p_y$(O), are orientated so as to point directly at the hydrogen atoms. This is shown in Figure 3.18; a bond angle of 90° is predicted. The oxygen $2p_x$ orbital, which is perpendicular to the plane of the molecule, is non-bonding, precisely the role which we found it to play. The two non-bonding orbitals are, therefore, 2s(O) and $2p_x$(O), which are of A_1 and B_2 symmetries, just as found above (their relative energies – one very stable A_1 orbital and one high energy B_2 orbital – also agree with the symmetry-based model). It is easy to show that the two bonding orbitals which result from the overlap of the $2p_z$(O) and $2p_y$(O) orbitals (those in the plane of the molecule) with the hydrogen 1s orbitals give rise to a reducible representation with A_1 and B_1 components. Arguments analogous to those developed above for β_1 and β_2 in the previous model demonstrate that the qualitative forms of the corresponding A_1 and B_1 molecular orbitals are those deduced earlier in this chapter. It is perhaps pertinent to comment that this particular model gets closer to the results of accurate quantum mechanical calculations than does any other. It predicts a low-lying non-bonding 2s orbital, one pure 2p non-bonding oxygen orbital and two bonding molecular orbitals involving oxygen 2p orbitals. Yet this is a model which fell into disuse in the 1950s!

Problem 3.10 Show that the $2p_z(O)$ and $2p_y(O)$ orbitals shown in Figure 3.18 transform together as $A_1 + B_1$.

Problem 3.11 In the text a variety of alternative arguments, all leading to the same conclusion, have been used to arrive at the general form of Figure 3.15. Select and rehearse a single set of arguments leading to this figure.

3.10 A *rapprochement* between simple and symmetry models

It is useful at this point to review the development of the arguments in this book. In Chapter 1 it was concluded that simple models of molecular bonding cannot be expected to be infallible predictors of molecular geometry. However, in the present chapter it has been shown, at least for the case of the water molecule, that these simple models may usefully be reinterpreted. It is possible to recast them and to show that the bonding descriptions they present are equivalent, qualitatively, to a symmetry-based description. Symmetry-based descriptions show that there can be different relationships between molecular geometry and the contribution to the bonding from the various bonding molecular orbitals. The stabilization resulting from one interaction may be independent of geometry; others may be very sensitive to geometry. Different interactions may make a maximum contribution to molecular stability at quite different patterns of bond angles. Although it is not required from the discussion so far, it transpires that these conclusions have a general validity.

It is here that the circle opened in Chapter 1 closes. Simple pictures of molecular bonding are perhaps more reliable predictors of relative energies of molecular orbitals than they are of molecular geometries (although more used for the latter rather than the former). When a simple picture fails to give a correct molecular orbital energy level pattern it is usually because there are some interactions in the molecule involving orbitals other than those considered in the simple model. In such cases the simple models are, nonetheless, usually good starting points for a detailed discussion. Finally, we note that, despite their apparent differences, when there is a variety of simple approaches to the bonding in a molecule they commonly lead to the same qualitative energy level diagram. Again, the only exceptions occur when different models include different interactions, but here the differences are themselves illuminating.

What is the particular attraction of a symmetry-based approach which leads us to refer all other models to it? A computational advantage has already been mentioned – interactions are only non-zero between wavefunctions of the same symmetry species so that the size of the problem, the number of interacting orbitals, is reduced. There is another important reason. Whenever excited electronic states or ionized species are considered it becomes essential to use a symmetry-based approach. This is because it is the only one which allows a simple connection between the discussion of the ground and excited (or ionized) states of a molecule. One illustration will make the point. Suppose an electron in the water molecule is excited from a bonding orbital to some high-lying, non-bonding, orbital and suppose that the excited electron comes from a single O—H bond. According to all of the simple models of the bonding in the water molecule those electrons associated with one bond are not associated with the other and so such an ionization would seem entirely possible. In the excited state the two O—H bonds would differ – one has only one bonding electron whilst the other has two. This is in contradiction to the observation that in all stable excited states of the water molecule the O—H bonds are equivalent (excluding unstable states from which dissociation into H + OH occurs). For a symmetry-based description, in which the bonding electron comes from a molecular orbital spread equally over both hydrogen atoms, the observed equivalence of the two hydrogen atoms in excited or ionized states follows naturally. A symmetry-based description is thus to be preferred because it can be applied to both ground and excited states.[14]

Problem 3.12 A student was heard to complain that 'symmetry arguments make difficult problems even harder by adding another complicating consideration'. Write a one-page document assessing this point of view.

3.11 Summary

The irreducible representations which appear in character tables are expressions of nodal patterns; these patterns/irreducible representations are orthonormal (p. 47) – each component is independent of the others and carries equal weight. This property enables reducible representations (p. 51) to be reduced systematically to their irreducible components (p. 52). In the context of molecular bonding this enables the interactions between orbitals of each symmetry type to be discussed separately (p. 55). Such discussions, together with simple nodal-plane criteria (p. 66), enable qualitative molecular orbital energy level diagrams to be constructed (p. 66) and the angular variation of each bonding interaction assessed (p. 66). This latter information may be conveniently represented as a Walsh diagram (p. 69). A symmetry analysis of simple pictures of molecular bonding reveals that they have similarities with each other and with the symmetry-based approach (p. 71).

[14] A paper which makes a direct connection between the models considered above and the photoelectron spectrum of the water molecule is: J. Simons, 'Why equivalent bonds appear as distinct peaks in photoelectron spectra'. *J. Chem. Educ.* **69** (1992) 522.

4 Vibrational spectra of the water molecule

In this chapter we will develop in some detail a topic that was introduced as something of an aside in the introduction to the previous chapter, the vibrations of the water molecule. Having found these vibrations, their symmetries and forms, the next question that arises is 'how can these results be checked?' or, in reality, 'in which spectroscopies are these vibrations to be seen?'. This will lead us to spectroscopic selection rules and all of the development that goes before them. The reader will perhaps be delighted to learn that, when the end of this chapter has been reached, all of the key aspects of the application of group theory to chemistry will have been covered. Although, of course, there is more – otherwise the remainder of the book would not have been written!

There is a problem with the water molecule. It is too simple. We met this problem in a different form in the last chapter, where we introduced four general points in space around the water molecule (in the guise of the ethylene oxide molecule) to overcome the difficulty. This time is different. The answer to our question is too obvious. What are the vibrations of the water molecule? Two bond stretches and a bond angle change. Simple, too simple. If one is interested in a molecule of any size the answer is by no means so obvious. In order to obtain the number of independent vibrations of a non-linear molecule one applies the $3N - 6$ rule (a rule that we shall be looking at in some detail later in this chapter). Here, N is the number of atoms in the molecule and $3N - 6$ is the number of vibrations. Consider a triatomic molecule, $N = 3$; it follows that $3N - 6$ is 3, just the number listed for the water molecule above. But even the $N = 3$ case can pose problems. Suppose that the three atoms are in a ring, that they lie at the corners of an equilateral triangle. There are three bond stretching vibrations and three angle change vibrations, a total of 6. But $N = 3$ and so the answer should be 3, just as for water. Something is wrong. Here, the answer is not difficult to see; each bond angle change does much the same thing as the bond stretch opposite it, as shown in Figure 4.1. But the same problem arises whenever there is a ring (which can be puckered) of atoms – and seldom is the explanation so obvious or so simple as in the example of a triangle. In such cases one can use a different approach, a general approach which is systematic – unlike the 'look and see' method that we used above for water. Fortunately, the water molecule provides an excellent example for the systematic method and it will be worked through in detail.

Symmetry and Structure: Readable Group Theory for Chemists Sidney F. A. Kettle

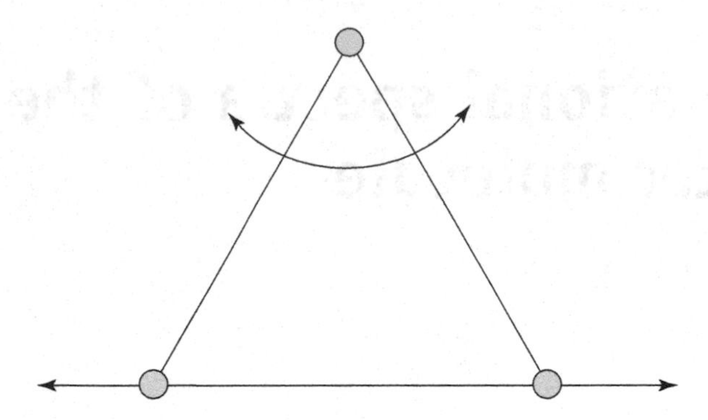

Figure 4.1 A bond angle change (top) in an equilateral triangular arrangement of atoms leads to atom motions which are very similar to those of the bond stretch opposite to it (lower)

4.1 Vibrations of the water molecule: Part 1 (easy)

Although the 'look and see' approach to the vibrations of the water molecule is simple, it is not trite. The reason is that it is often of value to apply it to much larger molecules, or rather to fragments of larger molecules. Consider a not-very-much-larger molecule: 1,1-dichloroethylene, shown in Figure 4.2. It can be thought of as a composite of three 'bits', a CH_2 fragment, a CCl_2 and a C=C bond. The first two bits can be treated in the same way as the water molecule; each is expected to have two bond stretches and a bond angle change vibration, a total of 6. Add the C=C bond stretch and we have a total of 7. This is not the whole story (the $3N - 6$ rule leads to a prediction of 12), but it is a good start. Further, we have learnt something about the forms of the vibrations – something that the $3N - 6$ rule gives no information on. It is when applied to really large molecules, when a small fragment is – mentally – abstracted and considered, that the method becomes very valuable, and far from trite. Quite often, several fragments can be treated in this way, simplifying an otherwise difficult problem.

We now return to the water molecule and consider first the two O—H stretch vibrations, shown in Figure 4.3. This diagram is deceptively simple. What it really represents is the oscillation of each H atom about its equilibrium position, the bond alternately lengthening and contracting. Diagrams in which the bond is shown lengthening are favoured because the arrows can often be drawn into empty space on the paper. A bond contraction would mean the arrows would overlap with the line representing the bond. If the bond stretching is a simple harmonic motion (something always assumed) then two arrows of the same length, one a bond contraction and the other a bond lengthening, would be the opposites, negatives, of each other. This relationship is something that we will be using shortly. Figure 4.3 seems

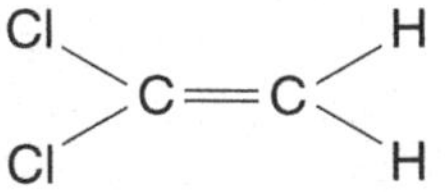

Cl H
C=C
Cl H

Figure 4.2 1,1-Dichlorethylene

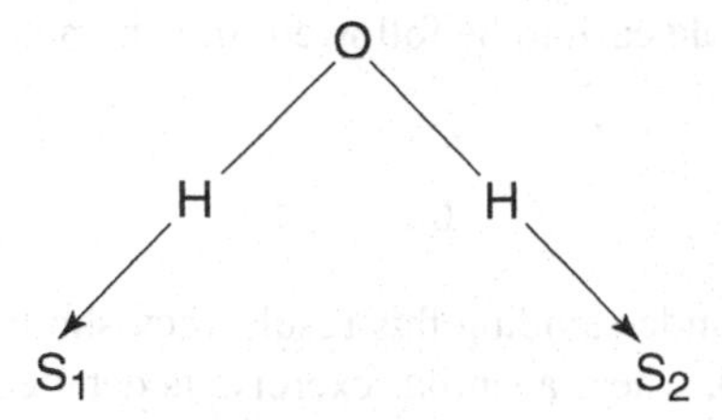

Figure 4.3 The two O—H stretch vibrations of the water molecule, labelled s_1 and s_2. These are two independent motions, not an in-phase combination

to show the O—H bonds stretching at the same time and to the same extent. As we will see, it is convenient if this is so, but it is not required. Figure 4.3 shows two *independent* O—H bond stretches. They could be at quite different points in their own motions – but it is very convenient to set things up so that the operations of the C_{2v} group can be applied to them without difficulty. One final defect in Figure 4.3: for clarity, the motions shown are far too big. In reality, the hydrogens would move by a maximum of something like one-tenth of the bond length.

Given that there are two independent motions shown in Figure 4.3, the next step is the standard one: to determine how they transform under the operations of the group, the C_{2v} group in our case. These transformations are shown in Figure 4.4, where the stretches are

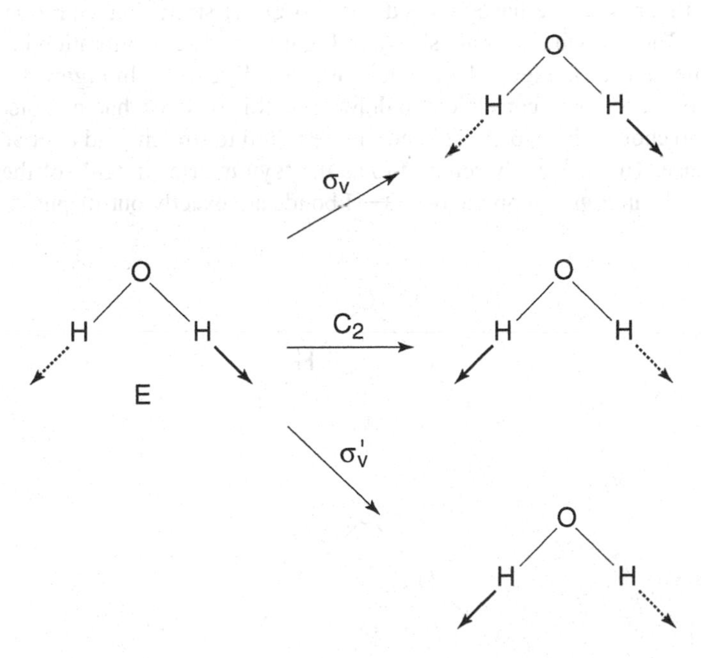

Figure 4.4 The transformation of s_1 and s_2 (Figure 4.3) under the operations of the C_{2v} point group; they are drawn differently to enable the transformation of each to be more readily followed

indicated differently, to enable each to be followed; the characters that result are:

E	C_2	σ_v	σ_v'
2	0	2	0

If the reader has problems understanding this result, they should look back at Section 3.3 (and the end of Section 3.2), where a similar exercise is detailed for the two 1s orbitals of the hydrogen atoms in water. On the other hand, if he or she has a sense of *déjà vu*, they are to be congratulated. The reducible representation we have just obtained is the same as that generated by the transformations of the hydrogen 1s orbitals. Quite different things can give the same result! Because it is the same reducible representation, its irreducible components must be the same. That is, $A_1 + B_1$. And, even better, we already know the form of these vibrations, vibrations in which the two H atoms move in concert. No longer do they move independently, as in Figure 4.3. One talks of 'group vibrations', where by 'group' we mean the two O—H stretch vibrations. It is potentially confusing to be talking about two different groups (the other is C_{2v}), but in practice no problems arise. The forms of the group vibrations can be taken from the hydrogen 1s discussion of the last chapter and are:

$$\psi(A_1) = \frac{1}{\sqrt{2}}(s_1 + s_2)$$

$$\psi(B_1) = \frac{1}{\sqrt{2}}(s_1 - s_2)$$

where, as in Figure 4.3, we have labelled the two O—H stretching vibrations s_1 and s_2 (s for stretch). The two vibrations are shown in Figure 4.5. The A_1 vibration in Figure 4.5a looks much the same as in Figure 4.3 but there is a key difference. In Figure 4.3 the arrows were similar because it was convenient to draw them this way; we had a choice. In Figure 4.5a there is no choice, the two O—H bonds are required to stretch (and contract) together, exactly in phase. This is usually referred to as the 'symmetric stretch' (of the two O—H bonds). Similarly, in Figure 4.5b the two O—H bonds are exactly out-of-phase.

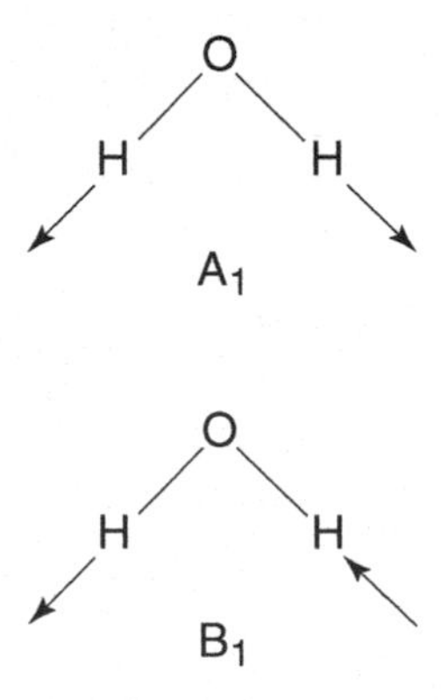

Figure 4.5 The A_1 (a) and B_1 (b) O—H bond stretch motions of H_2O

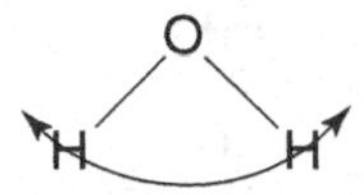

Figure 4.6 The H—O—H bond angle change vibration (bond deformation) of the water molecule

As the one stretches the other contracts (and the problem of an arrow overlapping with the bond has been circumvented). It is usually called the 'antisymmetric stretch' vibration. We will have more to say about Figure 4.5 shortly, but first we have to remember that there is another vibration of the water molecule, the bond angle change vibration. To put a symmetry label on it, we have to study its transformations under the operations of the C_{2v} point group. Hopefully, by now the reader will agree that this is a trivial problem. The bond angle change vibration is shown in Figure 4.6, but the author refuses to waste space giving pictures of its transformations under the operations of the C_{2v} point group. They would all look the same as in Figure 4.6. The vibration is turned into itself under all of the operations of the group; it is a totally symmetric vibration, A_1.

Problem 4.1 Check, by studying their transformations under the operations of the C_{2v} point group, that the vibrations in Figure 4.5 have the A_1 and B_1 symmetries indicated and that the vibration of Figure 4.6 has A_1.

The next step is to remind the reader that we are discussing vibrations. This is a less trivial comment than it seems. The point is that we are NOT considering the translation or rotation motions of the water molecule. They comprise a separate discussion, one that we shall embrace shortly. Put another way, the motions shown in Figure 4.5 must contain not the smallest trace of a translation or rotation of the water molecule. But they seem to. If the motion shown in Figure 4.5a were correct, then the centre of gravity of the water molecule would oscillate, translate, just a bit, up and down the C_2 axis as the hydrogen atoms moved up and down. Similarly, in Figure 4.5b the centre of gravity would oscillate back and forth, sideways. The answer is simple. The problem arises from yet another defect in Figure 4.3. The arrows in Figure 4.3 show O—H bond stretches but we have pictured these as motions of the H atoms. In reality, the O atom moves too, in such a way that the centre of gravity of each O—H bond does not move. Because the hydrogen atom is so much lighter than the oxygen, the amplitudes of oscillation of the H atoms are much greater than those of the O atom – so the diagrams are not too bad – but the O atom motion should not be forgotten. Figure 4.7 shows the upper part of Figure 4.5 again but with the oxygen atom motion included (as for the H atoms, the O atom motion is exaggerated for clarity). One consequence of the neglect of O atom motion in the corresponding diagram in Figure 4.5 is the preservation of the HOH bond angle. It did not change. It follows that in Figure 4.7 the HOH bond angle *is* changing throughout the vibration. But the HOH bond angle change vibration was considered separately above (Figure 4.6). Evidently, this separation was not exact; the A_1 bond stretching vibration contains a bit of the A_1 bond angle change vibration. Not surprisingly, the converse is also true. If the bond angle change vibration shown in Figure 4.6 is to leave the centre of gravity of the molecule unmoved then the

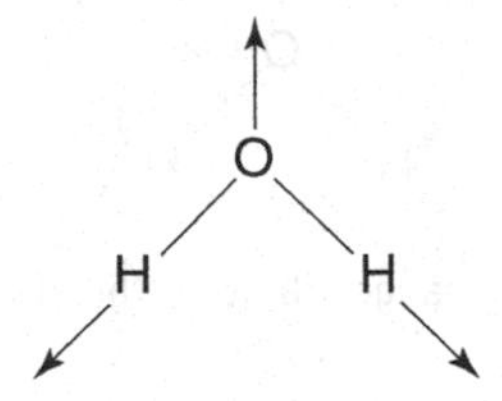

Figure 4.7 The A_1 vibration of Figure 4.5 but with the motion of the O atom included

oxygen atom has to oscillate, just a bit, along the twofold axis. But this motion means that the O—H bond lengths must oscillate, change, too – just a bit. Again, bending and stretching motions are mixed, just a bit. What of the B_1 combination shown in Figure 4.5? When modified to include the oxygen atom motion, Figure 4.8 is obtained. There are two comments to be made about this figure. The first is that it shows a motion in which *the HOH bond angle is rigorously preserved.* The reason is that Figure 4.8 shows a motion of B_1 symmetry whilst the HOH bond angle change is of A_1 symmetry. The two symmetries are different and so the two motions cannot be mixed. Second, it may seem strange – because the individual H atoms in Figure 4.8 have motions such that each is oscillating, just a bit, up and down the z axis – that the motion of the oxygen atom is rigorously perpendicular to that axis. Again, the explanation has to do with symmetry – but it can only be given after we have looked at things in more detail, as we do in the next section.

4.2 Vibrations of the water molecule: Part 2 (less easy!)

The discussion in the previous section depended on our use of chemical knowledge. We knew that the water molecule has O—H bonds and an HOH angle which are fairly fixed – and we exploited this insight. But, earlier, we saw that when there is a ring of atoms problems arise, and these problems are seldom as easily resolved as for the triangle-of-atoms case that we discussed. Another problem arises with angle changes. Consider a planar AB_3 molecule, all the B atoms being equivalent, so that there are three B—A—B bond angles of 120°. If we were to do the group theory on this (it will not be until Chapter 7 that the reader will be equipped for it) we would find a totally symmetric combination of these bond angles, one that requires all three angles to increase simultaneously. But this is impossible – they already have the maximum angle physically possible. We have to throw something away. But we not only discard vibrations, we throw away the symmetry labels that are attached to them. How can we be confident that we have chosen correctly and that we still have the symmetries of the vibrations correctly listed? The present section provides the answer

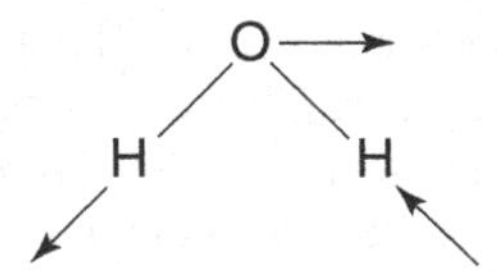

Figure 4.8 The B_1 vibration of Figure 4.5 but with the motion of the O atom included

to this question. It does so by ignoring any chemical knowledge of the problem. Instead, it elaborates the $3N - 6$ rule, first by looking at the $3N$ and then the 6. Once we have determined the symmetry labels to be associated with the $3N$, we simply subtract from them those in the '6' list. What is left is a complete list of the irreducible representations spanned by the vibrations. Hopefully, we can then add a bit of chemical knowledge to get some insight into what the vibrations look like.

First, the $3N$ – where N is the number of atoms in the molecule. Ignoring any bonding constraints (remember, no chemical knowledge) each atom can move independently in three mutually perpendicular directions, just like any other physical object. So, N atoms, together, have $3N$ independent motions, or, as it is usually put, $3N$ degrees of motional freedom. The next step is to determine the symmetry species, the irreducible representations, generated by the transformation of these $3N$ motions under the operations of the appropriate group – and for us that means the C_{2v} group. Key to doing this correctly is a good diagram. In the present case, we are lucky. We can draw three separate diagrams. This simplicity arises because of the simplicity of the C_{2v} group; generally, the three diagrams that we will draw would have to be combined into one, and it is this that can make life difficult unless the diagram is a good one.

The three diagrams that we will draw concern motion along three separate axes. When talking about the bonding in the water molecule it was pointed out that the 2p orbitals of the oxygen atom could be orientated in any direction, as long as they were mutually perpendicular. The same is true of the directions of the tiny displacement vectors that we will attach to each atom to indicate its movement in space. All choices are equal but some are more equal than others. Since we are looking for an easy life, we will make an educated choice. We will be asking how the vectors behave when rotated or reflected, under the operations of the C_{2v} point group. The ideal would be to get simple answers like 1, -1 or 0, and to do this the vectors have to be orientated in a way which enables such answers, one which recognizes the existence of the corresponding symmetry elements. So, if at all possible, let them lie along, or parallel to, rotation axes; in, or parallel to, mirror planes; perpendicular to mirror planes. If this sounds rather complicated, all the points are illustrated in the three diagrams which follow. For the water molecule we attach three tiny arrows to each atom, to indicate its three degrees of motional freedom. We will consider each set of arrows in turn, one member of each set on each atom. First, one on each atom is chosen to point either along or parallel to the C_2, z, axis. These three are shown in Figure 4.9. The other two arrows on each atom must be perpendicular to the first, and so perpendicular to the C_2, z, axis. For each atom, place one in the σ_v mirror plane, the plane of the molecule. They are shown in Figure 4.10. Finally, in Figure 4.11, the last arrow on each atom. To be parallel to the first two on each atom, this one has no choice but to lie perpendicular to the first two, for each atom. This third displacement vector either lies in or is parallel to the σ_v' mirror plane.

Now comes the key step, that of using the arrows in Figures 4.9–4.11 to generate reducible representations of the group. The experienced worker will do this by simply looking at the diagrams and writing down the answers, but this book is not written for experienced workers! Instead we will proceed in two ways, in parallel. The left-hand frames in Figures 4.12–4.14 repeat Figures 4.9–4.11 but with labels attached to the vectors. Taking advantage of the fact that we are ignoring chemistry, the atoms have been labelled as 1, 2 and 3. The displacement vectors are labelled by the direction in which they point (remember, in the water molecule,

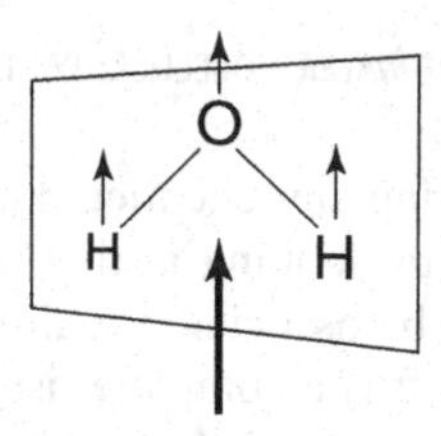

Figure 4.9 The movement of each atom parallel to the molecular *z* axis, a total of three independent motions

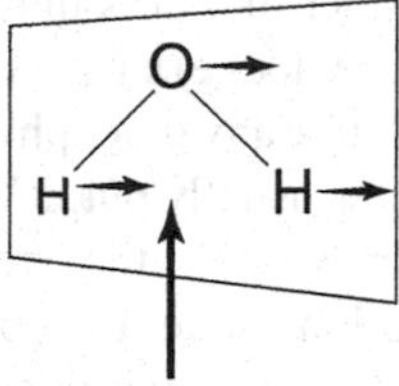

Figure 4.10 The movement of each atom parallel to the molecular *y* axis, a total of three independent motions

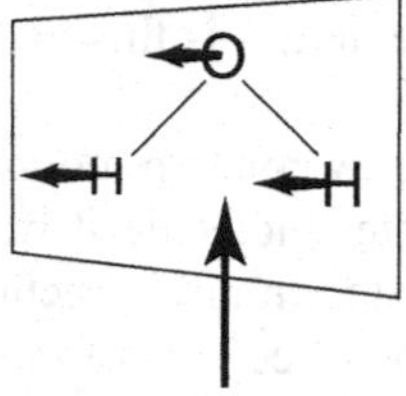

Figure 4.11 The movement of each atom parallel to the molecular *x* axis, a total of three independent motions

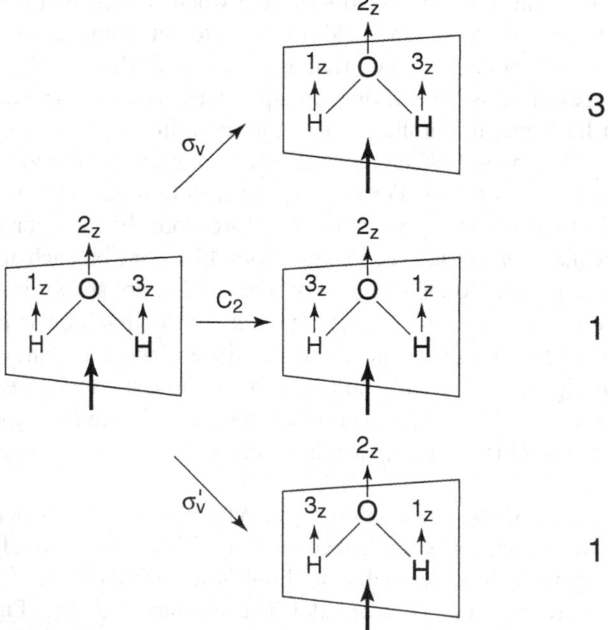

Figure 4.12 The characters generated by the transformations of the arrows of Figure 4.9 under the operations of the C_{2v} point group. Each character (right) is obtained from a comparison of the diagram closest to it with the original (left)

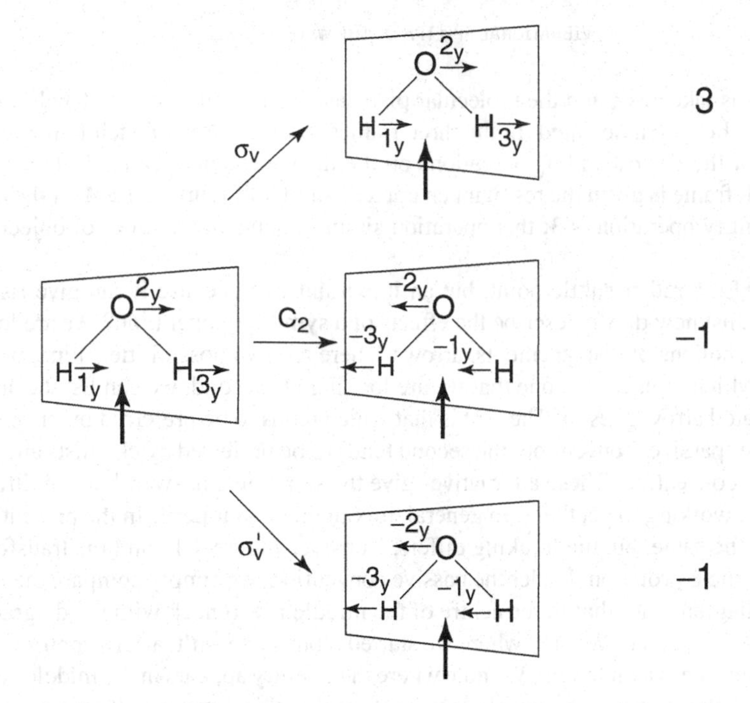

Figure 4.13 The characters generated by the transformations of the arrows of Figure 4.10 under the operations of the C_{2v} point group. Each character (right) is obtained from a comparison of the diagram closest to it with the original (left)

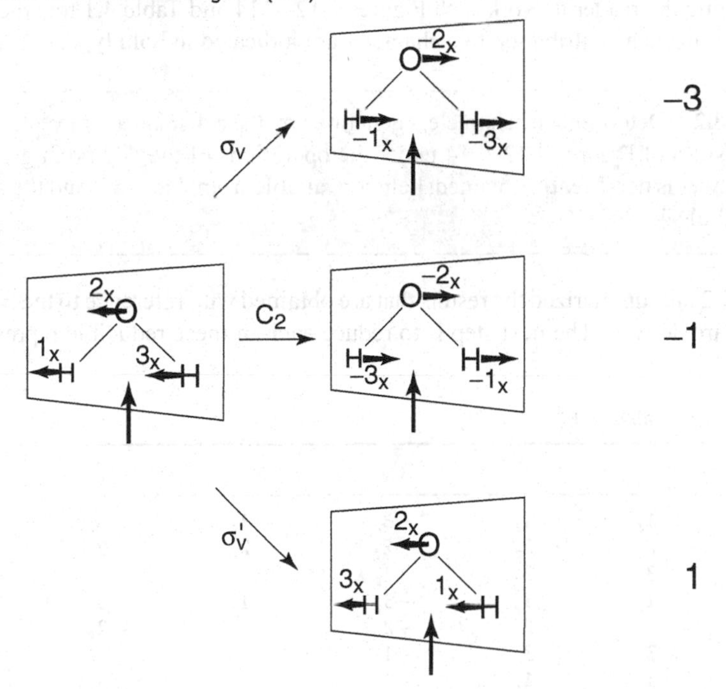

Figure 4.14 The characters generated by the transformations of the arrows of Figure 4.11 under the operations of the C_{2v} point group. Each character (right) is obtained from a comparison of the diagram closest to it with the original (left)

the C_2 axis is taken as z and the molecular plane as yz) as well as by the label of the atom with which they are associated. In the three frames in the middle of each figure are shown the effect of the C_2, σ_v and σ_v' operations on the displacement vectors. At the right-hand side of each frame is given the resultant character (for all of Figures 4.12–4.14 the character for the identity operation is 3; this operation simply counts the number of objects we are studying).

We now hit a rather subtle point, but an important one because it can give rise to real confusion. Just how do we describe the effects of a symmetry operation? We are looking at the transformations of displacements, arrows. There are two possibilities. First, we can list the arrow which replaces the one that we are looking at. Second, we can list the arrow that our nominated arrow goes to. The first is that which tends to be preferred by physicists and is called the 'passive' convention; the second tends to be preferred by chemists and is called the 'active' convention. These alternatives give the same final answer, but can differ in the detail of the working to get there. In general, mix them at your peril! In the present case the working is the same, but the thinking differs. Consider Figure 4.12 and the transformation of 1_z under the C_2 rotation. Under the passive convention, we simply compare the left-hand (starting) diagram with that in the centre of the middle row (check with the diagram): 1_z is replaced by 3_z; that's it. We stay where we started – but 1_z doesn't, a zero contribution. The active convention is a bit longer. We note where the 1_z entry appears in the middle centre and refer back to the starting diagram (left) to confirm that this entry is at the point previously occupied by 3_z. We move with the object under study. The actual transformations of the displacement vectors are given in Table 4.1; the resultant characters are not given (in the hope of forcing the reader to work with Figures 4.12–4.14 and Table 4.1 together!) but all entries which make a contribution to a character are indicated in bold type.

Problem 4.2 Determine the characters generated by the z displacement vectors on the left-hand sides of Figures 4.12–4.14 under the operations of the C_{2v} point group. If a wrong answer is persistently obtained, help is available from Table 4.1 and the answers given in Table 4.2.

In Table 4.2 are summarized the results that are obtained with reference to the figure from which they are derived. The next step is to reduce each of these reducible representations

Table 4.1

	E	C_2	σ_v	σ_v'
1_z	$\mathbf{1_z}$	3_z	$\mathbf{1_z}$	3_z
2_z	$\mathbf{2_z}$	$\mathbf{2_z}$	$\mathbf{2_z}$	$\mathbf{2_z}$
3_z	$\mathbf{3_z}$	1_z	$\mathbf{3_z}$	1_z
1_y	$\mathbf{1_y}$	-3_y	$\mathbf{1_y}$	-3_y
2_y	$\mathbf{2_y}$	$\mathbf{-2_y}$	$\mathbf{2_y}$	$\mathbf{-2_y}$
3_y	$\mathbf{3_y}$	-1_y	$\mathbf{3_y}$	-1_y
1_x	$\mathbf{1_x}$	-3_x	$\mathbf{-1_x}$	3_x
2_x	$\mathbf{2_x}$	$\mathbf{-2_x}$	$\mathbf{-2_x}$	$\mathbf{2_x}$
3_x	$\mathbf{3_x}$	-1_x	$\mathbf{-3_x}$	1_x

Table 4.2

	E	C_2	σ_v	σ_v'
Figure 4.12	3	1	3	1
Figure 4.13	3	−1	3	−1
Figure 4.14	3	−1	−3	1

into its irreducible components. Because it is so important a technique, the results derived from Figure 4.14 will be worked through in detail (for help, look back at Section 3.3) but the answers for the results from Figures 4.12 and 4.13 are just given. As an aid, the C_{2v} character table is given again in Table 4.3.

Consider the reducible representation:

	E	C_2	σ_v	σ_v'
	3	−1	−3	1
Test for A_1	1	1	1	1
Multiply	3	−1	−3	1

Addition gives 0, so there is no A_1 component in the reducible representation.

	E	C_2	σ_v	σ_v'
	3	−1	−3	1
Test for A_2	1	1	−1	−1
Multiply	3	−1	3	−1

Addition gives 4, so division by the order of the group, 4, shows that there is one A_2 in the reducible representation.

	E	C_2	σ_v	σ_v'
	3	−1	−3	1
Test for B_1	1	−1	1	−1
Multiply	3	1	−3	−1

Addition gives 0, so there is no B_1 component in the reducible representation.

	E	C_2	σ_v	σ_v'
	3	−1	−3	1
Test for B_2	1	−1	−1	1
Multiply	3	1	3	1

Addition gives 8, and division by 4 shows that there are two B_2 in the reducible representation.

We conclude that the reducible representation contains $A_2 + 2B_2$ components.

Table 4.3

C_{2v}	E	C_2	σ_v	σ_v'
A_1	1	1	1	1
A_2	1	1	−1	−1
B_1	1	−1	1	−1
B_2	1	−1	−1	1

The above paragraph detailed the reduction of the representation given in the last row of Table 4.2 into its components. The next step is to do the same for the other two rows. We only give the results and they are:

	E	C_2	σ_v	σ_v'
Figure 4.12	3	1	3	1
Irreducible components: $2A_1 + B_1$				
Figure 4.13	3	−1	3	−1
Irreducible components: $A_1 + 2B_1$				

> **Problem 4.3** Check the results just obtained for the transformations shown in Figures 4.12 and 4.13.

Summing the results for all three diagrams (Figures 4.12–4.14), we get

$$3A_1 + A_2 + 3B_1 + 2B_2$$

Had we been brave enough to have combined Figures 4.9–4.11 into a single diagram and worked with that, we would have obtained a reducible representation which would have been the aggregate of our three:

E	C_2	σ_v	σ_v'
9	−1	3	1

It may look more daunting, but reducing it must give the sum of the irreducible representation that we have obtained:

$$3A_1 + A_2 + 3B_1 + 2B_2$$

> **Problem 4.4** Reduce the representation just given and show that it has the components listed immediately above.

Before we proceed, there is a relevant question. Suppose we had chosen, or been forced, to do the job of generating a reducible representation the hard way, all vectors together, and made a mistake? (it happens to the most experienced of workers). No problem. Invariably (strictly, one should say 'almost invariably', but in practice the 'almost' can be omitted), if a mistake is made, the reducible representation will fail to reduce. That is, the numbers generated at the end of testing for one or more irreducible representation will not be exactly divisible by the order of the group. Discovering that a mistake has been made is easy; correcting it may be more difficult!

We have now dealt with the $3N$ part of the problem; we know the symmetries of all of the collective motions of the atoms in the water molecule. They are of $3A_1 + A_2 + 3B_1 + 2B_2$ symmetries. Notice the use of the word 'collective'. The thing that distinguishes the motions of this set and those with which we started (the sum of those contained in Figures 4.9–4.11) is that the latter were quite individual: individual displacements of individual atoms. The collective motions involve (in principle) all of the atoms of the molecule, even if the motion associated with the individual symmetry species has a form which does not require the participation of each and every one of the arrows in Figures 4.9–4.11. Buried

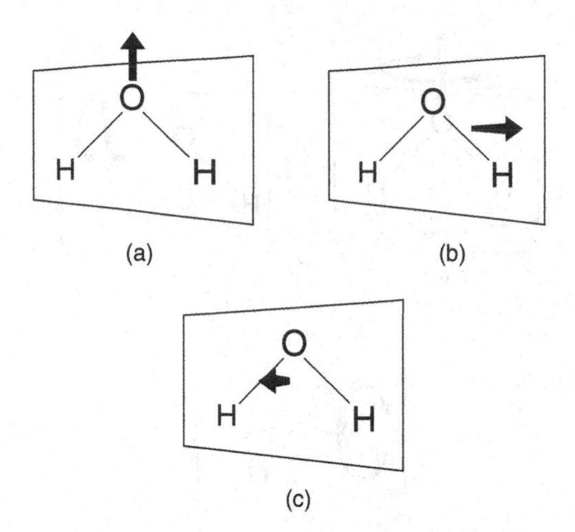

Figure 4.15 The translation motions of the water molecule, the molecule being considered as a solid, rigid, body: (a) along z, A_1; (b) along y, B_1; (c) along x, B_2

in the $3A_1 + A_2 + 3B_1 + 2B_2$ set, of course, are the vibrations of the water molecule. The vibrations are those motions in which one part of the water molecule moves relative to another part. All the other modes in the set must then be such that the three atoms are in fixed positions: they do not move relative to each other. These motions are the three translations and the three rotations of the water molecule. We have dealt with the $3N$ and now it is time to consider the 6.

Just like any solid body, there are three independent axes along which translation of the water molecule can occur, and about each of these axes the rotation of the entire molecule can occur too. So there is a total of six things that we have to consider. Of these, the three translations are simple, trivial almost. Normally, they would all be drawn together, but we have taken advantage of the freedom offered by the water molecule to draw them separately and do this in Figure 4.15. The next step is to study the transformations of each and to determine the symmetry species that they generate. Since each is a single unit, reducible representations cannot be generated and each transforms as an irreducible representation of the C_{2v} point group. This is a straightforward task and we do not detail it; the answers are given in the caption to Figure 4.15. If the reader finds that this problem is difficult, they should turn back and re-read Section 2.2. The explanation given there is exactly that applicable to the translations if $2p_z$ is replaced by T_z, $2p_x$ by T_x and $2p_y$ by T_y, where T_z, T_x and T_y represent translations along the z, x and y axes respectively.

Now we turn to the rotations of the water molecule. If the translations were easy, they are well compensated for by the rotations, which are not – unless, that is, the reader prefers an easy life, for there *is* an easy way but one that can only be appreciated after we have first done the job the hard way. The first problem is that of how to draw a rotation. Translations of the water molecule were easy; a magnified version of the tiny displacement vectors that we placed on atoms was an obvious way of showing them. But with rotations we are looking for something to represent a rotation of the *entire* water molecule. Some authors draw these rotations by showing the motions of the individual atoms that comprise the rotation. But this means using several arrows when talking about a single thing, and this risks confusion.

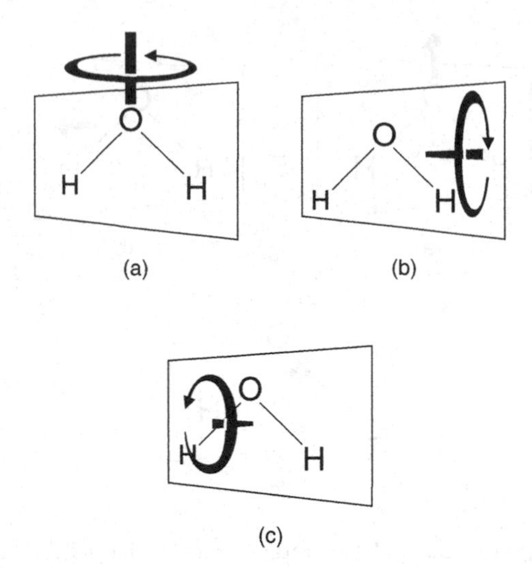

Figure 4.16 The rotation motions of the water molecule, the molecule being considered as a solid, rigid, body: (a) around *z*; (b) around *y*; (c) around *x*

The author's preference is to use a single circular arrow ↻, and this is how the rotations are shown in Figure 4.16.[1] But this is only the first problem. The next problem is that of showing the effect of the symmetry operations of the water molecule on the circular arrows. For this, it may be helpful to become momentarily blind to the existence of the water molecule and to think of the arrows themselves. This simplification has been adopted in Figure 4.17, where the effect of the operations of the C_{2v} point group on the arrows is shown (as a help, the relevant symmetry element has also been shown). Even this is not the end of the difficulty. The curved arrows sometimes move to positions in which, previously, there was no arrow. When this happens, it is important to remember that the arrows represent *rotations*, and the real question is whether the *rotation* is the same as before and, if not, how it is related to the original. Here, the relative orientation of the arrows is vital. A turned-round arrow means that a clockwise rotation has become an anticlockwise one, or vice versa – the negative of a clockwise rotation is an anticlockwise rotation. With these comments, the reader should be able to work through Figure 4.17 and check that they can obtain the characters given there (of course, the character for the identity operation is always 1). If they can, they will have demonstrated that R_z transforms as A_2, R_x as B_2 and R_y as B_1, where R_z, R_x and R_y represent rotations of the entire water molecule about the z, x and y axes, respectively.

Problem 4.5 Carefully study Figure 4.17a–c and generate the characters given there. Show that they are those of the irreducible representations listed immediately above.

[1] An incomplete circle is preferred to a complete circle because use of the latter means that the only thing that apparently changes as a result of an operation is the arrow attached to the circle.

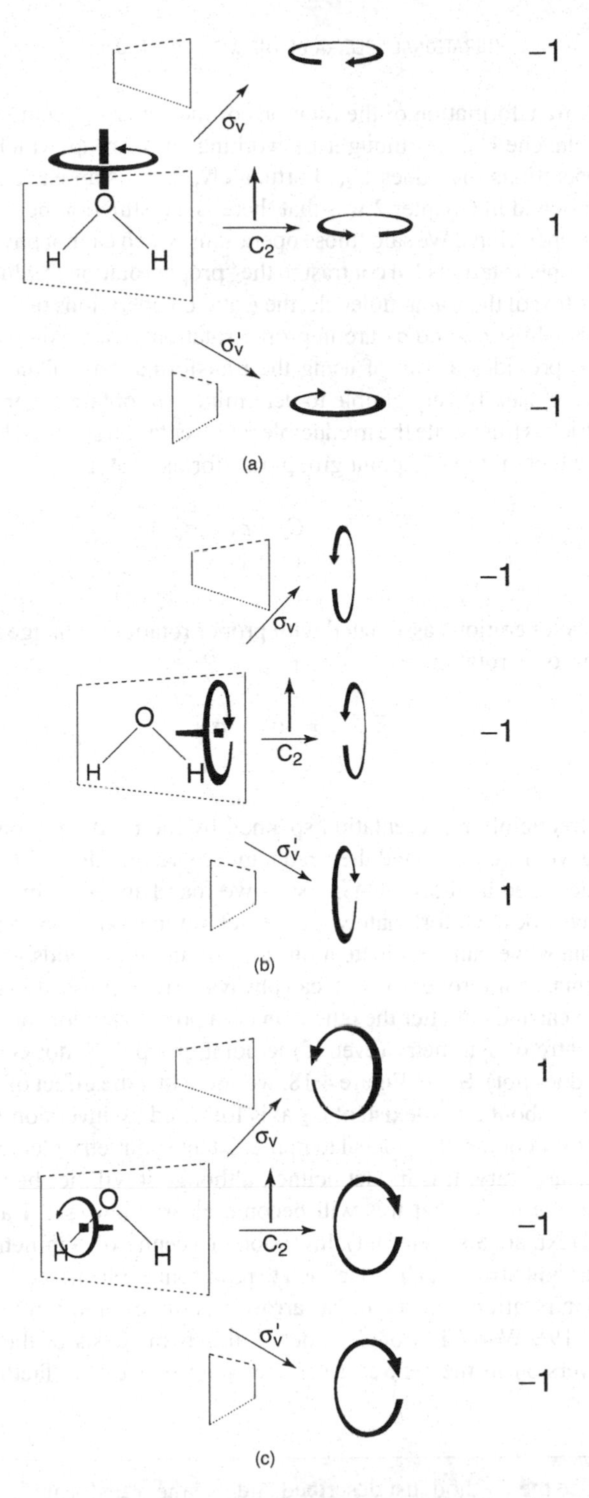

Figure 4.17 The effect of the symmetry operations of the C_{2v} group on the rotations of the water molecule. Each character at the right arises from a comparison of the nearest central arrow with the original (left). As an aid, the symmetry operation is symbolically shown next to the symbol representing it. Figure 4.17a shows rotation around z, 4.17b around y and 4.17c around x

The study of the transformation of the rotations of the water molecule was not a simple task; many have blanched at the thought of working in a group which contains many more symmetry operations than does C_{2v}. Fortunately, there is an easier way. One of the earliest things mentioned in Chapter 2 was that there is a distinction between 'proper' and 'improper' rotation operations. We said 'those operations which cannot physically be carried out are called "improper rotations" in contrast to the "proper rotations" which are physically possible'. In the context of the water molecule, the E and C_2 operations of the C_{2v} point group are proper rotations whilst σ_v and σ_v' are improper rotations. Surprisingly, this distinction between operations provides a way of using the transformations of the translations of a molecule – which are usually very simple to determine – to obtain the transformations of the rotations. The trick is this; write the irreducible representation spanned by the translation, in full. So, for T_z, which in the C_{2v} point group transforms as A_1:

	E	C_2	σ_v	σ_v'
T_z	1	1	1	1

Doing nothing to the operations associated with proper rotations, change the sign of those associated with improper rotations:

E	C_2	σ_v	σ_v'
1	1	−1	−1

The result is the irreducible representation spanned by the rotation about the same axis, which here is z, so we have generated the irreducible representation of R_z. The characters are immediately identified as those of A_2, just as we found above. Why does this method work – and it always does? Unfortunately we are not yet in a position to fully answer this question – but perhaps we can anticipate a bit, leaving the loose ends to be tied up later. Surprising as it seems, an improper rotation can always be re-expressed as a combination of two operations, one carried out after the other. One is a proper rotation, and the other is that of inversion in a centre of symmetry (even if the point group does not contain one, which C_{2v} most certainly does not). So, in Figure 4.18, we show that the effect of a σ_v reflection is equivalent to rotation about a non-existent C_2 axis followed by inversion in a non-existent centre of symmetry. Of course, this appeal to non-existent symmetry elements appears both highly artificial and arbitrary. It is in fact neither, although it will not be until Chapter 10, where we return to the topic, that this will become clear (Figures 5.4 and 8.6, together with the associated text, are also relevant). Inversion in a centre of symmetry always turns a translation – or a straight arrow, such as we used to represent a translation – into its negative. On the other hand, a rotation – or a circular arrow – is turned into a rotation in the same direction (Figure 4.19). Worked through in detail, this is the basis of the method that we have used. The inversion in the centre of symmetry introduces the factor of -1 that we invoked.

Problem 4.6 Use the method just described and use the transformation properties of T_x and T_y to show that R_x has B_2 and that R_y has B_1 symmetry in C_{2v}.

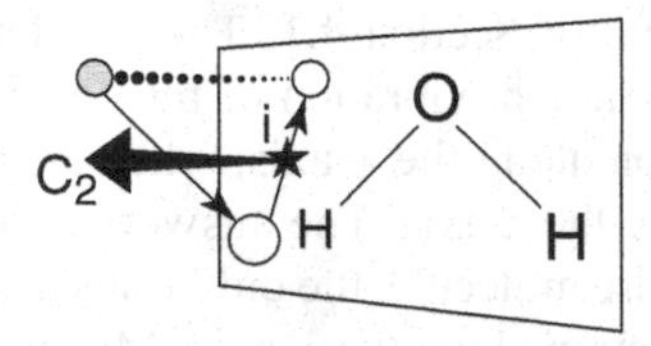

Figure 4.18 Reflection in the mirror plane of the water molecule moves the dark circle into the open circle of the same size (the two are joined with a dashed line). The same result can be obtained by, first, a rotation about a C_2 axis, placed anywhere provided that it is perpendicular to the mirror plane (the large open circle results; the action is indicated by an arrow). This is followed by inversion in a centre of symmetry, placed at the point at which the C_2 axis intersects the mirror plane. The centre of symmetry is indicated by a black star and the action by an arrow

We are now in a position to use the $3N - 6$ rule to obtain the vibrations of the water molecule. The $3N$ gave us:

$$3A_1 + A_2 + 3B_1 + 2B_2$$

We have not brought the symmetry species spanned by the three translations together, but this is easy to do. The sum is:

$$A_1 + B_1 + B_2$$

For the three rotations, the sum of their symmetry species is:

$$A_2 + B_1 + B_2$$

To obtain the vibrations, we simply have to subtract the irreducible representations spanned by the translations and the rotations from the first set listed. This is just applying the $3N - 6$ rule in detail. The, unsurprising, result is:

$$2A_1 + B_1$$

exactly as we obtained before. It is in difficult cases that this method becomes essential, and the final arbiter. But the simplicity of the method that was described in Section 4.1, together with the fact that it rapidly gives the form of the actual vibrations, means that it is always the first method to apply. Incidentally, we are now in a position to give the

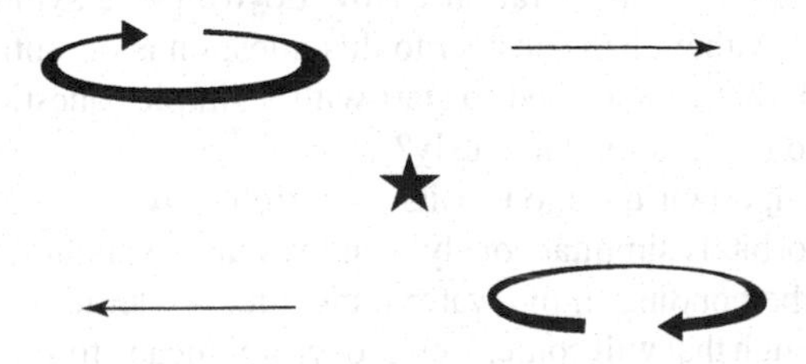

Figure 4.19 Inversion in a centre of symmetry (symbolically represented by a star) converts a rotation into a rotation in the same sense but converts a translation into a translation in the opposite direction

explanation promised at the end of Section 4.1. The problem left unanswered is that of why, in the B_1 (antisymmetric stretch, vibration of the water molecule), the oxygen atom is rigorously confined to motion along the y axis, whereas the two hydrogen atoms have motions which take them along the z axis. The answer is seen in the translation motions discussed above. As for the entire molecule, the only translation of the oxygen atom which is of B_1 symmetry is that of motion along the y axis. Movements along z (A_1) and x (B_2) are forbidden by symmetry in a B_1 vibration. The reason why the two hydrogens can move along z is that their motions along this axis *exactly* cancel each other. The resultant motion of the two hydrogens is entirely along the y axis.

Problem 4.7 Explain the fact that the motion of the oxygen atom shown in Figure 4.7 is rigorously along the z axis.

The next problem is that of the spectral activity of the vibrations of the water molecule that we have been exploring. Although there are other possibilities, by far the most important methods of studying molecular vibrations are infrared spectroscopy and Raman spectroscopy. We shall confine our discussion to these. So, the question to be answered is that of which of the vibrations of the water molecule are expected to be seen, to be active, in the infrared spectrum and which in the Raman. But before we can answer this question there is some serious foundation work to be done.

4.3 Product functions

It is clear that our discussion so far is inadequate if we wish to understand spectra. Spectroscopy involves three things. First, there is the starting point of a molecule. One normally thinks and talks of its ground state, be it a vibrational ground state, electronic ground state, or whatever. Then the spectroscopy involves an excited state, be it vibrational, electronic or whatever – the state to which the molecule is promoted by the spectral technique being used. And, finally, there is the technique itself, which provides some mechanism for interacting with the molecule and exciting it from its ground to excited state. Three things – ground state, excited state and spectroscopic mechanism – and any spectroscopy involves all three, simultaneously. And there is our problem. So far in this book we have only spoken about individual things and how they transform. We have been concerned with putting symmetry labels on things like orbitals and vibrations. How do we put a symmetry label on three things simultaneously? Although the answer to this question is our ultimate goal, prudence suggests that we would be well advised to start with a simpler question: how do we put a symmetry label on two things simultaneously?

In fact, this is an important question in its own right. Overlap integrals, for instance, typically involve two orbitals simultaneously – and we talked about the overlap of orbitals when we considered the bonding in the water molecule. So, let us talk about overlap; not overlap integrals, although that will come. Here, 'overlap' means forming the product of two quantities, normally orbitals (to get an overlap *integral* we have to carry out an additional summation). Specifically, we form the product of the numerical values, magnitudes, of two orbitals at a chosen point in space. For the present purpose, there is no need to specify much

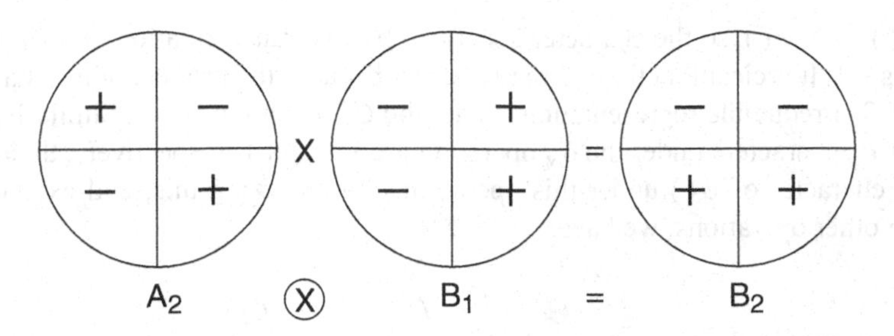

Figure 4.20 When the phases in the corresponding quadrants of the A_2 and B_1 nodal patterns of Figure 3.1 are multiplied together they give the B_2 nodal pattern. One says that 'the direct product of A_2 and B_1 is B_2' and a special multiplication symbol is used (bottom row)

further, although the reader may find it helpful to refer back to Figures 3.8 and 3.9, and the associated text, where the discussion is about overlap and overlap integrals. In Figures 3.8 and 3.9 the symmetry species involved were A_1 and B_1 and it would make sense to use these as an example as we proceed. Unfortunately, they are not sufficiently general and so instead we will consider the pair A_2 and B_1; we continue working in the C_{2v} point group. That is, we consider a product of (unspecified) orbitals $\varphi_1\varphi_2$ where φ_1 has symmetry A_2; we write this $\varphi_1(A_2)$. Similarly, we take φ_2 to be of B_1 symmetry and write it $\varphi_2(B_1)$. We have to determine the symmetry of their product ψ; that is, we have to fill the empty parentheses in:

$$\psi() = \varphi_1(A_2) \cdot \varphi_2(B_1)$$

where a dot has been placed between the two functions on the right to facilitate separate consideration of these two functions. The simplest, and perhaps most fundamental, way of tackling this problem is to apply the nodal pattern method used in the previous two chapters. Figure 4.20 shows the nodal patterns of the A_2 and B_1 irreducible representations together with their product (multiply the corresponding phases in the two left-hand figures to obtain that in the right-hand figure). The product is the pattern of the B_2 irreducible representation. We have our answer! Unfortunately, for large groups the nodal patterns can be rather convoluted and so the method that we have just used is difficult to apply. A non-pictorial method has to be used, so it is sensible to work it through for the present case too. In principle, the method is simple – it is the one used many times before in this book when working with single objects. We describe it as if we did not already know the answer, for this will help in the application of the method to more complicated examples.

In the equation $\psi() = \varphi_1(A_2) \cdot \varphi_2(B_1)$ we subject ψ to all of the operations of the group and obtain a set of characters by relating the transformed function to the original. The application of the C_2 operation, for example, to ψ means that we are really applying it to $\varphi_1(A_2)$ and to $\varphi_2(B_1)$, simultaneously. Using the C_{2v} character table (Table 4.3), under this operation $\varphi_1(A_2) \rightarrow \varphi_1 A(_2)$ because the A_2 irreducible representation has a character of 1 for this operation. Similarly, $\varphi_2(B_1) \rightarrow -\varphi_2(B_1)$ and a character of -1. Putting these two results together, we have that under the C_2 rotation operation

$$\varphi_1(A_2) \cdot \varphi_2(B_1) \rightarrow -\varphi_1(A_2) \cdot \varphi_2(B_1)$$

That is, $\psi() \to -\psi()$, so the character generated by the transformation of $\psi()$ under this operation is -1. It is clear that this -1 really occurs because the product of the characters of the A_2 and B_1 irreducible representations under the C_2 operation is -1. Similarly, because the A_2 and B_1 characters under the σ_v operation are -1 and 1, respectively, their product, -1, is the character of $\psi()$ under this operation. Summarizing this, and extending it to include the other operations, we have:

	E	C_2	σ_v	σ'_v
Characters generated by the transformation of $\varphi_1(A_2)$ (i.e. the A_2 irreducible representation)	1	1	-1	-1
Characters generated by the transformation of $\varphi_2(B_1)$ (i.e. the B_1 irreducible representation)	1	-1	1	-1
Characters of the transformation of $\psi()$, i.e. the products of the two rows of characters above.	1	-1	-1	1

The representation generated is the B_2 irreducible representation. That is, $\psi()$ can now be identified as $\psi(B_2)$, a result that we knew from the application of the nodal pattern method.

Problem 4.8 Using the procedure described above, fill in the empty parentheses in the following product functions:

$$\psi() = \varphi_1(B_1) \cdot \varphi_2(B_2)$$
$$\psi() = \varphi_1(B_2) \cdot \varphi_2(A_2)$$
$$\psi() = \varphi_1(A_2) \cdot \varphi_2(A_1)$$

It is now easy to see why the subject of Figure 3.8 was not a good example – it included the A_1 irreducible representation and so would have meant multiplying throughout by the number 1, which is scarcely the best way of seeing what is happening!

In the example just worked through, the general method of determining the symmetries of product functions was used: multiply together the characters of the irreducible representations which describe the transformation of the individual functions. The act of multiplying two irreducible representations in this way is said to give rise to the *direct product* of the two individual representations; if we multiply three irreducible representations we form a triple direct product, and so on.

As will be seen in the remainder of this chapter, direct products are very important in the application of symmetry to chemistry. For these applications, all that is needed is a list – a table – of two-function direct products. Triple and higher direct products can readily be deduced from such a table. The (two-function) direct product table for the C_{2v} point group is given in Table 4.4. In this table an obvious, and conventional, symbolism has been used. The entry at a particular point in the table is the symmetry of the direct product of the species which label the column and row in which the entry falls.

Table 4.4 Direct products of the irreducible representations of the C_{2v} group

C_{2v}	A_1	A_2	B_1	B_2
A_1	A_1	A_2	B_1	B_2
A_2	A_2	A_1	B_2	B_1
B_1	B_1	B_2	A_1	A_2
B_2	B_2	B_1	A_2	A_1

Problem 4.9 Check that Table 4.4 is correct – this will provide useful additional practice in the formation of direct products if done numerically and additional insight if done using nodal patterns.

Note that Table 4.4 is symmetric about the leading diagonal (top left to bottom right). Thus, the result obtained for the example considered above is

$$A_2 \otimes B_1 = B_2$$

(where the symbol $\otimes$, which is that conventionally used to denote the direct product between two irreducible representations, has been used in preference to the $\times$ which might have been expected).

It is equally true that:

$$B_1 \otimes A_2 = B_2$$

This equivalence follows because sets of numbers are being multiplied together and the result obtained is independent of the order in which they are multiplied – the origin of the diagonal symmetry of Table 4.4 is at once evident. Further, it is evident that the process of forming direct products is not limited to just two irreducible representations. We can have as many as we like – or, in reality, as many as we need. And because we are simply multiplying numbers, the order in which they appear is not important. So, in the C_{2v} group

$$(B_1 \otimes A_2) \otimes B_2 = (B_2 \otimes B_2) = A_1$$

or

$$B_1 \otimes (A_2 \otimes B_2) = (B_1 \otimes B_1) = A_1$$

or, rearranging,

$$A_2 \otimes (B_1 \otimes B_2) = (A_2 \otimes A_2) = A_1$$

and so on. This will become important in the section following the next, where this sort of rearranging proves helpful.

Problem 4.10 Use Table 4.4 to obtain symmetry labels for the following product functions:

$$\psi() = \varphi_1(A_1)\varphi_2(B_1)\varphi_3(B_2)$$
$$\psi() = \varphi_1(A_2)\varphi_2(B_1)\varphi(B_2)$$
$$\psi() = \varphi_1(A_2)\varphi_2(B_2)\varphi(B_1)$$
$$\psi() = \varphi_1(A_2)\varphi_2(B_1)\varphi(A_1)\varphi(B_1)$$
$$\psi() = \varphi_1(A_2)\varphi_2(B_1)\varphi(A_2)\varphi(B_1)$$
$$\psi() = \varphi_1(A_2)\varphi_2(A_2)\varphi(B_1)\varphi(B_1)$$

4.4 Direct products and quantum mechanical integrals

Lest the title of this section appear too daunting, let the reader be assured that not a single integral will be evaluated in it. Although, perhaps surprisingly, we will determine the value of many! And further, the approach which will be developed will lead to a reduction in the number of integrals that arise in quantum chemistry. This is not a topic that we will pursue; rather we will use group theory to show which integrals are zero, without any need to evaluate them! Such zero integrals reappear as spectroscopic selection rules, the topic of the next section. It is all too easy for quantum mechanics to appear formidable because of the large number of rather unpleasant looking integrals which it seems to involve. In practice, these integrals are found to be rather less objectionable because if they cannot be evaluated algebraically they can be evaluated numerically, invariably by a computer. Even so, a great deal of work can be saved by the intelligent use of group theory. Let us first consider what is meant by an integration over all space (which is the integration that is usually involved in quantum mechanics). Integration may be pictorially regarded as the adding together of an infinite number of infinitesimally small fragments. As a consequence of this it is sometimes possible to see the result of an integration without actually carrying out the calculation. Consider the p_z orbital shown in Figure 4.21. What is the value of the integral over all space of the p_z orbital? That is, what is the value of $\int p_z \delta v$ where δv is an infinitesimally small volume element? Treat this integral as $\Sigma p_z \delta v$, where the summation is over an infinity of minute volume elements. In order to perform this summation – this integration – one has to collect into one box, as it were, all of the infinitesimally small fragments which comprise this wavefunction. We must pay due regard to the signs of the fragments – some fragments are from that part of space in which the wavefunction has a positive amplitude and others are from that part in which it has a negative amplitude. From the shape of the orbital it is evident that for every volume element that makes a positive contribution to the integral, there is a corresponding volume element which makes a negative contribution. A pair of such mutually cancelling volumes is shown in Figure 4.21, the positive contribution from the top volume being cancelled by the negative contribution from the bottom. By adding

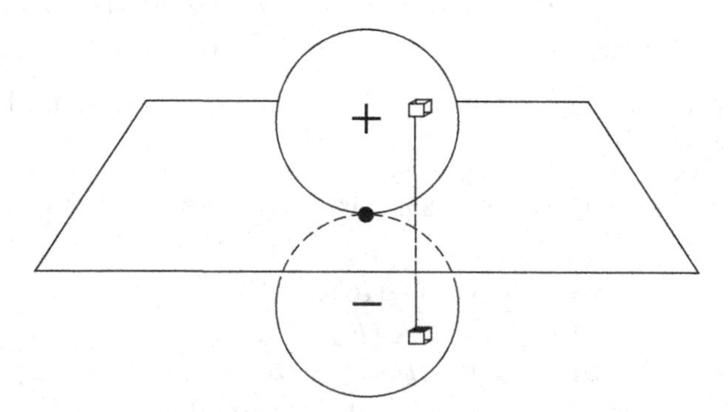

Figure 4.21 There is an exact cancellation of the contributions of the two boxes (at equivalent positions in the lobes of the p_z orbital shown) to an integral over all space of the p_z orbital. Summed over all space, the resultant is zero

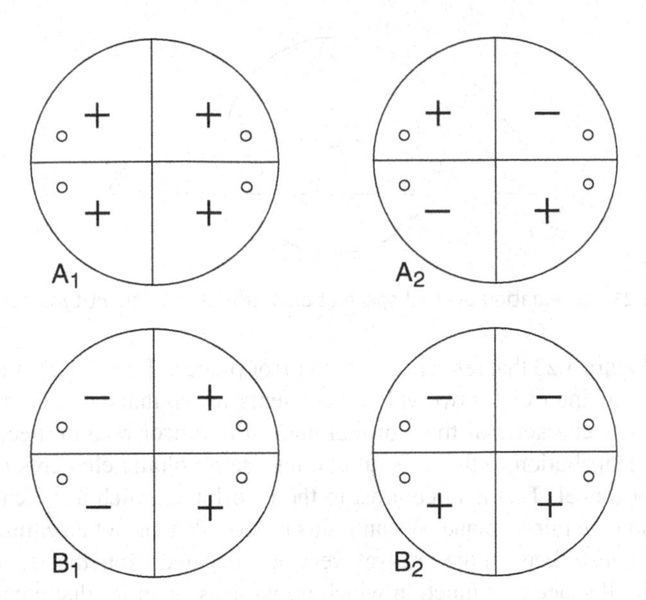

Figure 4.22 Nodal patterns of the irreducible representations of C_{2v}. Each circle is given the phase of the quadrant in which it falls and then all four circles in a diagram are summed. Just as in Figure 4.21, the result is zero for all but the A_1 pattern

together pairs of points in this way until the whole of space is included, it is seen that the value of the integral $\int p_z \delta v$ is zero – even though it has not been explicitly evaluated. It is the fact that arguments such as this can be cast, very simply, in the language of symmetry that makes group theory so valuable. Thus, an alternative way of stating the above argument is to recognize that the 'top' and 'bottom' of the p_z orbital are:

(a) shape-wise, related by reflection in a mirror plane, and
(b) of opposite phase.

These two facts, taken together, establish that the integral must be zero. Can this procedure be generalized? Is there a general rule to replace the two specific points made above, which are relevant only to the p_z orbital (and any other functions that behave similarly)? For the C_{2v} group the nodal patterns of the irreducible representations provide an immediate answer. They are shown again in Figure 4.22, each with four symmetry-related points indicated. Now, we have to sum over all four such points and then add all these individual sums together. From this it is evident that summation over all space leads to a mutual cancellation, an integral of zero, for all except the A_1 irreducible representation. Is this an indication of a general result? It is, but to see this it is helpful to look at another example from the C_{2v} group.

To establish the general rule it is helpful to qualitatively consider the integral over all space of an s orbital:

$$\int s\delta v$$

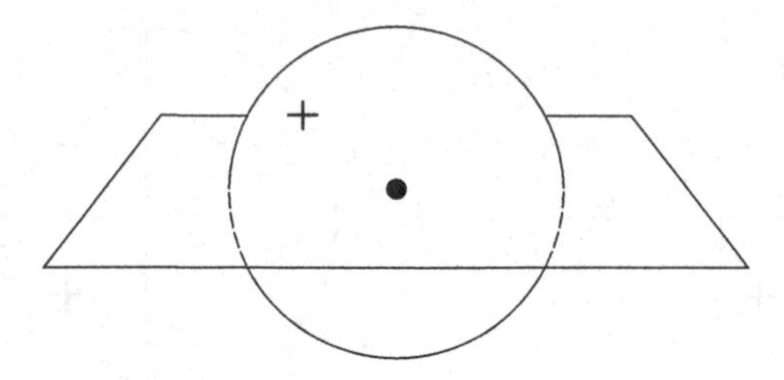

Figure 4.23 Integration over all space of an s orbital-type quantity is non-zero

It is clear from Figure 4.23 that reflection in the mirror plane in Figure 4.21 (there shown but not mentioned) now interrelates two volume elements which make *identical* contributions to the integral. The character of the s orbital under the mirror plane reflection operation is 1 and so the contribution to the integral coming from volume elements related by this operation do not cancel. This is in contrast to the p_z orbital, which has a character of -1 under reflection in the mirror plane. Because this mirror plane is not a symmetry operation of the C_{2v} group, its effects on the integral were not revealed. Nonetheless, it is clear that the integral over all space of a function which transforms as an irreducible representation that has all its characters equal to $+1$ will not be equal to zero by symmetry (although, as p_z shows, it may be zero for another reason – which can be a symmetry property not revealed by the point group being used). Figure 4.22 of course demonstrates the converse, that (for C_{2v}, but the result is general) integrals over non-totally symmetric functions give an answer of zero. We have actually already shown this in a different way. In Section 3.1 we met the orthonormal properties of irreducible representations. Another way of stating this property is to say that the overlap between functions transforming as different irreducible representations is zero. The fact that only an integral over all space of a totally symmetric function can be non-zero requires that the corresponding integral for all other irreducible representations be zero. Orthonormality does the work for us!

The next step is to apply this pattern to direct products; these are composed of individual symmetry species. So, one can, effectively, carry out an integration over a direct product. This is a most important step, because all quantum mechanical integrals of importance involve integration over more than one function. Can we say anything of importance about the value of overlap integrals between two functions (and, by extension, more than two)? Indeed we can. The point becomes rather clear for the C_{2v} group if we combine the content of Figure 4.22 with that of Table 4.4. Figure 4.22 shows that the only non-zero integrals are those over the totally symmetric irreducible representation and Table 4.4 shows that for two product functions the totally symmetric irreducible representation only occurs when the two functions transform as the same irreducible representation. We have reached an important and general conclusion:

> The only non-zero integrals in quantum mechanics are those over totally symmetric functions. An integral over a function which does not transform as the totally symmetric irreducible representation is zero. If an integral can be cast in a form such that it is the product of two quantities then, for the integral to be non-zero, these quantities must transform as the same irreducible representation.

Although the C_{2v} example that we have used has A_1 as its totally symmetric irreducible representation, the label used may differ from this and so the conclusion in the box above has been worded accordingly.

This and the previous section, on direct products, were not as new as it seems. In fact, both were used, without real explanation, in Chapter 3. There, they were the basis for the method presented for reducing a reducible representation into its irreducible components. There, overlap integrals were evaluated between the reducible representation and each irreducible representation in turn. Different irreducible representations gave an overlap integral of zero, whilst when an irreducible representation overlapped with itself it gave an overlap of 1 (to get this as the result we divided by the order of the group). Clearly, the results we have obtained in this section are of value. This conclusion is reinforced in the next section, where they are shown to give rise to the selection rules that are essential if spectroscopic data are to be interpreted.

4.5 Spectroscopic selection rules

Earlier in this chapter it was recognized that in spectroscopy there are three things that are relevant. The ground state, an excited state and some mechanism appropriate to the spectroscopy. If we can place symmetry labels on all three then we can use the results of the previous two sections and apply group theory, symmetry, to spectroscopy. None of these tasks is trivial when explored in detail. Fortunately, we do not have to go to great depth in the present discussion. Here, we are concerned with principles, not detail. Suffice to note that the wavefunctions associated with ground and excited states will themselves usually be product functions. After all, when we think of a molecule being in its ground state we are thinking of a single thing, the ground state. This notwithstanding the fact that we will be talking about a multi-electron molecule (if we are concerned with the electronic ground state; analogous arguments would apply to the vibrational ground state, if vibrations were our concern). The link between the two is that the ground state wavefunction will be a product of the wavefunctions of the individual electrons. We know how to place symmetry labels on the wavefunctions of individual electrons (they are basically the s, p_z and so on, with which we are familiar) and we know how to form products of these symmetry species (Section 4.3). So, in principle, at least, we know how to obtain a symmetry label for the ground state. The excited state symmetry is found in an analogous way, except that (for an electronic excited state) at least one of the electron wavefunctions will be different from one in the ground state. All we need is a symmetry label which, somehow, describes the spectroscopic process (and it was to cover this that in the box above the word 'quantities' was used when 'function' might have been expected from the earlier discussion).

Fortunately, the task of working out the symmetry species of the mechanism (the word 'operator' is that commonly used, indicating that it will have some mathematical form) appropriate to a particular form of spectroscopy is a very simple task compared with that of determining the detailed form of the operators themselves. In most forms of spectroscopy a beam of electromagnetic radiation is allowed to interact with the system under study (the term 'electromagnetic radiation' rather than the word 'light' is used because the wavelength of the radiation may be far from the visible region of the spectrum). In the simplest (Maxwell) picture, electromagnetic radiation is regarded as being composed of two

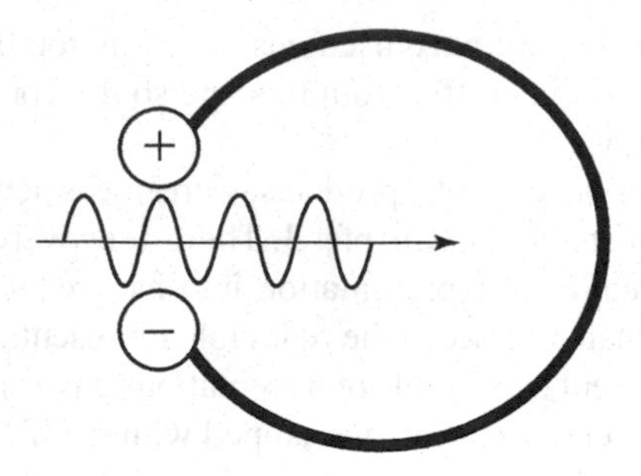

Figure 4.24 Passage of the electric field of a light wave (here shown moving from left to right) induces alternating charges in a – stationary – molecule (here represented as a loop, so that both poles of the field can be simultaneously sampled). One half-wavelength further on, the charges in the molecule will be the opposite of those shown

mutually perpendicular oscillating fields, one an electric field and the other magnetic; both fields are perpendicular to the direction of propagation of the radiation (we will shortly look at this in more detail). The most evident way, then, in which such radiation can interact with matter is by virtue of either or both of the electric and magnetic fields associated with it, so that the most common spectroscopic observations are of transitions which are either 'electric dipole' allowed or 'magnetic dipole' allowed. The oscillating electric field associated with the light wave – think of it as an oscillating voltage between two parallel metallic plates – induces an oscillating electric dipole [(+ −) which alternates with (− +)] in atoms and molecules lying between the metallic plates. When this oscillation matches a natural frequency of the atom or molecule, resonance occurs and energy is transferred from the light wave to the atom or molecule (Figure 4.24). Similarly, the magnetic field – think of some large electromagnet, energized by an oscillating current, with the atom or molecule in the middle – induces an oscillating magnetic dipole [(N S) which alternates with (S N)] which can also cause excitations by a resonance phenomenon (Figure 4.25). Magnetic dipole allowed transitions are of particular importance in nuclear magnetic resonance and paramagnetic resonance spectroscopies. To determine the symmetries that we need for spectroscopy, then, it is only necessary to determine the symmetry species associated with an electric dipole or a magnetic dipole; we can forget the mathematical form of the operators.

In order to obtain an electric dipole it is necessary to separate charges of opposite sign along an axis. In our three-dimensional world there are only three independent directions in which one may bring about such a charge separation and so there are just three electric dipole operators, one corresponding to the x, one to the y and one to the z axes. Further, because the Cartesian axes are dipolar – they also have + and − regions – the transformations of the electric dipole operators mimic – are isomorphous to – those of the Cartesian axes of a molecule. Equally, they are isomorphous to the translations of the molecule along x, y or z axes. Here, the normal presentation of character tables in compilations is of great help. It is standard for the symmetries of x, y and z to be given in character tables, in an additional column to the right of the character table itself. Each one is listed against the irreducible representation under which it transforms. The work is done for us!

Just as an electric dipole corresponds to a movement of electric charge along an axis, so a magnetic dipole corresponds to a rotation of charge about an axis (as in a solenoid

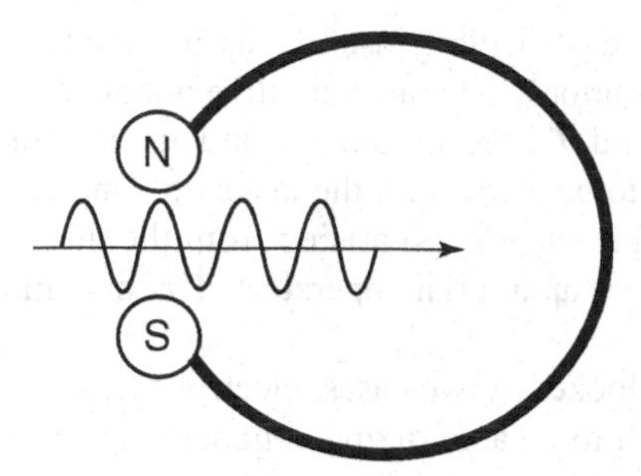

Figure 4.25 Passage of the magnetic field of a light wave (here shown moving from left to right) induces alternating poles in a – stationary – molecule (here represented as a loop, so that both poles of the field can be simultaneously sampled). One half-wavelength further on, the poles induced in the molecule will be the opposite of those shown. The electric and magnetic fields shown here and in Figure 4.24 are mutually perpendicular, so this diagram would best be thought of as lying in a plane perpendicular to the page

carrying a current). There are three magnetic dipole operators, one for each of the three Cartesian axes. The symmetry species of the magnetic dipole operators will be the same as those of the rotations about these axes. These rotations (usually denoted R_x, R_y and R_z in a character table) were met and used in the discussion of molecular vibrations in Section 4.2. The subsequent discussion there – in which the transformation of the rotations was derived from that of the corresponding translations – becomes relevant when only the latter are shown (as x, y and z) in a character table, as happens. If one is lucky, the entries R_x, R_y and R_z are also included at the right-hand side of a character table and indicate how the three magnetic dipole operators transform.

We are now in a position to make some general statements about whether or not an integral related to the intensity of a transition is required to be zero. That is, to state general selection rules for electric dipole and magnetic dipole allowed processes. This rule is derived from the integral given towards the end of the previous section by replacing wavefunctions and operator with the appropriate symmetry species.

> A transition is electric dipole allowed if the triple direct product of the symmetry species of the initial and final wavefunctions with that of the symmetry of a translation contains the totally symmetric irreducible representation. Similarly, a transition is magnetic dipole allowed if the triple direct product of a rotation with the symmetry species of the initial and final wavefunctions contains the totally symmetric irreducible representation.

These rules are rather a mouthful and triple direct products can be rather tedious to work out, so it is convenient to recast them into a simpler form by making use of the fact that the totally symmetric irreducible representation only arises in the direct product of an irreducible representation with itself. In compiling the triple direct product, first form the direct product of the symmetry species of the initial and final wavefunctions. The transition will only be allowed if this direct product contains within it the same symmetry species as that of the operator. This is a particularly useful way to state the selection rule because, as has been seen, there are commonly several alternative operators – dipole moment operators corresponding to T_x, T_y and T_z, for instance – and in this form a choice between the alternatives does not have to be made until the last step. One simply looks for a matching between the irreducible representation(s) arising from the direct product of wavefunction symmetries with those of the appropriate operators. If a matching exists, the transition is allowed.

Although we have only looked at two cases, electric and magnetic dipole transitions, in detail we have seen enough to be able to give a general spectroscopic selection rule – of which all others are particular cases. This is:

> A transition is allowed only if the direct product symmetry species of the initial and final wavefunctions contains the symmetry species of the operator appropriate to the transition process.

All that is needed in order to make use of this rule is a list of those spectroscopic processes which normally arise as a result of electric dipole transitions and those which normally arise from magnetic dipole transitions. This we will do shortly, but there is one thing more. What about Raman spectroscopy? After all, the present discussion has the aim of predicting something about the vibrational spectra of the water molecule and one would hope that this will include Raman predictions. Raman is a bit different. In a typical experimental set-up, a laser beam is incident on a sample and the light emitted is collected. The collection can be at any angle, and many have been used, although one of 90° between incident and scattered light is perhaps that favoured (although the incident light is strictly uni-directional – it comes from a laser – light is scattered at all angles). Related is the fact that the scattered light is of a different wavelength from the incident light, not one but several different wavelengths. Indeed, the quantities of interest *are* the differences in wavelength between incident and emitted light, because these are the vibrational energies. So, we have two coupled processes, each electric dipole in nature. If each is represented in the normal way by x, y or z, then the coupling between them is well represented by taking a product, any simple product: x^2, xy, xz, y^2, yx, yz, z^2, zx, zy are all acceptable, although it is not surprising that xy and yx, for instance, are normally equivalent. So the symmetry species appropriate to the Raman process are those of these product functions in the character table. But a word of warning to those who might now think that they are trained in the interpretation of Raman spectra (and not just Raman, although the problem is more acute there). One has to be careful to distinguish between laboratory axes and molecular axes. So, in laboratory axes it is easy to have polarized light, but it is much more difficult in molecular axes. In this chapter we have worked in molecular axes, but for Raman spectroscopy the symbols above may well be used to denote laboratory axes (after all, laser light is polarized in laboratory axes). We shall return to the distinction between molecular and laboratory axes, but for the moment we revert to the former.

We are now in a position to summarize the above discussion and give a list of the operators which are appropriate to the most common spectroscopies. This is done in Table 4.5.

Table 4.5

Form of operator and spectral region	Form of operator	Symmetry properties of the operator are the same as those of:
Electronic (visible and ultraviolet)	Electric dipole	T_x, T_y, T_z (or more simply x, y, z)
Vibrational (infrared)		
Rotational (microwave)	Magnetic dipole	R_x, R_y, R_z
NMR (radiofrequency)		
EPR (microwave)		
Raman (visible)	Polarizability (this resembles 'electric quadrupole' but is a bit wider)	$x^2, y^2, z^2, xy, yz, zx$ (or combinations of these)

Problem 4.11 (a) Confirm that the following transitions are electric dipole allowed:

Point group	Ground term symmetry	Excited term symmetry
C_{2v}	A_1	B_2
C_{2v}	B_1	B_1
C_{4v}	A_2	E
C_{4v}	E	E

(b) Confirm that the following electric dipole transitions are forbidden:

Point group	Ground term symmetry	Excited term symmetry
C_{2v}	B_2	B_1
C_{4v}	A_1	B_1
C_{4v}	B_1	B_2
C_{4v}	A_2	A_1

4.6 The vibrational spectroscopy of the water molecule

We now bring together two main themes of this chapter, the vibrations of the water molecule and spectroscopic selection rules. First, the vibrations. It will be recalled that there are three vibrations, or to give them their correct name, normal modes, of the water molecule. Two, largely bond stretch vibrations, have A_1 and B_1 symmetries, whilst the – largely – angle change ('deformation') vibration is A_1. Our problem then is a simple one – are A_1 vibrations infrared active and are they Raman active? And we ask the analogous questions for the B_1 vibration. Next, the selection rule. In order to apply this, we need three things. The symmetry of the ground state, the symmetry of the excited state and the symmetry of the relevant operators. The last of these was the subject of discussion in the preceding section; the answers have been summarized in Table 4.5. What of the ground state? Well, since we are interested in vibrational spectra, we have to talk about the vibrational ground state. Here, we follow what must now be recognized as a standard procedure. We apply the operations of the C_{2v} point group to the vibrational ground state and so determine its symmetry. Of course, in the ground state there are no vibrations excited; there is nothing to change the size or shape of the molecule. So, the ground state is of A_1 symmetry. Suppose we excite an A_1 vibration? Although the size and shape of the molecule will change (bond stretches do the former and bond angle changes the latter), at every point in the vibration the symmetry operations will turn the molecule into itself. Exactly. So, the symmetry of the excited state is A_1 when it is an A_1 vibration which is excited. The B_1 is different. The molecule is distorted during a B_1 vibration (indeed, it is only totally symmetric vibrations, here A_1, that do not distort the molecule in a way that temporarily destroys some of its symmetry). The relevant vibration was shown in Figure 4.5b; the focus of our attention is the relationship between the transient distortions of the molecule. We know the answer: the oscillating distortions of the molecule have B_1 symmetry, so the vibrational excited state when this vibration is excited is of B_1 symmetry. We have assembled all the information that we need in order to apply the selection rule, but we will wait. The reason is that there is a simplification which we can apply, one that has already been mentioned – but without the conclusion we shall reach. We have to form a triple direct product and, since this involves multiplying numbers, we can evaluate the triple product in any order. Choose to evaluate the direct product of the ground and excited states first. Now, the argument used above shows that the vibrational ground state of a non-vibrating molecule must always be totally symmetric. Its contribution to the direct product will be to multiply by a series of 1's. It changes nothing. The direct product of ground and excited states will always be the symmetry of the excited state. Instead of a triple direct product we now only have to evaluate a simple direct product, that between the excited state and operator symmetries, and look for the cases that give the totally symmetric irreducible representation.

But we know the answer. We will only get the totally symmetric irreducible representation as the direct product when the two irreducible representations that we are multiplying are the same. This means that all we have to do is to look at the character table, here C_{2v}, and ask 'does the vibration excited have a symmetry which is the same as that of an infrared/Raman operator?'. If the answer is 'yes' the transition is allowed. Table 4.6 is the C_{2v} character table, yet again, but this time with the transformations of the relevant operators indicated. Some of the latter seem to come from nowhere, but this is not so. We determined them in Chapter 2 (Section 2.2). There, we looked at orbitals like the $2p_x$ orbital of the oxygen atom in water – but the argument used there applies equally well to the axis, x, and those for the other 2p orbitals to the y and z axes. Similarly, talk about a d_{xy} orbital was really about the product xy. Their symmetries are the same. And we can now make it even easier. The symmetry of a product function like xy is given by the direct product of the symmetry species of x and y. All we really need in a character table is the transformations of x, y and z. As we have seen, from them, everything else, products, rotations, can be worked out quite

Table 4.6

C_{2v}	E	C_2	σ_v	σ'_v	
A_1	1	1	1	1	z, x^2, y^2, z^2, T_z
A_2	1	1	−1	−1	xy, R_z
B_1	1	−1	1	−1	y, yz, T_y, R_x
B_2	1	−1	−1	1	x, zx, T_x, R_y

easily. Nonetheless, in Table 4.6 we give the C_{2v} character table with everything explicitly shown.

We now address the key question: what are the spectral activities of A_1 and B_1 vibrations? If A_1 is infrared active then a dipole, either an axis label or a translation, will be shown in the right-hand box. Of course, they go together and we see both z and T_z; the A_1 vibrations are infrared active. Similarly, the B_1 vibration is infrared active because both y and T_y are alongside it in the right-hand box. Specifically, if we were to be able to devise a method of orienting water molecules (so that the molecular axes were fixed relative to the laboratory), then using polarized infrared radiation we could decide whether or not to excite the A_1 or B_1 vibrations. Infrared polarized along molecular y will excite B_1 but not A_1, for instance. Much ingenuity has been exercised in trying to devise suitable experiments to exploit this: for example, trapping molecules within a sheet of plastic which is then stretched to force the trapped molecules into a common orientation. We return to this point below.

What of Raman activity? We repeat the process, but this time look for products of axes. A_1 has x^2, y^2 and z^2 listed; it is a Raman active vibration, three times over! Similarly, the B_1 vibration is Raman active because yz is given alongside it. Because all of the vibrations of the water molecule are both infrared and Raman active, these techniques will not distinguish between them. Indeed, Table 4.6 shows that the only sort of vibration of a C_{2v} molecule that the techniques could distinguish is an A_2, which is Raman (xy) active but not infrared. Even this is not 100 %. A vibration, whilst allowed in a particular spectroscopy, might nonetheless give rise to a very weak band (non-zero can mean something very small), possibly too weak to be detected – in which case you might reasonably but mistakenly think that it is forbidden. In the past, when structural methods were less well developed, vibrational spectroscopy was often used to attempt to determine the structure of a molecule. Work through the predictions for all the likely symmetries and see which agree(s) with experiment. It was often referred to as a 'sporting' method. You could win, be right, or lose, be wrong: the latter when one or more vibrations gave rise to an allowed but invisibly weak peak. In the case of the water molecule you would be right (the only alternative geometry is that of a linear molecule, for which the predictions are different). The water molecule has vibrations at about 3756, 3652 and 1595 cm^{-1}. Bending vibrations occur at about half of the frequency of the corresponding stretches, so the peak at 1595 cm^{-1} can immediately be identified as the A_1 angle change vibration. We cannot immediately distinguish between the other two (although we will be able to by the end of this chapter). It turns out that the 3652 cm^{-1} peak is also of A_1 symmetry and the 3756 the B_1. We do not show the spectra themselves because they do not show the simplicity that the above discussion leads us to expect. We have ignored the possibility that the water molecule might change its rotation as well as vibrate. In fact, the water molecule has low moments of inertia and its vibrational spectra are complicated by rotational fine structure.

One last word. When we found that the A_1 vibrations of the water molecule are allowed because z and T_z are to be found in the right-hand box of Table 4.6, what did it mean? What would have been different if, say, x and T_x had been there instead? Remember something mentioned earlier: we have been talking about molecular axes, not laboratory axes (although the latter have to be important because our spectrometers use them!). The fact that an A_1 vibration is z active means that the dipole moment change associated with the mode has to be along the z axis (here, of course, the C_2 axis). Not x, not y, only z. So, if we were able to hold the molecule fixed in space (so that its axes become laboratory axes and so under our control) and send a beam of polarized infrared light onto it, we would get important results. The fact that the infrared beam is polarized means that the direction of the electric vector is specified, and so therefore is the direction of the dipole moment change that it can induce. If the beam were polarized in the z direction, the spectrum would show only the A_1 vibrations. If it were polarized in the y direction the spectrum would show only the B_1 vibration. If it were polarized in the x direction, it would show no vibrations at all (we talk of an ideal experiment; the real world is a bit less obliging). Such experiments would give us real spectral information! Alas, they are seldom possible; is is not possible to hold all the water molecules in a sample in parallel fixed positions in space. But all is not lost. We still have to understand the selection rules of Raman spectroscopy.

Take an A_1 vibration. It is Raman allowed because, for instance, zz appears in the right-hand column of Table 4.6. What does this mean? It means that if a beam of polarized radiation (really, a laser beam) is incident on the molecule which happens to be such that its axis of polarization (which has to be in laboratory axes) coincides with the molecular z, then there is a chance that an A_1 mode will be excited and light of a longer wavelength emitted which *will also have its axis of polarization along the molecular z axis*. 'There is a chance' because Raman scattering is not a very likely event. If the polarization of the light beam is inclined to z then the chances of zz scattering decrease until they become zero in a plane perpendicular to z. But what of xx and yy – they both appear alongside A_1 in Table 4.6? The answer is as for zz, but with the axis labels changed. We conclude that when A_1 modes scatter they will do so with the axis of polarization of the incident light maintained in the scattered light. Is this true of the B_1 mode? Against B_1, we see 'yz' (or it could be 'zy', the way that we formed direct products tells us this). Light incident and which happens to be polarized along the molecular y axis may excite a B_1 vibration and be re-emitted polarized along the molecular z axis. Or, if it happens to be polarized along the molecular z axis, it may excite a B_1 vibration and be re-emitted polarized along the molecular y axis. The light emitted associated with a B_1 vibration does *not* retain the polarization of the incident laser beam. We have cracked the problem! Only the light emitted after an A_1 vibration has been excited will retain the original polarization. So, if we put a sheet of polaroid film in the scattered beam and arrange its polarization to be perpendicular to that of the incident laser beam (which we can do because both are in laboratory axes), the A_1 mode will largely disappear from our spectrum. The B_1 will remain, a bit weaker, but still there. We cannot play this game in reverse and delete just the B_1 mode, but no matter. We can identify the A_1 mode. In general, this trick can be used to identify totally symmetric vibrational modes.[2]

We have now come to the end of our discussion of the water molecule and our use of it to explore the application of group theory to chemistry. To move on we need molecules of a higher symmetry and these will form the subjects of most of the following chapters (in Chapter 12 we will return to the water molecule in a discussion of direct products that follows on from that in the present chapter). So, this is a convenient point at which to explore

---

[2] Inevitably, the discussion is simplified. In fact, the magnitude of the polarization effect depends on the differences between x, y and z (and so, xx, yy and zz). It is greatest when these differences are zero – and so is most dramatic for cubic molecules, such as those which are the subject of Chapters 8 and 10, where a mode can be completely eliminated from the spectrum. Then, the Raman effect is greatest for molecules which have several centres of high electron density. So, water, which might be regarded as basically a distorted oxygen atom, is a poor Raman scatterer – indeed, it is for this reason that water is often used as a solvent for Raman studies of dissolved molecules. Finally, in that the discussion concerns isolated molecules, it deals with the Raman scattering of gases, the worst phase for the study of a weak phenomenon.

a topic for which the symmetry of the water molecule is too high, even though we shall refer to it for one last time in this chapter!

4.7 Optical activity

Classically, the signature that a molecule is optically active is that it rotates the plane of polarized light. Although this is a simple feature which can be experimentally tested, it is a surprisingly unhelpful way to describe the phenomenon. It is much easier to talk of two beams of circularly polarized light which combine to give plane polarized light. The reason why it is easier is that circularly polarized light samples the whole 360° space, and so it is easy to move from one angle to another, and that is the important feature of the phenomenon that we want to describe. Figure 4.26a shows, schematically, the connection between plane polarized and circularly polarized light. In order to rotate the plane of polarized light we simply have to go further around one circularly polarized component than the other. The general idea is shown in Figure 4.26b. Put another way, if a molecule is to be optically active, then it must be meaningful to say that light going through the molecule traces out something like the alternative corkscrew (helical) paths shown as the difference between

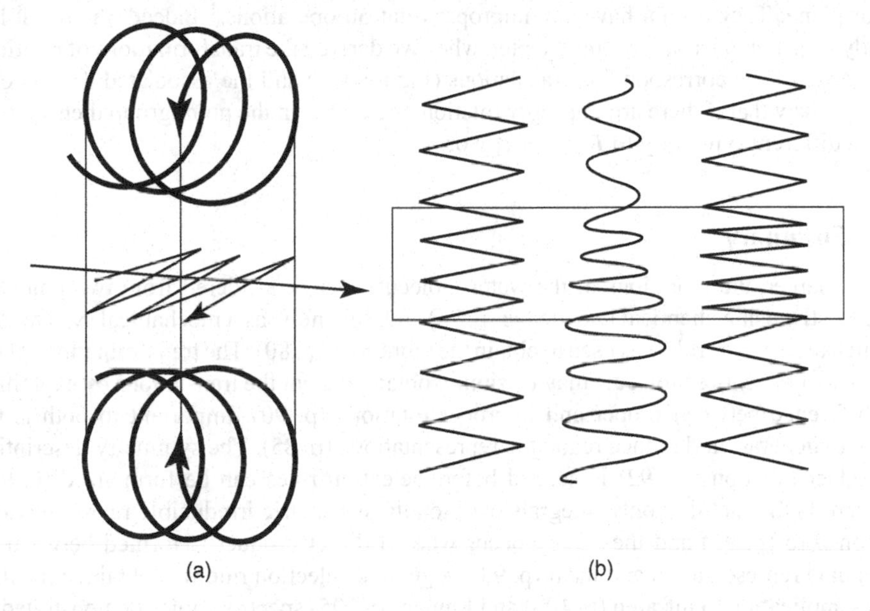

Figure 4.26 (a) The helixes at top and bottom represent light which is circularly polarized in opposite directions. Their sum is the plane polarized light (centre). This has been shown by joining (vertical lines) the two at their extremities, which correspond to the maxima in the centre. (b) A symbolic representation of the way that the plane of polarization of plane polarized light (centre) is rotated passing through optically active material. At left and right are shown the two circularly polarized beams that, at the bottom, combine in the way shown in Figure 4.26a to give the plane polarized beam. The central rectangle represents optically active material. This means that it contains an internal helical structure and so it reacts differently to the two incoming helical beams (one may be thought of as being of the same sense as that of the material, the other as of the opposite sense). The different interactions slow the two circularly polarized beams differently (centre) so that the emerging beams give rise to a rotated plane polarized beam (top)

Figure 4.26b left and right. How is such a corkscrew path to be defined? The corkscrew path, at the same time, encompasses both a translation and a rotation about the same axis. Both, at the same time. It cannot be a symmetry-forbidden process for the two to go together. In the water molecule they do not go together: T_z has a different symmetry to R_z, T_x has a different symmetry to R_x and T_y has a different symmetry to R_y. For a molecule to be optically active we need T_z to have the same symmetry as R_z; T_x as R_x and T_y as R_y. One can think of the possibility that any one pair would do, of course, but it turns out that this does not happen. It is all or none. Either a molecule is optically active or it is not. That is,

> Molecules[3] may be optically active when they have a symmetry such that T_α and R_α ($\alpha = x, y$ or z) transform as the same irreducible representation.

Comparison of this rule with the data given on the right-hand side of the compilation of character tables in Appendix 3 confirms the applicability of the commonly stated criteria for optical activity; optically active molecules possess neither a centre of symmetry nor a mirror plane. They do not have any improper rotation operations.[4] Indeed, the result has already been demonstrated in this chapter, when we derived the transformations of rotations from those of the corresponding translations (Figure 4.19 and the associated discussion). There we saw that if there are improper rotation operations in the point group then T_a must be of a different symmetry to R_a (a $= x, y$ or z).

4.8 Summary

In this chapter, the vibrations of the water molecule were considered from two points of view. Firstly, using chemical knowledge (p. 76) and, secondly, as a mechanical system, one in which the $3N - 6$ rule was used to obtain the vibrations (p. 80). The transformations of the rotations of the entire molecule may be simply obtained from the translations by exploiting the difference between proper and improper rotations (p. 90). Important to both is the ability to generate and reduce reducible representations (p. 85). The symmetry description of product functions (p. 92) is needed before selection rules can be formalized (p. 99). Key here is the fact that only integrals over totally symmetric irreducible representations are non-zero (p. 97) and these only occur when a direct product is formed between an irreducible representation and itself (p. 98). A general selection rule was obtained (p. 102) and its application to infrared (p. 105) and Raman (p. 105) spectroscopies demonstrated. It can also be used to explain molecular optical activity (p. 107).

---

[3] Note the word 'molecules' in this statement. It does not apply to crystals which, under some circumstances, can contain mirror planes of symmetry and yet be optically active.

[4] As an alternative general statement, one can say that optically active molecules do not have any S_n axis, where n can assume any value ($n = 1$ corresponds to a mirror plane and $n = 2$ to a centre of symmetry). S_n axes will be introduced in Chapter 8.

5 The D_{2h} character table and the electronic structures of ethene (ethylene) and diborane

The present chapter has several objectives, in addition to those indicated by its title. First, to introduce a new symmetry group and its character table. The group has been chosen because it is related to the C_{2v} group with which the reader is now familiar. It is not the simplest group that could have been used to discuss the bonding in ethene and diborane – the simplest would be the group D_2, a group which will be met shortly – but this discussion itself is only part of the objective of the present chapter. Use of the more complicated, D_{2h} (pronounced 'dee two aich') group will enable a start to be made to an exploration of the relationships between groups, the corresponding character tables and the pictorial representation of these character tables in nodal diagrams. A second objective is to present and to use the rather important technique of *projection operators*. Projection operators occur throughout quantum chemistry and are quite important. Despite their somewhat offputting name they provide a very simple method of obtaining the mathematical form of functions transforming as a particular irreducible representation.

5.1 The symmetry of the ethene molecule

The effect of bringing two CH_2 units, each of C_{2v} symmetry, together to form an ethene molecule has the effect of generating additional symmetry elements. All of the symmetry elements of the ethene molecule are shown in Figure 5.1. As shown in this figure, each CH_2 fragment is turned into itself by the operations that were met when discussing the C_{2v} point group – that is, there are two perpendicular mirror planes, the intersection between them defining a twofold axis, common to the two CH_2 units. As Figure 5.1 shows, the union of the two CH_2 units to form ethene has the effect of generating two new C_2 rotation axes, the three twofold axes being mutually perpendicular. This immediately suggests their use as Cartesian coordinate axes, a suggestion which we shall follow. Note that each of the three C_2 axes is unique: no two are similar. This is rather important because later in this book sets of twofold axes which are not unique will be met. In ethene, each twofold axis is unique because there is no operation in the group which interchanges any pair of them. This assertion can be checked after reading the next few paragraphs and a complete list

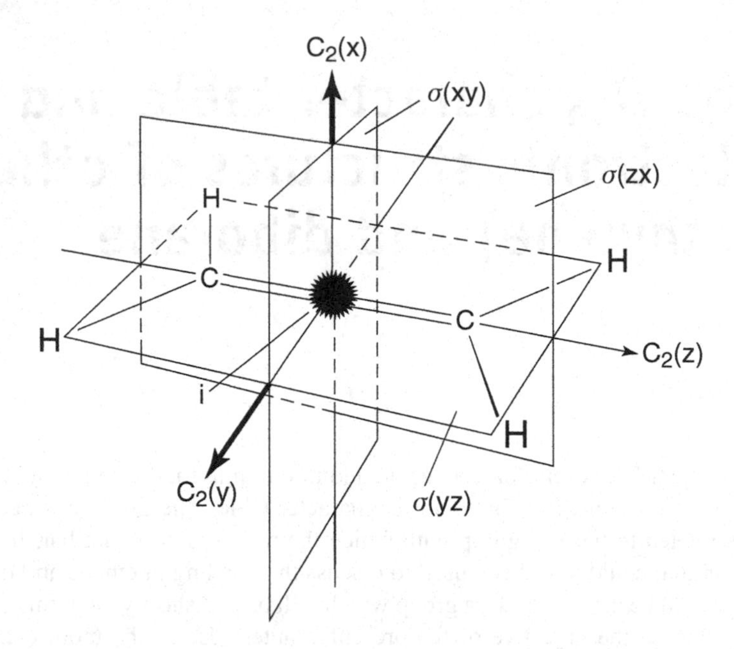

Figure 5.1 The symmetry elements of the ethene molecule; the list is complete except for the identity element, which has to be added. Here, the centre of symmetry, *i*, is shown as a furry circle. All of the other symbols have already been met

of the ethene symmetry operations has been obtained. In high-symmetry molecules it is quite common for there to be a set of rotation axes (or mirror planes or other symmetry elements) which are interchanged by other operations of the group. In such cases there is a corresponding complication in the character table and it is this complication which we seek to avoid here – it will be met in the next chapter – by working with another example of an Abelian[1] point group (the C_{2v} group of Chapters 2, 3 and 4 is Abelian, something which makes it particularly easy to work with).

When, as in this case, there are several apparently equally good choices for the z axis, it is usual to choose that axis which contains the largest number of atoms and so we shall take as z that twofold axis which passes through the two carbon atoms. It follows that the (local) z axis of each CH_2 fragment (of C_{2v} symmetry) is coincident with the molecular z axis. A helpful accident. Just as in the C_{2v} case, the labels of the other coordinate axes are determined by the convention that the planar molecule lies in the yz plane.

As is evident from Figure 5.1 the mirror planes of each CH_2 fragment persist as symmetry elements in the complete molecule. A third mirror plane exists in the ethene molecule. This passes through the mid-point of the carbon–carbon bond and is perpendicular to that bond. Figure 5.1 shows that each of the mirror planes is perpendicular to one of the coordinate axes; equally, it lies in the plane defined by the other two coordinate axes. Rather than use a σ_v notation for the mirror planes, each mirror plane is conventionally labelled by the

[1] It is not essential that the meaning of 'Abelian' be fully comprehended but this meaning was outlined at the end of the Chapter 2.

molecular coordinate axes which it contains: thus $\sigma(xy)$, $\sigma(yz)$ and $\sigma(zx)$. There will be more to say about these labels shortly. One of the difficult – and irritating – things about group theory to the newcomer is the way that the systems of notation that it defines are not always followed. Mirror planes, and more importantly the operation of reflection in them, are particularly prone to this. Eventually, one becomes hardened.

All of the symmetry elements listed so far are similar to those encountered in the C_{2v} group. Additionally, however, the ethene molecule contains a centre of symmetry, a point such that inversion of the whole molecule through it gives a molecule which is indistinguishable from the starting one. This centre of symmetry is indicated by the star-like point at the centre of Figure 5.1. More strictly, a centre of symmetry is such that inversion of any point of the molecule in it gives an equivalent point. Pictorially, if a straight line is drawn from any point (the starting point) in the molecule to the centre of symmetry and then extended an equal length beyond the centre of symmetry, the terminal point of the line is symmetry-equivalent to the starting point. This element and the corresponding operation are conventionally denoted by the lower case symbol i. A centre of symmetry, if there is one, is always at the centre of gravity of a molecule. A molecule may possess several rotation axes and several mirror planes but it can never possess more than one centre of symmetry, any more than it can have more than one centre of gravity.

The symmetry elements of the ethene molecule provide a better example of the use of the word 'point' when talking about a point group than do the symmetry elements of the water molecule. As is evident from Figure 5.1, all of the symmetry elements have one point in common: all pass through a common point, located at the centre of gravity of the molecule (in this context, the identity element is best thought of as corresponding to a C_1 rotation axis). Of course, in this example the point is also a centre of symmetry but this is not a requirement.

In summary, then, and talking now in terms of symmetry operations rather than symmetry elements, the symmetry operations which turn the ethene molecule into itself are:

$$E \quad C_2(z) \quad C_2(y) \quad C_2(x) \quad i \quad \sigma(xy) \quad \sigma(zx) \quad \sigma(yz)$$

This group of symmetry operations is commonly given the shorthand label D_{2h}. A detailed discussion of such shorthand labels will have to be deferred until Chapter 9 because it will not be until then that all of the symmetry operations on which the classification is based will have been met; up to that point they will have to be discussed individually. However, it is clear that the label D_{2h} requires some immediate explanation. Point groups which contain a principal C_n axis and, perpendicular to this principal axis, n twofold axes are called *dihedral point groups* and hence carry the label D_n (D for *d*ihedral). If one CH_2 group of the ethene molecule were to be slightly rotated about the z axis (so that the molecule becomes non-planar) then all of the mirror planes would be destroyed, as would the centre of symmetry. The resulting molecule would be of D_2 symmetry (Figure 5.2). If, perpendicular to the principal rotation axis – that of highest n value in C_n – in a molecule there is a mirror plane, that is, a plane *h*orizontal with respect to the C_n axis, then this mirror plane is denoted σ_h (recall that in Chapter 2 a mirror plane which is *v*ertical with respect to the principal rotation axes was denoted σ_v). If a D_n group also contains a σ_h mirror plane then the point group is labelled D_{nh}. The present point group falls into this category once a difficulty has been overcome. This is that in the present group, D_{2h}, no mirror plane has been called σ_h (or σ_v, for that matter). The reason for this is that in D_{2h} there are three

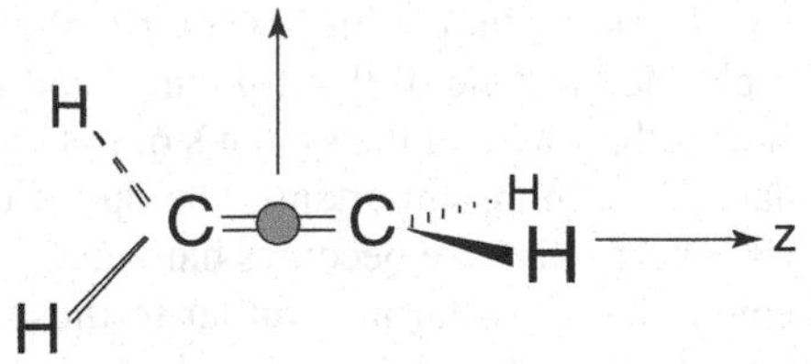

Figure 5.2 If the two ends of the ethene molecule are twisted in opposite senses around the z axis, the molecular mirror plane is destroyed and all other mirror planes too. Three C_2 rotation axes remain (one is shown here head-on)

C_2 axes, any one of which might equally well be chosen as the principal axis (our choice of z axis was determined by convention, nothing more fundamental). The mirror plane which should be labelled σ_h would depend upon which particular C_2 axis is nominated as principal axis. In this particular case, where all three mirror planes are equally good candidates for being labelled σ_h (or, indeed, σ_v), the egalitarian solution is to give none of them either label but, rather, designate them as has been done above. However, egality cannot alter the claim of the group to be recognized as one of the D_{nh} type; accordingly it is labelled D_{2h}.

5.2 The character and multiplication tables of the D_{2h} group

In order to proceed further we must obtain the character table of the D_{2h} point group. The procedure which was adopted for the C_{2v} case – considering the transformations of a variety of basis functions – could be used to generate the D_{2h} character table; the procedure is entirely analogous. For this reason, space will not be devoted to it. Rather, the reader is invited to use this method himself or herself in a problem which follows. The D_{2h} character table is given in Table 5.1. Perhaps the first reaction is to be horrified by its size. Any such horror is misplaced. The bigger the table the greater the number of distinctions that can be made using group theory and the more useful the table is. So, the size should be welcomed. Actually, it is not as bad as it first appears. Table 5.1 is divided into quadrants. Three of these

Table 5.1

D_{2h}	E	$C_2(z)$	$C_2(y)$	$C_2(x)$	i	$\sigma(xy)$	$\sigma(zx)$	$\sigma(yz)$	
A_g	1	1	1	1	1	1	1	1	s, d_{z^2}, $d_{x^2-y^2}$
B_{1g}	1	1	−1	−1	1	1	−1	−1	d_{xy}
B_{2g}	1	−1	1	−1	1	−1	1	−1	d_{zx}
B_{3g}	1	−1	−1	1	1	−1	−1	1	d_{yz}
A_u	1	1	1	1	−1	−1	−1	−1	f_{xyz}
B_{1u}	1	1	−1	−1	−1	−1	1	1	p_z
B_{2u}	1	−1	1	−1	−1	1	−1	1	p_y
B_{3u}	1	−1	−1	1	−1	1	1	−1	p_x

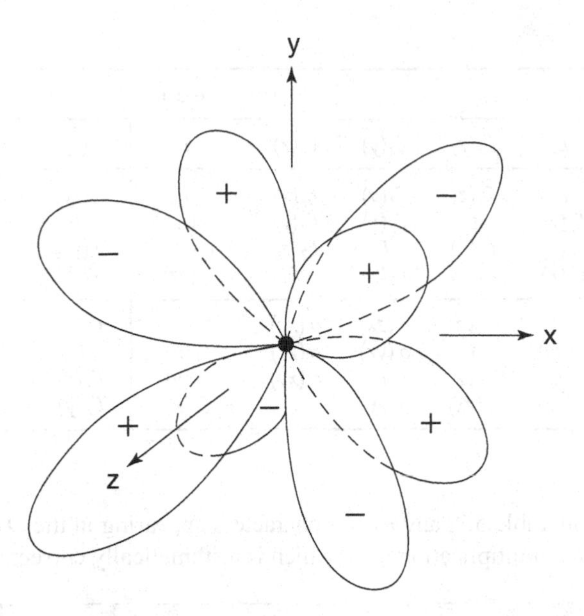

Figure 5.3 The f_{xyz} orbital of a hypothetical atom placed at the centre of gravity of the ethene molecule. Note that the phase of the orbital is positive in those regions of space in which the product *xyz* is positive

are familiar – they are just a repetition of the C_{2v} character table, even if the surrounding labels are different. The fourth is also the C_{2v} character table, but this time with all signs reversed. Not surprisingly, we shall have more to say about this.

In Chapter 2 the irreducible representations of the C_{2v} character table were generated by considering the transformations of the orbitals of a unique atom (the oxygen in H_2O). In order to use this technique in the present problem it is necessary to first have a unique atom. This can be done by placing a hypothetical atom at the centre of gravity of the ethene molecule. Using just the familiar s, p and d orbitals it is not possible to generate the A_u irreducible representation of the D_{2h} point group. This irreducible representation can be generated using one of the f orbitals, the f_{xyz} orbital, a diagram of which is shown in Figure 5.3. The orbitals of the hypothetical atom which generates each irreducible representation are shown at the right-hand side of Table 5.1. Care has to be taken to use the coordinate axis system shown in Figure 5.1.

Problem 5.1 Derive as much as you can of the character table of the D_{2h} point group (Table 5.1). Place an atom at the centre of gravity of the molecule and consider the transformations of the orbitals of this atom.

For every group there exists a group multiplication table; that for the D_{2h} group is given in Table 5.2. Its derivation is analogous to the derivation of the C_{2v} group multiplication table (Table 2.1), although more tedious and error-prone. Just as in the C_{2v} case it will be found that when the appropriate substitution of characters for the corresponding symmetry

Table 5.2

		First operation							
	D_{2h}	E	$C_2(z)$	$C_2(y)$	$C_2(x)$	i	$\sigma(xy)$	$\sigma(zx)$	$\sigma(yz)$
Second operation	E	E	$C_2(z)$	$C_2(y)$	$C_2(x)$	i	$\sigma(xy)$	$\sigma(zx)$	$\sigma(yz)$
	$C_2(z)$	$C_2(z)$	E	$C_2(x)$	$C_2(y)$	$\sigma(zx)$	i	$\sigma(yz)$	$\sigma(zx)$
	$C_2(y)$	$C_2(y)$	$C_2(x)$	E	$C_2(z)$	$\sigma(zx)$	$\sigma(yz)$	i	$\sigma(xy)$
	$C_2(x)$	$C_2(x)$	$C_2(y)$	$C_2(z)$	E	$\sigma(yz)$	$\sigma(zx)$	$\sigma(xy)$	i
	i	i	$\sigma(yz)$	$\sigma(zx)$	$\sigma(xy)$	E	$C_2(x)$	$C_2(y)$	$C_2(z)$
	$\sigma(xy)$	$\sigma(xy)$	i	$\sigma(yz)$	$\sigma(zx)$	$C_2(z)$	E	$C_2(x)$	$C_2(y)$
	$\sigma(zx)$	$\sigma(zx)$	$\sigma(yz)$	i	$\sigma(xy)$	$C_2(y)$	$C_2(x)$	E	$C_2(z)$
	$\sigma(yz)$	$\sigma(yz)$	$\sigma(zx)$	$\sigma(xy)$	i	$C_2(x)$	$C_2(y)$	$C_2(z)$	E

operation is made in Table 5.2, any row of characters appearing in the D_{2h} character table turns Table 5.2 into a multiplication table which is arithmetically correct.

Problem 5.2 By combining (multiplying) pairs of operations of the D_{2h} character table show that Table 5.2 is correct. Some help in this problem is provided by Section 2.3. A further tip on how to do this problem is provided by Figure 5.4. Take a general point in space (indicated by the solid star). Perform the first operation (in Figure 5.4, $\sigma(zx)$) to give the cross-hatched star, follow it with the second operation (in Figure 5.4, i) to give the open star. Then ask 'what single operation turns the solid star into the open one' (in Figure 5.4, $C_2(y)$). One concludes that $\sigma(zx)$ followed by i is equivalent to $C_2(y)$.[2] Of course, this is not new, it is just Figure 4.18 applied to a case where all of the symmetry operations exist in their own right.

Problem 5.3 Take any four of the irreducible representations of Table 5.1 and by substituting the appropriate character for each operation in Table 5.2 show that in each case an arithmetically correct multiplication table is obtained. If needed, Section 2.4 will provide guidance on this problem.

5.3 Direct products of groups

There are several interesting features of Table 5.2; for example, it is symmetric about either diagonal. Another is the way that it may be broken into four smaller blocks, pairs of which are identical. Similarly, as has been seen, the D_{2h} character table (Table 5.1) may also be broken into four blocks but now three of the blocks are identical and in the fourth the same set of characters appear, but with all signs reversed. There is a simple reason for these patterns. As is evident from Table 5.2 (and Figure 5.4), the operation $\sigma(xy)$ is equivalent to $C_2(z)$ followed by the inversion i. Similarly, $\sigma(yz)$ equals $C_2(x)$ followed by i and $\sigma(zx)$

[2] The reader may have spotted a similarity between Figure 5.3 and Figure 4.18. Indeed, they are variants on a common theme. There is one other variant – a reflection in a mirror plane followed by a rotation about an appropriately placed C_2 axis is equivalent to inversion in a centre of symmetry.

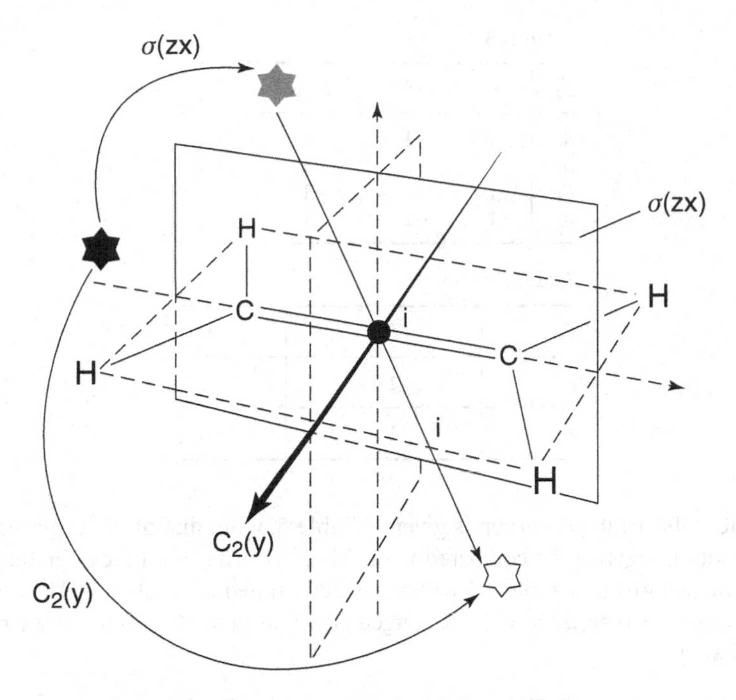

Figure 5.4 An illustration that $\sigma(zx)$ followed by i is equivalent to $C_2(y)$. In a similar way a C_2 can always be treated as σ combined with i; i can always be represented as C_2 combined with σ. Figure 4.18 is relevant

equals $C_2(y)$ followed by i. It follows that the operations of the D_{2h} group may be rewritten as follows:

$$\begin{array}{cc} E \quad C_2(z) \quad C_2(y) \quad C_2(x) & i \quad \sigma(xy) \quad \sigma(zx) \quad \sigma(yz) \\ \text{are equivalent to} & \text{are equivalent to} \\ \underbrace{E \quad C_2(z) \quad C_2(y) \quad C_2(x)}_{\text{followed by } E} & \underbrace{E \quad C_2(z) \quad C_2(y) \quad C_2(x)}_{\text{followed by } i} \end{array}$$

That is, the operations of the D_{2h} group may be obtained by forming all possible products of members of the set E, $C_2(z)$, $C_2(y)$, $C_2(x)$ which, as has already been seen, form the D_2 group – with members of the set E, i – two operations which together form a group called the C_i group (pronounced 'cee eye'), to which we shall return shortly. Technically, one says that 'the D_{2h} group is the *direct product* of the D_2 and C_i groups' – where 'direct product' means 'form all possible products of one group of symmetry operators with all of the symmetry operators of the other'.[3] When the operations of a group may be expressed as direct products in this way, so too, may the corresponding character tables. That is, the character table of the D_{2h} group is the direct product of those of the D_2 and C_i groups (and the phrase 'direct product' now refers to characters but its meaning is otherwise unaltered).

[3] In this section an *aufbau* approach is used, a larger group being built up from subgroups. An important notation appears if the inverse pattern is studied and the decomposition of a group into subgroups is considered. Because both C_i and D_2 may be obtained from D_{2h} in one way, and only one way, they are both said to be *invariant subgroups* of D_{2h}. In contrast, some other subgroups of D_{2h} – one is C_{2v}; another is C_2 – are not invariant subgroups because there is more than one way that they can be generated. This topic is dealt with more fully in Section 9.3; it is a topic which will become particularly important in Chapter 14.

Table 5.3

D_2	E	$C_2(z)$	$C_2(y)$	$C_2(x)$
A	1	1	1	1
B_1	1	1	−1	−1
B_2	1	−1	1	−1
B_3	1	−1	−1	1

Table 5.4

C_i	E	i
A_g	1	1
A_u	1	−1

The character table of the D_2 group is given in Table 5.3 and that of the C_i group in Table 5.4. They should, together, be compared with Table 5.1. The four blocks in Table 5.1 are just the characters given in Table 5.3 with signs determined by Table 5.4. Thus in three of the blocks Table 5.3 reappears with unchanged sign and in the fourth all of the characters are multiplied by −1.

Problem 5.4 Multiply the character Table 5.3 by the character Table 5.4 and thus generate Table 5.1. Note that 'multiply' here means different things for operations and characters. For the latter it means simple arithmetic multiplication but for the former it means 'carry out the operations one after the other'. The way that the relevant operations multiply is indicated earlier in this section; in order to generate Table 5.1 it is necessary to maintain the correct correspondence between products of characters and products of operations.

As we have already noted, an interesting thing about Table 5.3 is that the sets of characters that appear in it are the same as those of the C_{2v} point group (Table 2.4), although the operations and irreducible representation labels are not the same in the two groups. Groups which have character tables containing identical corresponding sets of characters are said to be *isomorphous* groups.[4] Isomorphous groups need have no operation in common – except, of course, the identity operation, which appears in all groups. However, isomorphism between character tables means that there is a close connection between the groups. Thus, something true in one group has a counterpart in an isomorphous group. An illustration of this is given in Section 12.1.

In the C_i character table (Table 5.4) the only distinction between the irreducible representations is the behaviour of the quantities they describe under the operation of inversion in the centre of symmetry, i. Both irreducible representations are denoted by A but something which is transformed into itself (i.e. is symmetric) under the inversion operation is

[4] Strictly, their multiplication tables must also show an analogous similarity but, in practise, the definition in the text is adequate.

distinguished from one which is turned into minus itself (i.e. is antisymmetric) by the subscripts g (from gerade, German for 'even') and u (from ungerade, German for 'odd') respectively; it is always true that a g suffix indicates an irreducible representation which describes something which is symmetric with respect to inversion in a centre of symmetry, whilst the suffix u describes something which is antisymmetric. The reader will find it helpful to compare the use of g and u suffixes in the irreducible representation labels of Table 5.1 with the corresponding characters under the i operation.

Problem 5.5 For each irreducible representation in Table 5.1 which carries a g suffix (e.g. A_g) list the character under the i operation. Repeat this exercise for each irreducible representation carrying a u suffix (e.g. A_u). Compare your result with Table 5.4.

Two final points. Lines have been included in Tables 5.1, 5.2 and 5.4 to clarify the discussion in the text. Normally they are omitted and, indeed, columns in these tables are sometimes permuted, concealing the pattern which is apparent from the way that these have been written here. In the compilation of character tables in Appendix 3 those that are direct products involving C_i have lines included (and, just for good measure, a few which are direct products involving a group which is isomorphous to C_i have lines included too). Finally, the choice of labels for the irreducible representations in Table 5.1. For the C_{2v} point group, it will be recalled that the irreducible representations that were symmetric with respect to rotation by C_2 (a character of $+1$ under this operation) were labelled A and those that were antisymmetric (a character of -1) were labelled B (Section 2.4). Exactly the same convention is used in D_{2h}. The difference, of course, is that there are three times as many distinct C_2 rotations in the latter, and so more B's, than in C_{2v}. In C_{2v}, there are equal numbers of irreducible representations labelled A and B; in D_{2h}, there are three times as many B's as A's. The suffixes 1, 2 and 3 on the B's in D_{2h} indicate under which C_2 rotation the character is $+1$ (there always is one). Commonly, 1 indicates z, 2 indicates y and 3 indicates x. There is the potential for confusion between different authors, and it happens. Perhaps, in retrospect, labels such as B_z, B_y and B_x might make some sense, but the 1, 2, 3 convention is firmly established. The distinction between the g and u suffixes has already been discussed.

5.4 Nodal patterns of the irreducible representations of the D_{2h} group

Although the reader may have been persuaded that the size of the character table of the D_{2h} group is helpful rather than forbidding, Table 5.1 probably remains somewhat overpowering. It is the purpose of the present section to draw pictures of the irreducible representations in Table 5.1 and so, hopefully, make them more familiar. A convenient starting point is with Tables 5.3 and 5.4; we start by drawing their irreducible representations as nodal patterns and then combine them, in the direct product fashion of the previous section, to obtain those of Table 5.1. First we look at Table 5.3; its similarity to the C_{2v} character table means that the nodal patterns are also similar. They are given in Figure 5.5. The patterns of $+$ and $-$ in the diagrams are the same as those of Figure 2.17, but the symmetry operations

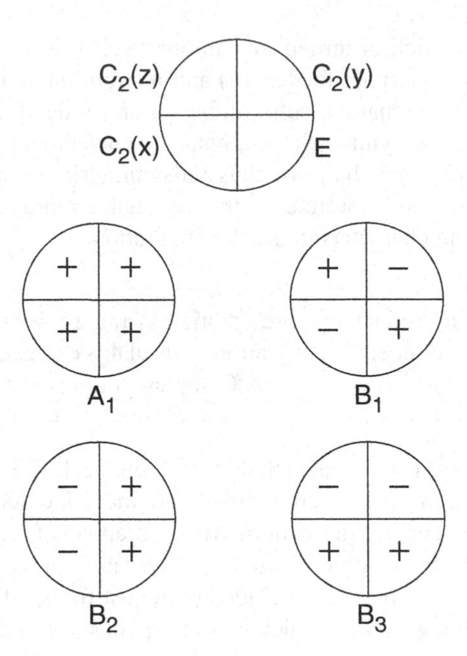

Figure 5.5 Nodal patterns of the irreducible representations of the D_2 point group. The symmetry relationships between the quadrants is shown at the top. Although the patterns are the same as those of the C_{2v} group (Figure 3.1) the symmetry labels differ. The significance differs too. In C_{2v} it was possible to carry out all operations of the group on a point and the z coordinate of the point did not change. D_2 is much more three-dimensional

to which they refer are different. These diagrams do not contain a centre of symmetry. We must remember this after we have worked with Table 5.4 and, subsequently, turn to its direct-product combination with Table 5.3. Now to Table 5.4. It is simple, and so too must be the corresponding nodal diagrams of the irreducible representations. They are drawn in Figure 5.6 (note that the horizontal line is a phase line, not an indication of a mirror plane operation). The nodal patterns of the D_{2h} group are combinations of those of Figures 5.5 and 5.6. Although the combinations are those of a direct product (i.e., every entry in Figure 5.5 is combined with every entry in Figure 5.6), care has to be taken. Consider Figures 5.7 and 5.8. Figure 5.7 shows the B_2 pattern of Figure 5.5 combining with the A_u pattern of Figure 5.6, everything being placed in somewhat arbitrary positions and orientations. A more sensible arrangement is that in Figure 5.8.[5] The two separate layers shown at the top are repeated at the bottom but in counter-perspective, so that the 'bottom' layer is bigger than the 'top', so that it can be seen. But in Figure 5.7 the bottom part may not seem consistent with the upper (the very bottom circle having the wrong phases). The reason is that two of the C_2 rotation operations of D_2 appear on the 'bottom' layer of the diagrams for D_{2h} (Figure 5.9). So, nodal diagrams have to be combined with care.

Figure 5.9 gives real insight into the character table, Table 5.1. Apart from the first, all of the irreducible representations of Table 5.1 contain the same number of $+1$ and -1's.

[5] Figure 5.7 was included in the hope that it would make Figures 5.8 and 5.9 more understandable.

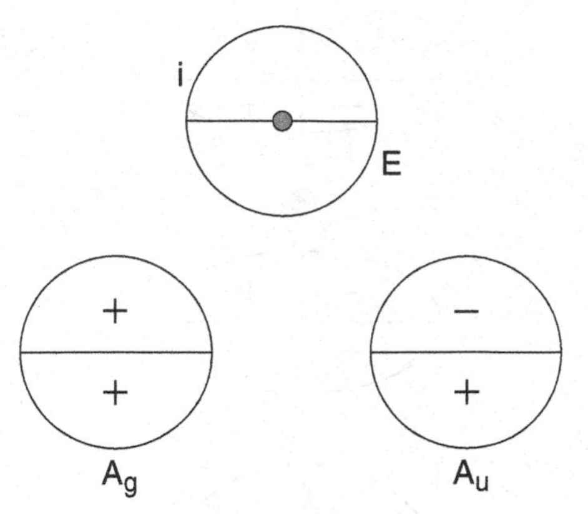

Figure 5.6 Nodal patterns of the irreducible representations of the C_i point group. The effect of the symmetry operations is shown at the top, where, for convenience, the centre of symmetry is shown as a dark circle

One might reasonably think that they are therefore rather similar. Figure 5.9 shows that this is not the case. The B_{1u}, B_{2u} and B_{3u} irreducible representations describe objects, orbitals, mathematical functions, whatever, which are inherently single-noded. The B_{1g}, B_{2g} and B_{3g} describe functions which are double-noded, whilst A_u describes triple-noded functions (which is why it had to be introduced using an f orbital) and A_g describes zero-noded functions. We begin to see the pattern. The bigger the character table, the greater the range

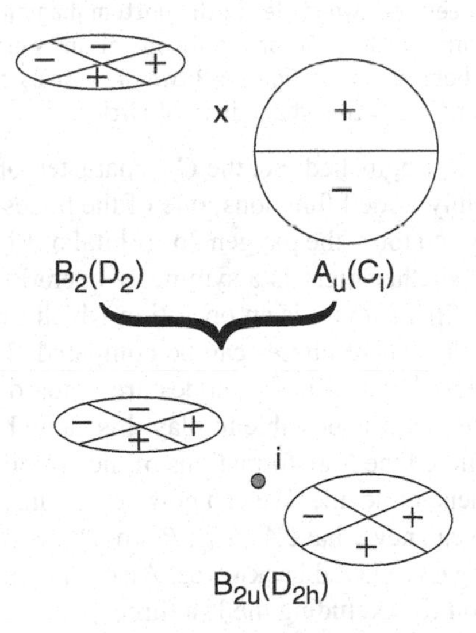

Figure 5.7 The nodal pattern of the B_{2u} irreducible representation of the D_{2h} point group, obtained as the direct product of the B_2 of D_2 with the A_u of C_i. The B_2 nodal pattern has been twisted relative to that shown in Figure 5.5 and no attempt has been made to bring the A_u into coincidence with it. With this strange combination, the resulting B_{2u} pattern is shown at the bottom, where the centre of symmetry of C_i has been added, for convenience.

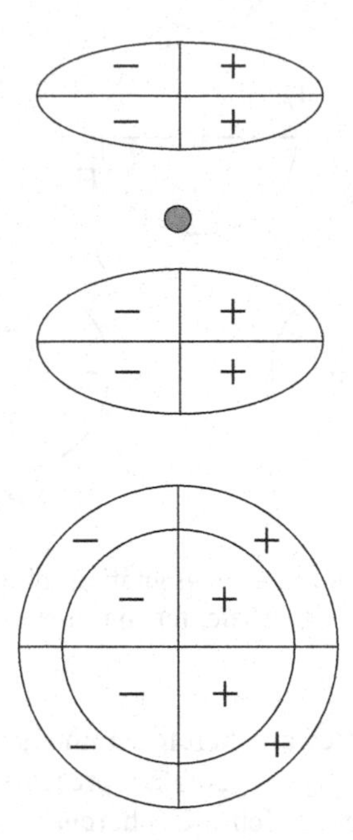

Figure 5.8 The B_{2u} nodal pattern of Figure 5.7 in a more symmetrical arrangement (top), again showing the centre of symmetry, and a more convenient way of representing this combination (bottom). The latter is drawn in counter-perspective, so that the upper layer in the top diagram is contained within the smaller circle of the bottom diagram. The lower layer in the top diagram is represented by the areas between the two circles in the bottom diagram. There is a hidden subtlety in this diagram, a hidden rearrangement. All of the C_2 operations were contained in the very top part of Figure 5.7, but in the bottom part of Figure 5.8 two of these C_2 connect the inner circle with the outer, only one connects points within the individual circles.

of nodalities which can be distinguished. So, the C_{2v} character table of Chapters 2, 3 and 4 offered no sensitivity to triply-noded functions; one of the nodes had to be ignored. Other nodes were sometimes ignored too – the oxygen $2p_z$ orbital of Figure 2.15c has had a node ignored. The key thing is whether there is a symmetry operation which interchanges the lobes of an orbital such as $2p_z$. If there is an operation which interchanges them, there is a mechanism by which their relative phases can be compared. The greater the number of symmetry operations the less likely it is that nodes are ignored. The size of Table 5.1 is potentially beneficial! Note that it is possible to play this game backwards. Hopefully, the reader will have already studied the transformations of the orbitals of an atom placed at the centre of gravity of the ethene molecule. We can now see, by inspection of Figure 5.9, that a p orbital on such an atom can never have A_u, B_{1g}, B_{2g} or B_{3g} symmetries – it does not have enough planar nodes. The game played backwards. Also, of course, it is inherently 'u', so there is an associated reason for excluding the last three.

Problem 5.6 The nodal patterns in Figure 5.9 are really arranged in three-dimensional space. Use Figure 5.1 – and the eight symmetry-related volumes between the mirror planes shown there – to sketch diagrams of the three-dimensional patterns for which Figure 5.9 is a shorthand.

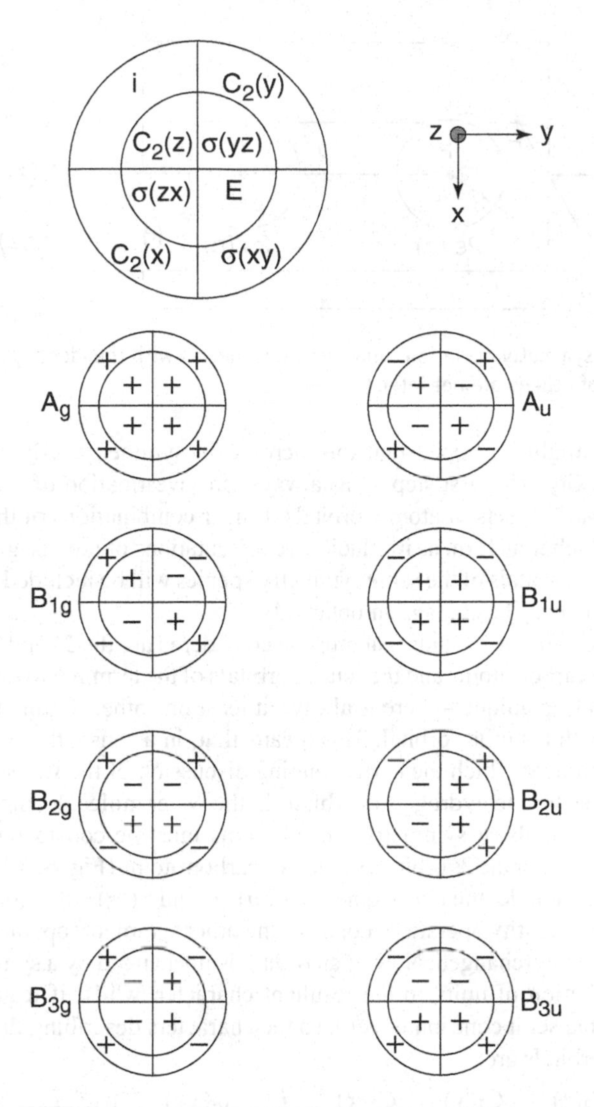

Figure 5.9 The nodal patterns of the irreducible representations of the D_{2h} point group. The association with symmetry operations is given in the top diagram. The arrangement of the symmetry operations is the 'natural' one; that is, it is one that displays the characteristics of the irreducible representations. So, B_{1u}, B_{2u} and B_{3u} are inherently dipolar and this is evident in the patterns

5.5 The symmetries of the carbon atomic orbitals in ethene

The character table of the D_{2h} group given in Table 5.1 will now be used in a qualitative discussion of the electronic structure of the ethene molecule. At first sight one might expect that this discussion would be more complicated than that for the water molecule because we now have six atoms to consider. On the other hand, instead of working with a group with only four irreducible representations we now have eight, so, as the reader should already

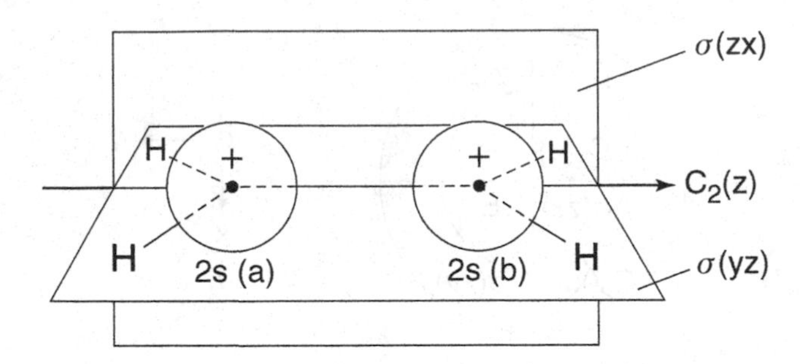

Figure 5.10 Those symmetry operations under which (together with the identity operation) the two carbon 2s orbitals of ethene are not interchanged

be suspecting, it might be hoped that the increase in symmetry will offset the greater molecular complexity. The first step is, as always, an investigation of the transformation properties of the various sets of atomic orbitals. Linear combinations of these orbitals will then be formed which transform as irreducible representations of the D_{2h} group. Finally, the interaction between orbitals of the same symmetry species will be included and a qualitative molecular orbital energy level diagram obtained.

The valence shell atomic orbitals that must be considered are the 2s and $2p_x$, $2p_y$ and $2p_z$ orbitals of the two carbon atoms and the four 1s orbitals of the terminal hydrogen atoms. Not one of these orbitals is unique – there is always at least one other, symmetry related, atom in the molecule with a similar orbital. This means that, in a sense, the present discussion must start at the point at which the corresponding discussion of the water molecule ended in Chapter 3. Just as for the hydrogen 1s orbitals in the water molecule, the transformations of corresponding orbitals of symmetry-related atoms must be considered together. As a simple example, consider the 2s orbitals of the two carbon atoms (Figure 5.10). Each of these orbitals remains itself under the $C_2(z)$ rotation, the $\sigma(zx)$ and $\sigma(yz)$ reflection operations and, of course, under the identity operation. For all of the other symmetry operations of the group the two orbitals are interchanged. Now, if an orbital is unchanged by a symmetry operation it makes a contribution of unity to the resultant character, whilst if it goes into another member of the same set it contributes zero, so the characters describing the transformation of the carbon 2s orbitals are:

E	$C_2(z)$	$C_2(y)$	$C_2(x)$	i	$\sigma(xy)$	$\sigma(zx)$	$\sigma(yz)$
2	2	0	2	0	0	2	2

Either by trial and error, or by systematic use of the group orthonormal relationships which were met in Chapter 3 (Section 3.1) and, briefly, given a more precise meaning in Chapter 4 (Section 4.4), it is concluded that this reducible representation has $A_g + B_{1u}$ components.[6]

The 2s orbitals of the two carbon atoms in the ethene molecule, considered in isolation, resemble the two hydrogen 1s orbitals in the hydrogen molecule (or in the water molecule). It is reasonable, therefore, to anticipate that the linear combinations of these orbitals which transform as A_g and B_{1u} will be similar to those which were obtained when discussing the

[6] Note a useful trick here, one sometimes used by experienced workers to impress the inexperienced. The character under the i operation is 0. This can only arise if the number of g irreducible representations spanned equals the number of u. So, if at any point there is an imbalance, one immediately knows in which direction to move to redress the balance.

water molecule (Figures 3.2, 4 and 6). That is, if we call the carbon 2s orbitals 2s(a) and 2s(b), as shown in Figure 5.10, then the correct linear combinations are of the form:

$$\frac{1}{\sqrt{2}}\left(2s\,(a) + 2s\,(b)\right) \quad \text{and} \quad \frac{1}{\sqrt{2}}\left(2s\,(a) - 2s\,(b)\right)$$

Later in this chapter a systematic way of deriving such linear combinations will be obtained and these functions can then be checked. Actually, whenever there are just two symmetry-related orbitals to be considered, the correct combinations are sum and difference combinations – like those above – irrespective of the details of the symmetry. Of the two combinations given above it is easy to demonstrate that the first has A_g symmetry and the second B_{1u}. There is a rather subtle aspect of this. Suppose that instead of choosing in Figure 5.10 to give the two carbon 2s orbitals the same phase they had been given opposite phases. The first combination above would, in this case, be an out-of-phase combination of the two orbitals, notwithstanding the + sign in the mathematical expression. The solution to this paradox is that in this case the first combination would have had B_{1u} symmetry and not A_g whilst the second would be the A_g combination. The systematic method of obtaining such functions takes account of our arbitrary choices of orbital phases and corrects for them. It is important to note that one cannot work with combination functions like those given above unless the phases chosen for the component atomic orbitals are known. One might think that the simple way would be to choose all orbitals to be of the same phase. Unfortunately such a simplification is not always possible. Thus, there are two alternative ways of drawing the $2p_z$ orbitals on the two carbon atoms; these are shown in Figure 5.11. In Figure 5.11a the $2p_z$ orbitals are chosen so that the phasing of the $2p_z$ orbitals coincides with that of the molecular coordinate axis system – the positive lobes point towards positive z and the negative lobes towards negative z. In Figure 5.11b the phase on one centre is reversed. This latter choice of phases has the advantage that under, say, the $C_2(x)$ rotation operation the $2p_z$ orbitals are simply interchanged. In the choice of Figure 5.11a they are

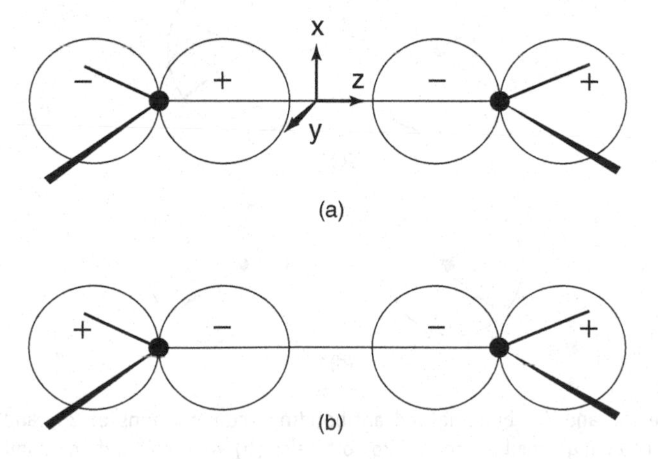

Figure 5.11 Alternative phase choices for the $2p_z$ orbitals of the carbon atoms in ethene. See the text for a discussion

not only interchanged by this operation but the phases of their lobes are also reversed. Here there is no simplification offered by convention; different people may, with equal validity, choose the phases differently. It follows that care has to be taken to check the basic choice of phases used by each person writing on the subject. If we choose the phases indicated in Figure 5.11a then the sum and difference combinations

$$\frac{1}{\sqrt{2}}\left(2p_z(a) + 2p_z(b)\right) \quad \text{and} \quad \frac{1}{\sqrt{2}}\left(2p_z(a) - 2p_z(b)\right)$$

are, respectively, the B_{1u} and A_g (C—C σ antibonding and bonding respectively) combinations of carbon $2p_z$ orbitals. If were to choose the phases of Figure 5.11b then these would, respectively, have been the A_g and B_{1u} combinations. The simplest way of avoiding the choice-of-phase problem is to state the symmetry of the combination, A_g or B_{1u}. These combinations of carbon $2p_z$ orbitals are shown schematically in Figures 5.12a and 5.12b. Also shown are the A_g and B_{1u} combinations of carbon 2s orbitals as Figures 5.12c and 5.12d. In these diagrams the atomic orbitals on the two carbon atoms are shown as overlapping each other, although this overlap has been neglected in the expressions given

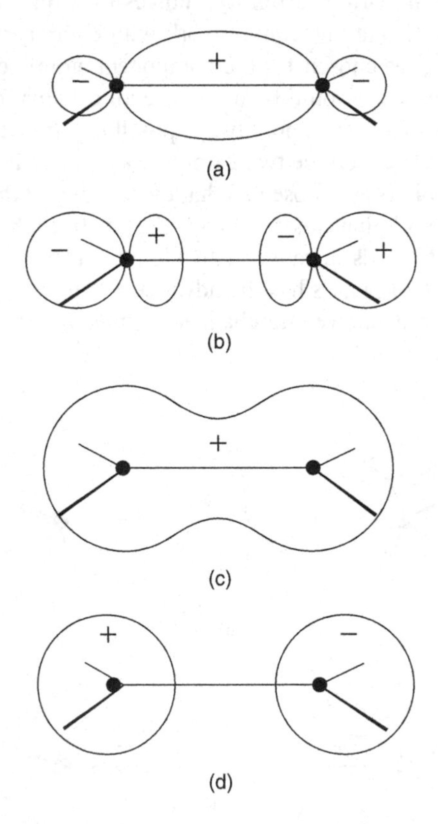

Figure 5.12 The B_{1u} and A_g (bonding and antibonding) combinations of $2p_z$ and 2s orbitals in ethene: (a) B_{1u} (bonding) combination of $2p_z$ orbitals; (b) A_g (antibonding) combination of $2p_z$ orbitals; (c) A_g (bonding) combination of 2s orbitals; (d) B_{1u} (antibonding) combination of 2s orbitals

above. This inconsistency is tolerated because it gives mathematical simplicity together with diagrammatic clarity.

Although detailed calculations did not fully justify it, in our discussion of the water molecule it was found to be convenient to mix together the oxygen 2s and $2p_z$ orbitals. Since they had the same symmetry this mixing is allowed and the resultant picture that emerged was closely related to simple ideas on the bonding in the water molecule – and was an advantage when later looking at the latter in more detail. For the same reason the carbon 2s and 2p orbitals will be mixed in the present example forming, effectively, carbon sp hybrids. If these hybrids had been formed as a first step, a simpler discussion would have resulted. Unfortunately, this was not permissible because at that stage it had not been established that the carbon $2p_z$ and 2s orbitals transform in a similar way (one says 'they transform isomorphously'). Instead of going back to the start of the argument and working with sp hybrids it is simplest to simply combine the A_g combinations of carbon 2s and $2p_z$ orbitals and similarly for the B_{1u} – the end result is the same. The result of mixing together – essentially, taking sum and difference combinations of – the two A_g orbitals of Figure 5.12 and (separately) the two B_{1u} orbitals is shown schematically in Figure 5.13. In each case both in-phase and out-of-phase combinations are shown. Two of these four orbitals will carry through, unmodified, into the final description of the ethene molecule. These are an A_g combination which is to be identified with the C—C σ bonding orbital (Figure 5.13a) and a B_{1u} combination which is the corresponding C—C σ antibonding orbital (Figure 5.13b). The other A_g and B_{1u} combinations (Figures 5.13c and 5.13d, respectively), which are largely directed away from the C—C bond, are involved in interactions with the terminal hydrogen atoms.

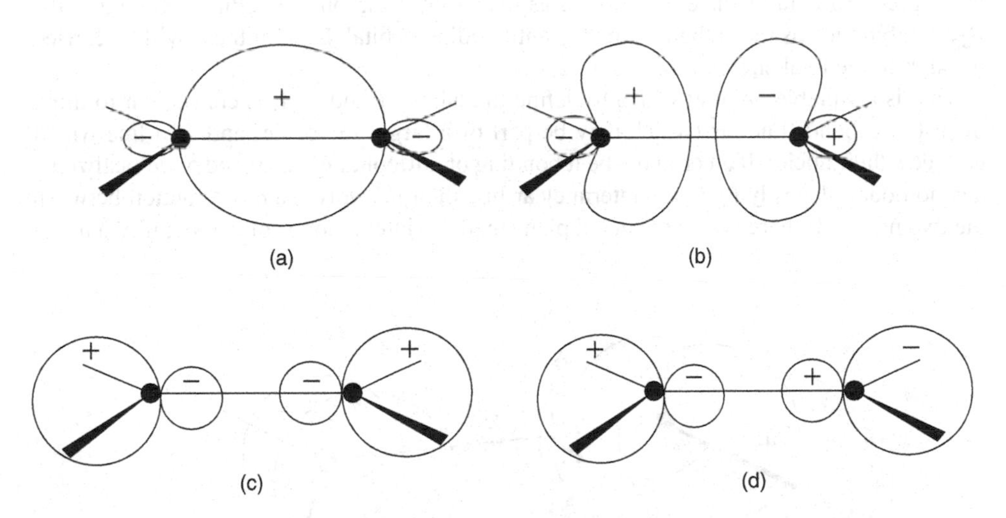

Figure 5.13 (a) The A_g C—C σ bonding orbital in ethene (this is, essentially, Figure 5.12a + Figure 5.12c). (b) The B_{1u} C—C σ antibonding orbital in ethene (this is, essentially, Figure 5.12b + Figure 5.12d). (c) The A_g carbon-based orbital involved in C-H bonding in ethene (essentially, Figure 5.12c–Figure 5.12a). (d) The B_{1u} carbon-based orbital involved in C-H bonding in ethene (essentially, Figure 5.12d – Figure 5.12b)

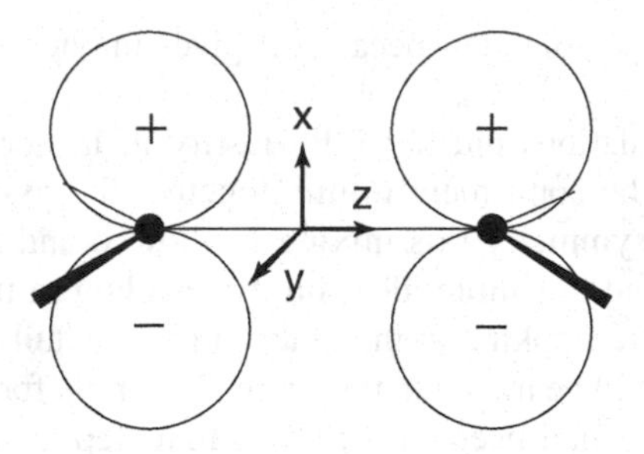

Figure 5.14 Carbon $2p_x$ orbitals in ethene. The phases of the orbitals have been chosen to be identical to those of the *x* axis

The other 2p orbitals of the carbon atoms are readily dealt with. For the pairs of $2p_x$ and $2p_y$ orbitals a similar phase ambiguity exists as for the $2p_z$ orbitals, although it is usually found to be less troublesome. In this chapter the phases shown in Figures 5.14 and 5.15 have been chosen. These orbitals transform as follows:

$$2p_x : B_{2g} + B_{3u}$$
$$2p_y : B_{3g} + B_{2u}$$

Symmetry-correct linear combinations transforming as the above irreducible representations are sum and differences of the carbon $2p_x$ and $2p_y$ orbitals of Figures 5.14 and 5.15 and are shown in Figures 5.16 and 5.17. The B_{3u} combination of carbon $2p_x$ orbitals shown in Figure 5.16 is immediately identified as the carbon–carbon π bonding orbital and the B_{2g} combination as the carbon–carbon π antibonding orbital. Both of these will be carried through to the final energy level diagram.

This is a suitable point at which to define the labels σ and π. It is convenient to think of just two bonded atoms (which may be part of a larger molecule) and of a line which connects their nuclei. If an orbital – be it bonding or antibonding, localized or delocalized – has no nodal planes lying in the internuclear line then it involves a σ interaction between the two nuclei. If there is a single nodal plane then the interaction is of π type; if two nodal

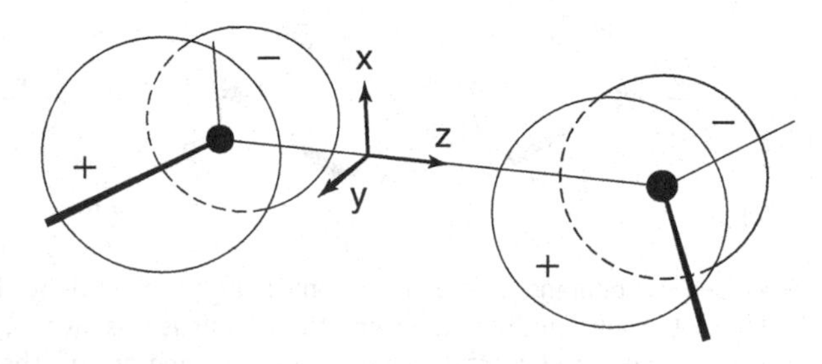

Figure 5.15 Carbon $2p_y$ orbitals in ethene. The phases of the orbitals have been chosen to be identical to those of the *y* axis

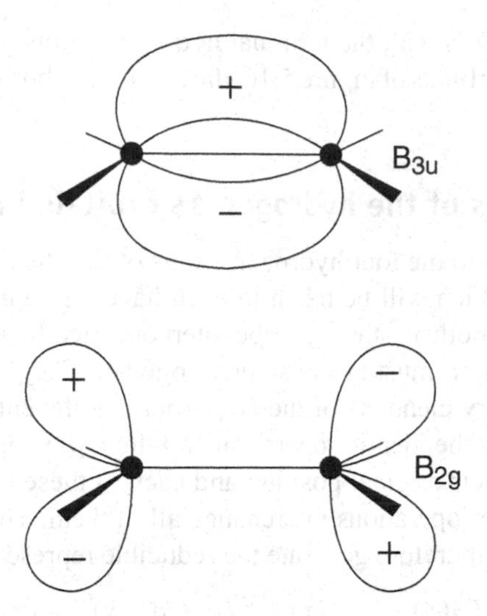

Figure 5.16 The B_{3u} (upper) and B_{2g} (lower) combinations of carbon $2p_x$ orbitals in ethene, shown in perspective

planes then it is δ. If a molecule is planar then the σ/π distinction extends over the entire molecule and one can correctly distinguish σ molecular orbitals from π molecular orbitals (the former are symmetric and the latter antisymmetric with respect to reflection in the molecular plane).

There is an element of inconsistency in Figures 5.16 and 5.17. The only difference between the 2p orbitals shown in Figures 5.14 and 5.15 is that the former are rotated through 90° relative to the latter. One would therefore expect to find that Figure 5.17 is identical to Figure 5.16 except for this same rotation. In anticipation that the primary interaction involving the

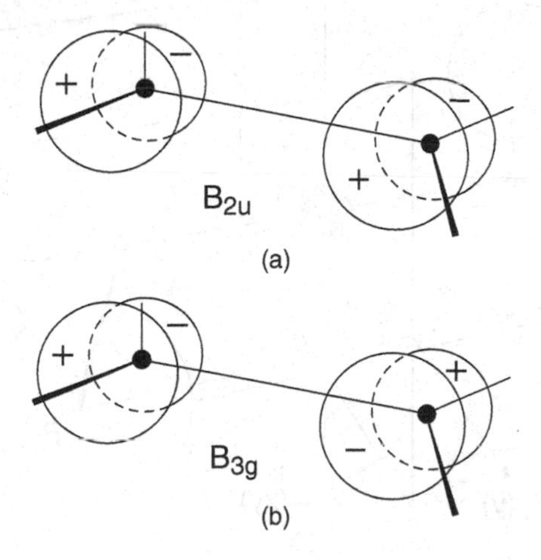

Figure 5.17 The B_{2u} (upper) and B_{3g} (lower) combinations of carbon $2p_y$ orbitals in ethene, shown in perspective

B_{2u} orbitals of Figure 5.17 is with the terminal hydrogen atoms, whereas there is no such interaction involving the orbitals of Figure 5.16, the carbon–carbon overlap has been ignored in Figure 5.17.

5.6 The symmetries of the hydrogen 1s orbitals in ethene

We now turn our attention to the four hydrogen atoms of the ethene molecule and consider the 1s orbital on each (which will be taken to each have the same phase). These orbitals are all equivalent to one another – they may be interconverted by the symmetry operations of the group – and so all four must be considered together. They are shown in Figure 5.18 together with the symmetry elements of the D_{2h} group. Of the entire set of corresponding symmetry operations only the identity operation and the $C_2(zx)$ operation leave any of the hydrogen 1s orbitals in their original position and each of these operations leaves all four orbitals unmoved; all other operations interchange all of them. The transformations of the four hydrogen 1s orbitals therefore generate the reducible representation:

E	$C_2(z)$	$C_2(y)$	$C_2(x)$	i	$\sigma(xy)$	$\sigma(zx)$	$\sigma(yz)$
4	0	0	0	0	0	0	4

This representation provides a useful illustration of the use of the method described in Section 3.3 – reducing it by trial and error could be a bit tedious. First select an irreducible representation of the D_{2h} group and multiply each character of the above reducible representation by the corresponding character of our selected irreducible representation. Add these products together and then divide by the order of the group (8 in the present

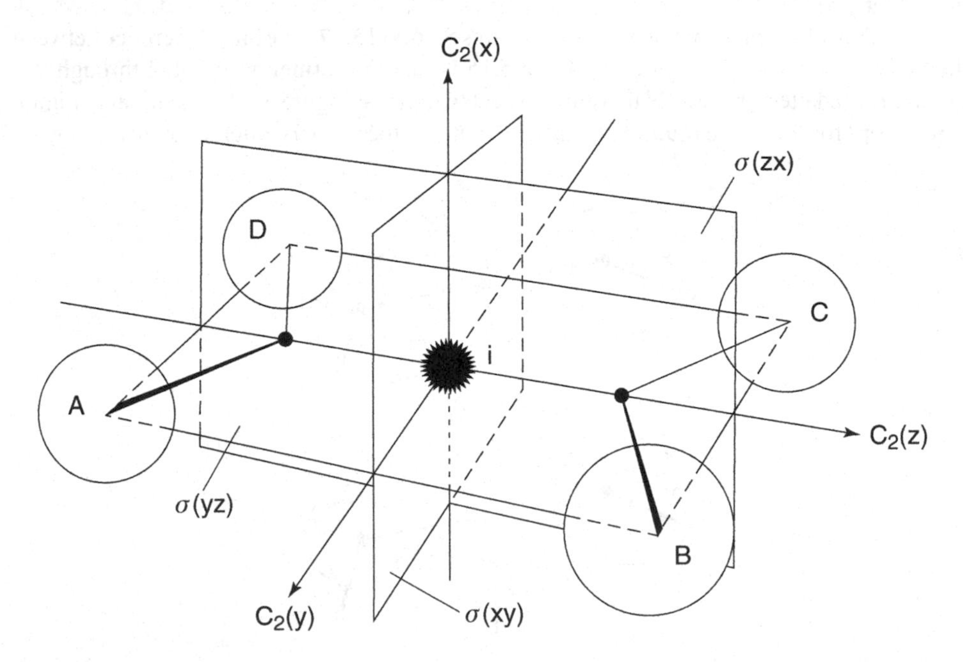

Figure 5.18 The four hydrogen 1s orbitals in ethene together with the symmetry elements of the D_{2h} group (excluding the identity)

case). The integer which results[7] is the number of times the selected irreducible representation appears in the reducible representation. Thus, if the A_g irreducible representation is selected the calculation proceeds as follows:

	E	$C_2(z)$	$C_2(y)$	$C_2(x)$	i	$\sigma(xy)$	$\sigma(zx)$	$\sigma(yz)$
	4	0	0	0	0	0	0	4
A_g	1	1	1	1	1	1	1	1
Products	4	0	0	0	0	0	0	4

so that the sum of products is 8. Division by the order of the group yields the result that the A_g irreducible representation appears once. Proceeding in this way with each irreducible representation selected in turn from the character table, it is concluded that the reducible representation has components

$$A_g + B_{3g} + B_{1u} + B_{2u}$$

That is, it is possible to form a linear combination of the hydrogen 1s orbitals which has A_g symmetry, another which has B_{3g} and so on. What are these combinations? This is the next problem which we have to solve.

Problem 5.7 Show that the above reducible representation contains $B_{3g} + B_{1u} + B_{2u}$ components in addition to the A_g deduced in the text.

5.7 The projection operator method

Although not essential to a qualitative discussion of the bonding in the ethene molecule, it is very useful at this point to seek combinations of the four hydrogen 1s orbitals, which, separately, transform as A_g, B_{3g}, B_{1u} and B_{2u} (combinations such as these are often referred to as 'symmetry-adapted functions'). This will provide a relatively simple introduction to the important *projection operator* method. As a bonus, some idea of the form of the C–H bonding molecular orbitals will be obtained. It would perhaps help to make the task simpler if we knew the answers in advance. The nodal patterns of Figure 5.9 provide a simple solution. Because they lie in a plane which contains the centre of symmetry, the four hydrogen 1s orbitals, effectively, lie in a plane half-way between the two shown in all of the diagrams in Figure 5.9. It follows that only those diagrams with matching phases 'top' and 'bottom' are acceptable. The hydrogen 1s orbitals must have the same phase in each. There are only four diagrams which satisfy this requirement, the A_g, B_{3g}, B_{1u} and B_{2u} – and these, of course, are the same irreducible representations spanned by the four hydrogen 1s orbitals. So, the linear combinations of hydrogen 1s orbitals which we will obtain by the projection operator method must have the A_g, B_{3g}, B_{1u} and B_{2u} phase patterns of Figure 5.9. This method is a bit more powerful than it seems. Suppose we had not generated the reducible representation and decomposed it into its irreducible components; we did not know that we were looking for A_g, B_{3g}, B_{1u} and B_{2u} patterns. The method still works! One simply asks the question 'which

[7] If a nonsense answer is obtained (for example, a fraction) then either an arithmetical mistake has been made or the reducible representation has been wrongly generated – this is one way in which such mistakes are commonly discovered.

Table 5.5

	E	$C_2(z)$	$C_2(y)$	$C_2(x)$	i	$\sigma(xy)$	$\sigma(zx)$	$\sigma(yz)$
Under the operation A becomes	A	D	B	C	C	B	D	A

projection patterns are compatible with the hydrogen 1s orbital set?'.[8] By 'compatible' one means that the pattern does not demand non-existent nodal characteristics inherent in the basis set. So, as we have seen, the hydrogen 1s orbitals cannot be antisymmetric with respect to reflection in the mirror plane containing the molecule. In the present case, just A_g, B_{3g}, B_{1u} and B_{2u} satisfy this criterion. We have obtained both the symmetries and the form of the symmetry-adapted functions in one simple step! It is probably using a method such as this, perhaps subconsciously, that the expert can impress the beginner by their ability to write down symmetry-adapted functions without effort. But the beginner would probably be well advised to check their own answers with the projection operator method; even experts can make mistakes. And when a symmetry becomes complicated, that of a cube for instance, suitable nodal patterns may not be readily available and the projection operator method becomes that of choice. So, now, the projection operator solution to the problem!

We first consider the transformations of the individual hydrogen 1s orbitals in much greater detail. Previously, we have only been concerned with whether or not a hydrogen 1s orbital was turned into itself under a particular symmetry operation. If it did not do this the destiny of the hydrogen atom did not concern us. This is no longer the case. We shall now look in detail at one of the four hydrogen 1s orbitals and determine the precise effect of each symmetry operation on this chosen orbital. Label the hydrogen 1s orbitals as shown in Figure 5.18 and consider the transformation of the orbital which is labelled A. The following discussion will be made easier if an eye is kept on Figure 5.18 and another(!) on Table 5.1 and the individual characters that it contains. Under the identity operation, A remains itself; under the $C_2(z)$ rotation it becomes the orbital labelled D; under the $C_2(y)$ rotation it becomes B and so on. A complete list of its transformations is given in Table 5.5; it is important that the reader checks that this table is correct.

Problem 5.8 Use Figure 5.18 to check that Table 5.5 is correct.

We are now in a position to generate symmetry-correct linear combinations of the hydrogen orbitals. We know that the set A, B, C and D gives rise to a B_{1u} combination and we shall now generate this combination (we select it because its generation highlights all the important points). Consider orbital A and the effect of the $C_2(y)$ operation. Table 5.1 shows that under this operation a function transforming as B_{1u} changes sign. It follows, therefore, that orbitals A and B must appear in the B_{1u} linear combination in the form (A–B) since this expression changes sign under the $C_2(y)$ operation. Now consider the $C_2(x)$ and $C_2(z)$

[8] Here it is assumed that all of the orbitals in the set under consideration (which might be bigger than the four hydrogen 1s orbitals in the text) are symmetry-related. If not all of them are, then they must be broken up into sets which are. The method then has to be applied to each of these in turn.

operations – under which A interchanges with C and D respectively. Because a B_{1u} function changes sign under $C_2(x)$ but retains its sign under $C_2(z)$ it is evident that C and D must appear as $-C$ and $+D$. It follows that the B_{1u} combination is of the (normalized) form:

$$\tfrac{1}{2}(\mathrm{A} - \mathrm{B} - \mathrm{C} + \mathrm{D})$$

It is a simple matter to check that this combination does indeed transform correctly as B_{1u} under all of the operations of the group. The important thing to recognize is the way that the sign with which an individual orbital appears in the result is determined by the appropriate character of the irreducible representation (which, of course, is made explicit in Figure 5.9).

The general method is at once evident. In order to generate a required linear combination we simply take the entries in Table 5.5 and multiply each entry by the corresponding character. The sum of the answer so obtained is the desired linear combination (although it will not be normalized). As an illustration of this method let us generate the B_{3g} linear combination of hydrogen 1s orbitals by this, the *projection operator*, method:

	E	$C_2(z)$	$C_2(y)$	$C_2(x)$	i	$\sigma(xy)$	$\sigma(zx)$	$\sigma(yz)$
Under the operation								
A becomes	A	D	B	C	C	B	D	A
B_{3g}	1	−1	−1	1	1	−1	−1	1
Product	A	−D	−B	C	C	−B	−D	A
	Sum: 2A − 2B + 2C − 2D							

The linear combination generated by this procedure is 2A − 2B + 2C − 2D. This function is not normalized since the sum of squares of coefficients appearing is 16, not 1; to normalize we have to divide by $\sqrt{16} = 4$ and so obtain the normalized B_{3g} combination.

$$\tfrac{1}{2}(\mathrm{A} - \mathrm{B} + \mathrm{C} - \mathrm{D})$$

The A_g and B_{2u} combinations are obtained in a precisely similar way. All four linear combinations are given in Table 5.6, and shown in Figure 5.19. Such combinations are also often referred to as 'symmetry-adapted combinations'.

Problem 5.9 Use the projection operator method to obtain the (normalized) A_g and B_{2u} combinations of hydrogen 1s orbitals.

Table 5.6

Symmetry species	Linear combination of 1s orbitals of hydrogen atoms in ethene
A_g	½(A + B + C + D)
B_{3g}	½(A − B + C − D)
B_{1u}	½(A − B − C + D)
B_{2u}	½(A + B − C − D)

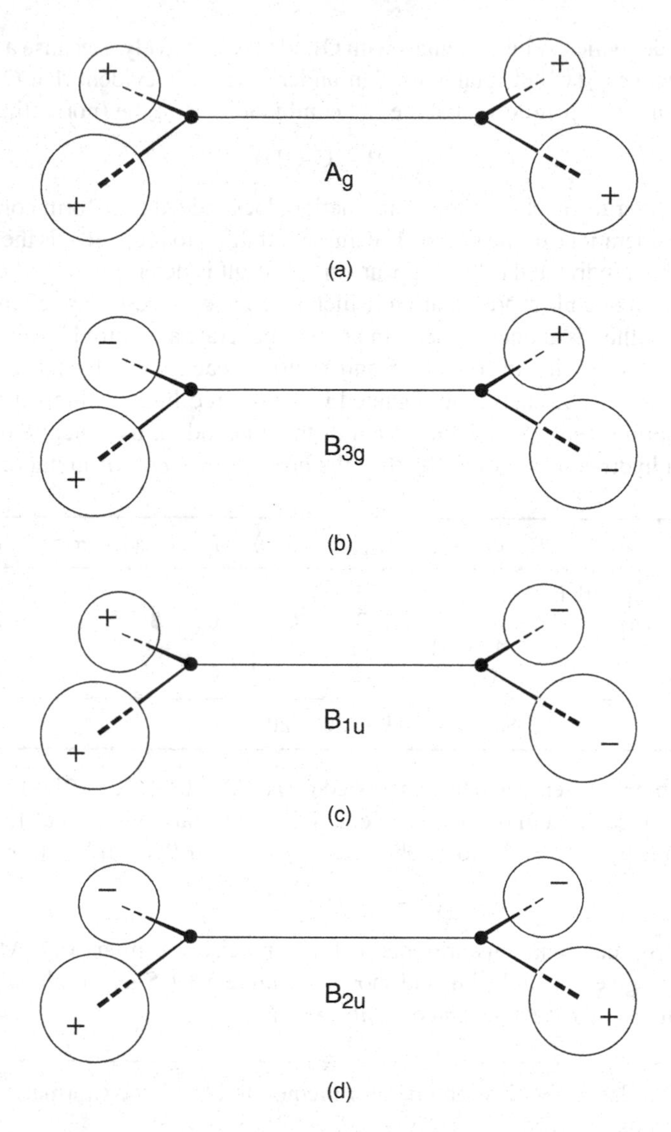

Figure 5.19 The symmetry-adapted combinations of hydrogen 1s orbitals in ethene. Note that the relative phases chosen for the individual hydrogen 1s orbitals are evident in the A_g combination. Here, all were chosen with identical phases but if one had been chosen with a phase opposite to all of the others then this would appear as a − phase in the A_g combination above. The sequence in which these combinations are shown is that in which they are listed in the text

It is to be emphasized that each of the four diagrams in Figure 5.20 pictures one orbital and *not* four. An instructive exercise at this point is to attempt to generate from the data in Table 5.5 a combination transforming as an irreducible representation which is absent (and so is not listed in either Table 5.6 or Figure 5.20) – for example B_{1g}. It will be found that the projection operator method has the great advantage of being self-correcting! The reason is

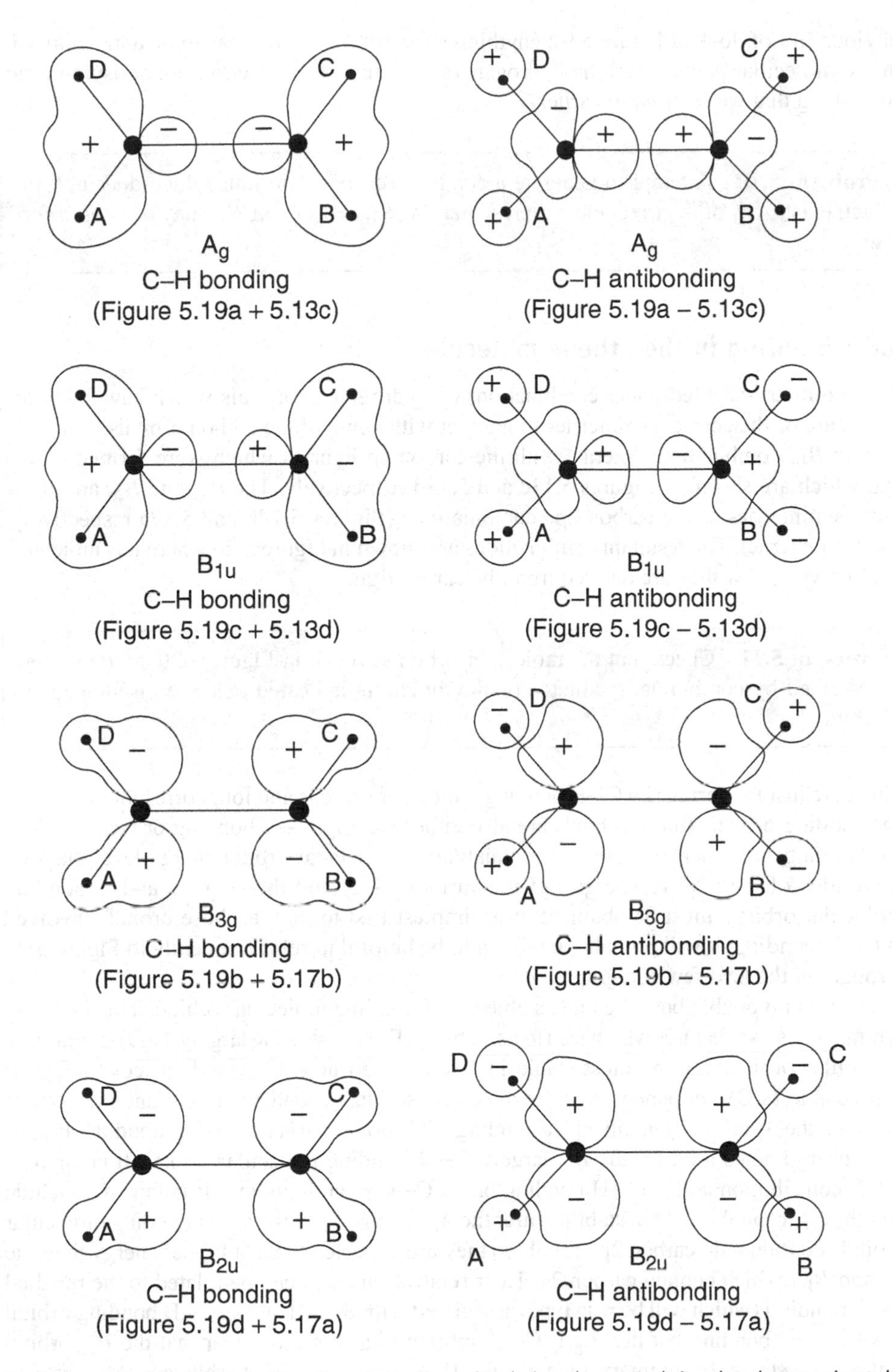

Figure 5.20 Bonding and antibonding molecular orbitals in ethene. Each is related approximately to those in earlier figures from which it is derived

obvious (if not, look at Figure 5.9); anything transforming as B_{1g} has to be antisymmetric in the molecular plane – and the hydrogen 1s orbitals are symmetric, not antisymmetric, something that we referred to earlier.

Problem 5.10 Attempt to generate a combination of 1s orbitals which does not, in fact, exist. Any of the irreducible representations B_{1g}, B_{2g}, A_u or B_{3u} may be chosen for this.

5.8 Bonding in the ethene molecule

The symmetry-adapted linear combinations of hydrogen 1s orbitals which have been obtained are of the correct symmetries to interact with some of the carbon orbitals. Thus, the A_g and B_{1u} combinations interact with the carbon sp hybrids which were formed earlier and which are shown in Figures 5.13c and 5.13d respectively. The B_{3g} and B_{2u} are of the same symmetries as the carbon $2p_y$ combinations (Figures 5.17b and 5.17a respectively) and also interact. The resultant combinations are shown in Figure 5.20 where it is indicated, qualitatively, how they are derived from the earlier figures.

Problem 5.11 Check that the molecular orbitals shown in Figure 5.20 are correctly described by combining, qualitatively, the diagrams indicated below each molecular orbital.

There are just four primarily C—H bonding molecular orbitals and four corresponding C—H antibonding orbitals (these orbitals are also either weakly C—C bonding or weakly C—C antibonding). In order to obtain even a qualitative molecular orbital energy level diagram some idea of the relative energies of the various C—H and the C—C σ and π bonding molecular orbitals must be obtained. It is simplest first to look at those orbitals involved in C—H bonding; it will probably be found to be helpful to refer frequently to Figure 5.20 throughout the next few paragraphs.

There is no doubt about the most stable C—H bonding molecular orbital. This is the A_g orbital. It has two features which lead to its stability. First, just as the largely 2s(O)-containing molecular orbital was the most stable in H_2O, so too here is the orbital containing an appreciable 2s(C) component expected to be very stable. Second, the important interactions in which the A_g orbital is involved are bonding – it is both C—H and C—C σ bonding. Rather similar arguments hold for the B_{1u} largely C—H bonding molecular orbital. It contains a 2s(C) contribution and is C—H bonding but is C—C σ antibonding. It is fair to conclude that the B_{1u} orbital is next in stability after the A_g. The B_{2u} and B_{3g} C—H bonding molecular orbitals contain only carbon 2p orbitals so they are expected to be at higher energy than the A_g and B_{1u}, which contain carbon 2s. Their relative energies can be related to the residual C—C bonding (which will be π in type) associated with each. The B_{2u} C—H bonding orbital is also C—C bonding but the B_{3g} is C—C antibonding. It seems clear that the B_{2u} orbital is the more stable. In summary, then, the C—H bonding molecular orbitals are expected to decrease in stability in the order:

$$A_g > B_{1u} > B_{2u} > B_{3g}$$

We now turn to the orbitals which are largely responsible for the carbon–carbon bonding. They are shown in Figures 5.13a and 5.16a. There is no doubt that the A_g largely carbon–carbon σ bonding molecular orbital will be more stable than the B_{3u} carbon–carbon π bonding molecular orbital because one contains 2s(C) whereas the other contains $2p_z$(C). However, it is not easy to unambiguously relate their energies to those of the C—H bonding molecular orbitals. The following argument is indicative. The bond energy of a single C—C σ bond is ca. 360 kJ mol^{-1}, although it is to be noted that this value is appropriate to a bond length slightly longer than that found in ethene. In contrast, the energy of an average C—H bond is ca. 420 kJ mol^{-1}. It seems reasonable, then, to anticipate that the stabilization resulting from the C—H bonding interactions should be somewhat greater than that of the A_g C—C σ interaction. This means that it would be reasonable to expect the C—H bonding molecular orbitals which have a carbon 2s component (those of A_g and B_{1u} symmetries) to be lower in energy than the carbon–carbon bonding orbital with a 2s component (that of A_g symmetry). We have, then, the stability order:

$$A_g \text{ (C—H bonding)} > B_{1u} \text{ (C—H bonding)} > A_g \text{ (C–C bonding)}$$

The next lowest C—H bonding orbital is B_{2u} and the question is whether its stability is sufficient to make it lower in energy than the A_g (C—C bonding). If we interpret the bond energy data given above as 'the centre of gravity of the energies of the four C—H bonding interactions should be below the energy of the single C—C bonding interaction' then the order

$$B_{2u} \text{ (C—H bonding)} > A_g \text{ (C—C bonding)}$$

seems probable, although not certain. All that can be said is that it seems likely that the two will be of similar energies with perhaps the B_{2u} the more stable. In fact, this is the pattern experimentally observed.

The carbon–carbon π bonding molecular orbital, of B_{3u} symmetry, is also best placed by appeal to experiment. A great deal of spectroscopic and other information on carbon–carbon π-bonded systems can be rationalized on the assumption that it is a carbon–carbon π orbital which is the highest occupied orbital. So, the B_{3u} (C—C bonding) is placed above the B_{3g} (C—H bonding). Together with the other arguments above, this leads to the molecular orbital energy level pattern shown in Figure 5.21. There are four valence electrons from each carbon and one from each hydrogen to be placed in these orbitals, a total of twelve. They occupy the six lowest orbitals in Figure 5.21; in this figure only one antibonding orbital, the lowest, C—C antibonding orbital of B_{2g} symmetry, is included.

Figure 5.21 can be checked in two ways. First, appeal can be made to detailed accurate calculations on this molecule. Second, the results of photoelectron spectroscopic measurements can be used. This theoretical and experimental work agrees on the energy level

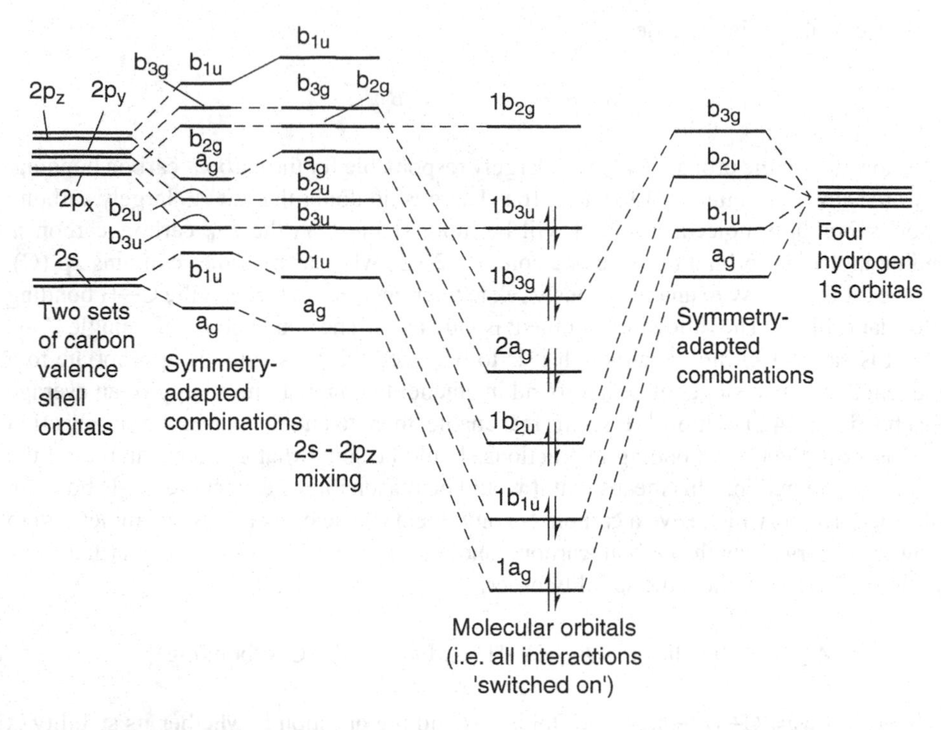

Figure 5.21 Schematic molecular orbital energy level diagram for ethene showing all relevant occupied orbitals and the lowest unoccupied orbital

sequence of ethene. The results are given below, the calculated values[9] being given in parentheses.

$1b_{3u}$	(C—C bonding)	10.51	(10.44) eV
$1b_{3g}$	(C—H bonding)	12.85	(13.04) eV
$2a_g$	(C—C bonding)	14.66	(14.70) eV
$1b_{2u}$	(C—H bonding)	15.87	(16.07) eV
$1b_{1u}$	(C—H bonding)	19.1	(19.44) eV
$1a_g$	(C—H bonding)	23.5	(26) eV

The agreement with the qualitative picture developed above is excellent, giving some confidence in the arguments that have been used. In particular, the hope that increased molecular symmetry would offset the greater molecular complexity compared with the water molecule seems to have been justified.

5.9 Bonding in the diborane molecule

Our discussion of the ethene molecule can be extended to another molecule, diborane. Diborane, B_2H_6, is of interest because it is the simplest of the boron hydrides (boranes).

[9] W. Von Niessen, G.H.F. Diercksen, L.S. Cederbaum and W. Domcke, *Chem. Phys.* **18** (1976) 469.

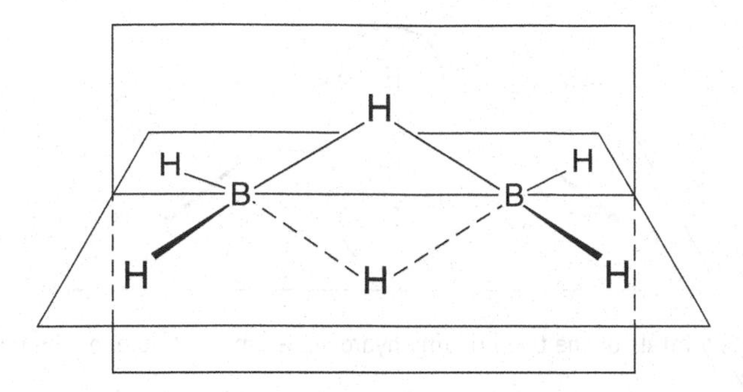

Figure 5.22 The structure of diborane, B_2H_6, shown in perspective

These, as a class, are often called 'electron deficient' because, whereas at least $(n - 1)$ electron pairs are regarded as necessary to bond n atoms, the boron hydrides all have fewer than $2(n - 1)$ electrons. Thus, there are only twelve valence shell electrons available in diborane to bond eight atoms. However, as will be seen, the term 'electron deficient' is a misnomer because the molecular structure is such that all bonding molecular orbitals are filled with electrons. Whereas diborane posed such a problem for simple bonding models that it appeared necessary to give it a separate classification, a symmetry-based discussion shows that there is no need to invoke new concepts.

The structure of diborane is shown in Figure 5.22, from which it can be seen that it has four terminal hydrogen atoms and two borons which together have the same symmetry, D_{2h}, as ethene (although the bond lengths and angles are different, of course). In addition, diborane has two hydrogen atoms out of, what is for ethene, the molecular plane. These two hydrogens are usually called the 'bridge' hydrogen atoms. It is the presence of these bridge hydrogen atoms in place of the C—C π bond of ethene that plays a major part in leading diborane to have a rather different chemistry to ethene.

Figure 5.22 does not show all of the symmetry elements of diborane. Comparison with Figure 5.1 shows that the bridge hydrogens, located on the $C_2(x)$ axis of Figure 5.1, in no way reduce the D_{2h} symmetry of the ethene-like B_2H_4 unit. Diborane, like ethene, has D_{2h} symmetry. It follows that apart from that involving the bridge hydrogen atoms, the bonding in the diborane molecule must, qualitatively, be similar to that given in the previous section for ethene since boron, like carbon, has 2s and $2p_x$, $2p_y$ and $2p_z$ valence orbitals. It therefore seems reasonable to expect the retention of the same energy level sequence:

$$A_g\,(\text{B–H}_t\text{ bonding}) < B_{1u}\,(\text{B–H}_t\text{ bonding}) < B_{2u}\,(\text{B–H}_t\text{ bonding}) < B_{3g}\,(\text{B–H}_t\text{ bonding})$$

where the suffix t has been added to distinguish terminally bonded hydrogens from the bridge hydrogens. There is little doubt that there is a substantial difference between the carbon–carbon bonding in ethene and the boron–boron bonding in diborane. This is shown by even a cursory study of the experimental data – the carbon–carbon bond length in ethene is 1.34 Å whilst the boron–boron bond length in diborane is 1.77 Å. The details of the B–B bonding will also be different from the C—C bonding in ethene because only the former has bridge hydrogens. Clearly, our discussion of the B–B bonding must start with these bridge hydrogens.

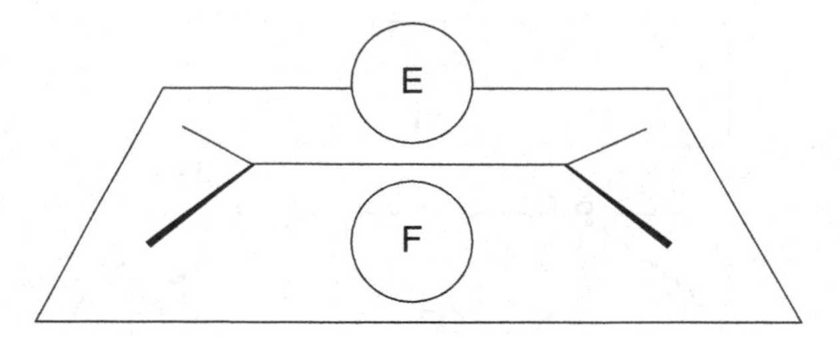

Figure 5.23 The 1s orbitals of the two bridging hydrogen atoms of diborane. The labels E and F will be used in the text

The transformations of the 1s orbitals of the two bridge hydrogen atoms in diborane (Figure 5.23) generate the following reducible representation:

E	$C_2(z)$	$C_2(y)$	$C_2(x)$	i	$\sigma(xy)$	$\sigma(zx)$	$\sigma(yz)$
2	0	0	2	0	2	2	0

a representation which has $A_g + B_{3u}$ components. As usual, the functions transforming as these irreducible representations are simply the sum and difference of the two 1s orbitals (which are labelled E and F, as shown in Figure 5.23, and taken to have the same phase). That is, they are:

Symmetry species	Linear combination of bridge hydrogen orbitals
A_g	$\frac{1}{\sqrt{2}}(E + F)$
B_{3u}	$\frac{1}{\sqrt{2}}(E - F)$

Problem 5.12 Check that the transformations of the two bridge hydrogen atoms in diborane are as given above. It is of particular importance to show that the two linear combinations of these orbitals transform as indicated.

The only orbitals shown in Figure 5.13 with which it is reasonable to expect any important interaction involving these bridge hydrogen orbitals are the boron–boron σ bonding orbital of A_g symmetry (which will be similar to that shown in Figure 5.13a but with boron atoms in place of carbon) and the boron–boron π bonding orbital of B_{3u} symmetry (which will resemble that shown in Figure 5.16a). The interactions between the bridge hydrogen orbitals and these boron–boron orbitals are shown qualitatively in Figures 5.24 and 5.25. Which of the bonding interactions shown in Figures 5.24 and 5.25 is the more important? For the B_{3u} (boron–boron π bonding) orbital the B_2H_4 plane is a nodal plane; its maximum amplitude must be out of this plane. In contrast, the maximum amplitude of the A_g (boron–boron σ bonding) orbital is in the B_2H_4 plane. Because the bridge hydrogens are above and below this plane it seems probable that the interaction will be greater with the B_{3u} boron combination than with the A_g. Whether this difference will lead to the orbital of B_{3u} symmetry being beneath that of A_g (in Figure 5.21 the B_{3u} is above the A_g) cannot

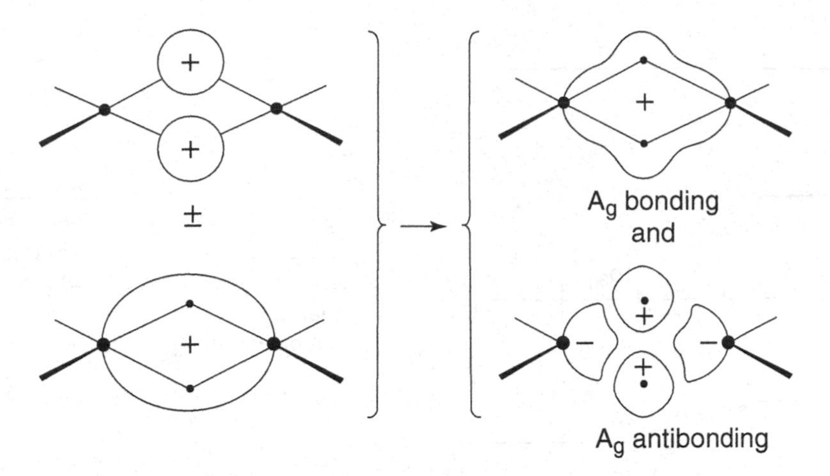

Figure 5.24 Schematic representation of the interactions of A_g symmetry involving the bridge hydrogens of diborane

be unambiguously predicted – in fact, it does. An additional reason for this pattern is the greater B–B bond length in diborane compared to the C—C bond in ethene. Because of this difference, it is likely that the A_g B–B σ bonding interaction is less than the C—C σ interaction in ethene.

In Figure 5.26 a schematic molecular orbital energy level diagram for the diborane molecule is given in which all of the above arguments are brought together. The left-hand side of this diagram shows schematically the ethene molecular orbital energy level pattern (Figure 5.21) which is then modified to take account of the bridge hydrogens. Qualitatively, the problem of the relative order of the B_{2u} (B–H$_t$ bonding) and A_g (B–H$_b$

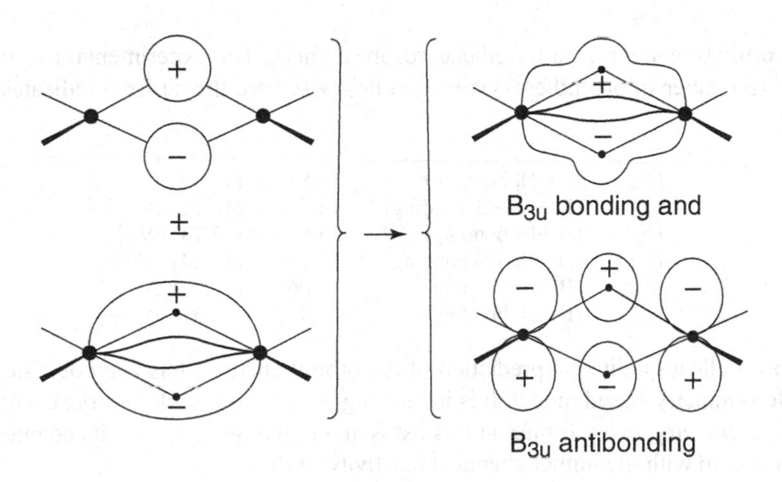

Figure 5.25 Schematic representation of the interactions of B_{3u} symmetry involving the bridge hydrogens of diborane

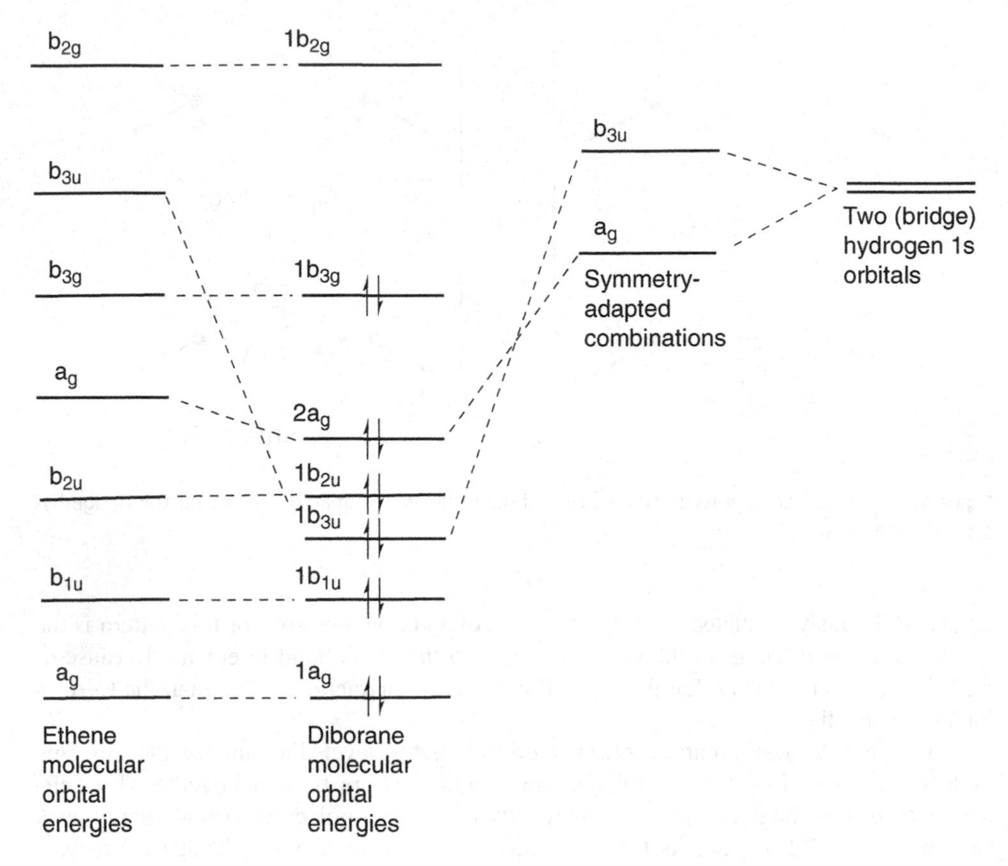

Figure 5.26 A qualitative molecular orbital energy level diagram for B_2H_6 and its relationship to that for C_2H_6

bonding) orbitals, encountered for ethene, reappears here. The experimental and theoretical data[10] (the latter in parentheses) are given below (where the suffix b indicates bridge hydrogens):

$1b_{3g}$	(B—H$_t$ bonding)	11.81	(11.95) eV
$2a_g$	(B—H$_b$—B bonding)	13.3	(13.12) eV
$1b_{2u}$	(B—H$_t$ bonding)	13.9	(13.73) eV
$1b_{3u}$	(B—H$_b$—B bonding)	14.7	(14.04) eV
$1b_{1u}$	(B—H$_t$ bonding)	16.06	(16.34) eV
$1a_g$	(B—H$_t$ bonding)	21.4	(22.57) eV

Again, an excellent qualitative prediction of the orbital energies has been obtained using our simple symmetry-based model. It is interesting to note that, with the sole exception of that of B_{3u} symmetry, every orbital in this list is at a higher energy than its counterpart in ethene, in accord with the higher chemical reactivity of diborane.

[10] D.R. Lloyd, N. Lynaugh, P.J. Roberts and M.F. Guest, *J. Chem. Soc. Faraday Trans. 2* **71** (1975) 1382.

Problem 5.13 The molecule N_2H_4, unlike B_2H_6 and C_2H_4, does not have D_{2h} symmetry (it has a low-symmetry structure which may be regarded as similar to ethane with one hydrogen removed from each nitrogen atom). Use Figure 5.21 to explain why a D_{2h} structure is not stable for N_2H_4. The discussion in the text associated with Figure 5.21 hints at the answer to this problem.

5.10 Comparison with other models

Most discussions of the electronic structures of the ethene and diborane molecules concern themselves almost exclusively with the carbon–carbon double bond and the bridge bonding respectively. Some of these descriptions appear rather different to those which have been given in the present chapter and it is the purpose of this section to discuss the relationship between the various models.

Consider ethene. Two models are commonly presented for this molecule. In the first, each carbon atom is sp^2 hybridized, two of three sp^2 hybrids being involved in bonding with the terminal hydrogen atoms whilst the third is responsible for the carbon–carbon σ bonding. A π bond is formed as a result of overlap between the 2p orbitals which were not hybridized. This model is pictured in Figure 5.27. The sp^2 hybrid orbitals on one carbon atom have been labelled a, d and e and those on the second carbon atom, b, c and f. The hybrids which are involved in carbon–hydrogen bonding are a, b, c and d. It is easy to show that the transformations of these orbitals under the operations of the D_{2h} point group follow (or, more precisely, are isomorphous to) those of the hydrogen 1s orbitals A, B, C and D which were considered earlier in this chapter (Section 5.5). It follows that this hybrid orbital model identifies the C—H bonding molecular orbitals as being of A_g, B_{3g}, B_{1u} and B_{2u} symmetries, a conclusion identical to that reached above. It is also straightforward to show that the hybrid orbitals e and f formthe basis for a reducible representation with A_g and

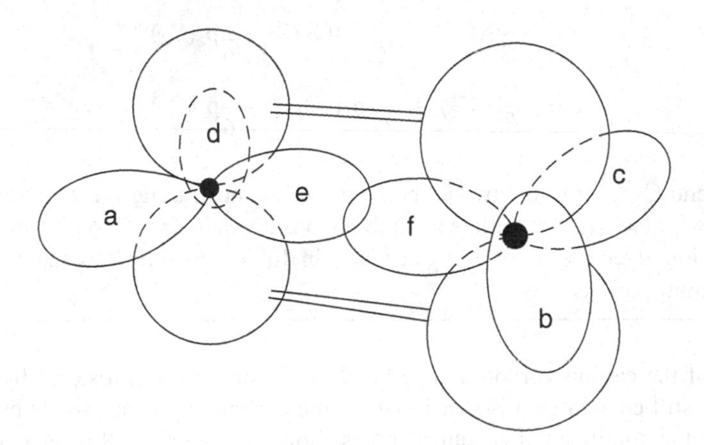

Figure 5.27 The '$sp^2 + p_\pi$' carbon atom model for the bonding in ethene. For simplicity the hydrogen atoms are omitted. The sp^2 hybrids carry the labels a–f

B_{2u} components, which must correspond to the C—C σ bonding and antibonding orbitals. Again, the qualitative description of the C—C σ bonding is identical to that of the symmetry-based model. The main differences can be seen when the two orbitals e and f in Figure 5.27 are compared with their counterparts in the model used above (Figure 5.13). These orbitals were regarded as equal mixtures of the carbon 2s and $2p_z$ orbitals whereas in the hybrid orbital model the s-orbital contribution is only one-third. However, it will be recalled that when the 2s and $2p_z$ orbitals were mixed in equal amounts it was mentioned that this was an arbitrary mixing, made on grounds of simplicity. The 1:1 ratio could, accidentally, have been correct. Equally, the hybrid orbital model could be right in its ratio of 1:2. Detailed calculations show that both are wrong – there are two A_g orbitals contributing to C—C bonding, one largely involving 2s(C) and the other $2p_z$(C). The aggregate 2s:$2p_z$ ratio is 1:1.3 so, from this viewpoint, the sp model adopted in the text is not too bad.

A point of apparent divergence between the two approaches is to be found in the carbon–hydrogen bonding orbitals of B_{3g} and B_{2u} symmetries. In the symmetry-based description these orbitals contain no contribution from the carbon 2s orbitals. In contrast, one might expect there to be such a contribution in the hybrid orbital description since each hybrid contains a 2s component. This is not the case. If the form of the hybrid orbitals is written out explicitly and the appropriate linear combinations of them are obtained using the projection operator method (these combinations are those given in Table 5.6 but with capital letters replaced by lower case letters) then it will be found that the carbon 2s orbital contributions also vanish in the hybrid orbital description.

Problem 5.14 The explicit forms of the relevant carbon sp^2 hybrid orbitals are:

$$a = \frac{1}{\sqrt{3}}s(C_1) + \frac{1}{\sqrt{2}}p_y(C_1) - \frac{1}{\sqrt{6}}p_z(C_1)$$

$$d = \frac{1}{\sqrt{3}}s(C_1) - \frac{1}{\sqrt{2}}p_y(C_1) - \frac{1}{\sqrt{6}}p_z(C_1)$$

$$b = \frac{1}{\sqrt{3}}s(C_2) + \frac{1}{\sqrt{2}}p_y(C_2) + \frac{1}{\sqrt{6}}p_z(C_2)$$

$$c = \frac{1}{\sqrt{3}}s(C_2) - \frac{1}{\sqrt{2}}p_y(C_2) + \frac{1}{\sqrt{6}}p_z(C_2)$$

where C_1 and C_2 refer to the two carbon atoms. By substituting these in the explicit expressions for the B_{3g} and B_{2u} linear combinations given in Table 5.6 (but substituting the expression given above for a in place of A in Table 5.6 etc.) show that the carbon 2s orbital contributions vanish.

A model of the carbon–carbon double bond in ethene which is historically important and which is still encountered is that in which the carbon atoms are sp^3 hybridized and each bond of the double bond is equivalent, as shown in Figure 5.28. It is a simple matter to show that the two carbon–carbon bonding orbitals labelled a and b in Figure 5.28

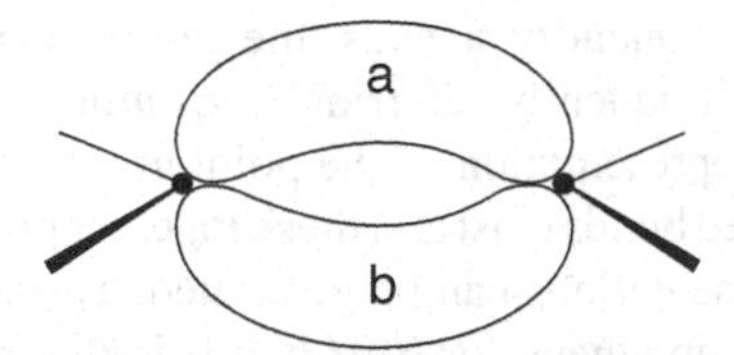

Figure 5.28 The sp^3 carbon atom model for the bonding in ethene. Hydrogens are omitted and the sp^3 hybrids to which they bond are represented by rods. Shown in the diagram are the bonding orbitals formed by the overlap of sp^3 hybrids on the carbon atoms

provide a basis for a reducible representation with A_g and B_{3u} components. These are the symmetries which have already been deduced as those of the carbon–carbon bonding orbitals. Indeed, if the projection operator method is used to obtain A_g and B_{3u} combinations of a and b (they are the sum and difference of the two) then orbitals are obtained which are, essentially, identical to the carbon–carbon bonding orbitals shown in Figure 5.13 and Figure 5.16. That is, the use of sp^3 hybrids at each carbon atom is also consistent with the model of C—C bonding derived in this chapter, although such a description pictures the orbital on each carbon atom which is involved in this bonding as being one-quarter composed of the carbon 2s orbital – the third value we have met! The use of sp^3 hybrids to explain the C—H bonding is also consistent with a symmetry-based discussion. Again C—H bonding molecular orbitals of A_g, B_{1u}, B_{2u}, and B_{3g} symmetries are obtained when the four C—H bonding sp^3 hybrids are used to generate a reducible representation of the group.

Perhaps the simplest description of the bonding of the bridging hydrogen atoms in diborane is the so-called banana bond picture. These bonds are shown in Figure 5.29; the close similarity with Figure 5.28 is immediately apparent. It is not at all difficult to show that the bridge bonds a and b in Figure 5.29 form the basis for two linear combinations, one of A_g symmetry and the other of B_{3u}. These symmetries are the same as those of the orbitals shown in Figure 5.25 as responsible for the bridge bonding. The similarity between the two descriptions follows at once.

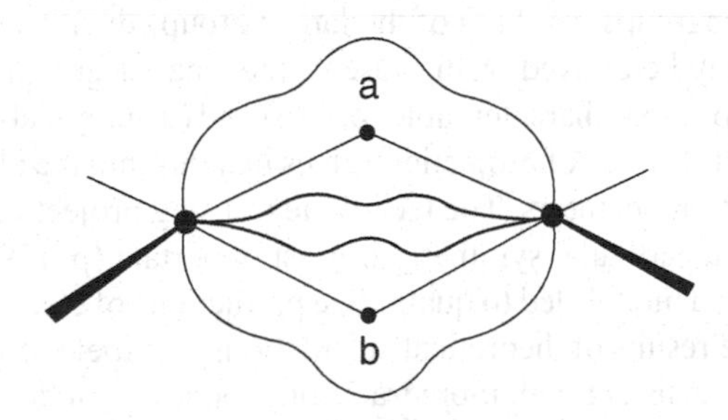

Figure 5.29 The 'banana bond' model for the bonding of the bridge hydrogens in diborane. The sp^3 hybrids on the boron atoms overlap with the 1s orbitals of the bridge hydrogens

When a chemist speaks of a quantity such as 'the carbon–carbon bond' or 'the carbon–hydrogen bond' he or she is frequently referring to quantities which do not, themselves, transform as an irreducible representation of the point group of a molecule. In such cases several other symmetry-related bonds exist and these together provide a basis set from which appropriate irreducible representations can be generated. Thus, in the present context one can say that 'the C—H bonds in ethene (or B–H bonds in diborane) can be combined into combinations which transform as irreducible representations of the D_{2h} point group'. Localized orbitals constructed so that they are equivalent to one another in this way and which can be used to derive symmetry-adapted combinations are often referred to as *equivalent orbitals*. The C—H bonding molecular orbitals shown in Figure 5.21 are all different; in contrast most chemists prefer to think of equivalent orbitals, although he or she would probably prefer to call them localized orbitals. As was recognized in the case of the water molecule (Chapter 3), and again in the present chapter for ethene and diborane, these two types of pictures are usually equivalent to each other. This is true for all of the simple models which have been shown to lead to results akin to those obtained by the symmetry-based approach. The orbitals which have been transformed to obtain reducible representations are all equivalent orbitals. That is, the approach developed in this chapter to the electronic structures of ethene and diborane is, fundamentally, no different from those with which the chemist is more familiar (the same is true of the discussion at the end of Chapter 3). On the other hand, the symmetry-based approach has considerable advantages. Thus, the observation that the C—H bonds in ethene are equivalent does not imply that the removal of any one C—H bonding electron requires the same energy as the removal of any other. The fact that there are several ionization potentials – as shown by photoelectron spectroscopy – only becomes clear in a symmetry-based description of the bonding. Despite this emphasis on symmetry it must be recognized that symmetry arguments, by themselves, tell us nothing about energy levels. It is only when these arguments are elaborated by including additional concepts, such as nodality, orbital composition and relative magnitudes of interactions, that relative energies begin to emerge.

5.11 Summary

In this chapter it has been seen that point groups may be related to each other. When a point group is the direct product of two smaller groups (the jargon is to refer to such smaller groups as 'invariant subgroups' (p. 115) of the larger group) then the multiplication tables of the larger groups may be derived from those of the smaller groups (p. 115), as may its symmetry operations (p. 115), character table (p. 116) and (usually) labels for its irreducible representations (p. 117). The use of projections gives insights into the ideas involved here, as they do also for projection operators. The technique of using projection operators to obtain linear combinations of a particular symmetry is most important (p. 129). As in the previous chapter, symmetry-based models led to qualitative predictions of electronic structure which were in accord with the results of theoretical calculations, photoelectron spectroscopic data (p. 136,140) and also consistent with more traditional bonding models (p. 141).

6 The electronic structure of bromine pentafluoride, BrF_5

Although the object of this chapter is a discussion of the electronic structure of bromine pentafluoride, this topic represents only about a third of its contents. The group theoretical methods that have been developed in the previous chapters must be extended to enable a discussion of almost any molecule, irrespective of its symmetry. This generalization is the major purpose of this chapter and takes up most of it. The key point which has to be studied is that of symmetry-enforced degeneracy. This is not as bad as it sounds. In a cube the three coordinate axes have to be equivalent. If we want to make one axis unique we can only do so by distorting the molecule so that this axis becomes distinct from the other two. The cubic symmetry and the equivalence of the axes go together. This equivalence is often expressed by using the word 'degeneracy', a word which makes obvious sense when we are talking about orbital energies, as we often will be. Hence 'symmetry-enforced degeneracy'. But, for the moment, the cube is too complicated, we need to start with something simpler.

As in the previous chapters, it is simplest to work with an example in mind and bromine pentafluoride is a very convenient one. The structure of the bromine pentafluoride molecule is shown in Figure 6.1. The bromine is surrounded by four fluorines at the corners of a square and by a fifth, unique, apical, fluorine situated so that the five fluorines form a square-based pyramid around the bromine atom. Perhaps surprisingly, the bromine is slightly beneath the plane defined by the four coplanar fluorines. A valence electron count shows that there are two non-bonding electrons on the bromine atom. These are presumably in an orbital directed towards the obvious 'hole' around the bromine, which if filled would mean that the bromine is surrounded by six groups at the corners of an octahedron (octahedral molecules will be the subject of Chapter 8 – and a cube has the same symmetry). The valence shell electron repulsion (Sidgwick–Powell–Nyholm–Gillespie) model (Chapter 1) suggests that lone-pair bond-pair repulsion will have a greater effect on the four co-planar fluorine atoms than will the repulsion between these bromine–fluorine bonds and the apical one. The consequence of this inequality will be that the co-planar B–F bonds will be bent towards the apical fluorine atom, giving the observed geometry of the molecule. As a simplifying assumption, however, in this chapter it will be assumed that the central bromine is coplanar with the surrounding four fluorine atoms. As we shall see, the symmetry is C_{4v}-basically, a fourfold axis in which lie four mirror planes.

Symmetry and Structure: Readable Group Theory for Chemists Sidney F. A. Kettle

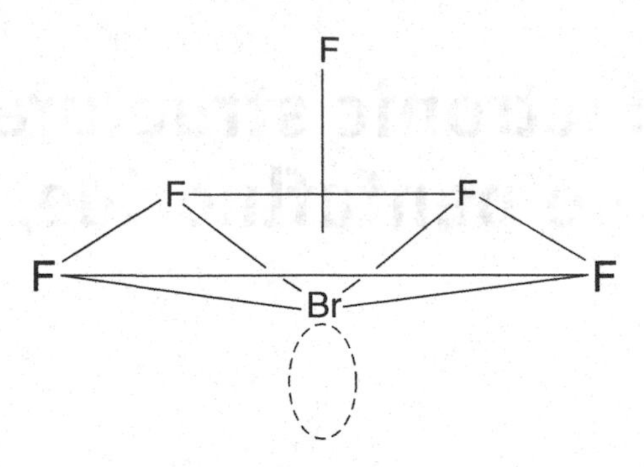

Figure 6.1 The actual structure of the BrF_5 molecule. For simplicity, in the text the bromine will be taken as coplanar with the four surrounding fluorines

The chapter will start in the way that most group theory problems start in chemistry. One looks up the relevant character table in a compilation of these (such as that given in Appendix 3). This is basically what we did for the D_{2h} character table of Chapter 5 (although we exploited its connection with the C_{2v} of earlier chapters). So, the character table for the C_{4v} group would simply be presented – the bromine pentafluoride molecule has this symmetry. Unfortunately it has no evident connection with any of the other character tables that we have met. Hopefully, this will make the reader unhappy with the idea of simply presenting the C_{4v} table. The character table will, in fact, be derived later in the chapter, using a method rather different to those used in earlier chapters for C_{2v} and D_{2h}. There are many differences between the C_{4v} character table and these two. Much of this chapter will be occupied by an exploration of these differences – this study is important because it will lead to the generalization of group theoretical concepts and techniques referred to above. In the C_{4v} group there may be more than one symmetry operation corresponding to a single symmetry element and, correspondingly, the character table contains numbers other than 1 and -1. The most important generalization will be of the orthonormality theorems. It is this generalization that will be used to generate the C_{4v} character table. This table is given in Table 6.1; it is helpful to see it at this point because the reader can then be made aware of the problems (and of their solutions!) before they are encountered.

In the C_{4v} character table, confusingly, E appears in the list of irreducible representation labels as well as in the list of operations. In this new usage it labels an irreducible representation which describes the transformation of two things simultaneously (its character under the 'leave alone' operation is 2). Irreducible representations which describe the

Table 6.1

C_{4v}	E	$2C_4$	C_2	$2\sigma_v$	$2\sigma_v'$
A_1	1	1	1	1	1
A_2	1	1	1	-1	-1
B_1	1	-1	1	1	-1
B_2	1	-1	1	-1	1
E	2	0	-2	0	0

properties of two things simultaneously are often called 'doubly degenerate' irreducible representations. The reason for this will become evident later in this chapter. We will start our discussion by looking in detail at the symmetry operations of the bromine pentafluoride molecule – of the C_{4v} group – and then return to the character table.

Problem 6.1 Both the D_{2h} and C_{4v} groups are of order eight – a total of eight operations is listed at the top of each table (compare Tables 5.1 and 6.1). However, the structures of the tables are rather different. Make a list of the qualitative differences between the two tables.

6.1 Symmetry operations of the C_{4v} group

The perspective shown in Figure 6.1 is not the best for seeing the symmetry of the BrF_5 molecule. This symmetry is most readily recognized by viewing the molecule along the bromine–axial fluorine bond as shown in Figure 6.2, from which it is clear that the four fluorine atoms lie at the corners of a square. Evidently, the bromine–axial fluorine bond coincides with a fourfold rotation axis (i.e. a C_4) of the molecule. This brings with it something new. In all of the symmetries previously considered there has always been a single symmetry operation associated with each symmetry element of a molecule. As a result, the same symbol has been used for operation and for element, leaving it to the context to make clear which was the subject of discussion. Although we shall persist with the latter convention it must now be recognized that there is not always a 1:1 correspondence between symmetry elements and symmetry operations. In the present case, although there is just one fourfold rotation axis in the BrF_5 molecule there are two corresponding symmetry operations. The molecule is turned into itself by a rotation of 90° in either a clockwise or an anticlockwise direction about the fourfold axis. These two operations have the effect of interchanging the fluorine atoms of BrF_5 in different ways and so are distinct operations. The clockwise and anticlockwise C_4 rotation operations associated with the C_4 axis are inseparable – one cannot have one without the other. Usually such pairs of operations are grouped together and written as $2C_4$, thus recognizing both their distinction and similarity. It will be seen that they are written this way in the C_{4v} character table (Table 6.1). Operations paired and written in this way are said to be 'members of the same class'. Although this is an adequate definition of 'class' for most purposes, the concept of class is an important

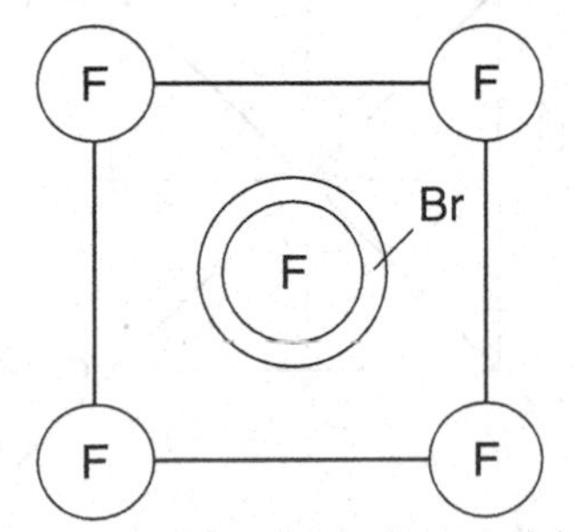

Figure 6.2 A view of the BrF_5 molecule looking down the apical (axial) F—Br bond. All bonds have been omitted but the square formed by the four equatorial fluorines is included in order to emphasize the fact that this is a view down a fourfold rotation axis

one in group theory and it is dealt with more formally and fully in Appendix 1. Because a fourfold axis exists it follows that a rotation of 180° (two steps of C_4 rotation) about this axis also turns the molecule into itself. However, this operation is *not* a C_4 rotation but a C_2, although, of course, you cannot have the former without also having the latter. Strictly, one should think of there being a C_2 axis coincident with the C_4 axis. The generality is clear – a high rotational symmetry may, automatically, imply the simultaneous existence of coincident axes of lower symmetry.

It is not trivial that the C_2 rotation operation may be regarded as a C_4 rotation operation carried out twice in succession (in the same clockwise or anticlockwise sense). Symbolically one can write

$$C_4 \times C_4 = C_4^2 \equiv C_2$$

where, following the discussion of Chapter 2, we have multiplied C_4 by C_4 to obtain C_2. In the same way it is easy to see that

$$C_4^3 \equiv C_4^{-1}$$

– carrying out three C_4 rotations in one sense, clockwise or anticlockwise, is equivalent to a single C_4 rotation in the opposite sense – and that

$$C_4^4 \equiv E$$

This is another point of difference with the groups met in previous chapters. For all of these groups it was found that any of their operations carried out twice in succession gave the identity, regenerated the original arrangement. For the C_4 operation it takes four steps in the same sense (clockwise or anticlockwise), and for a general C_n rotation it takes n.

The other symmetry elements (and associated operations) of the BrF_5 are fairly evident. In addition to the identity operation, the two C_4 rotation operations and the associated C_2 rotation operation (which, it should be noted, comprises a class of its own) there are four mirror planes which are indicated in Figure 6.3. It can be seen from this figure that these mirror planes are of two types. First, there are those which we have labelled $\sigma_v(1)$ and $\sigma_v(2)$, in each of which lie the bromine and three fluorine atoms. It is impossible to have one of

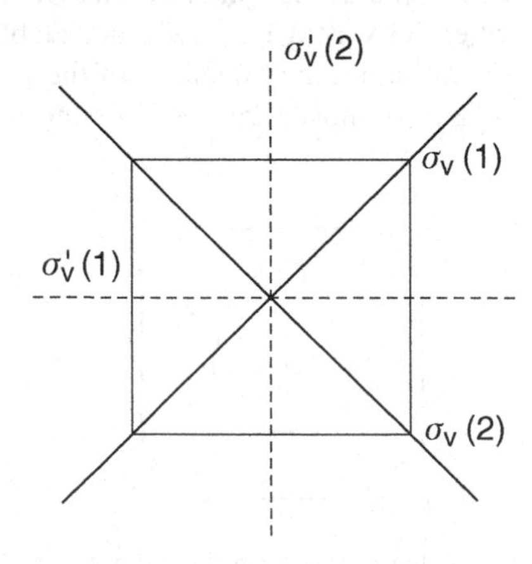

Figure 6.3 The two distinct pairs of mirror planes implicit in Figure 6.2. The square of Figure 6.2 is again shown

these operations without the other because of the C_4 axis. A C_4 operation rotates one of these σ_v mirror planes into the other. They are therefore inextricably paired together. These operations therefore comprise a class which is written as $2\sigma_v$ and it appears in this form in the character table. Second, there are those mirror planes which have been labelled $\sigma_v'(1)$ and $\sigma_v'(2)$ in Figure 6.3. Each contains the bromine and the axial fluorine atom and again are interrelated by the C_4 axis. They comprise the class $2\sigma_v'$. Several comments are relevant at this point. First, all four of the mirror planes contain the C_4 axis and so are σ_v mirror planes, as they have been labelled. Second, many authors prefer to give the mirror planes which we have called $\sigma_v'(1)$ and $\sigma_v'(2)$ the labels $\sigma_d(1)$ and $\sigma_d(2)$, or, as a class, $2\sigma_d$. This is because in a closely related group – that of the symmetry operations of a square – they carry this label. Strictly, however, the loss in symmetry in going from this group to C_{4v} forbids the use of the σ_d symbol (as will be seen in Section 8.1, this symbol has a rather precise meaning which forbids its use here). Third, a comment on the fact that $\sigma_v(1)$ and $\sigma_v(2)$ are interconverted by a C_4 rotation, as also are $\sigma_v'(1)$ and $\sigma_v'(2)$. When symmetry elements are interconverted by another operation of the group, it is a sure sign that the corresponding operations fall into the same class. Finally, it is the presence of the C_4 axis, together with the vertical mirror planes, that gives rise to the shorthand symbol for the group, C_{4v}.

Collecting together all of the symmetry operations of the C_{4v} group gives:

$$E \quad 2C_4 \quad C_2 \quad 2\sigma_v \quad 2\sigma_v'$$

and it is these operations that head the character table (Table 6.1). This is a convenient point at which to introduce the relationship between the operations of the C_{4v} group. They are given in Figure 6.4, which should be studied carefully since it will be used later to give diagrams of the irreducible representations. Any problems encountered should find a solution in the content of the previous paragraph. Note that the clockwise and anticlockwise C_4 rotations have been given a simpler notation than being denoted C_4 and C_4^3. It is helpful to use Figure 6.4 to show the class structure of the C_{4v} group. This is done in Figure 6.5, where the segments of the circle associated with members of the same class are similarly

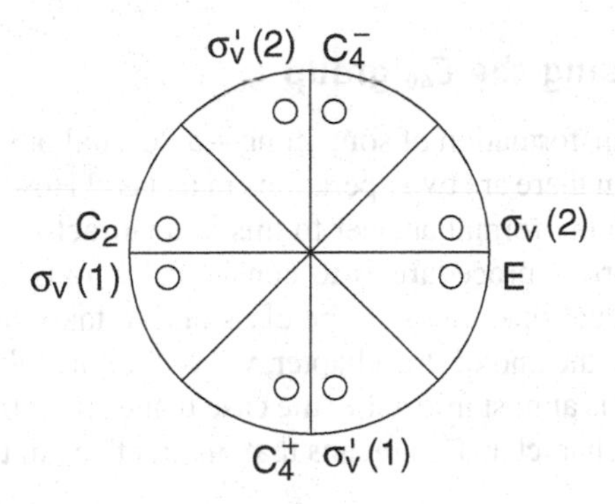

Figure 6.4 A projective view of the effects of the operations of the C_{4v} group. The two sets of mirror plane reflections are labelled σ_v and σ_v' and the members of each set distinguished as (1) and (2). Note that corresponding to the C_4 rotation axis there are two C_4 rotation operations and also a C_2 rotation operation. In this and similar diagrams the C_4 rotations have been labelled C_4^+ (clockwise rotation) and C_4^- (anticlockwise). Because it better fitted the arguments developed, they have usually been called C_4 and C_4^3, respectively, in the text

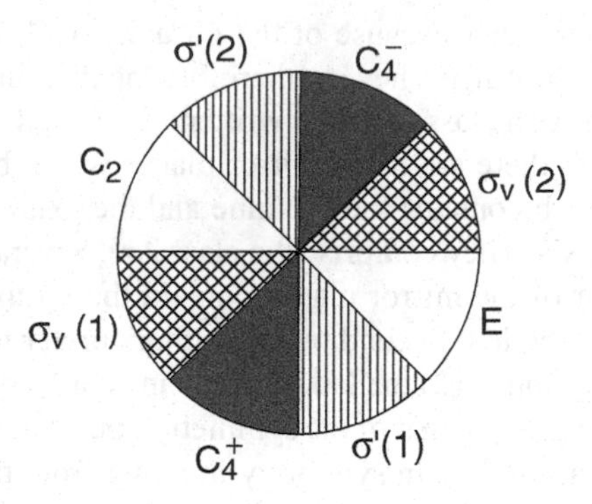

Figure 6.5 The fact that members of the same class are grouped systematically in Figure 6.6 is made evident if the quadrants corresponding to them are distinguished

marked. When we come to draw pictures of the irreducible representations, the + or − signs in the similarly marked areas do not vary independently, but are inextricably linked together in some way (often, but not always, they are identical). A point worth noting, but which will not be explored, is that if one had to compile the multiplication table for the C_{4v} group, the job would be made much simpler if Figure 6.4 were consulted.

Although the next major task is to derive the character table of the C_{4v} group, it is convenient first to consider a problem which will be encountered when using it.

> **Problem 6.2** Either draw a diagram or (better) make a model of the BrF_5 molecule and, by a study of this, make a list of the symmetry elements that it contains. Compare your list with that given above and explore the reason for any differences.

6.2 Problems in using the C_{4v} group

When considering the transformation of something – an orbital or set of orbitals, perhaps – what should be done when there are two operations in a class? How is a character generated in such a case? Although the formal answer to this is unattractive, the practical answer is simple. Formally, the correct procedure is to consider the transformation of each object under each of the individual operations in the class and to take the average of characters generated. Indeed, before the end of this chapter, we will do just this. But the need to take an average is very rare. It is almost invariably the case that each of the symmetry operations in a class gives the same character. This means that, in practice, all that has to be done is to:

> Select a single symmetry operation from a class (and it quite often happens that it is possible to set up the problem in such a way that there is one operation with which it is particularly easy to work) and take the character generated by this operation.

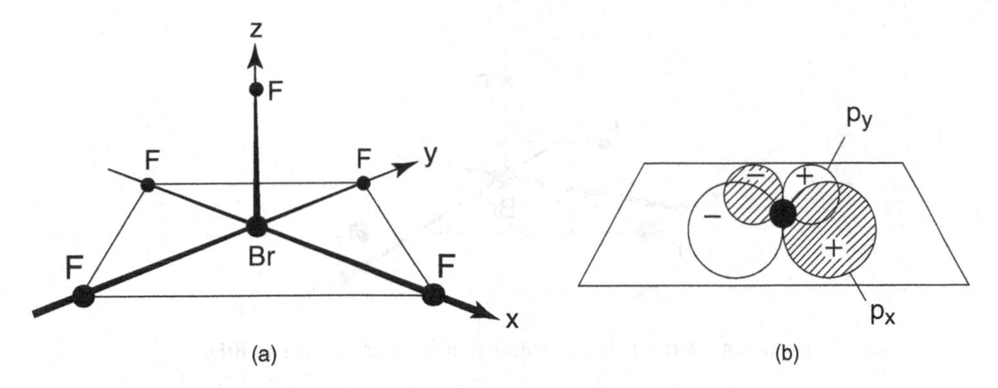

Figure 6.6 One choice of direction for x and y axes in BrF_5 (and consequent directions for the bromine $4p_x$ and $4p_y$ orbitals; the $4p_x$ orbital is shown cross-hatched)

There is yet one more problem which it is as well to consider before turning to the C_{4v} character table. As has been seen, the axis of highest rotational symmetry is conventionally chosen as the z axis so that the C_4 axis of BrF_5 is clearly to be taken as the z axis. However, we are left with the problem of where to place the x and y axes. Perhaps the most evident choice of directions is that shown in Figure 6.6, in which the positions of the four fluorines are taken to define the x and y axes – but what is wrong with the alternative choice given in Figure 6.7? The solution to this problem becomes clearer when it is noted that the x and y axes, just like the σ_v mirror planes in which they lie, are interchanged by the C_4 operations, irrespective of whether the orientation of Figure 6.6 or of Figure 6.7 is chosen for them. The orientations for x and y axes in these figures have an obvious attraction – they are choices which place the axes in mirror planes. A less attractive choice (but perfectly admissible one) such as that shown in Figure 6.8 still retains the property that x and y are interrelated by a C_4 rotation. Clearly, the x and y axes must be treated as a pair (if the choice for one is changed, so too must that for the other), just as the two σ_v and the two σ_v' mirror planes

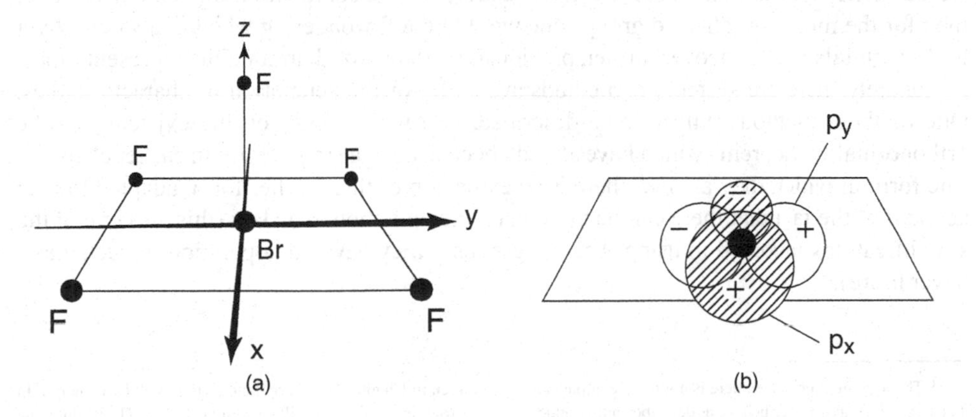

Figure 6.7 An alternative choice of direction for x and y axes in BrF_5 (and consequent directions for the bromine $4p_x$ and $4p_y$ orbitals)

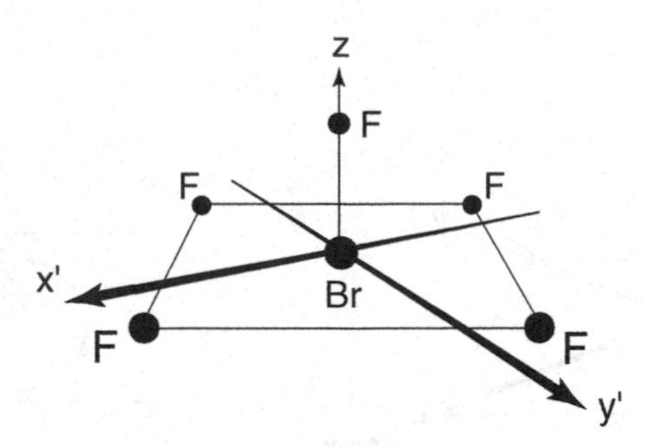

Figure 6.8 A third choice of direction for x and y axes in BrF_5

have to be treated as a pair. This intimate pairing of x and y axes is basic to the difference between the C_{4v} character table and the Abelian character tables of earlier chapters in this book.[1] The x and y axes are said to 'transform as a pair', a statement which will later be seen in the fact that these axes transform, together, as the E irreducible representation in Table 6.1. The choice of x and y axis directions – Figures 6.6, 6.7 and 6.8 – is ultimately unimportant; after all no physical property can in any way depend on the way we choose to place axes. Although this is an almost trivial statement, it is not so easy to work with axes placed in general positions. This elaboration is discussed in more detail in Appendix 2.

6.3 Orthonormality relationships

We now return to the problem of generating the character table of the C_{4v} point group. In principle, the procedure of Chapter 3 could be followed and the transformations of atomic orbitals of the bromine atom used to generate the irreducible representations of this group. Unfortunately, a complete set of irreducible representations could not be obtained even if f orbitals on the bromine atom were included – although if g orbitals were also included they would suffice! If this is the only method available for the compilation of a character table, then for the more complicated groups one would be left wondering whether a study of yet higher orbitals might uncover further, previously unrecognized, irreducible representations. Fortunately there are systematic methods available for the generation of character tables. One of these methods will now be described, one which relies on the existence of the orthonormality theorems which have already been used in a simple form in earlier chapters. The form in which they are used here is an extension of their earlier form, adapted to take account of the fact that there can be more than one operation in a class (this was one of the simplifications involved in using Abelian groups – they have one operation in each class, never more).

[1] The argument presented here is not quite complete, as will be seen in Chapter 11, where the C_4 group will be explored. In that group the x and y axes behave as described above and yet the C_4 group is Abelian. As will be seen in Chapter 11, the dilemma is resolved in that the group contains an E irreducible representation, in which the two components are said to be 'separably degenerate'

Of the theorems that follow, numbers 2, 3, 4 and 5 are those that are commonly called *the orthonormality theorems*.[2]

Theorem 1 In every character table there exists a totally symmetric irreducible representation.

Comment: The totally symmetric irreducible representation is the first given in any character table and has a character of 1 associated with each class of operation. It describes the symmetry properties of something which is turned into itself under every operation of the group. This really is a rather trivial theorem, introduced here for convenience. By definition, a molecule is turned into itself by every operation of the point group used to describe it. A totally symmetric irreducible representation must therefore exist for every point group.

Theorem 2 Take each element of any row of a character table (i.e. the characters of any irreducible representation), square each and multiply by the number of operations of the class to which the character belongs and add the answers together. The number that results is an integer which is equal to the order of the group (i.e. equal to the total number of symmetry operations in the group).

Comment: This theorem was first met in Chapter 3, when a systematic method of reducing a reducible representation into its irreducible components was obtained. Because an Abelian group was then involved, the number of operations in each class was one, so that there was no need to include the step of multiplying by the number of operations in a class. For a non-Abelian group, for which there is invariably at least one class containing more than one operation, care must be taken to include this additional step, otherwise the number obtained at the end of the summation will not be that of the order of the group.

Problem 6.3 Apply Theorem 2 to each of the irreducible representations of the C_{4v} point group (Table 6.1). The order of this group is eight.

Theorem 3 Take any two different rows of a character table (i.e. any two irreducible representations) and multiply together the two characters associated with each class. Then, in each case multiply the product by the number of operations in the class. Finally, add the answers together. The result is always zero.

Comment: This theorem, again, is one already met when reducing reducible representations. On that occasion, because there was only one operation in each class there was no need to

[2] In this chapter and all others up to Chapter 10, all the characters that will be met are real. The theorems that follow apply only to character tables with such real characters. When complex characters are encountered, changes to the theorems have to be made to deal with this. The changes that are needed are covered in Chapter 11.

explicitly include multiplication by the number of elements in the class. In general, however, this step must be included.

Problem 6.4 Apply Theorem 3 to at least five pairs of irreducible representations of the C_{4v} point group (Table 6.1). The E irreducible representation should be included in at least two cases.

As will be seen, Theorems 2 and 3 are at the heart of the method used to reduce reducible representations into their irreducible components. They are sometimes referred to as if the others did not exist and called 'the orthonormality relationships'.

Problem 6.5 Look back at Section 3.1 and read the discussion on orthonormality given there. Why are Theorems 2 and 3 referred to as 'orthonormality relationships'?

The fourth and fifth theorems are similar to Theorems 2 and 3 but relate to the columns of a character table instead of the rows. They are new to the reader but it can readily be checked that they are correct when applied to all of the character tables met so far.

Theorem 4 Consider any class (column) of a character table and square each of the elements in it; sum the squares and multiply the answer by the number of operations in the class. The answer is always equal to the order of the group.

Problem 6.6 Apply Theorem 4 to the columns of the C_{4v} character table.

Theorem 5 Consider any two different classes (columns) of the character table. This selects two characters of each irreducible representation. Multiply these pairs of characters of the same irreducible representation together, and sum the results. The answer is always zero.

Comment: In this case no explicit allowance has been made for the number of elements in a class. This is because multiplying by any factor which is common to all contributions to the sum would not change the final answer – it would still be zero.

Problem 6.7 Apply Theorem 5 to at least five pairs of columns of the C_{4v} character table.

Theorem 6 This states that a character table is always square – it has the same number of columns as it has rows; there are as many irreducible representations as there are classes of symmetry operations.

Comment: Yet again, it is easy to see that this theorem holds for all the character tables that have been encountered so far in this book.

Problem 6.8 As was hinted (but not elaborated) at the end of Section 2.4, some character tables contain complex numbers. Sometimes, authors of introductory texts attempt to protect their readers from such horrors by manipulation of the character table. The 'character table' for the group C_4 (the correct form of which contains complex numbers) taken from one such text is given below

C_4	E	$2C_4$	C_2
A	1	1	1
B	1	−1	1
E	2	0	−2

Show that this 'character table' does not fully obey Theorems 2, 4 and 5. (The correct character table will be discussed in detail in Chapter 11 and is given in Table 11.1.)

6.4 The derivation of the C_{4v} character table using the orthonormality theorems

In this section the C_{4v} character table is derived systematically, in contrast to the hit-or-miss method of previous chapters (where the examples were chosen to give hits, of course!). There are several methods of deriving character tables; that in this section is the easiest to understand, follow and use. The derivation of the C_{4v} character table starts by using Theorem 6. The total number of symmetry operations in the C_{4v} group is eight and it has already been seen that they fall into the five classes $E, 2C_4, C_2, 2\sigma_v$ and $2\sigma_v'$. Because the character table must be square (Theorem 6) it follows that there are just five irreducible representations. Theorem 4 requires that the sum of squares of characters lying in the column corresponding to the identity operation is eight (the order of the group). We therefore have to find five integers, the squares of which total eight. Further, because of the nature of the identity operation (it counts a number of objects), none of these integers can be negative or zero. The only set of integers which satisfies these conditions is the set 1, 1, 1, 1 and 2 ($1^2 + 1^2 + 1^2 + 1^2 + 2^2 = 8$). Including the totally symmetric irreducible representation (Theorem 1) we can write down the skeleton character table shown in Table 6.2, where the quantities a→p have yet to be determined.

Because an irreducible representation which describes the behaviour of a single object has characters which can only be 1 or −1 [3] (the object always goes into itself or minus itself, never into a different object under a symmetry operation), the entries a–l in Table 6.2 all have values of either 1 or −1.

Consider now the column corresponding to the C_2 rotation operation. Again, by Theorem 4, the sum of characters in this column has to equal eight and since from the last paragraph the squares of b, f and j are each +1 it follows that n^2 must be 4 so that n is either +2 or −2.

[3] More strictly, are always of modulus unity.

Table 6.2

E	$2C_4$	C_2	$2\sigma_v$	$2\sigma_v'$
1	1	1	1	1
1	a	b	c	d
1	e	f	g	h
1	i	j	k	l
2	m	n	o	p

From Theorem 4, and because each of the $2C_4$, $2\sigma_v$ and $2\sigma_v'$ classes have two operations in them, the elements m, o and p must each be equal to zero. If they had any other value, the sum of squares of elements in each column when multiplied by two, the order of the class, would give a number greater than eight. We are thus led to Table 6.3, in which all $\pm$ signs are to be regarded as independent of each other.

Consider the identity column in Table 6.3 together with one of the columns corresponding to any class of order two. For Theorem 5 to be satisfied (i.e. zero obtained when the products of corresponding characters are summed), the three characters listed for each class as ±1 must, in fact, contain one $+1$ and two -1's. Since at this point in the argument the middle three rows of Table 6.3 are identical, two -1's may be arbitrarily selected for any one class of order two. This we shall do for the $2C_4$ class. The new form of Table 6.3 could be written down but first it is convenient to apply Theorem 3 (the sum of the products of characters multiplied by class orders must be zero) using the double degenerate irreducible representation given in Table 6.3 together with the first (totally symmetric) irreducible representation. Theorem 3 is only satisfied if the ±2 entry under the C_2 class of the doubly degenerate irreducible representation is actually -2. The characters of the doubly degenerate irreducible representation have therefore all been obtained. These results are brought together in Table 6.4.

There are many ways of completing the generation of the character table. For example, apply Theorem 5 (the sum of products of elements of the two classes must be zero) to the columns headed by the E and C_2 operations (the E and C_2 classes) in Table 6.4. The theorem can only be satisfied if all of the characters in the C_2 class are $+1$. Remembering this result, consider the first two rows (irreducible representations) of Table 6.4 and apply Theorem 3 (the sum of the products of characters multiplied by class orders must be zero). The only way in which a sum of zero can be obtained is if the two ±1 characters in

Table 6.3

E	$2C_4$	C_2	$2\sigma_v$	$2\sigma_v'$
1	1	1	1	1
1	±1	±1	±1	±1
1	±1	±1	±1	±1
1	±1	±1	±1	±1
2	0	±2	0	0

Table 6.4

E	$2C_4$	C_2	$2\sigma_v$	$2\sigma_v'$
1	1	1	1	1
1	1	±1	±1	±1
1	−1	±1	±1	±1
1	−1	±1	±1	±1
2	0	−2	0	0

the second irreducible representation are actually −1. These results are summarized in Table 6.5.

Perhaps the most evident thing about the residual unknowns in Table 6.5 is that the characters associated with the third and fourth irreducible representations are the same. The application of either Theorem 3 or Theorem 5 readily shows that the four ±1 characters in Table 6.5 must be either

$$\begin{matrix} 1 & -1 \\ -1 & 1 \end{matrix} \quad \textit{or} \quad \begin{matrix} -1 & 1 \\ 1 & -1 \end{matrix}$$

Substitution of these sets of numbers alternately into Table 6.5 shows that they generate the same two irreducible representations; the alternatives merely differ in the order in which the irreducible representations are listed. The generation of the C_{4v} character table is complete!

Problem 6.9 The derivation of the C_{4v} character table has been explained in some detail. It is important that each step is followed closely because this will give valuable practice in the use of the orthonormality theorems. If it has not already been done in reading this section, carefully check each step in the derivation of the C_{4v} character table.

The final character table is given in Table 6.6 where the commonly adopted symbols for the irreducible representations have also been included. Note the difference between irreducible representations labelled B and those labelled A. Both are singly degenerate but the B's are antisymmetric with respect to a rotation about the axis of highest symmetry (C_4) whereas the A's are symmetric. This particular distinction may be compared with that discussed in Section 2.4, where the A's and B's in the C_{2v} point group were distinguished

Table 6.5

E	$2C_4$	C_2	$2\sigma_v$	$2\sigma_v'$
1	1	1	1	1
1	1	1	−1	−1
1	−1	1	±1	±1
1	−1	1	±1	±1
2	0	2	0	0

Table 6.6

C_{4v}	E	$2C_4$	C_2	$2\sigma_v$	$2\sigma_v'$	
A_1	1	1	1	1	1	z, z^2, x^2+y^2
A_2	1	1	1	−1	−1	
B_1	1	−1	1	1	−1	x^2-y^2
B_2	1	−1	1	−1	1	xy
E	2	0	−2	0	0	$(x, y), (zx, yz)$

by their behaviour under a C_2 rotation operation. The generalization is clear – for a group for which the highest rotational axis is C_n, A's are symmetric with respect to this operation whereas B's are antisymmetric.

Problem 6.10 A fragment of the C_{8v} character table is shown below. Complete this fragment.

C_{8v}	E	$2C_8$	$2C_4$	$2C_8{}^3$	C_2	...
A_1	1					...
A_2	1					...
B_1	1					...
B_2	1					...

Hint: Within this fragment there is no distinction apparent between A_1 and A_2 or between B_1 and B_2. Consider first behaviour under C_8; the other entries follow because C_4 and C_8^3 are multiples of C_8 and the characters under these operations must be consistent with that for C_8. This problem illustrates yet another approach to the compilation of character tables and the sort of relationships that exist within them. A discussion which parallels that required to answer this problem is to be found at the end of Section 10.5.

Problem 6.11 Use the second part of the hint in Problem 6.10 to explain why there are no B irreducible representations in the character table of the C_{7v} group (or, indeed, any group containing a C_n axis when n is odd).

We are now in a better position to discuss Table 6.6 than when it was first met as Table 6.1. There are five aspects of it on which it is appropriate to comment, all associated with the E irreducible representation. The first has already been mentioned, the label E itself. This is identical to the label used to describe the identity operation. Although this appears confusing, in practice it is not. This is because the contexts in which the two labels are used are always quite different; the context tells which is intended. In some texts, however, the ambiguity is avoided by a difference in typeface or, more simply but less frequently, by the use of the label I for the Identity operation. Second, the occurrence of the characters 2 and 0 in this irreducible representation is something new and requires comment. The appearance of the character 2 for the identity, leave alone, operation means that two things are being left alone, that the E irreducible representation describes the behaviour of a pair of objects simultaneously. The x and y axes of BrF_5, no matter which choice is made for them, are such a pair. Another, closely related, example which will be considered in detail shortly are the valence shell p_x and p_y orbitals of the bromine atom in BrF_5. The character 2 here, then, means the same as when it was met in Chapter 3; that two objects remain themselves. However, the way that the 0 appears is something new. Previously, in Chapter 3, this character was obtained because every object under consideration moved as a result of a symmetry operation. There is another way in which 0 can appear. This is when each object which remains unchanged is matched by one which changes its sign. The sum of 1 and -1 is, of course, 0. An example of this will be met when the p_x and p_y orbitals of the bromine atom in BrF_5 are considered.

Third, note the way in which the members of a basis set for the E representation are written (in the extreme right-hand column of the character table). The x and y axes are such a pair and are written (x, y) in contrast to the listing of functions which, separately and independently, provide a basis for a representation. The functions z and z^2 each, separately, forms a basis for the A_1 irreducible representation. The way that either can do this independently of the other is indicated by the way they are written: z, z^2. Fourth, this is a convenient point at which to formally introduce a piece of useful jargon (which we have already used!). Irreducible representations which describe the transformation of two objects simultaneously are said to be *doubly degenerate* whereas those describing the transformation of one are said to be *singly degenerate*. There are fundamental reasons for this usage but simplest is to note that if the objects that the irreducible representations describe are orbitals, then for E irreducible representations there *must* be two orbitals with exactly the same energy. Were they not the same, the act of carrying out, for example, a C_4 rotation would have the effect of changing energies (because it interchanges the bromine orbitals, p_x and p_y, for instance). The energy of an orbital would depend on whether or not we chose to do a C_4 operation and this clearly is ridiculous. This dilemma is only avoided by the orbitals having the same energy, being degenerate.

Fifth, if we were to construct a group multiplication table for the C_{4v} group we would find that only the characters of the various A and B irreducible representations could be substituted for their corresponding operations to give an arithmetically correct multiplication table. The substitution fails for the E irreducible representation. The reason is that we should really use 2×2 matrices to describe the E irreducible representation, not a simple number. When these *matrices* are substituted for the corresponding characters and multiplied by the laws of matrix multiplication then a correct multiplication table is obtained. This is explained in more detail in Appendix 2. At this point it is appropriate only to comment that ordinary numbers may be regarded as 1×1 matrices (whereupon the laws of matrix multiplication reduce to the ordinary laws of numerical multiplication) so that those irreducible representations containing only characters of value $+1$ or -1 may also be regarded as involving matrices. The connection between matrices and character tables is profound and important. Indeed, the name 'character' is given to the sum of the elements along the leading diagonal (top left to bottom right) of a matrix. This is no accident, as Appendix 2 makes clear. However, in almost all of the applications of group theory to chemistry there is no need to make explicit use of matrix algebra. Hence this book, in which there is no use of matrix algebra in the body of the text.

Two quite different methods of generating character tables have now been encountered, that of using the transformations of suitable basis functions and that of the use of the character table theorems. As has been indicated, there are yet other methods too but these will not be discussed. From this point on in the text, character tables will not be systematically derived. The procedure will be the one almost invariably used – a character table is taken from a compilation such as that in Appendix 3. When appropriate, however, comments on various aspects of those that are met will be included in the text.

6.5 Nodal patterns of the irreducible representations of C_{4v}

We are now in a position to draw the nodal patterns of the irreducible representations of Table 6.6. These pictures will show how the irreducible representations differ fundamentally from one another and, incidentally, show us why we would have needed to invoke g orbitals before we could obtain a complete set of orbital basis functions for the group. The patterns are shown in Figure 6.9, where, as an aid towards understanding, Figure 6.4 is included. A_1, as expected, is totally symmetric. A_2 is interesting because of its high nodality. It has four inherent nodal planes. One can immediately understand the need for a g orbital on the central atom if we are to have an orbital basis for the A_2 irreducible representation. The s orbitals have zero inherent nodal planes (they may have spherical nodal surfaces – these are what distinguish 1s from 2s from 3s etc.); p orbitals have one, d orbitals have two and f orbitals have three. To obtain four inherent nodal planes one needs to invoke g orbitals. The B_1 and B_2 patterns each have two nodal planes. The B_1 pattern is symmetric with respect to reflection in the σ_v mirror planes and antisymmetric with respect to reflection in the σ_v'. The B_2 pattern is the converse, antisymmetric in the σ_v and symmetric in the σ_v'. As we have met more than once before, care is needed. Interchange the labels on the mirror planes and you interchange the labels on B_1 and B_2. One author's B_1 may be someone else's B_2. Unfortunately, it happens that the choice is sometimes unspecified and one has to resort to some sort of working-backwards procedure in order to find out which choice has been made.

Now to the really interesting and informative nodal patterns. We have met irreducible representations with 0, 2 and 4 inherent planar nodes. What of the one-node case? These are characteristic of the E irreducible representation – as evidenced by the x, y entry in Table 6.6, and as shown in Figure 6.9. But in Table 6.6 the E irreducible representation has three characters of 0. Where are these in Figure 6.9? The answer is intimately related to the twin facts that in every case there are two operations in the class that has a 0 character and that there are two components of the E irreducible representation. Two operations, two components, a total of four combinations and all have to make an equal contribution to the character. Remember what was said when we first met the problem of obtaining a character when there is more than one operation in a class: 'the correct procedure is to consider the transformation of each object under each of the individual operations in the class and to take the average of characters generated'. Consider the σ_v operations. In the first E diagram, $E(1)$, the starting point, denoted by the E operation, with $+$ phase is converted into points with $+$ phase by $\sigma_v(1)$ and $-$ phase by $\sigma_v(2)$; an average of 0. In the second diagram, $E(2)$, the same mirror plane reflections give the phases $-$ and $+$ respectively; again an answer of 0. For the σ_v' mirror plane reflections the same answer is obtained, of course, but differently. In this case the phases are $+$ and $-$ in both diagrams. There is one other case to consider, that of the two C_4 rotation operations, but this is left as a problem for the reader. Finally, we again emphasize the fact that the irreducible representation labels in Table 6.6 are really

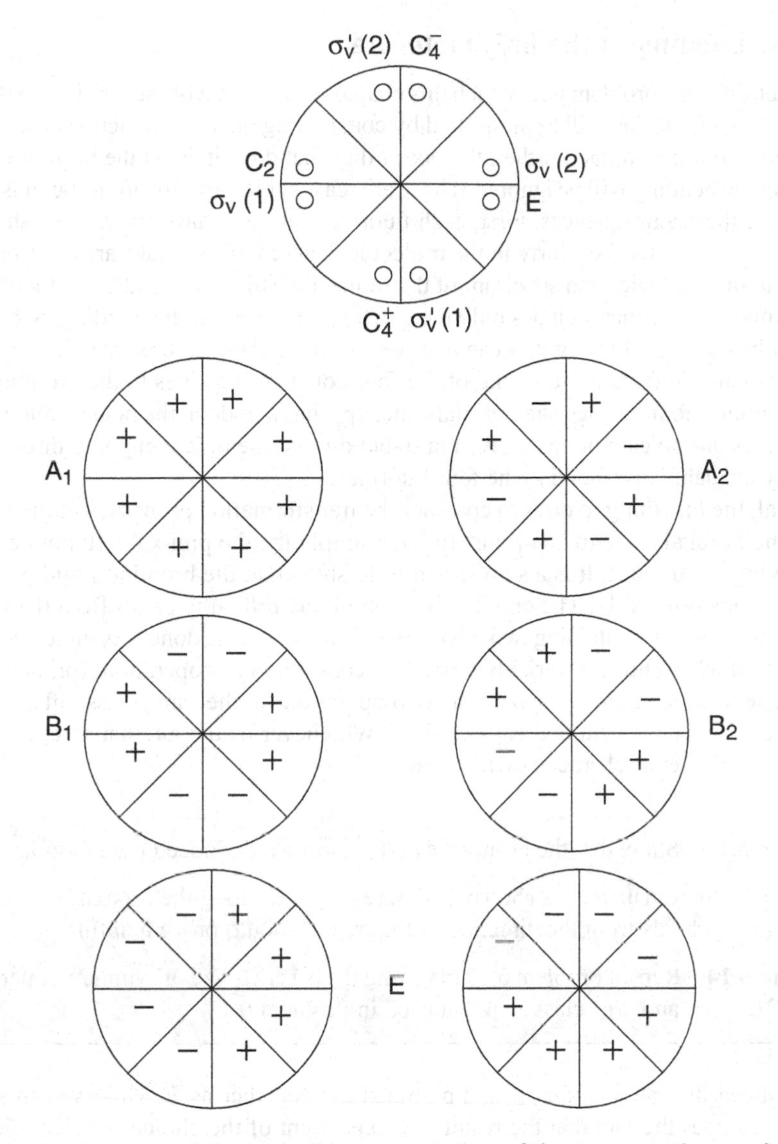

Figure 6.9 Nodal patterns of the irreducible representations of the C_{4v} group. The use of a 'natural' pattern in Figure 6.4 means that the inherent polarity of members of the E irreducible representation is evident

labels for different nodal patterns. Figure 6.9 shows these nodal patterns; clearly, it is very convenient to represent each by a simple label – but the labels should not be allowed to obscure the reality.

Problem 6.12 Show that the diagrams given for $E(1)$ and $E(2)$ in Figure 6.9 are consistent with a character of 0 for the $2C_4$ class.

6.6 The bonding in the BrF_5 molecule

We now return to the problem with which this chapter started, that of the bonding in the BrF_5 molecule. The discussion will be simplified by considering only σ interactions between the fluorine and bromine atoms. Further, the possibility that d orbitals on the bromine may be involved in the bonding will be ignored. These are reasonable simplifications but it is as well to anticipate their consequences. First, each fluorine atom will have six valence shell non-bonding electrons, a total of thirty in the molecule. There will be peaks arising from these electrons in the photoelectron spectrum of the molecule which may make it difficult to test our final model, since that includes only the electrons involved in the bonding. Second, the neglect of bromine d orbitals will mean that we will find, at most, three bonding molecular orbitals responsible for the σ bonding of the four coplanar fluorines to the bromine atom. Of the bromine's four valence shell orbitals one, $4p_z$, has a node in the plane containing the four fluorines and so cannot be involved in σ-bonding to them, leaving only three orbitals potentially available to σ-bond to the four fluorines.

As usual, the first thing to do is to consider the transformation properties of the bromine valence shell orbitals, $4s$, $4p_x$, $4p_y$ and $4p_z$ (for simplicity, the prefix 4 will not be used in the following discussion). It is a simple matter to show that the bromine s and p_z orbitals separately transform as A_1. The only likely point of any difficulty arises from the fact that there are three classes containing two operations. What has to be done? As indicated above, the answer which is almost invariably correct is 'consider either operation' (or, in the more general case in which there are *more* than two operations in the same class, often one will be a particularly convenient and easy choice). Whichever of the alternative operations is chosen, the same set of characters will result.

Problem 6.13 Show that the bromine s and p_z orbitals do indeed transform as A_1.

Hint: The z coordinate axis is shown in Figure 6.6 (it is along the C_4 axis, of course). Viewing the orbitals from the direction in Figure 6.2 should prove helpful.

Problem 6.14 Repeat Problem 6.13 but using the other choice of symmetry operation for the $2C_4$, $2\sigma_v$ and $2\sigma_v'$ classes to that used in Problem 6.13.

As has been indicated earlier, p_x and p_y transform together as E. However, this has to be shown, as does the fact that the result is independent of the choice of orientation of x and y axes (although the demonstration of the latter point will be incomplete because only the alternative axis sets of Figures 6.6 and 6.8 will be considered). The transformation of the p_x and p_y orbitals of the bromine atom under the eight symmetry operations of the C_{4v} group are detailed in Table 6.7 for the two choices of x and y axes. It is most important that this table should be worked through carefully. Note, in particular, that the detailed behaviour of the two sets of p orbitals under the mirror plane reflections depends on the choice of x and y axis directions. Despite these differences, the character resulting from the transformations is the same for either choice, which is a most important result. Similarly, the sum of characters generated by p_x and p_y is the same for all operations in any one class. The – consensus – characters generated by the p_x and p_y orbitals under the operations of the C_{4v} point group are given at the bottom of Table 6.7. The choice of x and y axes of Figure 6.8 (or any other arbitrary choice) would also lead to the same set of characters as

Table 6.7 The transformations of the bromine p_x and p_y orbitals in BrF_5. The table shows the orbital obtained when each operation operates on p_x and p_y. Its contribution to the aggregate character is given in parentheses after each orbital

	E	C_4	C_4^3	C_2	$\sigma_v(1)$	$\sigma_v(2)$	$\sigma_v'(1)$	$\sigma_v'(2)$
p_x (Fig.6.6) becomes	$p_x(1)$	$-p_y(0)$	$p_y(0)$	$-p_x(-1)$	$-p_x(-1)$	$p_x(1)$	$p_y(0)$	$-p_y(0)$
p_y (Fig.6.6) becomes	$p_y(1)$	$p_x(0)$	$-p_x(0)$	$-p_y(-1)$	$p_y(1)$	$-p_y(-1)$	$p_x(0)$	$-p_x(0)$
p_x, p_y together	2	0	0	−2	0	0	0	0
p_x (Fig.6.7) becomes	$p_x(1)$	$-p_y(0)$	$p_y(0)$	$-p_x(-1)$	$-p_y(0)$	$p_y(0)$	$-p_x(-1)$	$p_x(1)$
p_y (Fig.6.7) becomes	$p_y(1)$	$p_x(0)$	$-p_x(0)$	$-p_y(-1)$	$-p_x(0)$	$p_x(0)$	$p_y(1)$	$-p_y(-1)$
p_x, p_y together	2	0	0	−2	0	0	0	0
C_{4v}	E	$2C_4$		C_2	$2\sigma_v$		$2\sigma_v'$	
Representation generated by p_x and p_y together	2	0		−2	0		0	

those given in Table 6.7, although the truth of this is not self-evident. For the proof of the statement the use of matrix algebra is unavoidable. The reader who wishes to check out this particular aspect will have to turn to Appendix 2, where the proof is given. Comparison with Table 6.6 shows that the representation which has been generated using p_x and p_y as bases is the E irreducible representation of the C_{4v} group. Because the x and y axes transform similarly to p_x and p_y (just drop the p's in Table 6.7 to obtain the transformation of the axes) it follows that these too transform as E, as asserted earlier in this chapter.

The next task is to determine the irreducible representations spanned by the fluorine σ orbitals involved in bonding with the bromine. No attempt will be made to specify in detail the composition of the fluorine σ orbitals. It will be a mixture of s and p orbitals but the participation of each of these components is not symmetry determined and, in any case, the choice does not affect the qualitative conclusions that will be reached. For simplicity, in the diagrams in this chapter these hybrid orbitals will be drawn as spheres (in contrast, it is more convenient that they be drawn as pure p orbitals in Appendix 4). The next step is the usual one, a consideration of the transformation properties of these fluorine hybrid orbitals. That of the axial fluorine lies on all of the symmetry elements of the C_{4v} group. All of the corresponding operations turn the orbital into itself. It therefore transforms as the totally symmetric irreducible representation of the C_{4v} point group (A_1). The hybrid σ orbitals of the four symmetry-related fluorine atoms transform as a set and form a basis for a reducible representation which must be decomposed into its irreducible components. The generation of the reducible representation is straightforward but two comments are relevant. First, care has to be taken in the definition of σ_v and σ_v' mirror planes. In this chapter the choice shown in Figure 6.3 will be followed. That is, the fluorine atoms lie in the σ_v mirror planes. This choice is arbitrary but it is important to be consistent, otherwise meaningless results will be obtained. Second, as before, for the classes containing two symmetry operations either operation may be chosen to obtain the character. So, use that which is the easier (most people find that C_4 is easier than C_4^3, for example). Following this procedure it should readily be found that the reducible representation generated by the transformations of the fluorine σ orbitals is:

E	$2C_4$	C_2	$2\sigma_v$	$2\sigma_v'$
4	0	0	2	0

Next, this reducible representation has to be reduced to its irreducible components. Again, recognition has to be made of the fact that three classes contain two operations. The reduction of a reducible representation depends on the group theory orthogonality relationships given earlier in this chapter. In particular, Theorems 2 and 3 are relevant. So, the above representation has to be multiplied by the characters of each irreducible representation in turn. These products are then multiplied by the number of operations in the class and the results summed. If no mistake has been made, the sum is a multiple of 8 (the order of the C_{4v} group), the multiplication factor giving the number of times that the chosen irreducible representation appears in the reducible representation. This is worked out for the case of the B_1 irreducible representation below:

	E	$2C_4$	C_2	$2\sigma_v$	$2\sigma_v'$
Reducible representation	4	0	0	2	0
B_1	1	−1	1	1	−1
Multiply	4	0	0	2	0
Number of operations in class	1	2	1	2	2
Multiply last two rows	4	0	0	4	0

Add the entries in the last row; the sum = 8. We conclude that the B_1 irreducible representation occurs once in the reducible representation. Repetition of this process shows that the reducible representation has $A_1 + B_1 + E$ components. Alternatively, quicker but less evidently rigorous, one could ask the question 'which of the nodal patterns of Figure 6.9 are compatible with the four fluorine orbitals?'. The A_2 has to be excluded because it has four nodal planes and these are incompatible with the fluorine hybrids, each of which has to lie on a mirror plane and is symmetric with respect to reflection in it. Similarly, one of the B_1 and B_2 has to be excluded; interestingly, this approach bypasses the problem of 'which set of mirror planes is which?' (although there is a choice inherent in the answer obtained). As a bonus, the nodal patterns indicate that the correct linear combinations of fluorine orbitals transform as a particular irreducible representation (and, in the case of E, of the components). One simply adds the phase shown in Figure 6.9 to the corresponding hybrid; the resulting combination is correct (unless a silly mistake has been made!). Or one can get the same results the hard way, the way that we now outline.

Problem 6.15 Show that the reducible representation generated above has $A_1 + E$ components in addition to the B_1.

Labelling the fluorine orbitals as indicated in Figure 6.10, and proceeding as in Section 5.6, the normalized form of the linear combinations of fluorine hybrid orbitals which transform as the A_1 and B_1 irreducible representations – the symmetry-adapted combinations – are readily obtained. The discussion in Section 5.6 is so close to that needed at this point that it will not be repeated. The only thing new is to point out that each and every operation of the group must be considered (so, *both* C_4 and $C_4{}^3$). The character that is used to generate a particular symmetry-adapted combination is that in the character table(!), applied to *each*

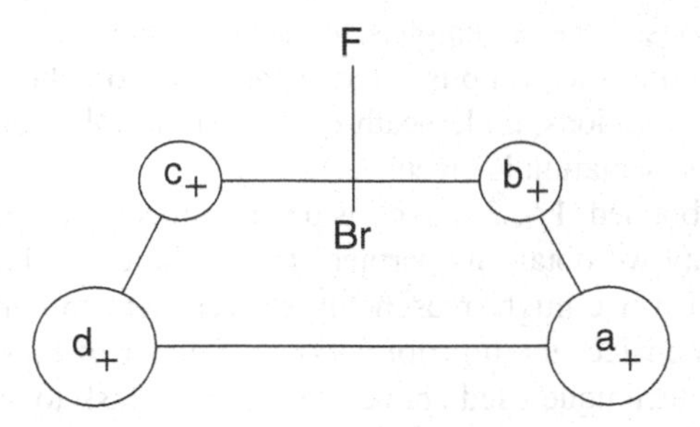

Figure 6.10 The labelling and phases of the σ hybrid orbitals of the four coplanar fluorines. For simplicity these hybrid orbitals are drawn as circles; the square of Figure 6.2 is shown in perspective in this and following diagrams in order to locate the fluorine atoms without including all of them

operation in a class. So, for the B_1 irreducible representation, *both* C_4 and $C_4{}^3$ are associated with a character of -1.

Symmetry species	Symmetry-adapted combinations of fluorine orbitals
A_1	$1/2(a + b + c + d)$
B_1	$1/2(a - b + c - d)$

Problem 6.16 Working with the labels of Figure 6.10, use the projection operator method to generate the A_1 and B_1 functions given above.

The generation of the two combinations which transform as E is a more difficult problem, and it will be considered in some detail. Those who had difficulties with Problem 6.16 should find that the following discussion enables them to be overcome. As for the A_1 and B_1 combinations the projection operator method described in the last chapter will be used. Using the fluorine hybrid orbital labelled a in Figure 6.10 as generating element and the mirror plane operations as labelled in Figure 6.3 the following transformations are found (they were probably generated in tackling Problem 6.16):

Operation	E	C_4	C_4^3	C_2	$\sigma_v(1)$	$\sigma_v(2)$	$\sigma_v'(1)$	$\sigma_v'(2)$
Under the operation orbital a becomes	a	d	b	c	c	a	b	d
The E irreducible representation	2	0	0	-2	0	0	0	0
Multiply	2a	0	0	-2c	0	0	0	0

(C_4 and C_4^3 are clockwise and anticlockwise rotations of 90° respectively.)

The sum of products is 2a − 2c which gives, on normalization:

$$\frac{1}{\sqrt{2}}(a - c)$$

as one of the E functions. Note, as emphasized several times already, that in the above derivation each operation of the group is listed separately. So, when there is a class comprising two symmetry operations, underneath *each* operation the corresponding character of the E irreducible representation is given.

The wave function obtained, $1/\sqrt{2}(a - c)$, is one member of the pair of functions transforming as E. How may we obtain its partner? In this function there is no contribution from the orbitals b and d; we might reasonably expect them to contribute to the orbital we are seeking. If we consider the transformations of either of these orbitals and follow the projection operator technique used above it is a simple task to show that the function $1/\sqrt{2}(b - d)$ is generated. This is the second function for which we have been looking.

Problem 6.17 Generate the second E function, $1/\sqrt{2}(b - d)$.

The functions $1/\sqrt{2}(a - c)$ and $1/\sqrt{2}(b - d)$ transform as a pair under the E irreducible representation of the C_{4v} group and are shown in Figure 6.11. The method used to obtain the second member of the degenerate pair was based on an enlightened guess. In the next chapter a more systematic method of generating such functions will be presented. One final word on these combinations. Whereas in the A_1 combination adjacent fluorine σ orbitals have the same phase – and so any interaction between them is bonding – in the B_1 they are always of opposite phase. Any interaction between them is antibonding. In each of the E combinations there is no interaction between adjacent σ orbitals (only *trans* orbitals appear in any one combination). This argument leads us to expect a relative energy order $A_1 < E < B_1$, a sequence which will be reflected in the presentation of the molecular orbital energy level diagram for BrF_5 (Figure 6.15).

We are now almost ready to consider the interaction between bromine and fluorine σ orbitals. First, however, recall that the s and p_z bromine orbitals separately transform as the A_1 irreducible representation. Analogous situations have been encountered in earlier chapters, when the corresponding orbitals were combined to obtain two mixed, hybrid, orbitals of the form $1/\sqrt{2}(s \pm p_z)$. This simplifying procedure will also be followed in the present case. One of the mixed orbitals is orientated in a way that should give good overlap with the σ orbital of the apical fluorine atom, an orbital which, as has been seen, is also of

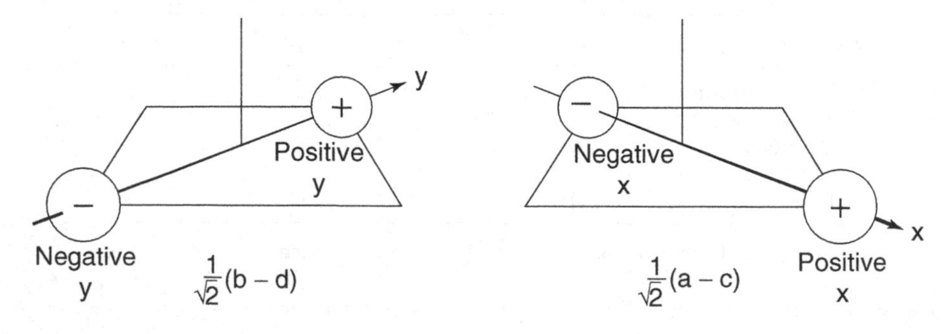

Figure 6.11 The two fluorine σ hybrid orbital combinations which transform as E in the C_{4v} point group and which are derived in the text

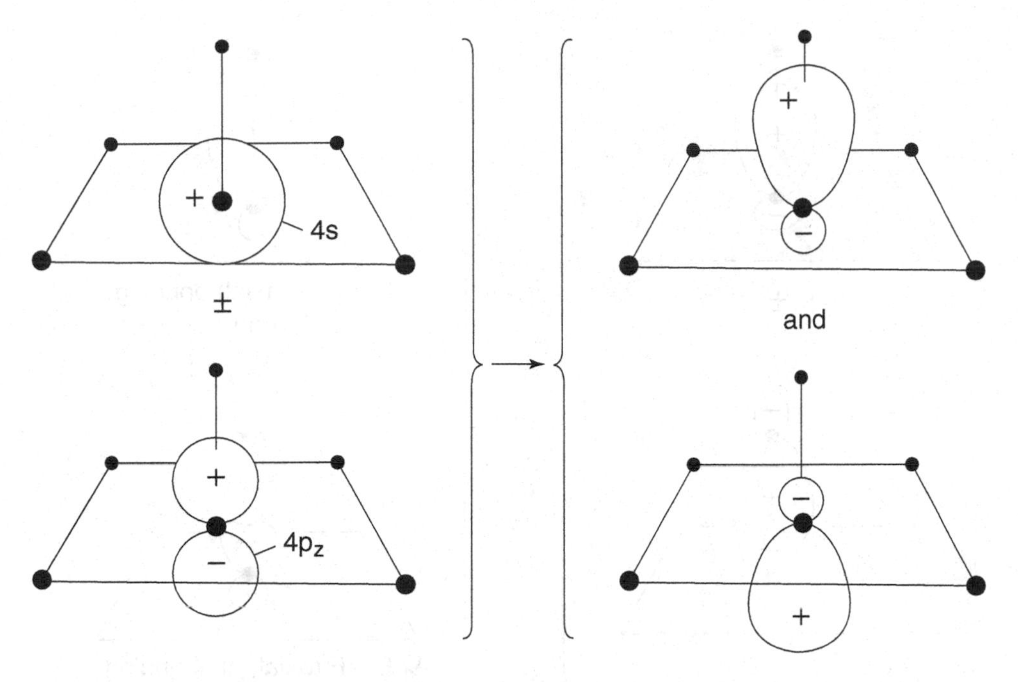

Figure 6.12 Hybrid orbitals (right) derived from bromine atomic orbitals (left) of A_1 symmetry in BrF_5

A_1 symmetry; these steps are shown schematically in Figures 6.12 and 6.13. The remaining s–p_z mixed orbital on the bromine atom points in the direction indicated by dashed lines in Figure 6.1 and therefore might be regarded as the orbital which (in the electron-pair repulsion model) causes the distortion of the molecule that was noted at the beginning of this chapter. Unfortunately, as will be seen, reality is perhaps more complicated than this. The complication arises from the fact that there is also a combination of σ orbitals from the planar fluorines which has A_1 symmetry. Clearly, it can interact with A_1 orbitals of the bromine. However, one of these latter A_1 orbitals is p_z and this has a nodal plane in which the fluorines lie (in our simplified geometry of coplanar bromine and fluorines). The basal plane fluorine A_1 combination interaction will therefore be almost entirely with the bromine s orbital. Correspondingly, it seems probable that the bromine s orbital involvement with the axial fluorine and in the basal lone pair will be rather less than assumed above.

The only other valence orbitals on the bromine atom are p_x and p_y which, together, are of E symmetry. They are shown in Figure 6.7b. They interact with the two fluorine σ orbital combinations of E symmetry which were generated earlier in this section and which are shown in Figure 6.11. Provided the p orbitals and the fluorine orbital combinations are properly chosen – and this means that the same set of coordinate axes is used for each – then each p orbital only interacts with one σ orbital combination. The orbitals in Figures 6.7b and 6.11 are properly chosen and the results of their interactions to give what have been represented as sum (bonding) and difference (antibonding) combinations are shown in Figure 6.14.

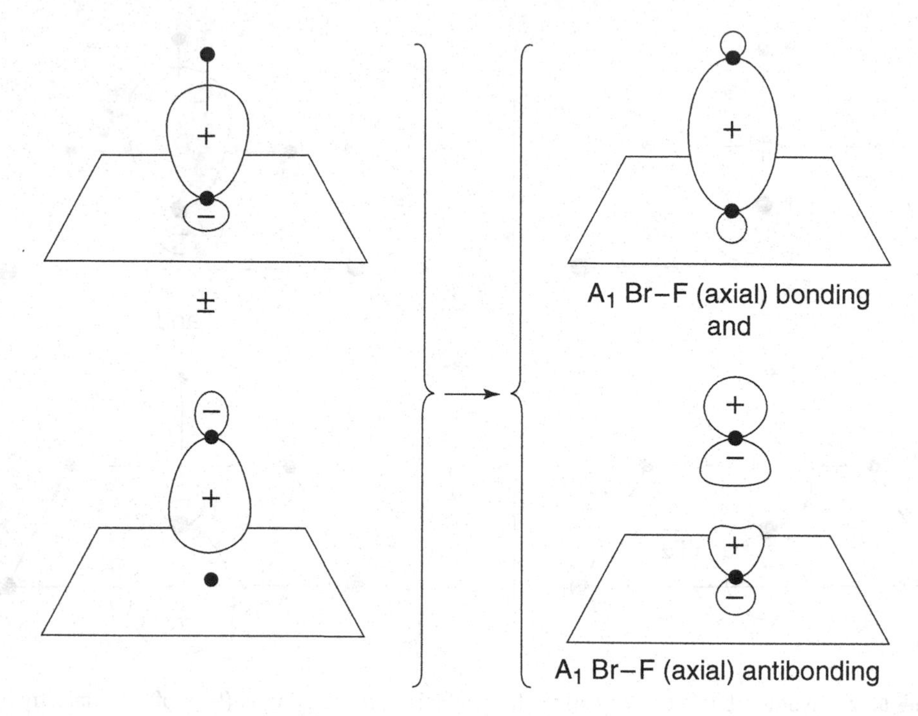

Figure 6.13 Bonding of the axial fluorine to the bromine atom in BrF_5; some form of sp hybrid is envisaged as involved on each atom

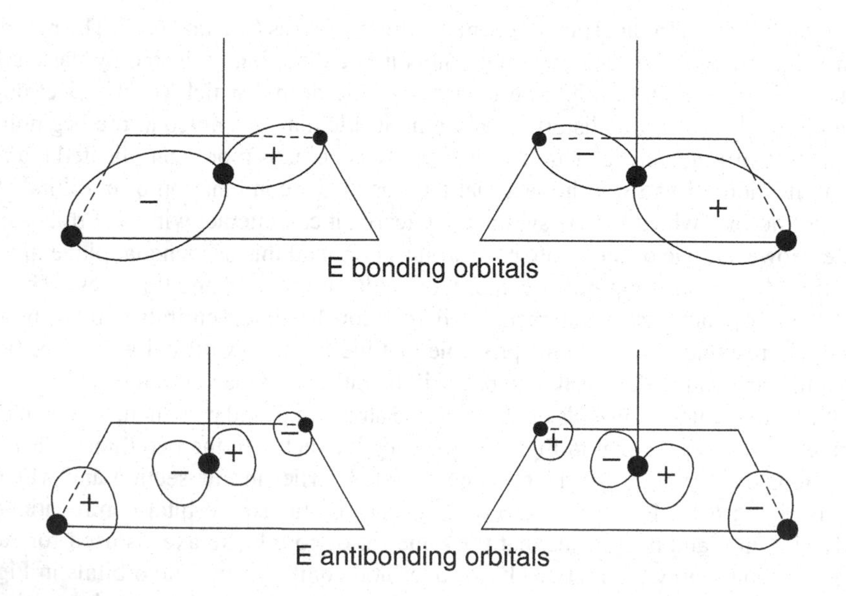

Figure 6.14 The two bonding orbitals and the two antibonding orbitals of E symmetry in BrF_5 arising from interactions of the four coplanar fluorines with the central bromine atom

Problem 6.18 Repeat the above discussion of the interactions of orbitals of E symmetry using the choice of coordinate axes shown in Figure 6.8 (use the bromine p orbitals shown in Figure 6.8b).

Hint: The projection operator method used in the text automatically selected the coordinate axis choice shown in Figure 6.7 because of an (implicit) choice to consider the transformation of an individual fluorine σ orbital (a). The method can be forced to give combinations appropriate to the axes of Figure 6.8 by considering, instead, the transformation of a pair of neighbouring σ orbitals. Thus, the pairs (a + b) and (a + d) are suitable pairs to use in tackling this problem.

The above discussion is summarized in the energy level pattern given in Figure 6.15, where, as has been recognized, there must be some uncertainty about the details of the positions of the orbitals of A_1 symmetry. Into the orbital pattern shown in this figure a total of twelve electrons (seven from the bromine and one from each fluorine orbital) have to be placed. It will be remembered that this diagram does not include the fluorine non-bonding electrons.

There are some interesting consequences of Figure 6.15 and of our discussion of the bonding of the BrF_5 molecule. We have suggested that one molecular orbital (of A_1 symmetry) is primarily involved in the bonding of the axial fluorine to the bromine. If this view is correct then this bromine–fluorine bond involves two electrons. In contrast, in the picture developed above, the strongly bonding molecular orbitals involving the planar fluorine atoms are of E symmetry (although there will be a smaller contribution from an A_1 orbital). That is, the four coplanar fluorine atoms are bonded to the central bromine atom by little more than two molecular orbitals. If this conclusion is correct, it suggests that the bonding between the bromine and each of the four coplanar fluorines is rather weak, a view

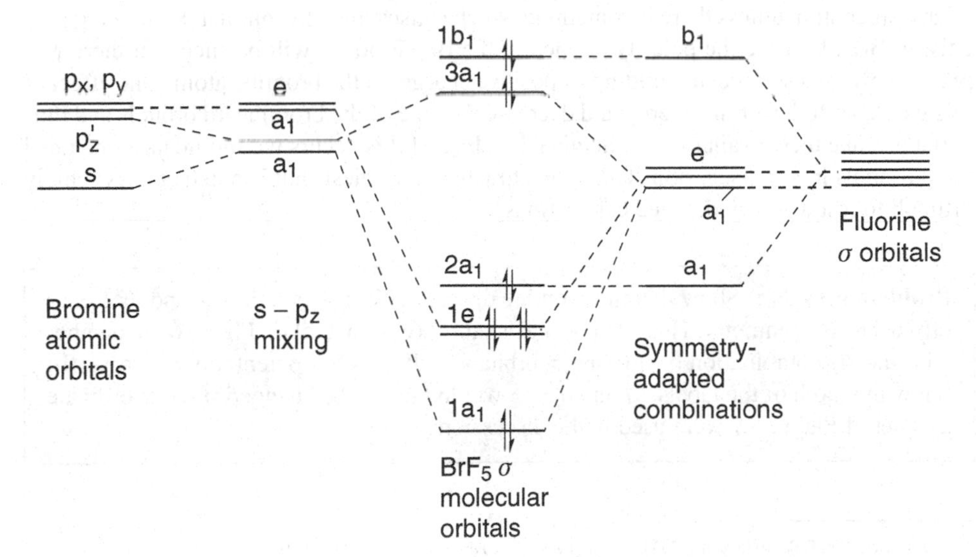

Figure 6.15 Schematic molecular orbital energy level diagram for BrF_5. The a_1 orbital shown in a central trio above the heading 'Symmetry-adapted combinations' is the σ orbital of the axial fluorine

supported by the fact that bromine pentafluoride is an extremely powerful fluorinating agent. Some further support for this difference between axial and planar fluorines is to be found in molecular structure determinations which show Br–F bond lengths of 1.68 Å (axial) and 1.78 Å (equatorial).

Both theoretical calculations and photoelectron spectroscopic data are available for BrF_5. For us they are complicated by the fact that in the above discussion all the electrons on the bromine and on the fluorine atoms which are not involved in σ bonding have been omitted. These electrons, then, have to be regarded as non-bonding and so are expected to be relatively easy to ionize. Fortunately, some of the symmetries they span are not included in the σ bonding set (A_2, for example) and this helps to identify them. The highest lying, most easily ionized, electrons (at 13.5 eV in the photoelectron spectrum)[4] are believed to be amongst those that we have omitted, a lone pair of electrons on the bromine atom. Then come at least three peaks (between 15 and 17 eV) corresponding to ionization of the fluorine 2p non-bonding electrons, included amongst which are the $1b_1$, $2a_1$ and $3a_1$ electrons of Figure 6.15. Between 18 and 22.5 eV are two peaks which are almost certainly composites but which have been reported as including ionization from the 1e and $1a_1$ Br–F σ-bonding molecular orbitals of Figure 6.15. Theoretical calculations are available for both ClF_5[5] (which has a structure similar to that of BrF_5) and BrF_5[6] itself. The two sets of calculations, which used somewhat different theoretical models, are in good qualitative agreement with each other and with the experimental data, although they suggest that perhaps the $1a_1$ orbital of Figure 6.15 is just a bit too low in energy to be seen in the photoelectron spectrum. For us the most important general conclusion is the promising result that, once again, our relatively simple symmetry-based arguments lead to an energy level pattern which is in good qualitative agreement both with experiment and with detailed theoretical calculations.

In our discussion of the bonding in BrF_5 we have ignored the presence of 4d orbitals on the bromine. The justification for this is that the 4d orbitals of the isolated bromine atom are so large and diffuse that they cannot overlap effectively with a valence shell atomic orbital of any other atom unless there is something which causes them to contract. Something may exist in BrF_5 because the polarity of each of the Br—F bonds will be such that there will presumably be a significant build-up of positive charge on the bromine atom. One effect of this would be to lower the energy and decrease the size of the bromine 4d orbitals and thus perhaps make them available for chemical bonding. If this occurs we should have included the d orbitals in our discussion. This is an attractive hypothesis but is one that is extremely difficult to test, even by detailed calculations.

Problem 6.19 (a) Show that the bromine d_{z^2} orbital has A_1 symmetry and its $d_{x^2-y^2}$ orbital has B_1 symmetry. This latter orbital is shown (contracted!) in Figure 6.16 together with the B_1 combination of fluorine σ orbitals with which it potentially interacts. (b) Show that both of the labels B_1 and $d_{x^2-y^2}$ would have to be changed if the coordinate axis set of Figure 6.7 were used in the discussion.

---

[4] R.L. De Kock, B.R. Higginson and D.R. Lloyd, *Chem. Soc. Faraday Discuss.* **54** (1972) 84.
[5] M.B. Hall, *Chem. Soc. Faraday Discuss.* **54** (1972) 97.
[6] G.L. Gutzev and A.E. Smolyar, *Chem. Phys. Lett.* **71** (1980) 296.

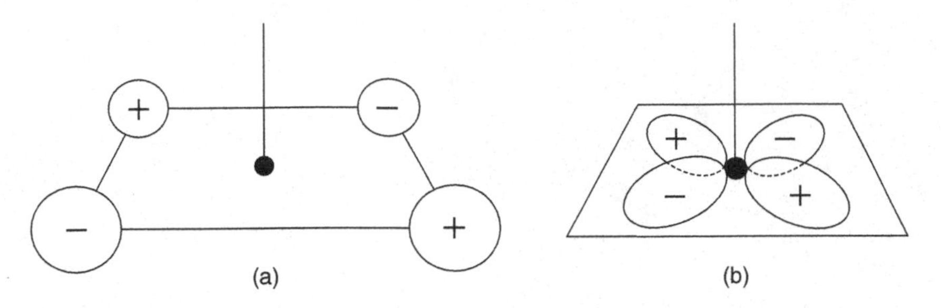

Figure 6.16 (a) The B_1 combination of fluorine σ orbitals. (b) The $d_{x^2-y^2}$ (B_1) orbital of bromine

6.7 Summary

In this chapter it has been found that operations may be divided into classes (p. 147) and that when some classes contain more than one operation the character table contains at least one degenerate representation (p. 146). The presence of a degenerate representation in the C_{4v} group enabled the orthonormality relationships to be presented in a more general form (p. 152). The procedures previously used to reduce a reducible representation have to be modified in the more general case although the projection operator technique is basically unchanged (p. 164). Application of these techniques to the problem of the bonding in BrF_5 suggested both a reason for its existence – polar Br—F bonds possibly enabling participation of bromine d orbitals in the bonding (Problem 6.19) – and for its reactivity – the four coplanar fluorines are not strongly bonded (p. 169).

7 The electronic structure of the ammonia molecule

As usual, the content of this chapter contains more than is evident from the title. It confronts the problem of unpleasant-looking characters. What, for instance, is to be made of a character of 0.6180? Of course, it is not as bad as at first appears. It is just 2cos 72° and it is likely to arise whenever a fivefold rotation axis is contained in a problem ($5 \times 72 = 360$). In fact all odd-fold (3, 5, 7, 9...) rotation axes give rise to such apparently strange characters and similar problems arise with all of them. It is the first, and simplest, of these, a threefold rotation axis, which is covered in this chapter. Almost all of the problems associated with odd-fold rotation axes will be encountered.[1]

In the first chapter of this book four different qualitative descriptions of the bonding in the ammonia molecule were discussed in outline. The symmetry-based approach has now been developed to a point at which this problem may be reconsidered in more detail. At the same time a problem encountered in the last chapter will reappear – that of the choice of directions of x and y axes. The form in which this problem appears is one which will lead to a general solution, a solution which will enable molecules of high symmetry, such as those which will be subject of Chapters 8 and 10, to be tackled.

7.1 The symmetry of the ammonia molecule

The structure of the ammonia molecule is given in Figure 7.1 which also shows the symmetry elements possessed by this molecule. The axis of highest rotational symmetry (which will therefore be taken as the z axis) is a C_3 rotation axis and has associated with it clockwise and anticlockwise rotation operations. To help the reader remember this distinction – because we will be working with them quite a lot – they will be called C_3^+ and C_3^-. In a more general notation they would be called C_3 and C_3^2, a type of notation which was actually that used in the last chapter (in fact, both of the notations were used there). In addition to the threefold rotation axis, there are three mirror planes each of which contains the threefold axis (and

[1] One problem which will not be covered is that of multiples of the basic rotation. So, when there is a fivefold rotation axis one is likely to encounter −1.6180, 2cos 144°, in addition to 0.6180, 2cos 72°. The extension of the content of the chapter to cover such cases is straightforward, but is best done on a case-by-case basis. So, for a C_5 axis it is possible to exploit some classic relationships based on $\tau = 1.6180$ ($-\tau$ and $\tau - 1$ are the cos values above). τ is the so-called 'golden section' and satisfies the equation $\tau^2 = \tau - 1$.

Symmetry and Structure: Readable Group Theory for Chemists Sidney F. A. Kettle

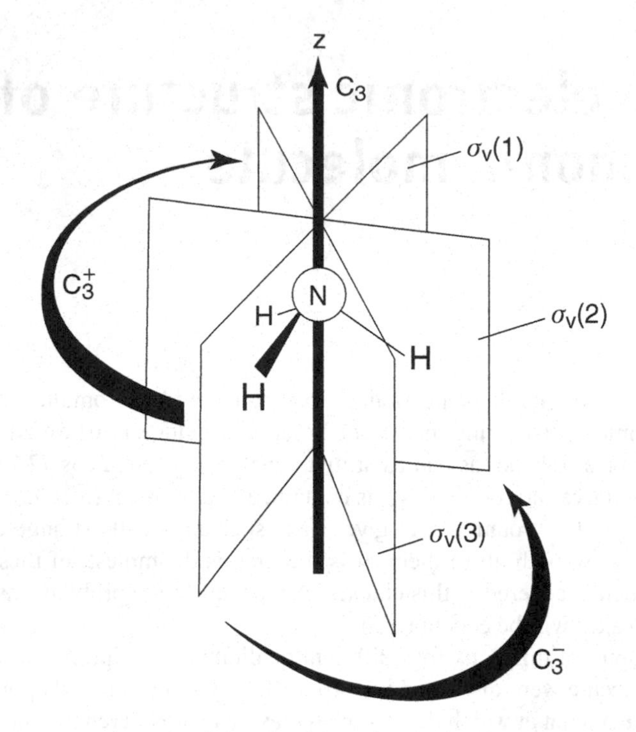

Figure 7.1 The symmetry elements of the ammonia molecule. One hydrogen atom is located in each mirror plane

are vertical with respect to it – that is, they are σ_v mirror planes), with one hydrogen atom lying in each mirror plane. The symmetry operations which turn the ammonia molecule into itself are therefore

$$E \qquad C_3{}^+ \qquad C_3{}^- \qquad \sigma_v(1) \qquad \sigma_v(2) \qquad \sigma_v(3)$$

This group is called the C_{3v} point group, the shorthand symbol C_{3v} indicating the coexistence of the C_3 axis and the vertical mirror planes.

Problem 7.1 Show that this set of operations comprises a group.

Hint: It will be found helpful to refer back to Problem 5.2. The group multiplication table for the C_{3v} group is given in Table 9.2 (and Appendix 2).

The class structure of the symmetry operations of the C_{3v} group is suggested from the similarities between the various operations and is

$$E \qquad 2C_3 \qquad 3\sigma_v$$

Alternatively, the formal methods described in Appendix 2 may be used to deduce this class structure (the $2C_3$ class is given as a worked example in Appendix 2). The character table of the C_{3v} point group is given in Table 7.1.

Table 7.1

C_{3v}	E	$2C_3$	$3\sigma_v$	
A_1	1	1	1	z, z^2, x^2+y^2
A_2	1	1	-1	
E	2	-1	0	$(x, y), (zx, yz), (xy, x^2-y^2)$

In this table we have followed Table 6.6 and given the usual presentation of character tables. On the right-hand side of the table are shown functions which are a basis for a particular representation. Thus the z axis, chosen following the convention which locates it along the C_3 axis, transforms as A_1 and the x and y axes, together, are a basis for the E irreducible representation.

Problem 7.2 Use the theorems of Section 6.3 to derive Table 7.1.

Note: This is a relatively short problem but one that gives excellent practice in the use of the orthonormality theorems.

There is one particular point about the C_{3v} character table which has to be discussed in detail. This concerns the axis pair (x, y) which, as shown in Table 7.1, transforms as the doubly degenerate irreducible representation E. Because they must be perpendicular to the z axis, the x and y axes lie in a plane perpendicular to the C_3 axis. But where in this plane do they lie? This problem is similar to one which was discussed in Chapter 6 where, in the C_{4v} point group, it was found that a variety of directions could be chosen for the x and y axes. So too, in the present problem there is no unique choice for the x and y axis directions. However, the present problem is more difficult than that encountered in the C_{4v} case and so it will be looked at in some detail. Suppose the x axis is chosen so that it lies in one of the σ_v mirror planes as is shown in Figure 7.2, which gives a view looking down the threefold axis.

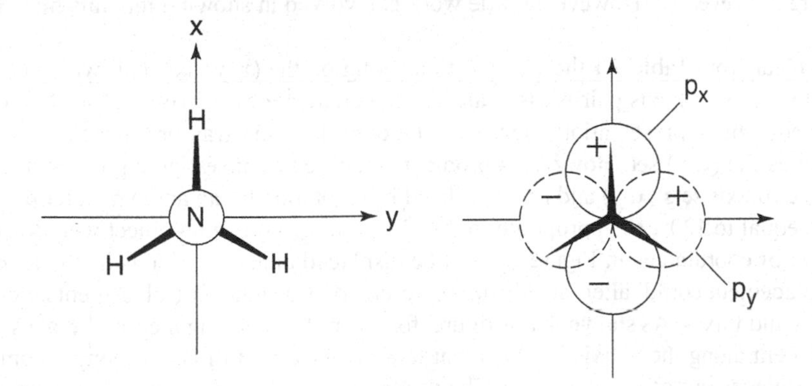

Figure 7.2 The choice of direction of x and y axes discussed in the text and consequent orientation of the p_x and p_y orbitals. The p_y orbital is shown dashed

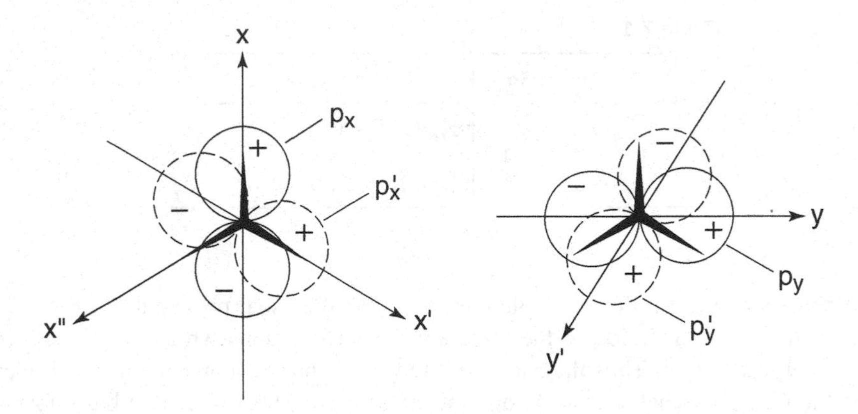

Figure 7.3 The x and y axes of Figure 7.2 together with an alternative set (x' and y') produced by a C_3 rotation of x and y. In both cases the corresponding p orbitals are also shown. The x'' axis will be referred to in the text

Two, related, problems at once arise. First, there is no evident reason why a particular mirror plane should be selected rather than one of the others. Second, the choice which has been made for the x axis means that the y axis is forced to be quite differently orientated in space. However, having made a choice we will stay with it and move on to the next problem, that of the effect of a C_3^+ rotation operation on these x and y axes, shown in Figure 7.3. It is seen from this figure that the x axis is rotated so that it lies along one of the directions which could have originally have been taken as the x axis but was not. Similarly, the y axis is rotated into a direction appropriate to this second choice of x axis. In Figure 7.3 the alternative x and y axes are indicated by primes (so that x is rotated into x' and y into y'). This is a quite new situation. So far in this book symmetry operations have turned objects into themselves or interchanged them. Here, a symmetry operation has generated something which did not previously exist, or so it seems. Well, the truth is that the x' and y' axes did previously exist – it is just that they were not revealed. However, a little work is involved in showing that this must be the case.

As is clear from Table 7.1 the C_3^+ rotation acting on the (x, y) axis pair which converts them into the (x', y') axis pair is associated with a character of -1 (this is the character of the E irreducible representation under C_3 rotations). In some way or other the (x', y') set is -1 times the (x, y) set. How? This problem is tackled by investigating the relationship between two axis sets (x, y) and (x', y') related by a rotation by an angle α (later α will be taken as equal to 120°, as appropriate to the C_{3v} point group). If an object were to start at the origin of coordinates in Figure 7.4 and be displaced along the x' axis it is evident that this displacement could, alternatively, be represented as a sum of displacements along the original x and y axes. As shown in this figure, for an angle α relating the x and x' axes, a unit displacement along the x' axis is equivalent to a displacement of $\cos \alpha$ along x combined with a displacement of $\sin \alpha$ along y. The rotated x axis, x', is a mixture of the original x and y, as too is the rotated y, y'.

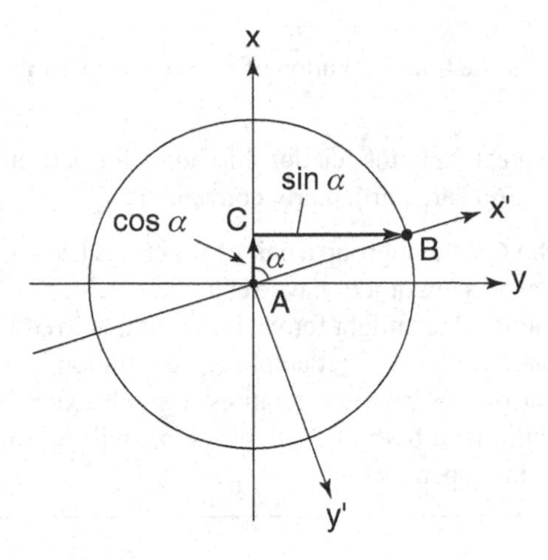

Figure 7.4 In this figure the circle is taken to be of unit radius. It follows that the unit displacement AB is the sum of the displacements AC and CB which, respectively, have magnitudes of cos α and sin α

When determining the contribution to a character made by the transformation of something such as an x axis, so far in this book we have asked the question 'is the x axis turned into itself, into minus itself or into something different', and we have associated the characters of 1, -1 and 0 with these three situations. We have now encountered a situation in which the x axis is rotated into an axis which may be described as in part containing the original x axis. Accordingly, our question must be modified to the simpler, but more general, form, 'to what extent is the old axis contained in the new?'. As is evident from Figure 7.4, and the discussion above, the numerical answer to this question is cos α, where α is the angle of rotation. An axis which is left unchanged by a rotation corresponds to $\alpha = 0$, so $\cos \alpha = 1$, the character that we have associated with this situation. Similarly, for a rotation of 180°, $\cos \alpha = -1$, which again is the character that the rotation of a coordinate axis by 180° gives. For $\alpha = 90°$ (when the x axis is rotated so that it becomes the y axis) $\cos \alpha = 0$, again the expected answer. The general rule (which applies to all C_n rotations, both those with n even and those with n odd) is clear:

> When an axis is rotated by an angle α by a symmetry operation its contribution to the character for that operation is cos α.

Comment: This statement holds for axes; for products of axes it has to be modified. Thus, because x contributes cos α it follows that x^2 contributes $\cos^2\alpha$ and that x^3 contributes $\cos^3\alpha$, and so on. Note that this rule applies to products of axes which are perpendicular to the axis of rotation. Thus, if the rotation axis is the z axis then the function xz will vary as cos α because only the x axis is perpendicular to the rotation axis – the z axis is left unchanged. However, the function xy will vary as $\cos^2\alpha$ because both x and y separately vary as cos α.

Problem 7.3 Show that the transformation of x^2 under a rotation of α about the z axis is given by the factor $\cos^2\alpha$.

Hint: It is sufficient to check that this relationship holds for particular values of α; $\alpha = 0°, 90°, 180°, 270°$ and $360°$ are particularly convenient.

Problem 7.4 (a) In the C_{5v} point group a pair of functions transforming as the doubly degenerate irreducible representation E_1 have a character of 2cos 72° under a C_5 rotation. Suggest a pair of functions which might form a basis for this irreducible representation.
(b) Repeat this problem for the E_2 irreducible representation, for which the character is 2 cos 144°. Solution of this problem requires a small extension of the argument developed above. (Solutions to both of these problems will be found in the character table for the C_{5v} group in Appendix 3)

Returning to the case of the C_{3v} point group, it is concluded that the x and y axes each make a contribution of $\cos 120° = -\frac{1}{2}$ to the character under the C_3, 120°, rotation operations. The sum of these two, −1, is indeed the character of the E irreducible representation under this operation. It was in Section 3.2 that it was first mentioned that two quantities, such as axes or orbitals, can be mixed by the operations of a group. We are now able to understand just what this means. The effect of a C_3 rotation on the original x and y axes is to rotate them to give new axes, each of which is a mixture of the original axes. In such cases the contribution that each axis makes to the character is always fractional. Everything that has been said about the x and y axes also holds for the $2p_x$ and $2p_y$ orbitals of the nitrogen atom in ammonia because the transformation of $2p_x$(N) is isomorphous to that of x, as is that of $2p_y$(N) to y (Figure 7.3). This parallel has already been anticipated by taking the molecular x and y axes to pass through the nitrogen atom – although they could be chosen to pass through any point along the C_3 axis – so that the above argument could be used as a basis for a discussion of the bonding in the ammonia molecule without the need to redefine axes.

7.2 Nodal patterns of the irreducible representations of C_{3v}

In previous chapters of this book we have given pictures of the nodal pattern associated with each irreducible representation. They seemed to be systematic. For the C_{2v} group, a group of order 4, there were patterns with 0, 1 or 2 nodes, a total of 4 different. For the D_{2h}, a group of order 8, patterns with 0, 1, 2 or 3 nodal planes were met, with a total of 8 different. Finally, for C_{4v}, also of order 8, the patterns had 0, 1, 2 or 4 nodal planes (none had 3), and a total of 6. The lower total arose because the doubly degenerate irreducible representation only gave two patterns (remember, in counting up to the order of the group the characters in the identity column have to be squared, so that the contribution of a doubly degenerate irreducible representation is 4). What of the C_{3v} group? It is of order 6, with two singly degenerate and one doubly degenerate irreducible representations. We expect one nodal pattern for each of the former and two for the latter. The patterns are shown in Figure 7.5, where the first diagram gives the interrelationship of the six symmetry operations viewed down the threefold axis.

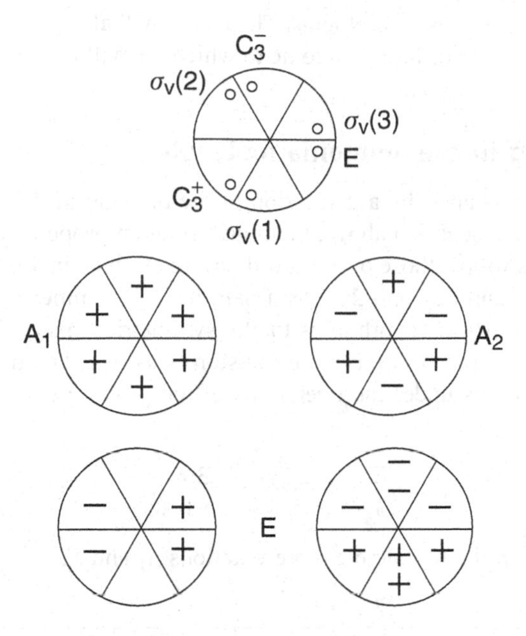

Figure 7.5 Nodal patterns of the irreducible representations of the C_{3v} group. Note the way that the accommodation of a nodal plane in one component of the *E* irreducible representation is compensated for by its partner

As we have come to expect, the pattern of the totally symmetric, A_1, is nodeless. Those of the *E* have one node each; we will look at these *E* pictures in more detail shortly. The last pattern, the A_2, has three nodes. There is a lesson to be learnt from the sequence C_{2v} (2 nodes maximum), C_{3v} (3 nodes maximum) and C_{4v} (4 nodes maximum); compare the point group symbol with the maximum number of nodes. But back to those E pictures. Take the first, *E*(1). Here, the nodal plane cuts through one of the six segments into which the circle is divided. There has to be a change of phase across the nodal plane, but each segment has to have its own, unique, phase. The only way out of this dilemma is to give the segment no phase at all. In contrast, the other *E* function, *E*(2), has a double amplitude in those segments left blank in *E*(1). No doubt all this seems somewhat strange to the reader, but hopefully not for long. In the next section we will use the projection operator method to obtain linear combinations of the hydrogen 1s orbitals in ammonia. There we will obtain combinations resembling those that we have just seen. Actually, the *E* combinations highlight an aspect which can prove to be a help in difficult problems: that all equivalent objects make equivalent contributions to all irreducible representations. In the present context, the objects can be taken to be the segments. So, for the A_1 and A_2 nodal patterns in Figure 7.5 each quadrant is represented with equal weight (although the sign may be either + or −). For the *E*'s the same is true *if they are treated as a pair*; that which one lacks is compensated by the other. Together, they have a weight of two. This points to a general property of reducible representations: for a triply degenerate set (and we will meet them in the next chapter) they, together, have a weight of three – but they can appear different, from one another. In fact, triply degenerate sets tend to be well behaved, simply because we live in a three-dimensional world. It is the others that present problems. The reader will almost certainly have met this with d orbitals; one, d_{z^2}, is 'different', a topic to which we will return in Chapter 8.

7.3 The bonding in the ammonia molecule

We now complete this chapter by a discussion of the bonding in the ammonia molecule. As is evident from the discussion above, the transformation properties of the nitrogen valence shell 2p orbitals follow those of the coordinate axes given in Table 7.1. The nitrogen $2p_z$ has A_1 symmetry and $2p_x$ and $2p_y$, as a pair, have E symmetry; it is a trivial exercise to show that the nitrogen 2s orbital is totally symmetric (this orbital is spherical and lies on all symmetry elements; it therefore transforms as A_1). The transformation of the three hydrogen 1s orbitals under the operations of the group gives rise to the reducible representation

E	$2C_3$	$3\sigma_v$
3	0	1

which is a linear sum of the irreducible representations A_1 and E.

Problem 7.5 Reduce the above reducible representation into its irreducible components.

Hint: An explicit solution to a similar problem has been given in the previous chapter.

Much of the discussion so far in this chapter has developed from the fact that the operation of rotation by 120° has the effect of mixing functions which provide a basis for the E irreducible representation. This same problem reappears again when we try to determine the symmetry-adapted combinations of hydrogen 1s orbitals in the ammonia molecule which

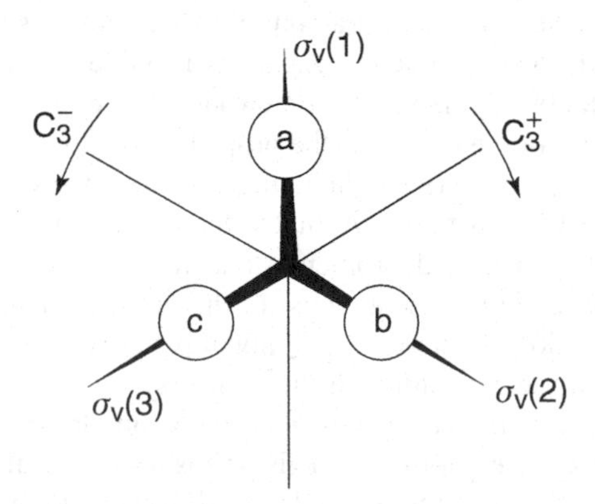

Figure 7.6 The labels used in the text for the hydrogen 1s orbitals of the ammonia molecule. For convenience, the operations are also shown

Table 7.2

Operation	E	C_3^+	C_3^-	$\sigma_v(1)$	$\sigma_v(2)$	$\sigma_v(3)$
a is turned into	a	b	c	a	c	b
Characters of the E irreducible representation	2	−1	−1	0	0	0
Multiply	2a	−b	−c	0	0	0
Sum	2a − b − c					
Normalize	$\frac{1}{\sqrt{6}}(2a - b - c)$					

transform as the E irreducible representation, a problem which will now be considered in detail. Labelling the hydrogen 1s orbitals as indicated in Figure 7.6 and considering the transformation of the orbital labelled a, the six symmetry operations of the group are found to lead to the following transformations, where the σ_v mirror planes are labelled as in Figure 7.1:

E	C_3^+	C_3^-	$\sigma_v(1)$	$\sigma_v(2)$	$\sigma_v(3)$
a	b	c	a	c	b

Application of the projection operator technique described in Section 5.6 shows the A_1 function to be:

$$\frac{1}{\sqrt{3}}(a + b + c)$$

There is no difficulty in obtaining one of the E functions. The steps involved are shown in Table 7.2 and lead to the function

$$\frac{1}{\sqrt{6}}(2a - b - c)$$

A problem arises when we try to obtain the second E function. A similar problem was met in Section 6.6 when discussing the four orbitals of the coplanar fluorine atoms in BrF_5. In that case the problem was relatively simple because the first E function contained contributions from only two of the σ orbitals; the projection operator technique applied to one of the other σ orbitals immediately gave the second E function. There is no such simple solution to the present problem; all three hydrogen 1s orbitals appear in the E function that has been generated, albeit with unequal weights. Following the procedure described in Chapter 5 the transformations of either the hydrogen 1s orbital b or c could be used as a basis for the projection operation method – but which? If b is used then the combination

$$\frac{1}{\sqrt{6}}(2b - c - a)$$

is obtained, whilst if c is used the function

$$\frac{1}{\sqrt{6}}(2c - a - b)$$

is obtained.

Problem 7.6 Show, by constructing tables analogous to Table 7.2, that the transformations of the hydrogen 1s orbitals b and c lead to the E functions

$$\frac{1}{\sqrt{6}}(2b - c - a) \quad \text{and} \quad \frac{1}{\sqrt{6}}(2c - a - b) \text{ respectively}$$

We have, apparently, obtained three quite different functions transforming as E – yet we know that, for a doubly degenerate irreducible representation, there can only be two. As indicated above, this problem is closely related to the three possible choices for the x axis that were discussed earlier and the solution to the problem is also similar. What we have, in fact, done in using a, b and c separately is to have generated E functions appropriate to the x, x' and x'' axes, respectively, of Figure 7.3. They are shown in Figure 7.7a. The functions corresponding to the x' and x'' axes, like these axes themselves, are mixtures of the functions appropriate to the original x and y axes. It is the latter pair that we are seeking. The first member of the pair we have pure, but the second we have only as part of a mixture (or, rather, as part of two mixtures).

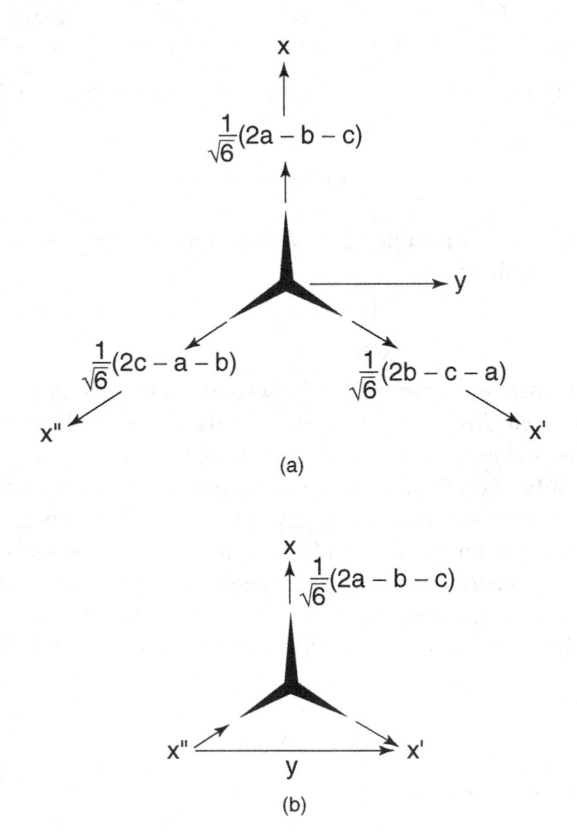

Figure 7.7 (a) Alternative symmetry-adapted combinations of hydrogen 1s orbitals in NH_3 corresponding to the axes x, x' and x'' of Figure 7.3. Just as one of these axes has to be selected so, too, does one of the three symmetry-adapted combinations. (b) The (vector) sum of displacements along $-x''$ and x' is a displacement along y

There are many ways of obtaining the second E function from the mixtures. Perhaps the simplest is to exploit the fact that if the first function we are seeking corresponds to the x axis, then the second corresponds to the y axis. This vector (axis-like) property of the functions is indicated by the arrows in Figure 7.7a. If, as shown in Figure 7.7b, the direction of the vector pointing in the direction x'' is reversed and added to that pointing in the x' direction a vector pointing in the y direction is obtained. These steps now have to be repeated using functions rather than vectors.

The negative of the function associated with x'' is

$$-\frac{1}{\sqrt{6}}(2\mathrm{c} - \mathrm{a} - \mathrm{b})$$

and adding it to the function associated with x'

$$\frac{1}{\sqrt{6}}(2\mathrm{b} - \mathrm{c} - \mathrm{a})$$

gives

$$\frac{1}{\sqrt{6}}(-2\mathrm{c} + \mathrm{a} + \mathrm{b} + 2\mathrm{b} - \mathrm{c} - \mathrm{a}) = \frac{1}{\sqrt{6}}(3\mathrm{b} - 3\mathrm{c})$$

That is, the second E function is of the form

$$(\mathrm{b} - \mathrm{c})$$

or, normalized,

$$\frac{1}{\sqrt{2}}(\mathrm{b} - \mathrm{c})$$

Problem 7.7 The sum of vectors pointing along x' and x'' of Figure 7.7a is the negative of a vector pointing along x. Show that an analogous statement is true for the corresponding E functions.

Problem 7.8 The fact that the two E functions which have just been obtained have quite different mathematical forms tends to be received with suspicion. Show that their forms are such that the orbitals a, b and c make equal total contributions to the E functions.

Hint: Sum the squares of coefficients in the normalized E functions.

Problem 7.9 Compare the two E functions just obtained with the E nodal patterns of Figure 7.5. Since they describe the same irreducible representation, they should be compatible. Are they? Could the nodal patterns of Figure 7.5 be used to generate the E functions above? If so, how?

The symmetry-adapted combinations of hydrogen 1s orbitals which have just been generated are shown in Figure 7.8 together with the nitrogen orbitals of the same symmetry with which they interact. Note that in this figure the unequal contribution of a, b and c

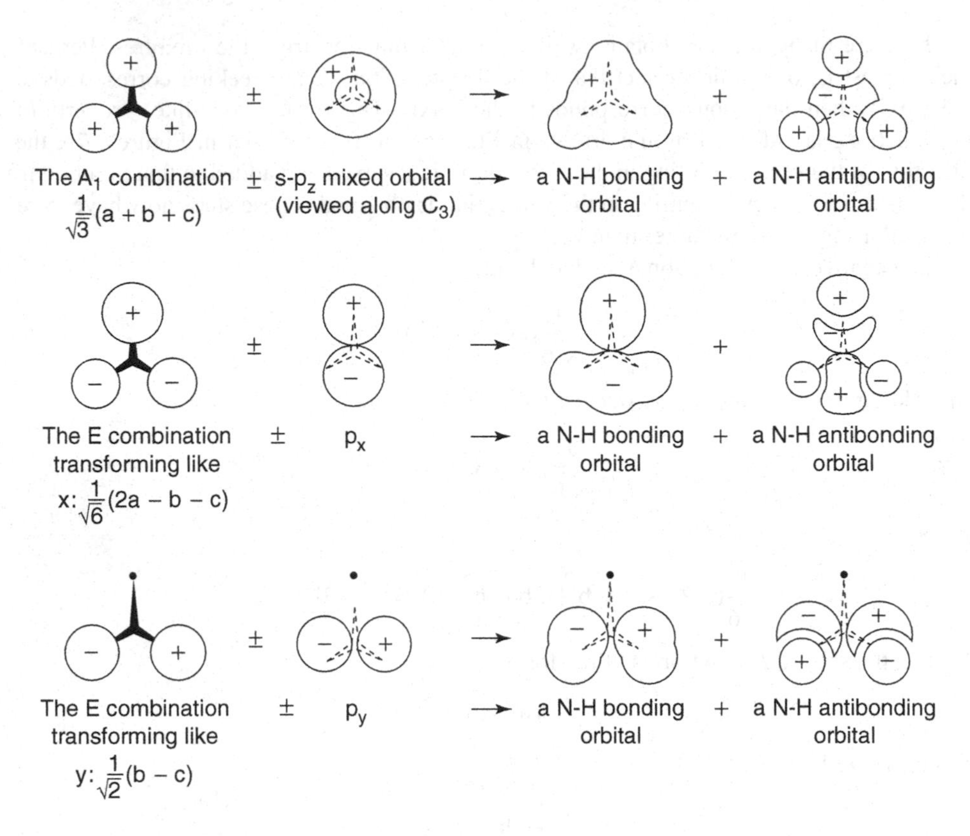

Figure 7.8 Schematic pictures of the bonding and antibonding molecular orbitals of A_1 and E symmetry in NH_3 and the way that they are derived from atomic and group orbitals

to each of the two symmetry adapted combinations of E symmetry is reflected in the diagrammatic representation of the orbitals. It will also be noted that in Figure 7.8 we have followed the approximate procedure of taking a combination of nitrogen 2s and $2p_z$ orbitals as the nitrogen orbital which interacts with the A_1 combination of hydrogen 1s orbitals. The resulting schematic molecular energy level diagram of ammonia is shown in Figure 7.9. There are eight valence electrons which have to be allocated to these orbitals (five from the nitrogen and one from each of the three hydrogens) and they are accommodated in the lowest molecular orbitals of A_1 and E symmetry, all of which are M—H bonding, and in the second A_1 orbital, which is essentially the nitrogen lone-pair orbital. These qualitative conclusions are to be compared with the results of detailed calculations and with the results of photoelectron spectroscopy.

Calculations[2] show that the A_1 orbitals of the ammonia molecule have energies of ca. −11.6 and ca. −31.3 eV and that of E symmetry ca. −17.1 eV. Of these, the more stable of the A_1 orbitals has, as experience leads us to expect, a major contribution from the nitrogen 2s atomic orbital. These data are in general agreement with the photoelectron spectroscopic results[3] which give energies of ca. 10.2 eV, 27.0 and 15.0 eV for these levels, respectively –

[2] C.D. Ritchie and H.F. King, *J. Chem. Phys.* **47** (1967) 564.
[3] A.W. Potts and W.C. Price, *Proc. R. Soc.* **326** (1972) 181.

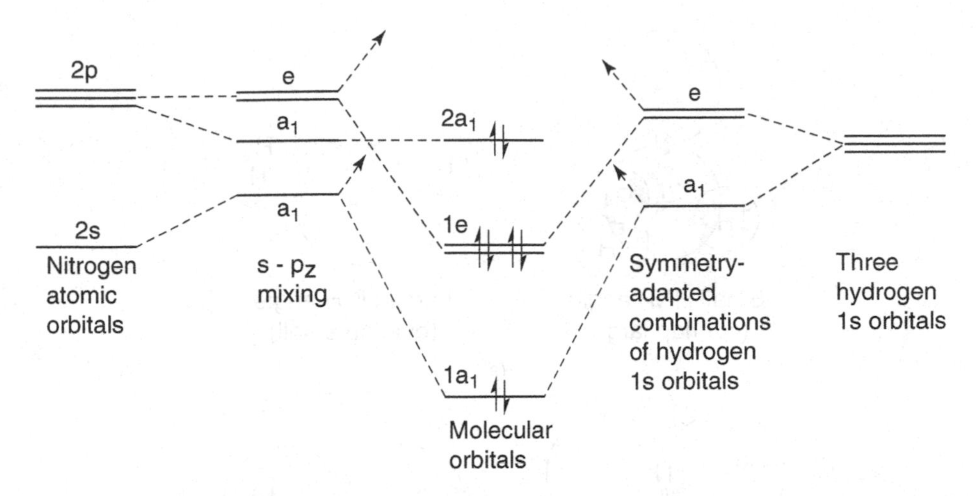

Figure 7.9 A schematic molecular orbital energy level diagram for NH_3

again, an encouraging result that a symmetry-based model is in good qualitative agreement both with detailed calculations and with experiment.

Ammonia is a molecule for which, like the water molecule, it is a simple matter to describe the angles at which the various contributions to the molecular bonding maximize. Using arguments entirely similar to those of Chapter 3 for the water molecule, it is concluded that the bonding interactions of A_1 symmetry involving the nitrogen $2p_z$ orbital (Figure 7.10a) maximize at small bond angles, whereas the interactions between orbitals of E symmetry maximize for the planar molecule (Figure 7.10b). These angular variations are conveniently summarized in a Walsh diagram, just as for the water molecule in Chapter 3. This diagram is given, qualitatively, in Figure 7.11.

Problem 7.10 Check that Figure 7.11 does, indeed, summarize the discussion of the above paragraph. What can be concluded about the non-bonding nature of the highest A_1 orbital from this diagram?

As indicated in Chapter 1, calculations show that the total bonding in the ammonia molecule is a maximum when the molecule is planar so it can be concluded that the E interactions dominate. However, this argument neglects the effects of repulsive forces on the molecular geometry and, as stated in Chapter 1, the same calculations show that it is these that – just – lead to the molecule adopting a pyramidal shape. At the observed bond angle there are both A_1 and E contributions to the bonding. Were we to remove an electron from the highest A_1 molecular orbital, which contains a large nitrogen $2p_z$ contribution, and which, despite our simplified discussion, makes a contribution to the molecular bonding, it would be reasonable to expect that a more nearly planar molecule would result. Experiment, indeed, indicates that in its ground state $NH_3{}^+$ is a planar molecule.

Although the bonding in planar NH_3 or $NH_3{}^+$ has not been discussed in this text, it is, nonetheless, of interest to consider a related planar species. This is the molecule

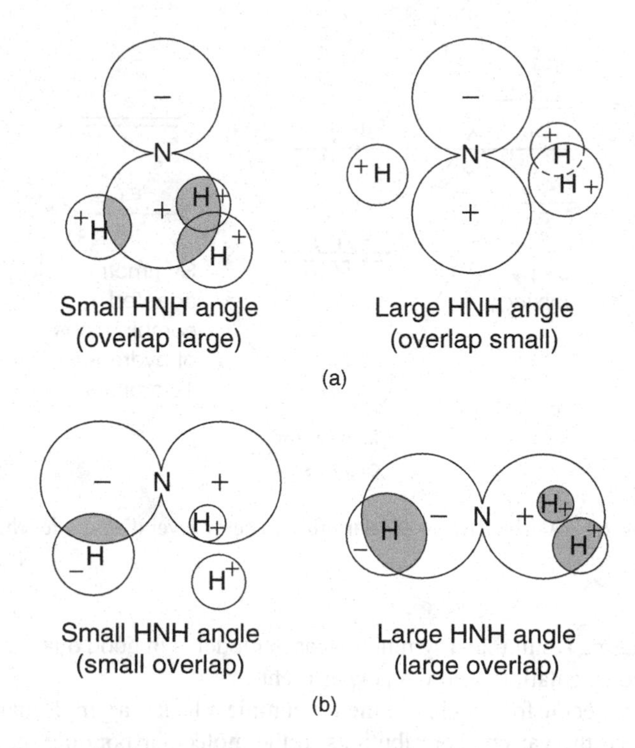

Figure 7.10 (a) The overlap between the A_1 symmetry-adapted hydrogen 1s combination and the nitrogen $2p_z$ orbital decreases as the HNH bond angle increases (this decrease is related to the fact that the hydrogen combination and the $2p_z$ orbitals have different symmetries in the planar molecule). (b) The overlap between an E symmetry-adapted hydrogen 1s combination and a nitrogen $2p_x$ orbital increases with the HNH bond angle

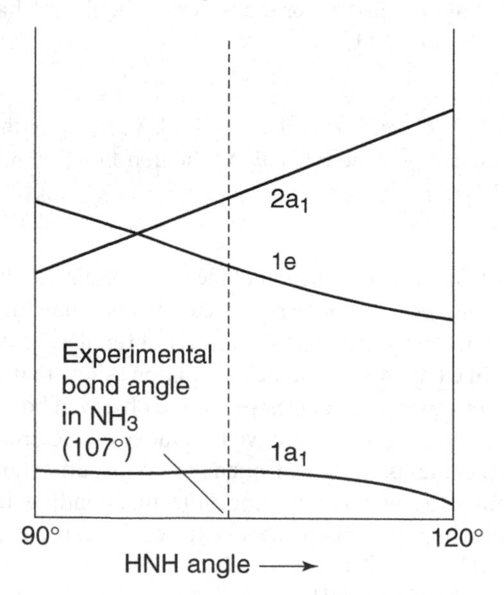

Figure 7.11 A Walsh diagram for NH_3

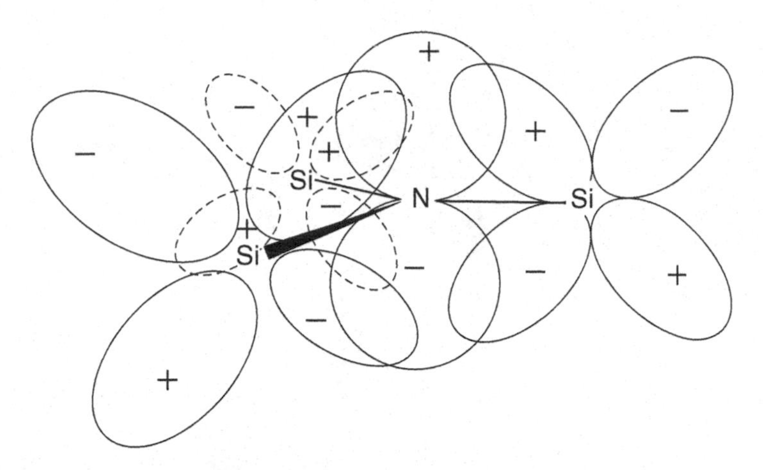

Figure 7.12 Postulated p_π/d_π bonding in the planar molecule trisilylamine, $N(SiH_3)_3$. In fact, the silicon $3d_\pi$ orbitals are much larger and diffuse than pictured here

trisilylamine, $(SiH_3)_3N$. In this molecule the Si_3–N framework is planar (unlike the C_3–N, skeleton in trimethylamine $(CH_3)_3N$, which is pyramidal like ammonia). The question of why trisilylamine should be planar has been widely discussed in the past and sometimes associated with Si–N π bonding involving the empty 3d orbitals of silicon accepting electrons from the lone pair on nitrogen (which in this geometry occupy a pure 2p orbital, Figure 7.12). Our discussion has indicated that the planarity of this molecule could arise if the delicate balance between bonding and repulsive forces found for ammonia – and which appears to occur for many such molecules – is such as to favour the planar form of trisilylamine. This argument, of course, does not require the existence of any Si–N π bonding. It could be, of course, that the presence of a small amount of π bonding is decisive in tipping a delicate balance. Equally, such an interaction might be important not for any – small – π bonding stabilization which results but because the resulting more diffuse electron distribution leads to a reduction in the destabilization resulting from electron repulsion. However, it is important to recognize that the observed planar geometry of trisilylamine does not of itself prove the existence of significant d–p π bonding in this molecule. Indeed, there are now (fairly complicated) organic species known which have a similar planar structure and for which it would be difficult to advance a π bonding argument.

7.4 Summary

In this chapter the problem of the transformation of functions has been discussed which form the basis for a degenerate reducible representation but which appear to be differently oriented with respect to the symmetry elements and may, indeed, have different mathematical forms (p. 183). Despite these superficial differences, the fact that they are mixed or interchanged by some operations of the group is sufficient to ensure their ultimate equivalence (p. 176). The differences are also found in the appropriate nodal patterns of the irreducible representations of the C_{3v} group (p. 179).

8 The electronic structures of some octahedral molecules

The methods developed so far in this book will be exploited to the full in the present chapter, where the electronic structure of the octahedral molecule SF_6 will be considered in detail and the results then extended to transition metal complexes. SF_6 is a quite large molecule but its symmetry is also considerable; enough to enable us to consider not only the bonding between sulphur and fluorine but also the non-bonding electrons on the fluorines. There are short-cuts which can be used in symmetry discussions and the present approach is such as to enable several of them to be introduced. Throughout this book new symmetry operations have been met in each chapter. This is also true for the present chapter; the operations will complete the types encountered in point groups and so a general review of point group classifications follows as the next chapter. This will prepare the way for Chapter 10, in which the relationships between point groups and spherical symmetry will be covered in some detail.

In Figure 8.1 is shown a cube, an octahedron and a tetrahedron. An octahedron is closely related to a cube. If the mid-points of faces of a cube are joined together the figure that is generated is an octahedron. The octahedron has eight faces but what is of more importance is the fact that it has six apices because when these apices are occupied by six atoms around an atom at the centre of the figure an octahedral molecule results (Figure 8.2). In the majority of octahedral ML_6 compounds the central atom is a metal ion whilst the surrounding atoms or ions are usually those of an electronegative element and are called ligands. Such species are referred to as 'octahedral complexes'. Although it is not convenient to start the discussion with such molecules they will be looked at in more detail later in the chapter.

A tetrahedron (Figure 8.1) is also derived from a cube, as was recognized in Chapter 1 (see Figure 1.4 and the discussion in Section 1.2.4). The fact that both the octahedron and tetrahedron are related to the cube means that it is possible to give a common discussion of the electronic structure of octahedral and tetrahedral transition metal complexes. In the present book we shall not embark on this discussion although the starting point will be indicated; the tetrahedral group will be considered in more detail in Chapter 10. First, however, a look at the symmetry of the octahedron.

Symmetry and Structure: Readable Group Theory for Chemists Sidney F. A. Kettle

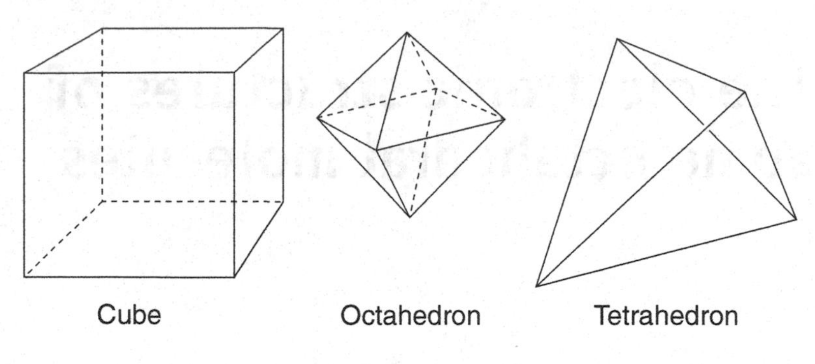

Figure 8.1

8.1 The symmetry operations of the octahedron

In Figure 8.3a are shown those pure rotational symmetry operations which turn an octahedral ML_6 molecule into itself (there are other operations, too, but it is convenient to start with the rotations). The octahedron contains three fourfold rotation axes and, of necessity, three coincident twofold rotation axes. There are also six twofold axes which are quite distinct from those that are coincident with the fourfold axes. Finally, there are four threefold rotation axes. Not surprisingly, Figure 8.3a shows a rather bewildering array of symmetry axes but there is a simple way of reducing the complexity. This is by associating symmetry elements with geometrical features. Thus, each C_3 axis passes through the mid-points of a pair of equilateral triangular faces on opposite sides of the octahedron. There are eight faces and so four pairs of opposite faces. It follows that there are four different C_3 axes. Similarly, the C_4 and coincident C_2 axes pass through opposite pairs of apices; there are six apices and so just three C_4's and C_2's. The other, C_2', axes pass through the mid-points of pairs of opposite edges. Because the octahedron has twelve edges there are six C_2' axes. The fact that the operations associated with each set of axes form separate classes is actually evident from the way that rotation axes are interchanged by other operations of the group (for instance, a C_3 operation interchanges the C_4 axes).

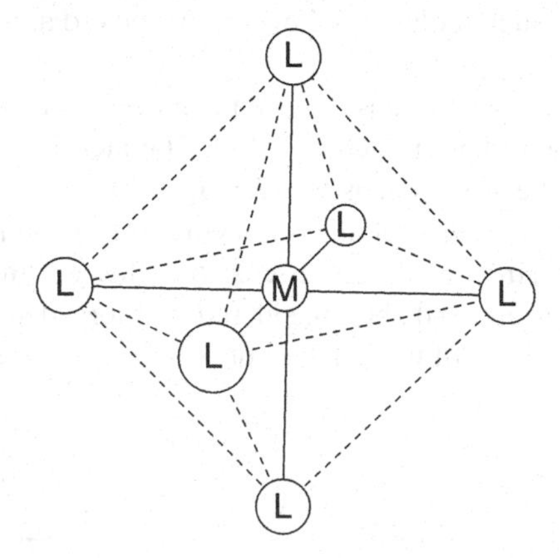

Figure 8.2 An octahedral molecule ML_6. The name derives from the fact that it has eight faces

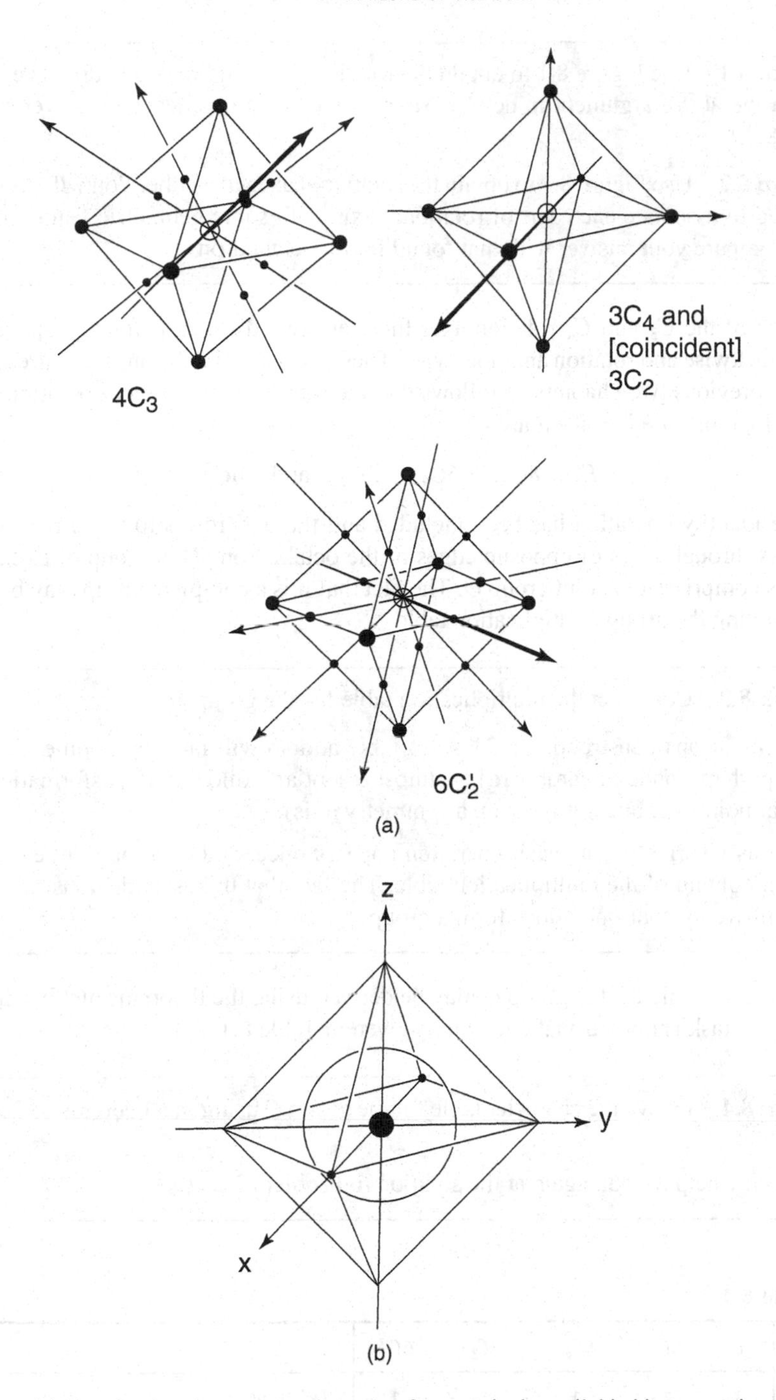

Figure 8.3 (a) The rotational symmetry elements of an octahedron, divided between three diagrams for clarity. (b) The conventional choice of coordinate axes for an octahedron

Problem 8.1 Use Figure 8.1 to obtain the rotational axes of an octahedron (i.e. work through the above argument in detail). Your answers can be checked by reference to Figure 8.3.

Problem 8.2 Use Figure 8.1 to obtain the rotational axes of a cube. You will probably first need to associate each sort of rotational axis with some geometric feature of the cube. Compare your answer with that found for an octahedron.

For each of the C_3 and C_4 rotation axes there are two distinct symmetry operations – rotation clockwise and rotation anticlockwise. These axes and operations have already been met in the previous two chapters. It follows that the rotational symmetry operations which turn an ML_6 molecule into itself are

$$E, \quad 8C_3, \quad 6C_4, \quad 3C_2 \quad \text{and} \quad 6C_2'$$

where the identity operation has been included and the $6C_2'$ refers to those twofold axes which pass through pairs of opposite edges of the octahedron. This group of twenty-four operations comprises the point group O. The fact that it is a complete group may be shown by constructing the group multiplication table.

Problem 8.3 Construct the multiplication table for the group O.

Note: This means constructing a 24 × 24 table and so will take some time. A good model (perhaps made of cardboard) is almost essential. Follow the transformations of a general point (i.e. one not lying on a symmetry axis).

Hint: As is invariably true, each operation appears once, and once only, in each row and each column of the multiplication table. The fact that this is so demonstrates that the set of twenty-four operations form a group.

The character table for the group O may be derived using the theorems met in Chapter 5 (although the task is not a trivial one) and is given in Table 8.1.

Problem 8.4 Derive the character table for the group O using the theorems of Section 6.3.

Hint: It may help to look again at the solution to Problem 7.2.

Table 8.1

O	E	$8C_3$	$6C_4$	$3C_2$	$6C_2'$	
A_1	1	1	1	1	1	$(x^2 + y^2 + z^2)$
A_2	1	1	−1	1	−1	
E	2	−1	0	2	0	$[1/\sqrt{3}\,(2z^2 - x^2 - y^2), (x^2 - y^2)]$
T_1	3	0	1	−1	−1	(x, y, z)
T_2	3	0	−1	−1	1	(xy, yz, zx)

There are several aspects of Table 8.1 which call for comment. For the first time triply degenerate irreducible representations are encountered; they are labelled T (with various suffixes). Their existence was implied earlier in this chapter when it was commented that 'octahedral molecules have x, y and z axes equivalent to each other'. Either these axes provide the basis for a reducible or an irreducible representation. In the event, it is irreducible and, as indicated by the basis functions given at the right-hand side of Table 8.1, they actually form a basis for the T_1 irreducible representation.

Problem 8.5 Show that the x, y and z axes, as a set, form a basis for the T_1 irreducible representation of the point group O.

Hint: Take the Cartesian axes to coincide with the C_4 axes (Figure 8.3b). For each class of operation select that individual operation which makes the transformation simplest to follow. The answer to this problem will be detailed – in an equivalent form – at the beginning of Section 8.2.

On the right-hand side of Table 8.1 are shown more basis functions than have previously been met. The reason is that the discussion of transition metal complexes later in this chapter will require a knowledge of how the d orbitals of the transition metal at the centre of the octahedron transform. Table 8.1 shows that the d orbitals d_{xy}, d_{yz} and d_{zx} are degenerate and transform as T_2 whilst d_{z^2} (or, more accurately, $d_{(1/\sqrt{3})(2z^2-x^2-y^2)}$) and $d_{(x^2-y^2)}$ are degenerate and transform as E. The function $x^2 + y^2 + z^2$ which, like an s orbital, has spherical symmetry transforms as A_1.

It is evident from Figures 8.1 and 8.2 that both an octahedron and a cube contain symmetry elements in addition to the rotations that have so far been listed. They possess a centre of symmetry, i, σ_h and σ_d mirror planes, and some rotation–reflection axes which are denoted S_n. They both contain S_4 and S_6 rotation axes. This type of element is not an easy one to fully appreciate and they will be looked at in detail shortly. All are shown in Figure 8.4. Of these, the i and σ_h (a mirror plane horizontal with respect to an axis of highest symmetry, here C_4) have been met in Chapter 5. The σ_d mirror plane is something new. Mirror planes that bisect the angle between a pair of twofold axes are called σ_d mirror planes, the suffix d being the first letter of the word dihedral (the same word which gives its initial letter to groups such as D_2, D_{2h} and D_{3h}, groups which have, respectively, two, two and three twofold axes perpendicular to the axis of highest symmetry). In the octahedron there are six σ_d mirror planes. Although they, indeed, bisect the angles between the C_2 axes it is easier to count them by noting that each σ_d mirror plane cuts opposite edges of the octahedron, just like the C_2' axes. There are six such pairs of edges and so six σ_d mirror planes.

Problem 8.6 Start with the definition of σ_d mirror planes as those that bisect the angle between pairs of C_2 axes and thus show that there are six σ_d mirror planes.

Hint: How many pairs of C_2 axes are there?

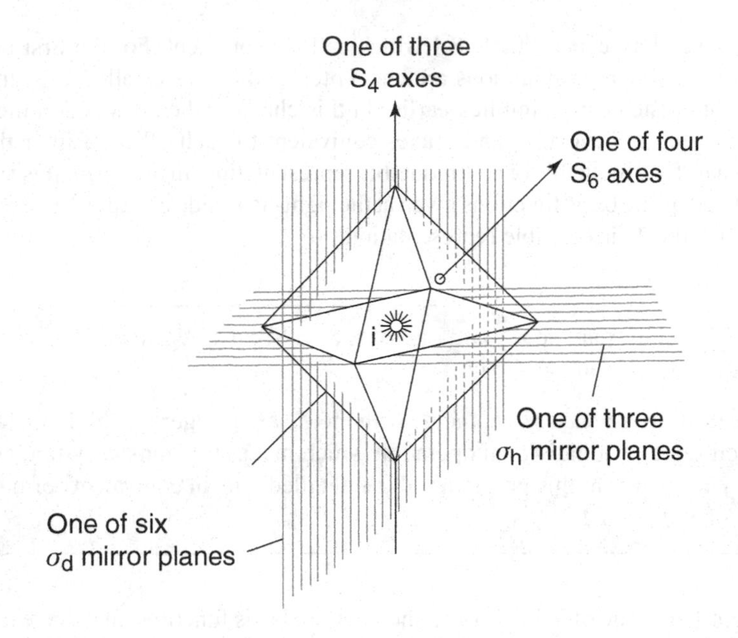

Figure 8.4 Some of the symmetry elements associated with improper rotation operations of an octahedron

Note that the mirror planes which have been labelled σ_h bisect the angles between pairs of C_2' axes. These mirror planes could have been labelled as σ_d'. However, convention dictates that the label σ_h takes precedence over σ_d whenever both are applicable.

Operations such as S_6 and S_4 are interesting because, as will be seen, they are two-part operations, conventionally taken as a rotation part and a reflection part. Hence they are called rotation–reflection operations. It has been seen that the cube and octahedron have the same rotational symmetry (Problem 8.2) and they have the same additional operations also. It follows that both have S_6 and S_4 axes. The S_4 axes are easier to see for the cube and are illustrated in Figure 8.5a. As this figure shows, the operation consists of a rotation by 90° (clockwise and anticlockwise rotations being associated with different S_4 operations) followed by reflection in a mirror plane perpendicular to the axis about which the 90° rotation was made. It is clear that this operation interconnects corners of the cube, but what is not so clear is that it is necessary-because the pairs of corners connected by the S_4 operations in Figure 8.5a are also connected by C_2 operations (the C_2 axes emerging through mid-points of the cube faces on the right- and left-hand sides of Figure 8.5a). The difference between the S_4 and C_2 operations is shown by the stars in Figure 8.5a. The star labelled 1 moves to the position occupied by star 2 under the S_4^- operation but these two points are not interconnected by a C_2 rotation. The S_6 operation (rotate by 60° and then reflect in a perpendicular mirror plane) is most readily seen for an octahedron standing on a face and is illustrated in detail in Figure 8.5b. In the case of the S_4 operations both the 90° rotation and reflection have an independent existence as C_4 and σ_h operations respectively. In the case of the S_6 operations the rotation and reflection do not exist in their own right as symmetry operations of the octahedron and cube.

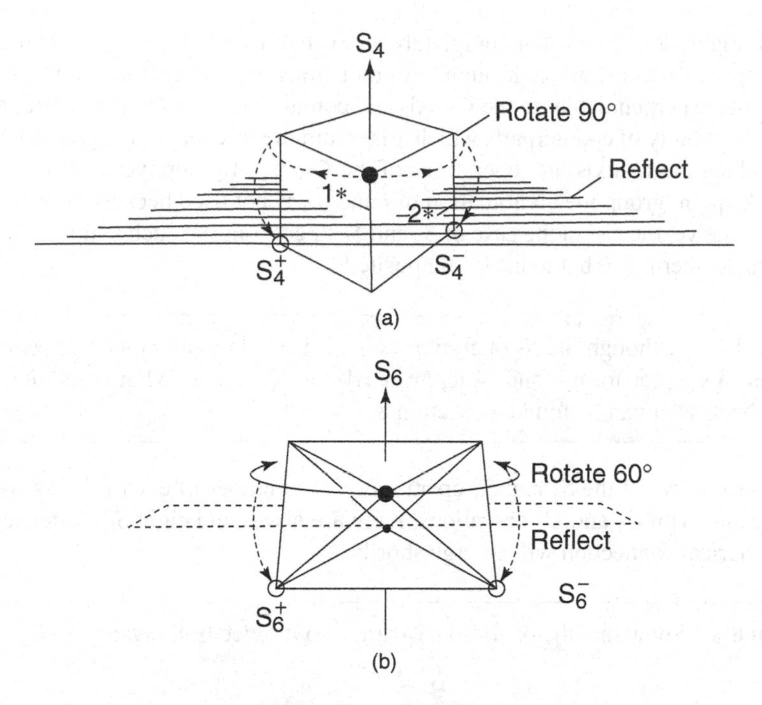

Figure 8.5 (a) An S_4 symmetry operation. (b) An S_6 symmetry operation

The S_n operations seem rather strange because each involves two operations, C_n and σ_h, which may or may not not have an independent existence. This apparently paradoxical situation may be made more acceptable by returning for the moment to the C_{2v} point group discussed in Chapter 2 and to a comment made several times in the following chapters (repeated because at first sight it can seem ridiculous). Figure 8.6 shows the water molecule and the C_2 operation which interrelates the two hydrogen atoms. Just like the σ_v reflection of Figure 4.18, the C_2 rotation can be expressed as the combination of two non-existent

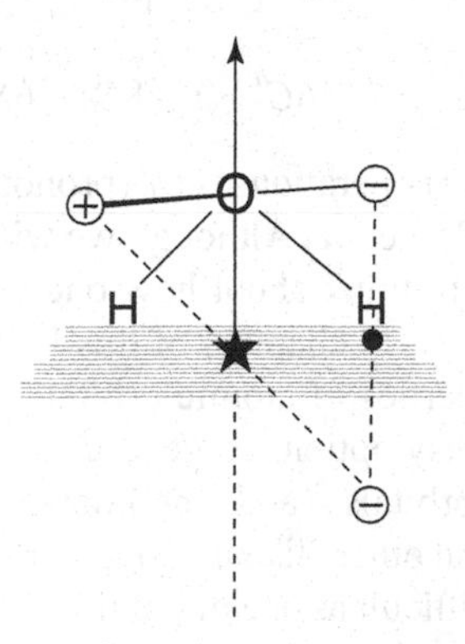

Figure 8.6 The C_2 rotation operation of the H_2O molecule (represented by a solid line connecting two C_2-related circles) is equivalent to an inversion at some point along this axis (indicated by a star) followed by reflection in a σ_h mirror plane containing this inversion centre (a sequence represented by a dashed line). The + and − signs indicate position relative to the plane of the paper.

operations. Figure 8.6 shows that completely equivalent to this single C_2 operation is the combined operation of inversion through *any* point along the C_2 axis followed by reflection in a mirror plane perpendicular to the C_2 axis and containing the inversion centre. Neither i nor σ_h (or the infinity of counterparts which arise from the freedom of the pair to be located anywhere along the C_2 axis) are operations of the C_{2v} point group, yet their combination is. In the C_{2v} point group the combination of i and σ_h is not used because there is a much simpler alternative, the C_2. In the case of S_4 and S_6 operations, no such simpler form exists and there is no alternative but to use a composite.[1]

Problem 8.7 Although the S_2 operation exists it is seldom mentioned as such. This is because a simpler form – and different label – exists for it. What is its alternative name? The answer can be found in Section 8.5.

It is necessary to count the S_4 and S_6 operations. The number of each follows from their correspondence with C_4 and C_3 operations; there are six S_4 and eight S_6. A deeper reason for this numerical connection will emerge shortly.

Problem 8.8 Show that the operation S_4 carried out twice is equivalent to C_2

$$S_4^2 = C_2$$

that S_6 carried out twice is equivalent to C_3

$$S_6^2 = C_3$$

and that S_6 carried out thrice is equivalent to i

$$S^3 = i$$

It is concluded that the complete list of symmetry operations of the octahedron (or cube) is:

$$E \quad 8C_3 \quad 6C_4 \quad 3C_2 \quad 6C_2' \quad i \quad 8S_6 \quad 6S_4 \quad 3\sigma_h \quad 6\sigma_d$$

The shorthand symbol for this set of operations is O_h (pronounced 'oh aiche'). The character table of the O_h group is given in Table 8.2. Although we will basically accept this character table as correct, it is interesting to think about how one might derive it. First, one might follow the procedure used in the early chapters of this book and attempt to derive it by a study of the transformation properties of atomic orbitals on an atom at the centre of the octahedron. Alas, this is not an easy option. To generate the A_{1u} irreducible representation we would need to invoke the m orbitals – and who knows what the m orbitals look like? [2] We will return to this question, and others like it – together with their answers – later, where we will find that they are not as difficult as one might think. Alternatively, one could hope to obtain the character table by generating a group multiplication table and, at least, substitute

[1] In fact, C_2, i and σ form a trio; any one can be expressed in terms of the other two, suitably positioned.

[2] To complete the picture, should anyone really want to know, the complete set of 19 m orbitals transform as $A_{1u} + A_{2u} + E_u + 3T_{1u} + 2T_{2u}$.

Table 8.2

O_h	E	$8C_3$	$6C_4$	$3C_2$	$6C'_2$	i	$8S_6$	$6S_4$	$3\sigma_h$	$6\sigma_d$		
A_{1g}	1	1	1	1	1	1	1	1	1	1		$x^2+y^2+z^2$
A_{2g}	1	1	−1	1	−1	1	1	−1	1	−1		
E_g	2	−1	0	2	−1	2	−1	0	2	−1		$\left[\frac{1}{\sqrt{3}}(2z^2-x^2-y^2), (x^2-y^2)\right]$
T_{1g}	3	0	1	−1	−1	3	0	1	−1	−1	(R_x, R_y, R_z)	
T_{2g}	3	0	−1	−1	1	3	0	−1	−1	1		(xy, yz, zx)
A_{1u}	1	1	1	1	1	−1	−1	−1	−1	−1		
A_{2u}	1	1	−1	1	−1	−1	−1	1	−1	1		
E_u	2	−1	0	2	−1	−2	1	0	−2	1		
T_{1u}	3	0	1	−1	−1	−3	0	−1	1	1	(T_x, T_y, T_z)	(x, y, z)
T_{2u}	3	0	−1	−1	1	−3	0	1	1	−1		

−1's to generate the singly degenerate irreducible representations (the others would require the substitution of matrices and these, themselves, would need to be derived).

A group multiplication table for the O_h group may be constructed with some considerable effort – and most will feel that the effort is not worth it (it is a 48 × 48 table). Finally, the character table may be derived using the methods of Section 5.3 exploiting the orthonormality theorems – but, fortunately, an easier method exists. This arises from the fact that the O_h group is the direct product of the groups O and C_i (the group containing E and i). Even this method means that we have to take the character table for O for granted. The concept of a group being the direct product of two other groups was met in Chapter 4, where the D_{2h} character table was seen to be the direct product of those of C_{2v} and C_i. In the same way, the character table for O_h, Table 8.2, is the direct product of Table 8.1 (the character table for O) and Table 5.4 (the character table for C_i). That this is so is evident from the way Table 8.2 is set out; it consists of four blocks containing the characters of Table 8.1 modulated by the signs of the four characters of Table 5.4 (which, for convenience, is repeated again as Table 8.3). In particular, the g and u suffixes contained in Table 8.3 reappear on the irreducible representation labels in Table 8.2.

Problem 8.9 Show that the labels used for the irreducible representations in Table 8.2 may be derived immediately from those of the character tables Table 8.1 and Table 8.3.

Direct product relationships, of course, apply both to operations and to characters; we met this for the D_{2h} group in Chapter 5. That is, they apply to the operations listed at the head of a character table as well as to the characters within the table itself. Because of the

Table 8.3

C_i	E	i
A_g	1	1
A_u	1	−1

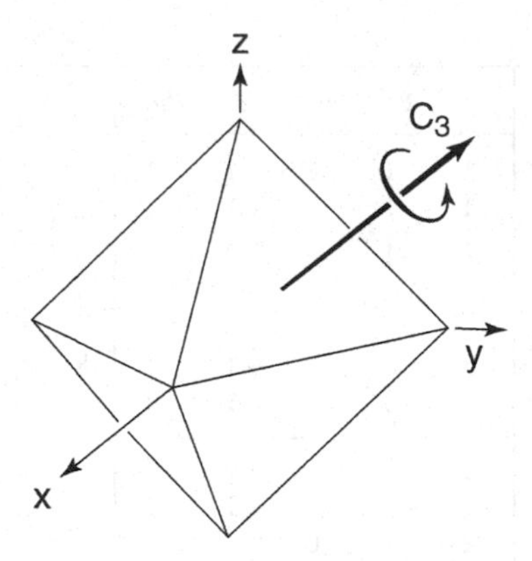

Figure 8.7 The choice of C_3 rotation operation (and consequent axis permutation) that will be used in this chapter

relationship between the groups O and O_h it is often possible to pretend that the symmetry of an octahedral molecule is O and then determine the g or u nature of the irreducible representations obtained by simply considering the effect of the inversion operation, i. Thus, a p orbital is ungerade – undergoes a change of phase – under the i operation. This, together with the knowledge that a set of p orbitals transform as T_1 in the group O (Table 8.1), is sufficient to establish that they transform as T_{1u} in O_h. Incidentally, but relevant to the set of p orbitals, when in this chapter the effects of a C_3 rotation operation are illustrated, the particular C_3 operation shown in Figure 8.7 will always be the one chosen. The effect of this choice is that coordinate axes, and thus labels, always permute as follows:

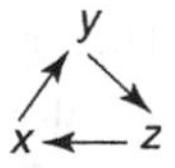

These permutations apply equally to products of axes. So, when Figure 8.31 is reached it will be seen that this C_3 operation turns $d_{x^2-y^2}$ into $d_{y^2-z^2}$ – which is just what the permutation gives.

The relationship between the groups O_h, O and C_i means that those operations possessed by O_h which are not present in O may be written in such a way that each is equivalent to some operation of O together with the operation i. The operations of O are *proper* (or *pure*) *rotation operations*, and the additional operations are *improper rotations*. As was seen at the end of Chapter 4, this latter name is used to denote any point group operation which is not a pure rotation. The precise correspondence between proper and improper rotations in the O_h group is evident from the way that Table 8.2 is written and is

E combined with i gives i
C_3 combined with i gives S_6
C_4 combined with i gives S_4
C_2 combined with i gives σ_h
C_2 combined with i gives σ_d

The existence of these relationships immediately explains why there is the same number of S_6 operations as C_3 and the same number of S_4 as C_4 – and so on – in O_h.

The above discussion indicates a different definition of S_n operations to that used earlier in this chapter. They may be defined as 'rotation–inversion' operations and, indeed, this is the way that they are described by crystallographers. This definition is 'Rotate by $(180 + \theta°)$, where $\theta = (360/n)°$, and follow by inversion in a centre of symmetry'. C_n [rotate by $(360/n)°$] and i may, or may not, exist in their own right as operations in a group containing S_n. The two definitions of S_n operations, rotate–reflect and rotate–invert, are entirely equivalent. A little thought will show that the duality exists because of the connection between i, C_2 and σ_h, detailed above for the C_{2v} point group; the connection is general.[3] But a word of warning. When a given C_3 (or C_4) operation is combined with either a σ_h reflection or i to give an S_6 (or S_4) operation, the alternatives do not give the same result. That is, the result of applying an S_6 (or S_4) operation depends on how the operation is defined. This would only pose a problem if the S_6 (or S_4) existed in isolation – there were no others. But it is one of a set and the *complete set* of operations always leads to the same *set* of results, no matter the definition. The 'complete set' in this context, of course, means all of the operations in a single class, which here is either $8S_6$ or $6S_4$.

Problem 8.10 Show that the operation S_4^- (defined as a rotation–reflection operation) is the same as the operation S_4^+ (defined as a rotation–inversion operation). Here, the superscripts denote the direction of 90° rotation (thus avoiding having to add on 180°).

Hint: Use Figure 8.5a.

8.2 Nodal patterns of the irreducible representations of the O_h group

This is the last time in this book that we will discuss in detail the nodal patterns which are associated with the irreducible representations of point groups (although they will reappear in Chapter 12, where some different groups are the subject). It is a section which is not necessary for an adequate understanding of the rest of the chapter and the reader can skip it if he or she wishes. However, it gives insights which are otherwise not available and so a study is likely to be worthwhile. But the reader should not take it too seriously. It will involve an appeal to orbitals they will never have heard of, and will probably never hear of again. The m orbital mentioned earlier is only the start; there are more to come. A question which the reader will ask of the author, regularly, is that of 'how do you know that?'. To which the answer is very simple – 'it is much easier than appears; all will be explained in Chapter 10'.

It is not surprising that the problem implicit in the heading to this section should be so difficult. We are looking for twenty nodal patterns (sum the characters listed under the E operation in Table 8.2), nodal patterns which are all different from each other and each of which is compatible with the forty-eight symmetry operations of the O_h group. Here, some

[3] So, the 180° in $(180 + \theta)°$ arises from the C_2.

Table 8.4

	Orbital									
	s	p	d	f	g	h	i	k	l	m
Value of l = number of planar nodes	0	1	2	3	4	5	6	7	8	9

strange orbitals will help. But first, something about orbitals in general. Consider s, p, d and f orbitals; the first three, at least, should be reasonably familiar to the reader. They have degeneracies of 1, 3, 5 and 7 respectively; degeneracies of $(2l + l)$, where $l = 0, 1, 2$ and 3 respectively. Here, l is a quantum number that may have been met with any one of several labels – 'orbital' is perhaps the most popular and best. Here, we will view it differently; l is the total number of planar nodes associated with each of the orbitals in the set to which it refers.[4] This pattern continues; so g orbitals have $l = 4$, are ninefold degenerate and have four nodal planes; h are 11-fold degenerate and have five nodal planes ($l = 5$) – and so on. All that could be of interest to us are listed in Table 8.4. The first four orbital labels have been met; their use originates in the history of the subject. The others are given in alphabetical order with some missing. Those missed have either already been used (like s and p, although their absence is not evident in Table 8.4) or have a significance somewhere else in quantum chemistry (like j).

In previous chapters we were able to give diagrammatic presentations of irreducible representations. This was possible because, with one exception (the D_{2h} group, Chapter 5), a view down a unique axis allowed the points interrelated by the symmetry operations to be pictured. The nodal patterns associated with each irreducible representation could then be shown in the same perspective; even D_{2h} was possible with a distorted perspective. Clearly, for a group like O_h, in which there is a plethora of symmetry elements, many at oblique angles to each other, such a simple presentation is not possible. Not that the problem is insoluble. So, in Figure 8.8 is given a diagram traditionally used by crystallographers as a solution. For us, it has several defects. Not least is that its use would force us to attempt to visualize three-dimensional arrays of nodal planes from a two-dimensional diagram.

More three-dimensional – and certainly more useful – is the arrangement of Figure 8.9. This figure shows a sphere with the corners of an octahedron as white balls on its surface. The nine mirror planes of the octahedron (three σ_h and six σ_d) are shown as black circles which project slightly from the surface of the sphere. The σ_h mirror planes are seen as joining the corners of the octahedron and the six σ_d intersect to give six-cornered stars above the middle of each face of the octahedron. This figure is useful because there is a total of forty-eight of the equilateral triangle-like areas in-between the mirror planes, which is the same as the number of operations in the O_h group. So, all of the complete nodal patterns associated with the irreducible representations of the group will assign a phase to each of these triangles. For the A_{2u} irreducible representation, for instance, adjacent triangular areas always have opposite phases. As met in the last chapter, some of the doubly degenerate irreducible representations have some faces left blank but such blanks are always compensated by the

[4] Sometimes, as with a d_{z^2} orbital, two planar nodes combine to give a conical nodal surface.

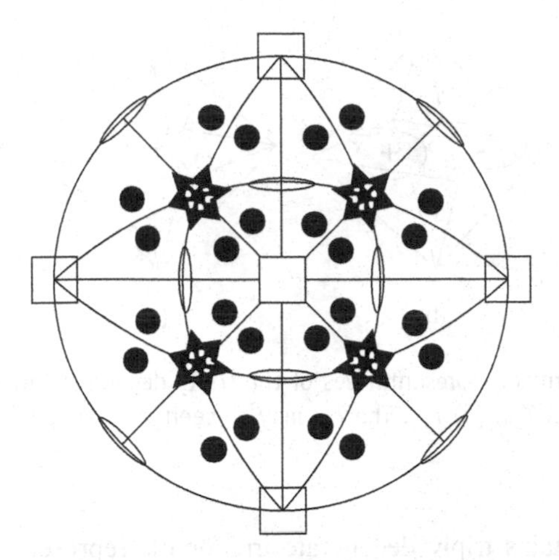

Figure 8.8 A traditional projection of symmetry elements of the O_h group. It is best thought of as similar to Figure 8.9 but viewed down a C_4 axis (white balls in Figure 8.9). The (curved) lines are mirror planes (also shown in Figure 8.9) and they divide the surface of the octahedron into 48 triangles. Only 24 of these can be seen from one side and these 24 are shown here. If Figure 8.9 is thought of as an orange, Figure 8.8 shows half the skin of the orange, flattened and stretched at the circumference to give a complete circle. The black dots are symmetry-related points; the squares represent fourfold axes. Twofold axes are shown as ellipses and threefold axes as six-cornered stars

other member of the pair, just as in Figure 7.5. Because in Figure 8.9 the interconnected points are projected onto the surface of a sphere, only half of them can be shown. The sphere could be made transparent but then it would be impossibly complicated. It is difficult to avoid the conclusion that a basic diagram which is both simple and comprehensive is somewhat elusive. This conclusion is reinforced by the recognition that there can be up to nine planar nodes to be added (as we have seen, the A_{1u} irreducible representation pattern has this number, each one coincident with a mirror plane)! But all is not lost. The fact

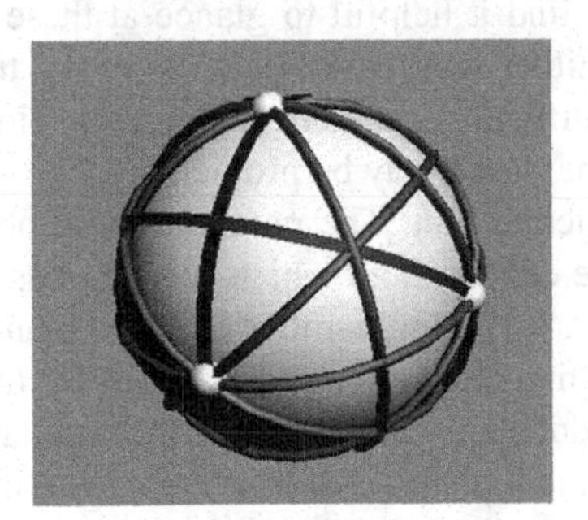

Figure 8.9 Whereas Figure 8.8 represented a projection of the octahedron onto a circular surface, Figure 8.9 represents a projection onto a spherical surface. The white balls are corners of the octahedron and the black lines mirror planes. In picturing the nodal patterns of the irreducible representations of O_h it is necessary to assign a phase to each triangular area shown in this figure. Although it is not obvious from the perspective adopted for this diagram (made worse by the fact that the mirror planes project from the surface), in principle 24 such triangles are visible

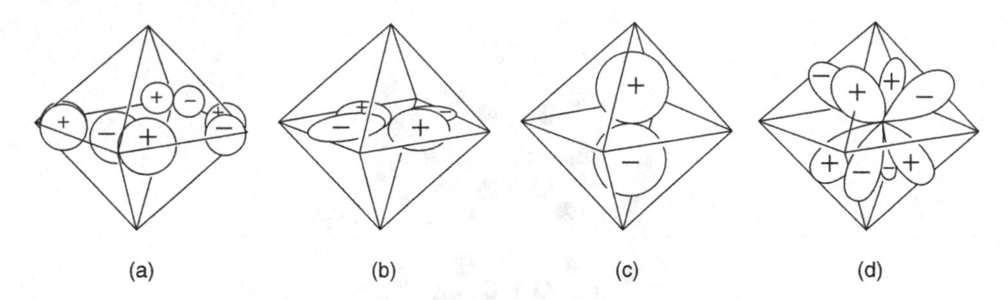

Figure 8.10 Nodal patterns of representatives of the triply degenerate irreducible representations of O_h: (a) T_{1g}; (b) T_{2g}; (c) T_{1u}; (d) T_{2u}. The T_{2u} may be seen most simply by adding an extra nodal plane to the T_{2g}

that the O_h group contains triply degenerate irreducible representations arises from the fact that the coordinate axes x, y and z are equivalent. For the triply degenerate irreducible representations there is always one function associated with x, an equivalent one with y and a third, also equivalent, associated with z. Any one will act as a representative for all three. In practice, this means that we can work with a single nodal pattern and this will be associated with one axis (although this can mean that it is perpendicular to it).

Simple diagrams are possible, although it may sometimes be convenient to change the viewpoint adopted. Such diagrams for the triply degenerate irreducible representations are given in Figure 8.10. Some of these diagrams contain more than an echo of atomic orbitals based on an atom at the centre of the octahedron. In terms of such orbitals which provide a basis, the relationships are the following, the number of nodal planes being given in parentheses: T_{1g}, g orbitals (4); T_{2g}, d orbitals (2); T_{1u}, p orbitals (1); T_{2u}, f orbitals (3). These are not difficult. Plots of p, d and even f orbitals are easy to find. The g orbital appears more of a problem but here life is made easy by the recognition that the three rotations of the octahedron, R_x, R_y and R_z, transform as T_{1g}. Picture a rotation and you have a picture of a T_{1g} nodal pattern. The T_{1g}, T_{2g}, T_{1u}, and T_{2u} nodal patterns reappear in Figures 8.22 to 8.26, and the reader may find it helpful to glance at these – they may seem different from those in Figure 8.10. Either way, nodal patterns of the triply degenerate irreducible representations are not too difficult to obtain. It is for the singly and doubly degenerate irreducible representations that there may be problems.

Consider the singly degenerate first. The potential for problems can be seen from the types of atomic orbital of the central atom which serve as bases for them. These are: A_{1g}, s(0); A_{2g}, i(6); A_{1u}, m(9); A_{2u}, f(3) – the number of nodal planes are given in parentheses. These are not as horrific as at first they might seem. Just as the total number of mirror planes in O_h $(3\sigma_h + 6\sigma_d)$ are the nodal planes of A_{1u}, so the three nodal planes of A_{2u} *are* the $3\sigma_h$.[5] For the A_{2g} the $6\sigma_d$ *are* the nodal planes. All four are shown in Figure 8.11, which adopts the perspective of Figure 8.9 but simplified as much as possible, with apologies for the A_{1u}, which really *is* difficult to draw! We are left with the two doubly degenerate irreducible representations, E_g and E_u. Here, we might expect problems. The reader may recall that in Chapter 7 the nodal patterns of the doubly degenerate irreducible representations of the C_{3v}

[5] In fact, this pattern has already been met in Figure 5.3, and repeated in Figure 8.11d in an octahedral environment.

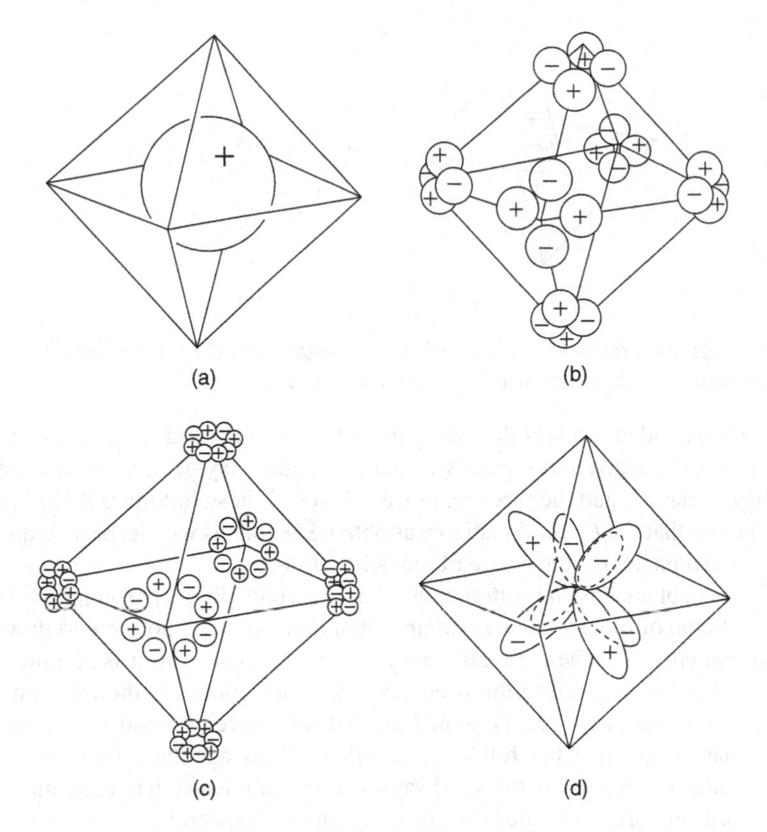

Figure 8.11 Nodal patterns of the singly degenerate irreducible representations of O_h: (a) A_{1g}; (b) A_{2g}; (c) A_{1u}; (d) A_{2u}

point group contained blank entries. These were a consequence of the problems associated with choice of the x and y axes when working with a C_3 axis. These axis-choice problems do not persist in O_h – but there are four C_3 axes, and these were the real source of the C_{3v} problem. Indeed, there are corresponding problems in O_h – but for the E_g irreducible representation there is a relatively simple solution, one that has something in common with that for C_{3v}. As for the other irreducible representations, we can find help in basis functions. The basis functions for the E_g irreducible representation are two of the d orbitals of the central atom, $d_{x^2-y^2}$ and d_{z^2} (but the latter is better written as $d_{2z^2-x^2-y^2}$); they are shown in Figure 8.12; later in the chapter they will be considered in much more detail. The reason is that these basis functions will play an important part in the crystal field theory developed there. What of the E_u irreducible representation? It is probably the most difficult. Easiest is to obtain it in stages. The functions of Figure 8.12 are centrosymmetric, g. We can use them to obtain corresponding u functions by adding a mirror plane (horizontal, in the xy plane), as shown in Figure 8.13. But these cannot be the final answers; in these diagrams the x and y axes lie in a nodal plane, that which has just been added, whereas the z axis most assuredly does not lie in a nodal plane. The molecule involved is not cubic! We can make it cubic by adding two more nodal planes, in the yz and zx planes, so that all three σ_h mirror

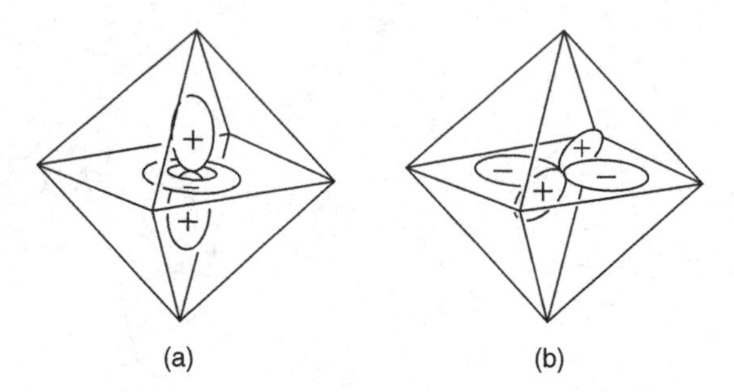

Figure 8.12 Nodal patterns of the E_g irreducible representation of O_h: (a) the function $2z^2 - x^2 - y^2$ (the d_{z^2} orbital); (b) the function $x^2 - y^2$ (the $d_{x^2-y^2}$ orbital)

planes have been added. The result is the pair of E_u functions that we have been seeking. Diagrams of them are given in Figure 8.14 but the reader may get a more understandable result if they, mentally, add the three σ_h mirror planes to those in Figure 8.12. Finally, we note that the way that the E_u combination was obtained explains why its basis is an h orbital (l = 5, five nodal planes) at the centre of the octahedron.

We have now obtained nodal patterns for all of the irreducible representations of the O_h point group. Some of them were very complicated and almost impossible to draw. Whilst these nodal patterns describe the reality, they also make clear why it is so much simpler to work with the labels given to the irreducible representations. In the label notation all irreducible representations are equal, something that can scarcely be said of the development above. One last word. In what follows, we will be drawing functions transforming as particular irreducible representations. Always – unless a mistake has been made – there must be a close similarity with the corresponding diagram(s) above.

8.3 The bonding in the SF_6 molecule

As is so often the case, the O_h point group is easier to use than to talk (or write!) about. To illustrate its use (or, more correctly, how its use may often be avoided), we now turn to a discussion of the bonding in the SF_6 molecule. The valence shell atomic orbitals of the sulphur atom will be taken as 3s and 3p, ignoring the 3d. The behaviour of d orbitals in octahedral molecules is of major importance in transition metal chemistry and this is the

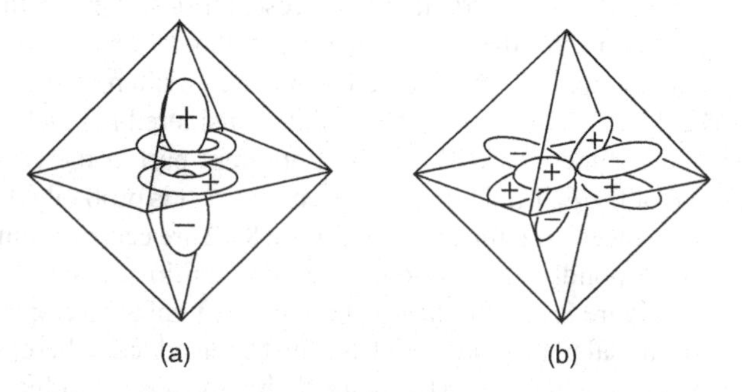

Figure 8.13 The first step in obtaining the nodal patterns of the E_u irreducible representation: (a) adding a node to Figure 8.12a; (b) adding a node to Figure 8.12b

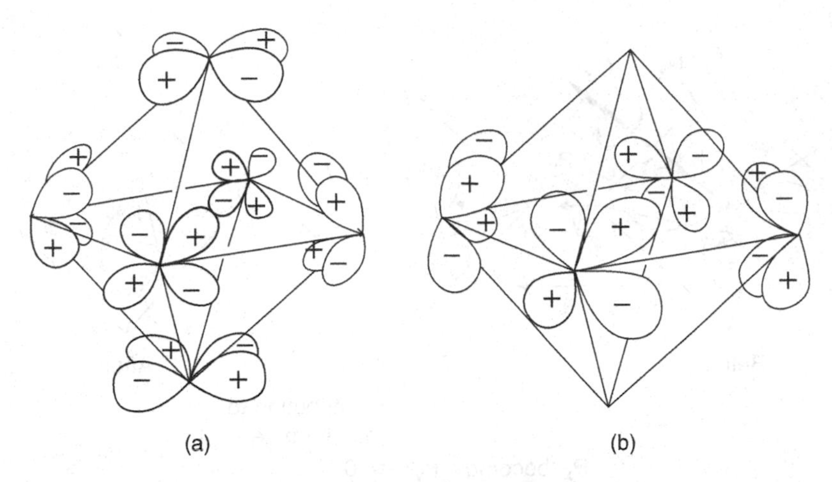

Figure 8.14 Nodal patterns of the E_u irreducible representations of O_h: (a) and (b) correspond to those labels in Figures 8.12 and 8.13; as an aid to clarity the actual functions shown here have been located at the corners of the octahedron – they can be thought of as d orbitals on atoms there

context in which they will be discussed later in this chapter. Throughout our discussion of SF_6 it will be convenient to work in the point group O rather than the correct group O_h. The reason for this lies in the structure of the O_h group. It is twice as large as O (48 operations compared with 24) and so is more cumbersome to handle. But, as has been seen, O_h is the direct product of O with C_i so that the only additional information that O_h has compared with O is that of behaviour under the additional operation introduced by the group C_i – that is, behaviour under the operation i. It is easier to ask of a basis function 'how does it transform under i' and to add either g (gerade = symmetric with respect to inversion in a centre of symmetry) or u (ungerade = antisymmetric respect to inversion in a centre of symmetry) as a suffix to the irreducible representation of O than to plough through the whole set of O_h operations. This pattern is easy when the function concerned is centred on the S atom, for it is therefore also centred on the centre of symmetry. When it is elsewhere, and in the present context that means on the F atoms, then it is neither inherently g or u (with respect to the centre of symmetry at the S atom) and will always be one of several (they go in multiples of six – there are six F atoms). In such a case, the several divide equally between g and u.

The following section will be made much easier if the reader can construct for themselves a suitable model, perhaps from expanded foam – and a cube may be easier to make than an octahedron. However, it has been written on the assumption that such an aid is not to hand and that, therefore, the high symmetry – and the numerous equivalent axes and alternative bases which it presents – may pose a problem. It therefore proceeds more slowly and with more detail than absolutely necessary. It is easy to show that the sulphur 3s orbital, shown in Figure 8.3b, transforms as the totally symmetric, A_1, irreducible representation of the point group O. In O_h, of course, it has A_{1g} symmetry. Figure 8.3a shows the axis system that will be used in the following discussion, although in many of the following figures the octahedron is drawn from a different viewpoint from that shown in Figure 8.3. Like the coordinate axes of Problem 8.5, the sulphur 3p orbitals transform together as the T_1 irreducible representation (T_{1u} in O_h). This particular problem will be looked at in some detail because it illustrates how to handle the sometimes bewildering task of working

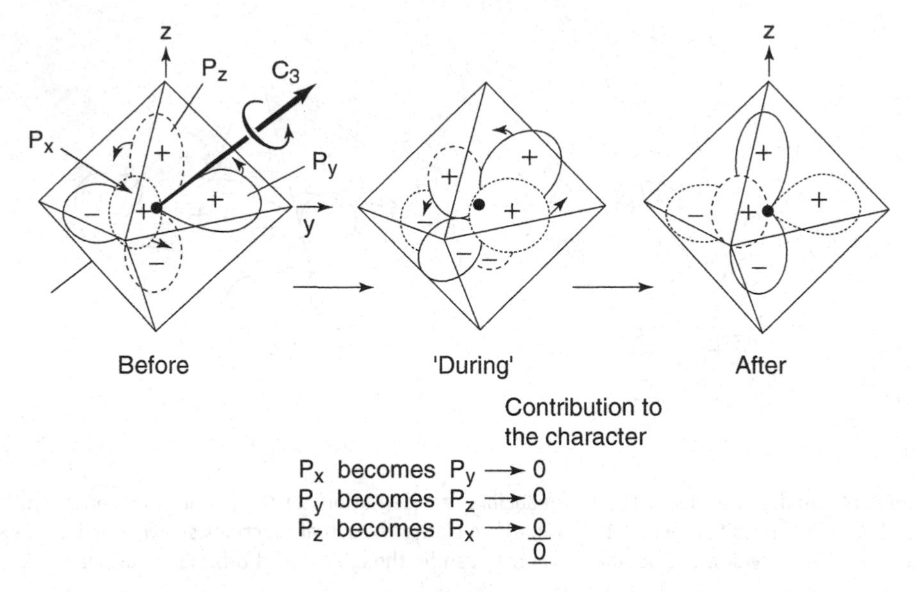

Figure 8.15 The transformation of a set of p orbitals of a central atom of an octahedral molecule under a C_3 rotation operation. The p orbitals are distinguished by solid, broken and dotted lines. Simplest is to concentrate on the behaviour of the positive lobes

with several equivalent objects in a high symmetry environment. Our discussion is largely diagrammatic because good diagrams – as much as a good model – are very important. It is therefore essential that each figure is studied carefully and the transformations that it shows are followed in detail. Figures 8.15–8.18 illustrate the transformations of the set of 3p orbitals under a representative operation of each class of the O point group (Figure 8.15 the C_3, Figure 8.16 the C_4, Figure 8.17 the C_2 and Figure 8.18 the C_2', this being the sequence in which they are listed in the character table). For clarity of presentation the lobes

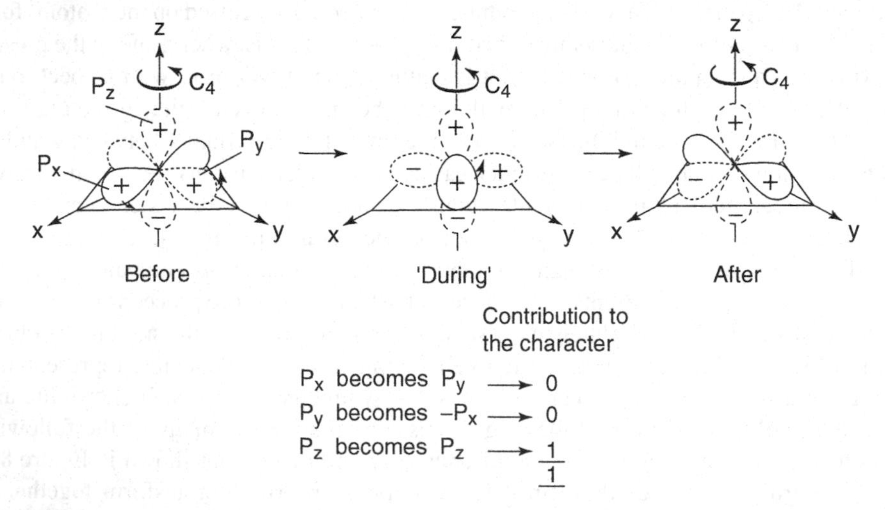

Figure 8.16 The transformation of a set of p orbitals of a central atom of an octahedral molecule under a C_4 rotation operation

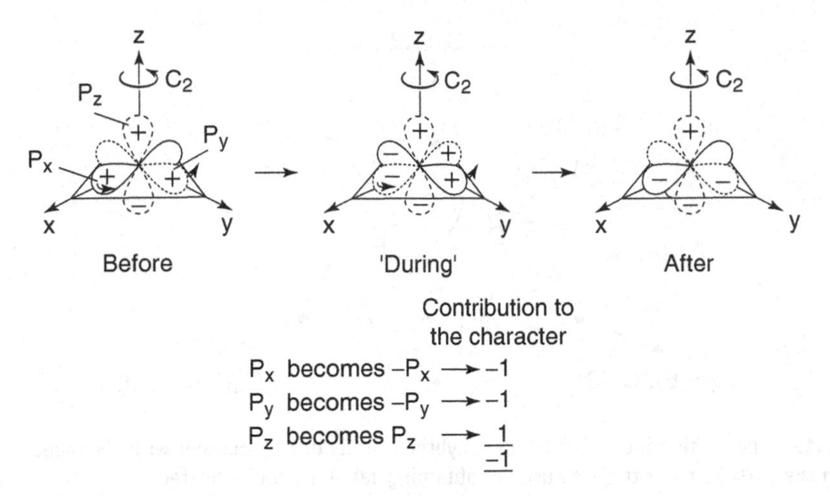

Figure 8.17 The transformation of a set of p orbitals of a central atom of an octahedral molecule under one of the $3C_2$ rotation operations

of the p orbitals are shown more as ellipses than the circles used so far in this book; the different p orbitals are distinguished by the way they are outlined. As an aid to visualizing the interconversions, these figures not only show the starting arrangement but also show them at some point whilst the operation is in progress as well as in the final arrangement. The actual transformations brought about by these operations are listed in the figures and the compilation of the corresponding characters is detailed. Figures 8.15–8.18 should be studied very carefully; the characters to which they give rise are:

E	$8C_3$	$6C_4$	$3C_2$	$6C_2'$
3	0	1	−1	−1

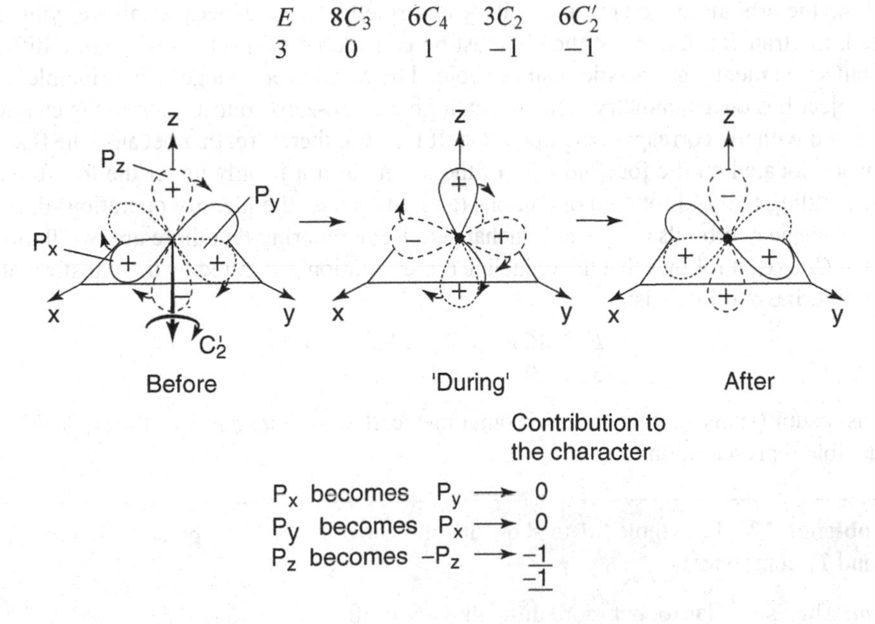

Figure 8.18 The transformation of a set of p orbitals of a central atom of an octahedral molecule under one of the $6C_2'$ rotation operations

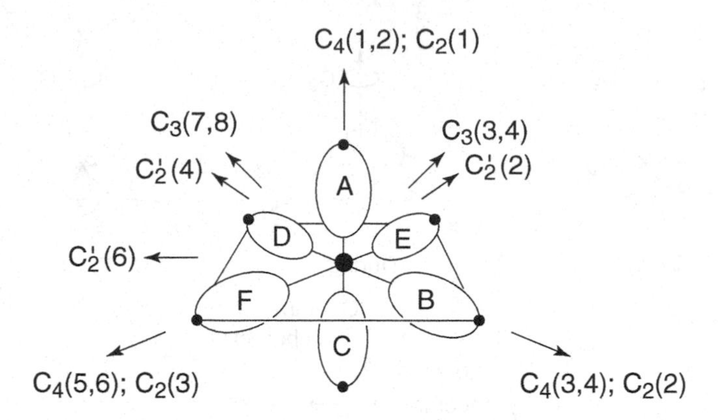

Figure 8.19 The six fluorine S–F σ-bonding hybrid orbitals of SF_6 together with the labels used for them in the text. Some of the axes used in obtaining Table 8.5 are indicated

Comparison with Table 8.1 confirms that this set of characters is the T_1 irreducible representation.

Problem 8.11 By a detailed study of Figures 8.15–8.18 derive the above set of characters.

The s and p orbitals of the sulphur atom in SF_6 bond with those fluorine orbitals that point towards the sulphur atoms; without defining their composition further they will simply be called the fluorine σ orbitals. This set of orbitals is shown schematically in Figure 8.19 (the labels on the orbitals and on the axes will be described later) and, because all are symmetry related, the transformations of the six must be considered as a set. This is not a difficult task but some mental gymnastics can be avoided by remembering a general principle; only if an object lies on a symmetry element can it give a non-zero contribution to the character associated with the corresponding operation. It follows, therefore, that because the fluorine atoms are located on the fourfold axes of the octahedron it is only under the fourfold and corresponding twofold rotation operations (and, of course, the identity operation) that any of the fluorine σ orbitals can be left unchanged. Remembering that there are two fluorines on each C_4 axis, it follows that the reducible representation generated by the transformation of the fluorine σ orbitals is:

E	$8C_3$	$6C_4$	$3C_2$	$6C_2'$
6	0	2	2	0

This is a sum (sums such as this are sometimes called a *direct* sum) of the $A_1 + E + T_1$ irreducible representations.

Problem 8.12 Use Table 8.1 to show that the above reducible representation has A_1, E and T_1 components.

Hint: This is similar to, but more difficult than, Problems 6.15 and 7.5.

This result was obtained in the group O. There are several ways in which we could proceed to obtain the g and u nature of the A_1, E and T_1 combinations in O_h – the most obvious way is to repeat the above sequence again but using the full O_h group. However, we would still be without the explicit forms of the combination of σ orbitals which transform as each irreducible representation, and this information is something that will be needed later. If these explicit forms were available at the present point in the text, a detailed study of them would show their g or u nature. It is therefore simplest next to use the projector operator method to obtain these combinations (working in the point group O) and subsequently to ask which are g and which are u in nature. To some extent the result may be anticipated. As we have already commented, the fluorine σ orbitals used to generate the reducible representation were neither inherently symmetric nor antisymmetric with respect to inversion in the centre of symmetry – because they were not located at this centre they were interchanged by the inversion operation. In such a situation this indifference is reflected by an equal number of combinations of g and u symmetries being generated. That is, in the present case there must be three linear combinations of fluorine orbitals which are g and three which are u. There are, then, two possibilities. Either we have $A_{1g} + E_g + T_{1u}$ in O_h or, alternatively, $A_{1u} + E_u + T_{1g}$. Physically, only the first choice makes any sense, because the irreducible representations generated by the transformation of the sulphur valence shell s and p orbitals are included in this set whereas they are not in the second. That is, if the first set is correct then there can be interactions between the fluorine σ orbitals and the sulphur orbitals – and so the existence of the molecule explained – whereas for the second set there would be no interactions and the molecule SF_6 would not exist!

As has just been said, in order to obtain the fluorine σ orbital combinations transforming as the A_1, E and T_1 irreducible representations the projection operator method will be used – a method that has been met several times before. Because the present case provides a particularly good example of the general method it will be given in detail, bringing together the techniques developed in previous chapters.

First, each ligand σ orbital is given a label, A–F, as shown in Figure 8.19. The transformation of one of these orbitals under the operations of the group is then considered in detail. In Table 8.5 are listed the twenty-four operations of the group O, and beneath each is the ligand σ orbital into which A is transformed by the particular operation. Within each set of operations, $8C_3$ for example, the order in which the operations are considered is unimportant; what matters is that all are included. The operations are indicated by the labels on the axes in Figure 8.19; they should allow the reader to check the individual transformations listed below.

Table 8.5

E	$C_4(1)$	$C_4(2)$	$C_4(3)$	$C_4(4)$	$C_4(5)$
A	A	A	F	E	B
$C_4(6)$	$C_2(1)$	$C_2(2)$	$C_2(3)$	$C_3(1)$	$C_3(2)$
D	A	C	C	D	E
$C_3(3)$	$C_3(4)$	$C_3(5)$	$C_3(6)$	$C_3(7)$	$C_3(8)$
E	B	B	F	D	F
$C_2'(1)$	$C_2'(2)$	$C_2'(3)$	$C_2'(4)$	$C_2'(5)$	$C_2'(6)$
D	B	E	F	C	C

Problem 8.13 Use Figures 8.3a and 8.19 to obtain Table 8.5.

Hint: Good diagrams are important – it may well be necessary to sketch out parts of Figure 8.3a several times to retain clarity in distinguishing the different effects of the various operations. Note that because there are just six fluorine σ orbitals but twenty-four operations, each orbital label appears four times in Table 8.5. To help with this problem as many axes as is consistent with graphic clarity have been indicated in Figure 8.19. Note that there is nothing sacred about the particular labels we have used; they are just a tool to get to the final result. If something different seems more natural to the reader, they should follow their instinct – and perhaps generate the orbitals in a different order. Nonetheless, they will arrive at the same result, unless they make a mistake.

For each of the irreducible representations in the direct sum obtained from the transformation of the sulphur orbitals, $A_1 + E + T_1$, the labels in Table 8.5 are multiplied by the character appropriate to the corresponding operation. The products are then added together. The sum is either the required ligand group orbital (the name used commonly in inorganic chemistry; more generally, it is called a 'symmetry adapted function') or is simply related to it. For the A_1 group orbital, multiplying each of the orbital labels by 1 (the value of each of the A_1 characters) and adding the products together gives

$$4\mathrm{A} + 4\mathrm{B} + 4\mathrm{C} + 4\mathrm{D} + 4\mathrm{E} + 4\mathrm{F}$$

On normalizing, the A_1 combination is obtained

$$\Psi(\mathrm{a}_1) = \frac{1}{\sqrt{6}}(\mathrm{A} + \mathrm{B} + \mathrm{C} + \mathrm{D} + \mathrm{E} + \mathrm{F})$$

Turning to the E orbitals, the sum obtained after multiplication is

$$4A + 4C - 2B - 2D - 2E - 2F$$

which after normalizing is

$$\Psi\mathrm{e}(1) = \frac{1}{\sqrt{12}}(2\mathrm{A} + 2\mathrm{C} - \mathrm{B} - \mathrm{D} - \mathrm{E} - \mathrm{F})$$

Problem 8.14 Derive ψe(1).

Hint: This problem is quite similar to that solved in Table 7.2 and the associated discussion.

Now a problem which is closely related to one met in Chapter 7 – how to obtain the second E function. Using either B or E as the generating orbital in Table 8.5 the (un-normalized) combinations which would have been obtained are:

$$\text{from B}: \quad 4\mathrm{B} + 4\mathrm{D} - 2\mathrm{A} - 2\mathrm{E} - 2\mathrm{C} - 2\mathrm{F}$$
$$\text{from E}: \quad 4\mathrm{E} + 4\mathrm{F} - 2\mathrm{A} - 2\mathrm{B} - 2\mathrm{C} - 2\mathrm{D}$$

Neither of these can be the second E function because they are different and the choice between them is arbitrary; they cannot both be correct and there cannot be three different

E functions. The method described in Section 7.2 may be used to systematically obtain the second function. Study of the results obtained there suggests that the difference between the functions given above should be taken. This difference is

$$6B + 6D - 6E - 6F$$

which on normalizing gives the second function

$$\Psi e(2) = \frac{1}{2}(B + D - E - F)$$

Problem 8.15 Show by squaring and adding the coefficients with which the fluorine σ orbitals appear in $\Psi e(1)$ and $\Psi e(2)$ that each orbital makes an equal contribution to the E set.

Hint: This problem resembles Problem 7.8. Note that the sum of squares of coefficients is equal to the ratio

$$\frac{\text{Number of } E \text{ functions}}{\text{Total number of } \sigma \text{ orbitals}} = \frac{2}{6}$$

In Section 7.2 the argument used leading to the generation of a second E function depended on the fact that the functions which we were seeking to generate had vector-like properties. Those that have just been obtained do not behave like axes (as the basis functions given at the right-hand side of Table 8.1 show; they transform like sums of products of axes). The method deduced in Section 7.2 clearly has a wider generality than could have been anticipated.

The T_1 functions are readily obtained. The transformation of A in Table 8.5 when multiplied by the T_1 characters and added gives 4A − 4C which, when normalized, gives

$$\Psi t_1(1) = \frac{1}{\sqrt{2}}(A - C)$$

Similarly, the transformations of B (or D) and E (or F) give

$$\Psi t_1(2) = \frac{1}{\sqrt{2}}(B - D)$$

and

$$\Psi t_1(3) = \frac{1}{\sqrt{2}}(E - F)$$

respectively. Because these three functions each involve different fluorine orbitals they are clearly independent of each other.

Problem 8.16 Derive the three T_1 functions listed above.

The complete list of symmetry-adapted functions is given in Table 8.6. In order to determine their symmetries in O_h the behaviour of these functions under the operation of

Table 8.6

Symmetry		
In the group O	In the group O_h	Symmetry-adapted function (ligand group orbitals)
A_1	A_{1g}	$\frac{1}{\sqrt{6}}(A + B + C + D + E + F)$
E	E_g	$\frac{1}{\sqrt{12}}(2A + 2C - B - D - E - F)$ $\frac{1}{2}(B + D - E - F)$
T_1	T_{1u}	$\frac{1}{\sqrt{2}}(A - C)$ $\frac{1}{\sqrt{2}}(B - D)$ $\frac{1}{\sqrt{2}}(E - F)$

inversion centre of symmetry has to be determined. Because this operation interchanges the fluorine σ orbitals as follows:

$$A \longleftrightarrow C$$
$$B \longleftrightarrow D$$
$$E \longleftrightarrow F$$

the effect of this operation is obtained by making these substitutions (A for C, C for A etc.) in the functions given in Table 8.6. When this is done, A_1 and E are left unchanged but each T_1 function changes sign. It is concluded that they transform in O_h as A_{1g}, E_g and T_{1u} respectively. These labels have been included in Table 8.6. We were correct in our earlier educated guess! This means that symmetries of the sulphur 3s and 3p orbitals (A_{1g} and T_{1u} respectively) are matched within the fluorine σ orbital symmetries. The bonding molecular orbital of A_{1g} symmetry is shown in Figure 8.20a, a representative T_{1u} bonding molecular orbital in Figure 8.20b and one of the two E_g in Figure 8.20c. Because the two E_g functions have different mathematical forms, it should perhaps be commented that it is the second that was generated which is shown in Figure 8.20c.

Problem 8.17 Sketch a diagram of the first-generated E_g function.

There is no doubt that the bonding orbitals have bonding energy stabilities in the order

$$A_{1g} > T_{1u} > E_g$$

This order of orbital energies is also the order in terms of the number of nodes. The A_{1g} bonding molecular orbital is nodeless, the T_{1u} has one nodal plane and the E_g has two. These nodal patterns are implicit in the expressions given in Table 8.6 and are also evident in Figure 8.20.

This discussion has assumed that only σ bonding is involved in the interaction between the central sulphur atom and the surrounding fluorines. Although this is quite a good approximation for SF_6 it is useful, nonetheless, to extend the discussion to include those p_π orbitals on the fluorine atoms which so far have been ignored. This is because they are of relatively high energy and they will be seen in the photoelectron spectrum. These p_π orbitals

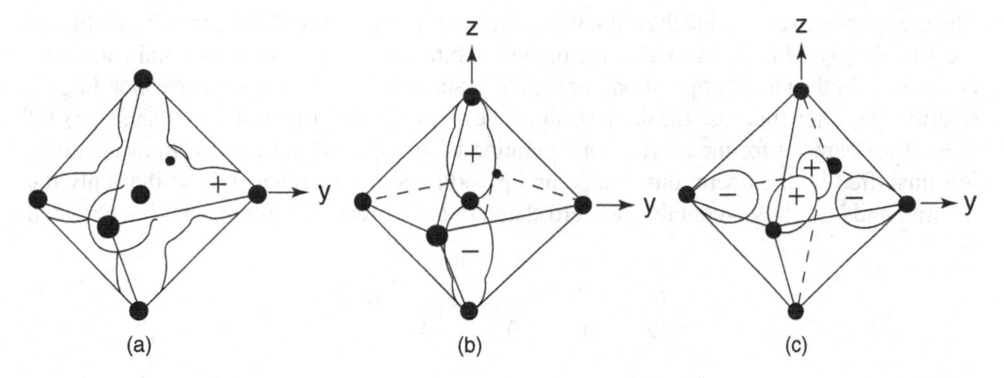

Figure 8.20 (a) The A_{1g} bonding molecular orbital. (b) One of the T_{1u} bonding molecular orbitals (that involving the sulphur p_z orbital). (c) One of the fluorine σ orbital symmetry-adapted combinations of E_g symmetry

transform as a degenerate pair of E symmetry under the local C_{4v} symmetry of each fluorine atom. It follows from the discussion of this symmetry in Chapter 6 that there is no unique specification of the direction of the local x and y axes but the choice and notation in Figure 8.21 prove to be convenient in practice.

Problem 8.18 Figure 8.21 appears rather complicated. Show that it has an internal consistency in that all p_x orbitals are all oriented in the same x direction, the p_y in the same y and the p_z in the same z direction.

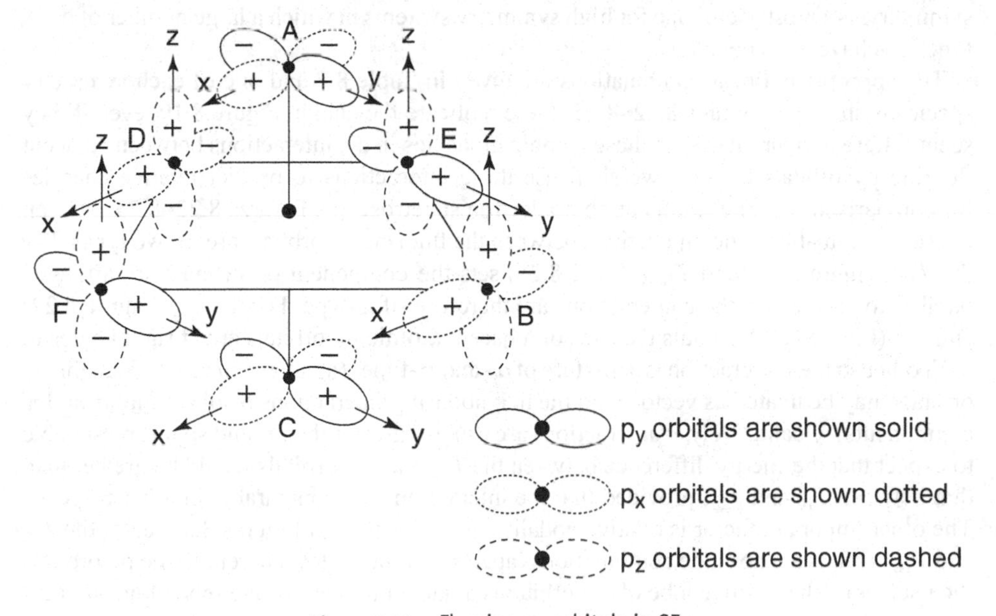

Figure 8.21 Fluorine p_π orbitals in SF_6

The next step is to determine the reducible representation generated by the transformation of these twelve p_π orbitals. As for the case of the σ orbitals it is only possible to obtain non-zero characters for the identity operation, for the C_4 rotation operations and the corresponding C_2 rotation operation (because the fluorine atoms lie on the fourfold axes of the octahedron). Of these, the character for the C_4 rotation operation is zero (because, for those fluorine atoms left unshifted by the operations, the p_x and p_y orbitals are interchanged) so that only the identity and $3C_2$ classes contain non-zero characters. The reducible representation obtained is

E	$8C_3$	$6C_4$	$3C_2$	$6C_2'$
12	0	0	−4	0

which has components $2T_1 + 2T_2$ (under O_h symmetry these become $T_{1g} + T_{1u} + T_{2g} + T_{2u}$).

Problem 8.19 Show that the twelve fluorine p_π orbitals of Figure 8.21 generate the above reducible representation and that it has $2T_1$ and $2T_2$ components in the point group O.

Appropriate linear combinations are obtained by the usual projection operator method but a difficulty arises because, in O symmetry, two quite independent sets of functions transform as T_1 and two other sets as T_2. The problem of distinguishing between them is readily solved by working, instead, in O_h symmetry – where all sets are symmetry-distinguished – but this is a rather tedious task because this group has forty-eight symmetry operations, each of which has to be considered separately. In Appendix 4 an alternative, short-cut, method of obtaining these linear combinations is described. This method, depending on an ascent-in-symmetry, is a most useful one for high symmetry systems in which a large number of basis functions have to be handled.

The appropriate linear combinations are given in Table 8.7 and one of each symmetry species is shown in Figures 8.22–8.25 (these replicate the data in Figure 8.10, even if they seem different). Our interest in these combinations lies in the interactions between adjacent fluorine p_π orbitals because we shall use these interactions to predict relative energies for comparison with the results of photoelectron spectroscopy. Figures 8.22–8.25 show an interesting situation. The interactions between the fluorine p_π orbitals are of two types. For the T_{1u} (Figure 8.23) and T_{2u} (Figure 8.25) sets the component p_π orbitals are arranged parallel to each other; their interactions are therefore of π-type. For the T_{1g} (Figure 8.22) and T_{2g} (Figure 8.24) orbitals the axes of adjacent atomic p_π orbitals are at right angles to each other so their interaction is a mixture of σ- and π-types (as shown in Figure 8.26 the p_π orbitals may be treated as vectors and the neighbouring interactions resolved into σ and π components). Qualitatively, σ interactions are usually greater than π and so it is reasonable to expect that the energy difference between the T_{2u} and T_{1g} orbitals would be greater than that between T_{1u} and T_{2g}, provided that the interactions are comparable in other respects. The other important factor is relative nodality. As evident from Figures 8.22–8.25, the T_{1u} and T_{2g} orbitals are no-node combinations (apart from the nodes inherent in the p_π orbitals themselves) – the positive lobe of a p orbital is adjacent to the positive lobe of an adjacent

Table 8.7 Symmetry-adapted combinations of fluorine p_π orbitals

T_{1g} orbitals
$t_{1g}(1) = \frac{1}{2}[p_x(A) - p_x(C) + p_z(E) - p_z(F)]$
$t_{1g}(2) = \frac{1}{2}[p_y(A) - p_y(C) - p_z(B) + p_z(D)]$
$t_{1g}(3) = \frac{1}{2}[p_x(B) - p_x(D) + p_y(E) - p_y(F)]$
T_{1u} orbitals
$t_{1u}(1) = \frac{1}{2}[p_z(B) + p_z(D) + p_z(E) + p_z(F)]$
$t_{1u}(2) = \frac{1}{2}[p_y(A) + p_y(C) + p_y(E) + p_y(F)]$
$t_{1u}(3) = \frac{1}{2}[p_x(A) + p_x(B) + p_x(C) + p_x(D)]$
T_{2g} orbitals
$t_{2g}(1) = \frac{1}{2}[p_x(A) - p_x(C) - p_z(E) + p_z(F)]$
$t_{2g}(2) = \frac{1}{2}[p_y(A) - p_y(C) + p_z(B) - p_z(D)]$
$t_{2g}(3) = \frac{1}{2}[p_x(B) - p_x(D) - p_y(E) + p_y(F)]$
T_{2u} orbitals
$t_{2u}(1) = \frac{1}{2}[p_z(B) + p_z(D) - p_z(E) - p_z(F)]$
$t_{2u}(2) = \frac{1}{2}[p_y(A) + p_y(C) - p_y(E) - p_y(F)]$
$t_{2u}(3) = \frac{1}{2}[p_x(A) + p_x(B) - p_x(C) - p_x(D)]$

p_y (A) means the p_y orbital on atom A as indicated in Figure 8.21.

p orbital – whereas the T_{1g} and T_{2u} orbitals each have two additional nodes – the positive lobe of one p orbital is adjacent to the negative lobe of its neighbour. We have:

T_{1g}: 2 nodes, σ and π interactions
T_{2u}: 2 nodes, π interaction only
T_{1u}: 0 nodes, π interaction only
T_{2g}: 0 nodes, σ and π interactions

Our discussion leads us to expect that the stability of these orbitals varies in the order

$$T_{2g} > T_{1u} > T_{2u} > T_{1g}$$

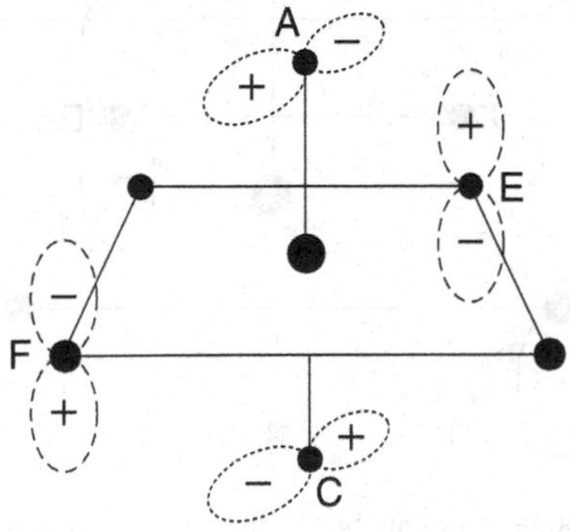

Figure 8.22 One of the T_{1g} fluorine p_π combinations

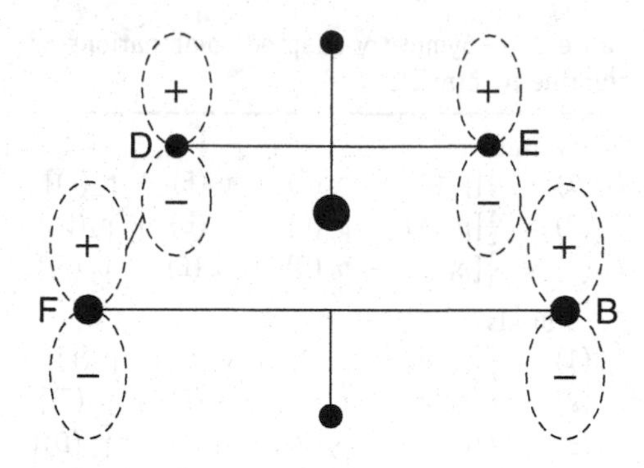

Figure 8.23 One of the T_{1u} fluorine p_{π} combinations

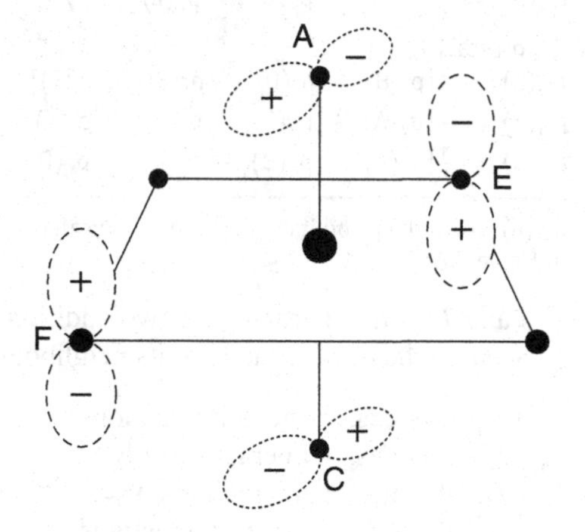

Figure 8.24 One of the T_{2g} fluorine p_{π} combinations

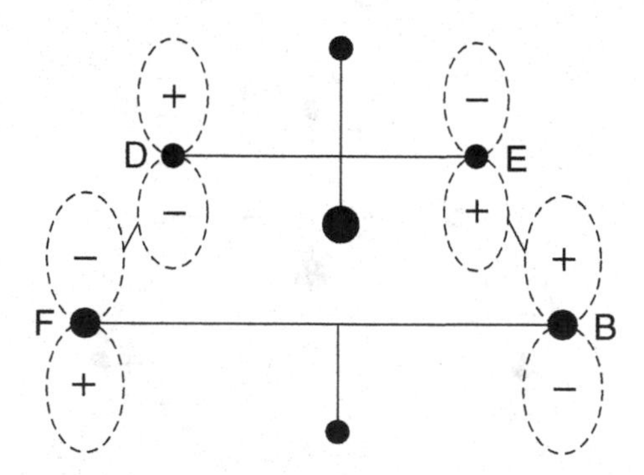

Figure 8.25 One of the T_{2u} fluorine p_{π} combinations

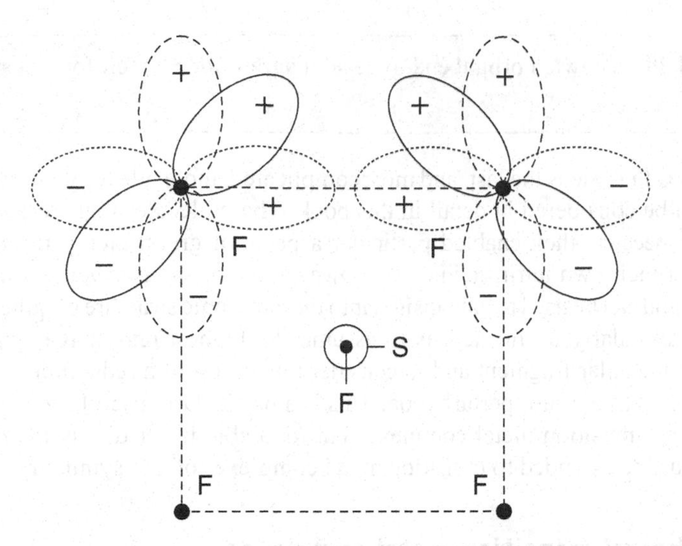

Figure 8.26 The overlap of 'coplanar' p_π orbitals on adjacent fluorine atoms (shown solid) may be expressed as a sum of a π overlap (between orbitals shown dashed) and a σ overlap (between orbitals shown dotted). This diagram shows SF_6 viewed down a C_4 axis

In arriving at this order any possible bonding of the fluorine p_π orbitals with the sulphur atom has been ignored. This defect is easily remedied. The only symmetry in common with the sulphur valence shell orbitals is T_{1u}. The latter are involved in S–F σ bonding and so interactions between the p_π-derived T_{1u} set and the (more stable) S–F σ bonding T_{1u} set have to be considered (these are the only ones that will affect the photoelectron spectrum). Any interaction will lead to a further stabilization of the S–F σ bonding set and a corresponding destabilization of the p_π-derived T_{1u} orbitals. This destabilization could change the p_π orbital energy sequence; although the stability order

$$T_{2g} > T_{2u} > T_{1g}$$

seems clear enough, the T_{1u} set could slot in between the T_{2g} and T_{2u} – as before – or between the T_{2u} and T_{1g} (assuming that the destabilization is not too great). In practice, the stability observed sequence for SF_6 seems to be[6]

$$T_{2g} > T_{2u} > T_{1u} > T_{1g}$$

or, including the orbitals associated with σ bonding ($A_{1g} > T_{1u} > E_g$):

$$A_{1g} > T_{1u}(1) > T_{2g} > E_g > T_{2u} > T_{1u}(2) > T_{1g}$$

This order is in good agreement with that given by the qualitative model developed above, although this was not able to predict that the σ-interaction energy level sequence would overlap with the π levels.

---

[6] W. von Niessen, W.P. Kraemer and G.H.F. Diercksend, *Chem. Phys. Lett.* **63** (1979) 65. The order given above is of calculated orbital energies; in the vertical ionization potentials (observed in the photoelectron spectrum) the T_{2u} and T_{1u} are identical.

Problem 8.20 Draw an orbital energy level diagram for SF_6 (cf. for instance, Figure 7.9).

Sulphur hexafluoride is the last, and most complicated, molecule for which the electronic structure will be considered in detail in this book. The molecules that were selected were chosen more because they enabled particular aspects of group theory to be introduced, rather than for their own intrinsic interest. However, the methods developed are of general applicability and can be used to gain insight into the electronic structure of quite complicated molecules. Particularly useful here is to assume the highest reasonable symmetry for a molecule or molecular fragment and to consider the effects of a reduction in symmetry to the real-life level as a minor perturbation. This is a particularly useful trick when working with octahedral transition metal complexes but to enable it our discussion of octahedral molecules must be extended to transition metal complexes of this symmetry.

8.4 Octahedral transition metal complexes

It is probably true that a majority of transition metal complexes have octahedral symmetry, at least approximately. Entire books have been written on this subject but only the more important features will be described here. At the simplest level an octahedral transition metal complex may be regarded as built up from a transition metal ion, M^{n+}, surrounded by six atoms or ions arranged at the corners of a regular octahedron. The six surrounding atoms may indeed be single atoms or they may be an atom through which a molecule is attached to the transition metal ion. In the simplest picture the metal ion is bonded to the six surrounding *ligands* (a collective noun covering both bonded atoms and molecules) by pure electrostatic attraction. This simple model leads to *crystal field theory* and it is this which will now be discussed in outline. Although simple, it provides the basis for all other, more detailed, models and so time spent studying it is time well spent, even if it is unrealistic. The most important thing about it is that it introduces all of the important symmetry-based arguments. Discussions of bonding can come later.

Transition metals are characterized by the fact that they exhibit variable valencies in their salts. The corresponding transition metal cations have different numbers of d electrons, the number of d electrons varying with the valence state of the cation. Loosely speaking, if a transition metal ion is oxidized then it loses a d electron; if it is reduced it gains one. Attention is therefore focused on the d electrons and on the d orbitals in which they are located. In an octahedral ML_6 molecule a set of d orbitals on the central metal atom divides into two sets. One, consisting of the d_{xy}, d_{yz} and d_{zx} orbitals, has T_{2g} symmetry, as indicated in Tables 8.1 and 8.2. Figures 8.27–8.30 illustrate the transformations of members of this set and their individual contributions to the resulting characters detailed.

Problem 8.21 Check the transformations of the d_{xy}, d_{yz} and d_{zx} orbitals shown in Figures 8.27–8.30 and thus show that these orbitals transform as T_2 in O. Because they are centrosymmetric orbitals, it follows that they transform as T_{2g} in O_h.

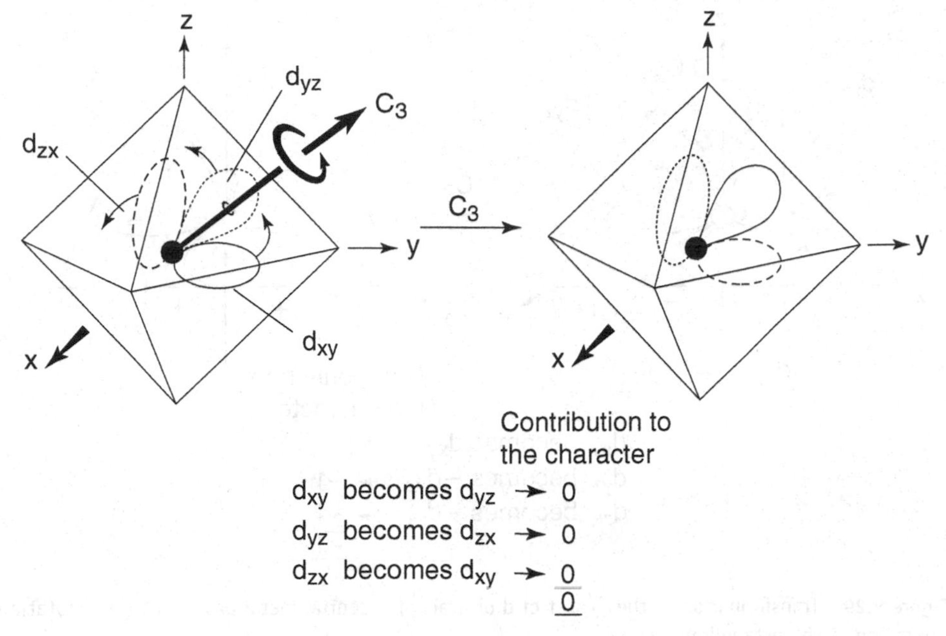

Figure 8.27 Transformation of the T_{2g} set of d orbitals of a central metal atom under a C_3 rotation operation of the octahedron

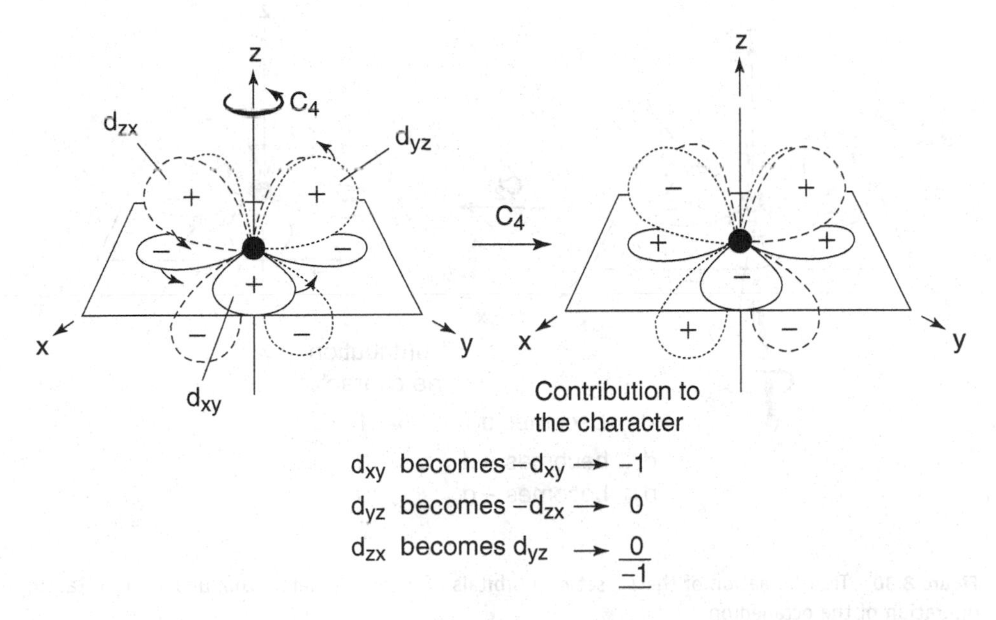

Figure 8.28 Transformation of the T_{2g} set of d orbitals of a central metal atom under a C_4 rotation operation of the octahedron. The d orbitals are distinguished in the same way as in Figure 8.27

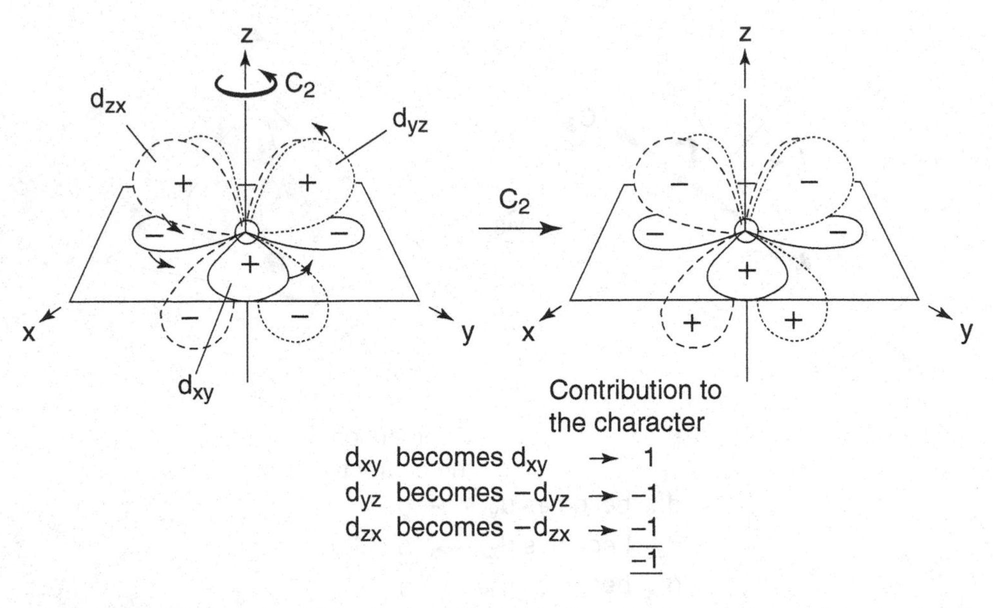

Figure 8.29 Transformation of the T_{2g} set of d orbitals of a central metal atom under a C_2 rotation operation of the octahedron

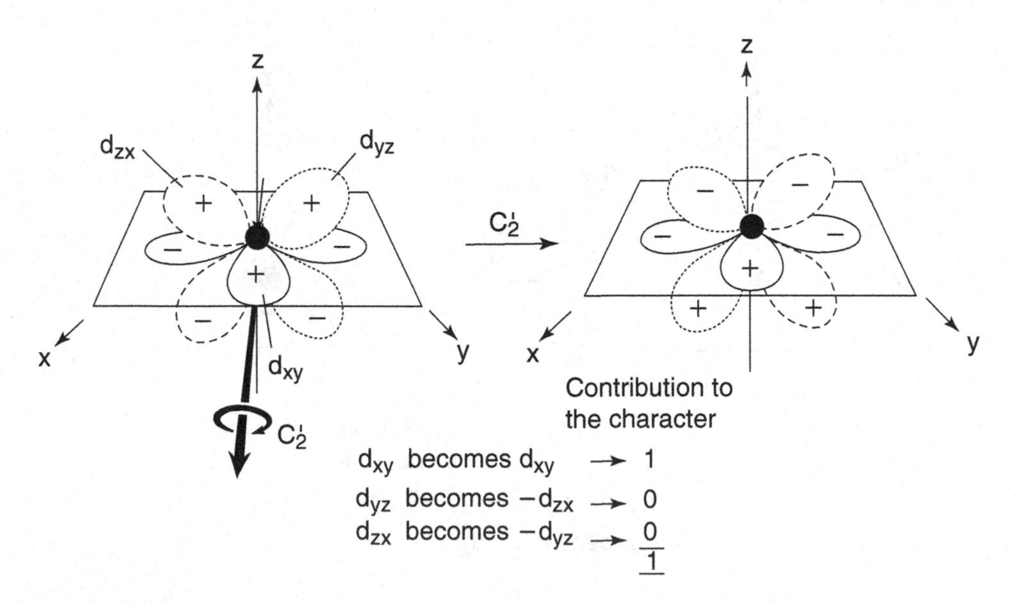

Figure 8.30 Transformation of the T_{2g} set of d orbitals of a central metal atom under a C_2' rotation operation of the octahedron

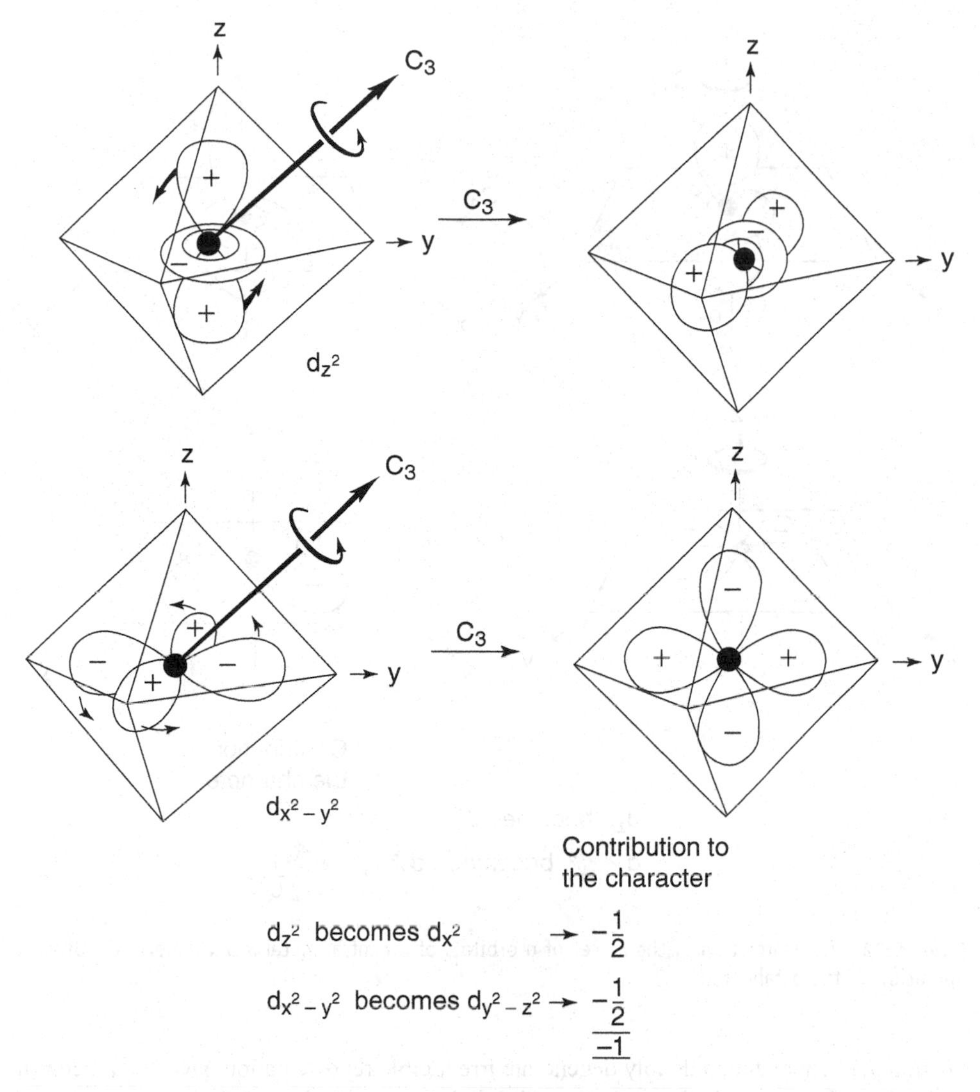

Figure 8.31 Transformation of the E_g set of d orbitals of a central metal atom under a C_3 rotation operation of the octahedron. It will help to understand this diagram if it is recognized that the C_3 operation shown has the effect of converting $z \rightarrow x$, $x \rightarrow y$ and $y \rightarrow z$ (Figure 8.7)

The second set of d orbitals,[7] $d_{x^2-y^2}$ and $d_{(1/\sqrt{3})(2z^2-x^2-y^2)}$, transform together as the E_g irreducible representation. Their transformations are illustrated in Figures 8.31–8.34 where their individual contributions to the characters are also given. The only point of difficulty arises in connection with the C_3 rotation operations and resembles that discussed in detail in

[7] These orbitals are usually called $d_{x^2-y^2}$ and d_{z^2}. In the present context we have to recognize that the label z^2 is a shorthand symbol for $(1/\sqrt{3})(2z^2 - x^2 - y^2)$. Note that the latter is cylindrically symmetrical around the z axis.

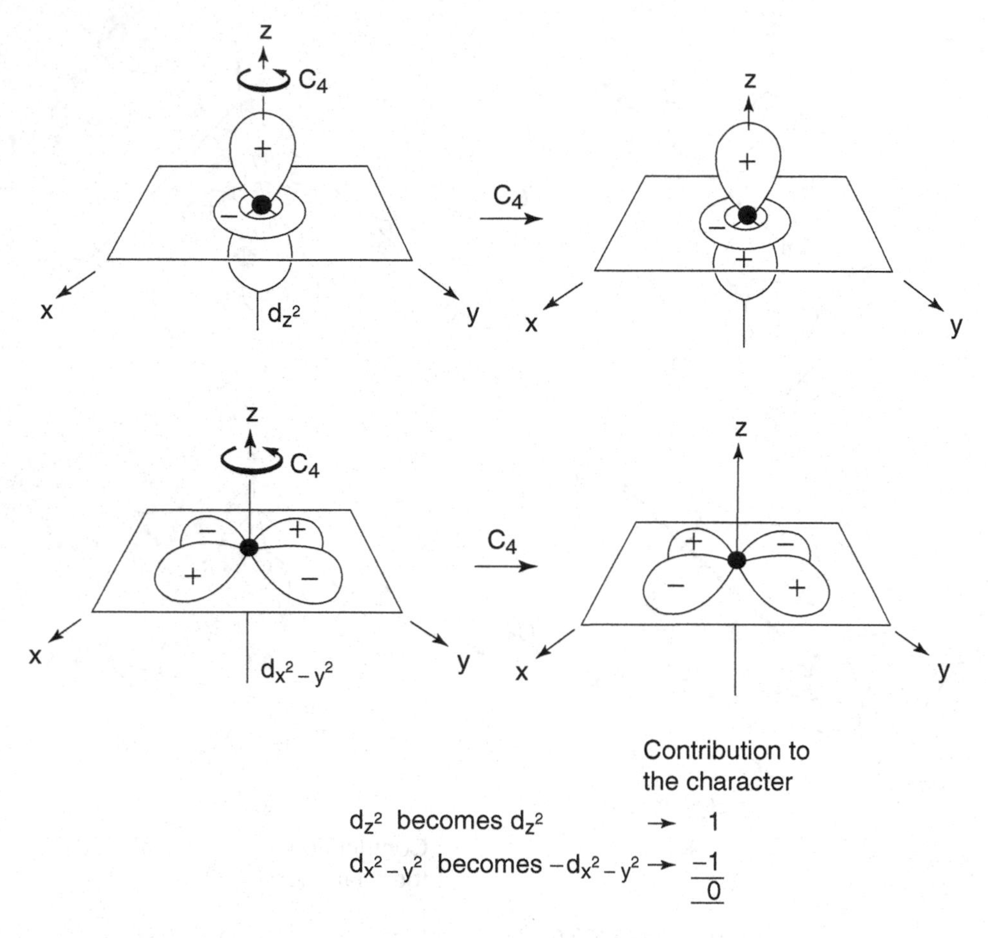

Figure 8.32 Transformation of the E_g set of d orbitals of a central metal atom under a C_4 rotation operation of the octahedron

Section 7.1. There, too, a doubly degenerate irreducible representation gave a character of -1 under a C_3 rotation. In the present case it is helpful to write the E_g orbitals as $d_{(x^2-y^2)}$ and $d_{(1/\sqrt{3})[(z^2-x^2)-(y^2-z^2)]}$, because this helps to demonstrate that rotation of the pair by 120° to give, for instance, $d_{(y^2-z^2)}$ and $d_{(1/\sqrt{3})[(x^2-y^2)-(z^2-x^2)]}$ (i.e. $x \rightarrow y \rightarrow z \rightarrow x$) leads to functions which may be expressed in terms of the original. It is easy to show by expansion of the coefficients that, for instance,

$$d_{(y^2-z^2)} = -\frac{1}{2}d_{(x^2-y^2)} - \frac{\sqrt{3}}{2}d_{(1/\sqrt{3})[(z^2-x^2)-(y^2-z^2)]}$$

so that the coefficient with which $d_{x^2-y^2}$, the 'starting' orbital, appears in this expression ($-1/2$) is its contribution to the character under the C_3 rotation. The contribution of $d_{(1/\sqrt{3})(2z^2-x^2-y^2)}$ to $d_{(1/\sqrt{3})(2x^2-y^2-z^2)}$ is similarly shown to be $-1/2$ so that the aggregate

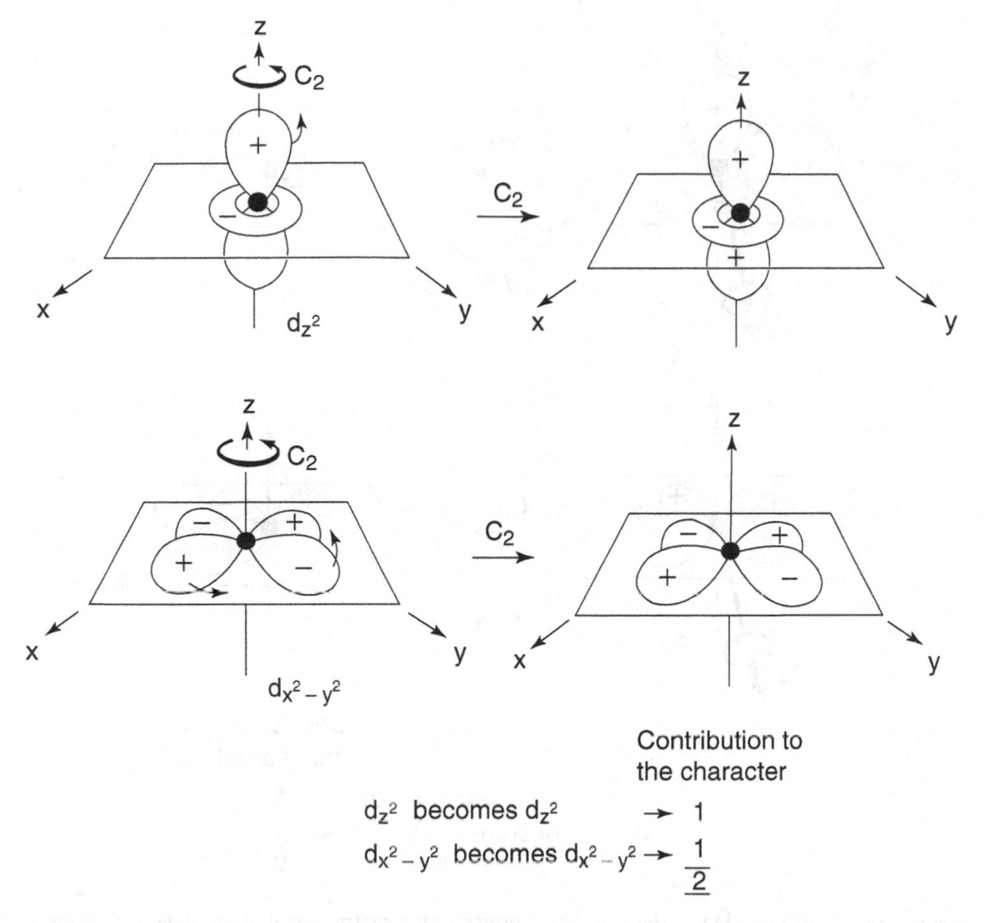

Figure 8.33 Transformation of the E_g set of d orbitals of a central metal atom under a C_2 rotation operation of the octahedron

character is −1. Of course, we have met all this before, in Section 8.2 (for those who had the strength to read it!), but there were no detailed arguments given.

> **Problem 8.22** Check the transformation of the $d_{x^2-y^2}$ and d_{z^2} orbitals given in Figures 8.31–8.34 and thus show that these orbitals transform as E in O. Because they are centrosymmetric orbitals it follows that they transform as E_g in O_h.

Crystal field theory, being a purely electrostatic theory which does not admit the existence of bonding and antibonding molecular orbitals, asserts that since the d electrons (like all other metal electrons) are non-bonding, they will occupy preferentially that arrangement in which electrostatic repulsion with the ligands (most simply represented as point negative charges) and with each other is a minimum. It is convenient to consider these two factors

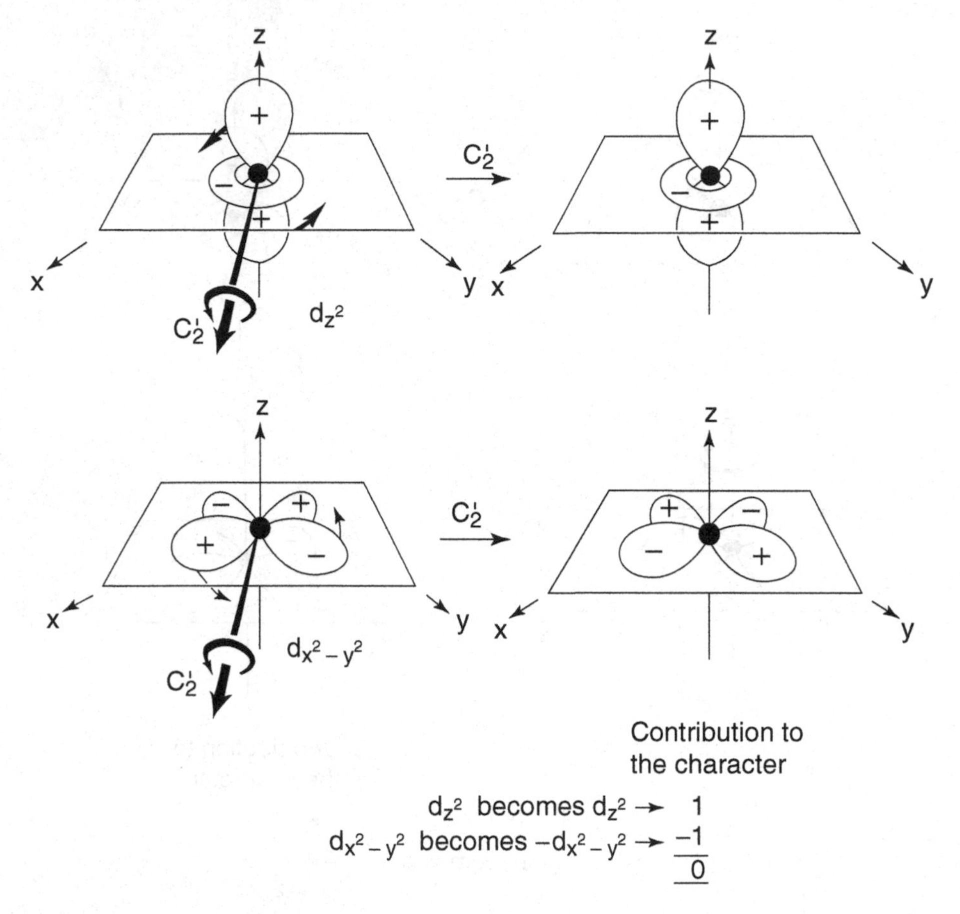

Figure 8.34 Transformation of the E_g set of d orbitals of a central metal atom under a C_2' rotation operation of the octahedron

separately. Consider first the requirement of minimum electrostatic repulsion between the metal d electrons and the negatively charged ligands. Figure 8.35 shows a representative E_g orbital (the $d_{x^2-y^2}$) and a representative T_{2g} (the d_{xy}). Symmetry ensures that whatever we conclude about these holds also for the other member(s) of their respective sets. As Figure 8.35 suggests, it is in the E_g orbitals that an electron gets closest to the ligands and so experiences the greatest electrostatic repulsion. This conclusion, which is confirmed by detailed calculations, means that the T_{2g} set of d orbitals has a lower (electrostatic) energy than the E_g set. The energy splitting between the two sets is usually denoted by either Δ or $10Dq$. The choice between the two symbols is personal – Δ is usually taken to be a quantity derived from experiment whilst $10Dq$ is theoretical in origin; in principle, both D and q can be calculated. If d electron–ligand repulsion were the only factor to be considered then the d electrons in octahedral transition metal complex ions would occupy the lower, T_{2g}, set of d orbitals until these were filled up. However, this preference is opposed by the effects of electron repulsion between the d electrons themselves. This electron repulsion is

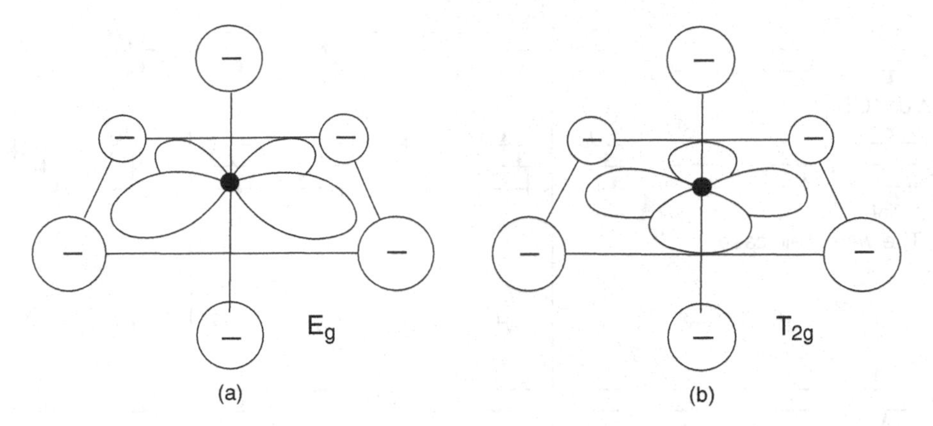

Figure 8.35 Representative (a) E_g and (b) T_{2g} orbitals of a central metal atom in an octahedral metal complex. In this figure – indicates negative electrical charge; the d orbitals are envisaged as also containing electron density so that electron–electron repulsion occurs

minimized if, as far as possible, the d electrons occupy different d orbitals with parallel spin. That is, occupation of the E_g orbitals will start as soon as the T_{2g} set is half-full. We have here a straight conflict between two opposing forces. When the d electron–ligand repulsion wins we have the so-called 'strong field' case; when the d–d electron repulsion dominates we have the so-called 'weak field' case. Alternative, but not quite equivalent (see below), names are to talk of 'low spin' and 'high spin' complexes, the names originating from the fact that weak field complexes have more unpaired electrons – a higher resultant spin – than do strong field complexes, when there is a choice.

In summary, in crystal field theory, the relative magnitudes of Δ ($10Dq$) and the d electron repulsion energies – the so-called 'pairing energy' – determine the way that the set of d orbitals are occupied. This is illustrated in Figure 8.36, where the clear difference between high spin and low spin electron arrangements for ions with between four and seven d electrons is evident (the names 'high' and 'low' spin really only apply to these electron configurations). Small orbital occupation differences also exist for ions with two, three and eight d electrons but these differences are rather subtle and are not manifest in obvious orbital occupancies. Consequent upon these differences between high and low spin cases are a variety of associated spectral, magnetic, structural, kinetic and thermodynamic differences.

Inclusion of covalent bonding, along the lines discussed earlier in this chapter for sulphur hexafluoride, in the interaction between metal ion and ligands in a transition metal complex leads to *ligand field theory*. It differs from crystal field theory in that quantities which are well defined in crystal field theory become less well defined in ligand field theory (and are generally treated as parameters to be deduced from experiment). Qualitatively, Figure 8.36 remains appropriate except that, as will be explained, the E_g set of d orbitals is now identified as the antibonding counterpart of the E_g set involved in metal–ligand σ bonding.

In contrast to the discussion of SF_6 earlier in this chapter, the valence shell of the central atom in transition metal complexes consists of s, p and d atomic orbitals. This means that the nine available metal orbitals span the A_{1g}, T_{1u}, E_g and T_{2g} irreducible representations. It will be recalled that the σ orbitals of the surrounding six atoms – be they fluorine in SF_6 or ligands in a complex – span $A_{1g} + T_{1u} + E_g$. In a transition metal complex these ligand

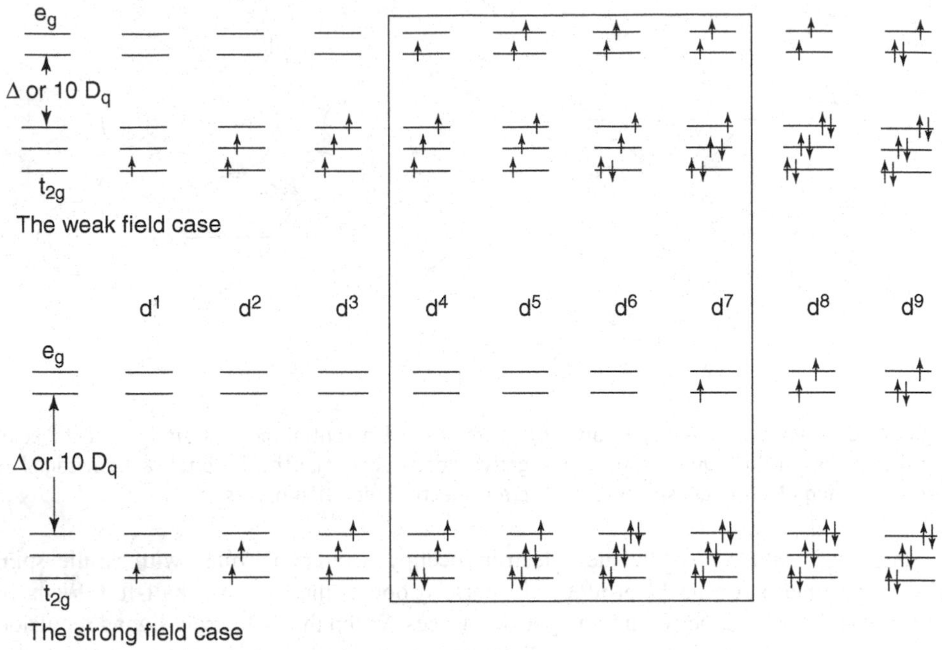

e_g
Δ or 10 D_q
t_{2g}
The weak field case

d^1 d^2 d^3 d^4 d^5 d^6 d^7 d^8 d^9

e_g
Δ or 10 D_q
t_{2g}
The strong field case

Figure 8.36 Differences in arrangement of electrons in the d orbitals of a metal atom in an octahedral complex occur for the d^4–d^7 configurations (those within the box)

orbitals are full. This is evident if the ligand is a closed shell anion such as F^-, Cl^- etc. and is equally true if it is a molecule such as H_2O or NH_3, where the ligand σ orbital is identified as a lone pair of electrons on the electronegative atom. This means that the interaction with the metal orbitals can be regarded as stabilizing the ligand orbitals – lowering their energy. In this case the metal orbitals are to be regarded as being correspondingly destabilized by virtue of the same interactions, and the e_g orbitals, which in crystal field theory are pure d orbitals, are to be regarded as antibonding combinations of ligand σ and metal d orbitals. Here the common practice of using lower case letters before the word 'orbital' is followed; thus, 'e_g orbitals' (this usage was first met in Section 3.6). To avoid possible disruption of the logistic sequence, this convention has not been applied earlier in this chapter but use of the labels e_g and t_{2g} is so common in transition metal chemistry that they have to be introduced at some point.

Before concluding this section on transition metal ions it is of interest to note that in ligand field theory the d orbitals of T_{2g} symmetry may also interact with ligand orbitals. It will be recalled that the fluorine p_π orbitals in SF_6 transformed as $T_{1u} + T_{1g} + T_{2u} + T_{2g}$. The p_π orbitals of any ligand, L, in an octahedral ML_6 complex will have the same symmetries. Evidently, in transition metal complexes the metal d orbitals of T_{2g} symmetry may interact with the T_{2g} set of ligand π orbitals. If the relevant ligand π orbitals are empty – and therefore of high energy – then the effect of ligand–metal T_{2g} interactions will be to depress (stabilize) the lower t_{2g} orbitals. These are those corresponding to the t_{2g} d orbitals shown in Figure 8.36 – and to raise the energy of the (empty) t_{2g} ligand π orbitals. The

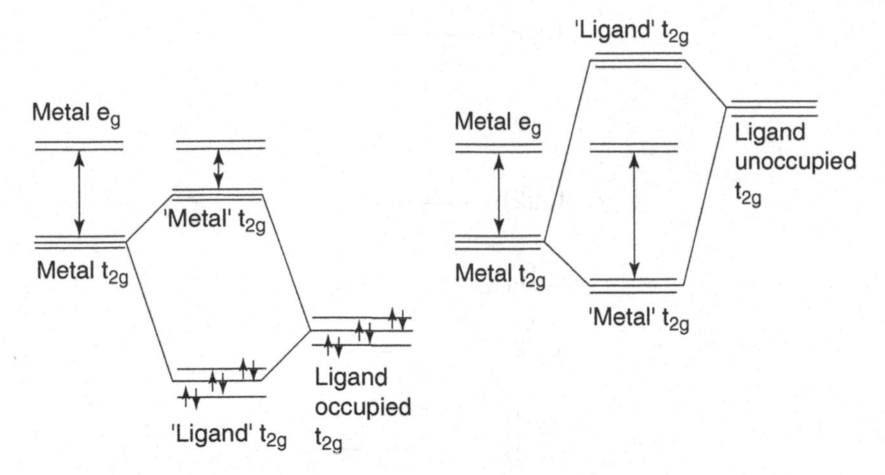

Figure 8.37 The effect of ligand π orbitals on the e_g–t_{2g} splitting (indicated by the double-headed arrows) depends on the relative energies of the metal and ligand t_{2g} orbitals

effect on the molecular orbitals corresponding to the d orbitals of Figure 8.36 will be to increase the splitting Δ. If the ligand π orbitals are filled – and therefore of relatively low energy – the effect will be to decrease the e_g–t_{2g} splitting. These two cases are illustrated in Figure 8.37. The π bonding that has just been described seems to be of importance because it is found that it is ligands with available but empty π orbitals that give large values of Δ, and thus strong field complexes, whilst those with filled π orbitals give small values of Δ and so weak field complexes. Examples of the former are the CN^- and CO ligands (the empty π orbitals being C–N or C–O π antibonding) and examples of the latter are Br^- and Cl^-. For these halide anions the filled π orbitals are the atomic p_π orbitals and several comments are relevant.

First, and perhaps most important, is the fact that the above argument can lead to misleading beliefs about energy levels. There is a subtle distinction between 'highest energy' and 'easiest to remove' when talking about electrons. It is a distinction which was recognized in Section 3.6, where it was unimportant. But there we were talking about a small molecule (H_2O) and here we are talking about a large molecule. When an electron is removed from a molecule, the electrons that are left rearrange to a new, most stable, pattern. This is an energy term in addition to the ionization energy. So, despite the above arguments, and although they may well be unpaired, d electrons may not always be the electrons of highest energy in a complex. In Figure 8.38 are shown the results of a recent, rather accurate, calculation on the ion $[Fe(CN)_6]^{2-}$. Although the highest four sets have symmetries which are those expected for the ligand π, the two which have to be non-bonding, T_{1g} and T_{2u}, are close together and the others well separated, indicative of important interactions. In fact, those sets labelled $t_{2g}(2)$ and $e_g(2)$ are largely d orbitals, as shown in Figure 8.39. For this molecule, the ligand π orbitals interleave the t_{2g}–e_g pair, to a first approximation. It is $t_{2g}(1)$ which has a high ligand π component. These calculations, therefore, are not in good agreement with Figure 8.37, although the pattern which this shows is generally assumed to be correct. Clearly, whilst group theory can give us an idea of what to look for and, with the addition of a reasonable model, make suggestions about the outcome, real life is more complicated.

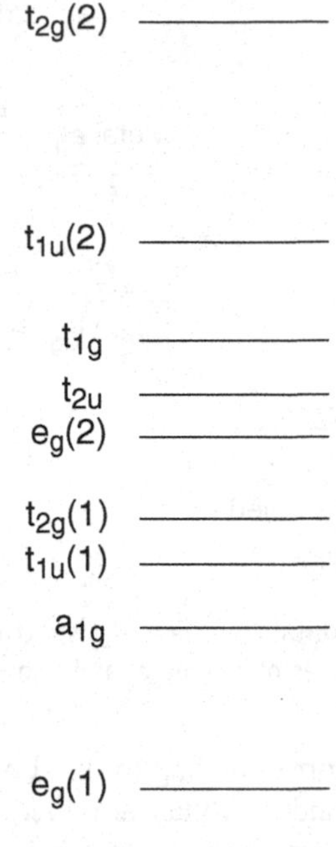

Figure 8.38 The (calculated) highest occupied orbitals of $[Fe(CN)_6]^{2-}$. This should be compared with Figure 8.37 but with the addition of ligand π orbitals of T_{1g}, T_{1u} and T_{2u} symmetries (the T_{2g} are already included) and ligand σ orbitals of A_{1g}, E_g and T_{1u} symmetries (pointing away from the metal atom)

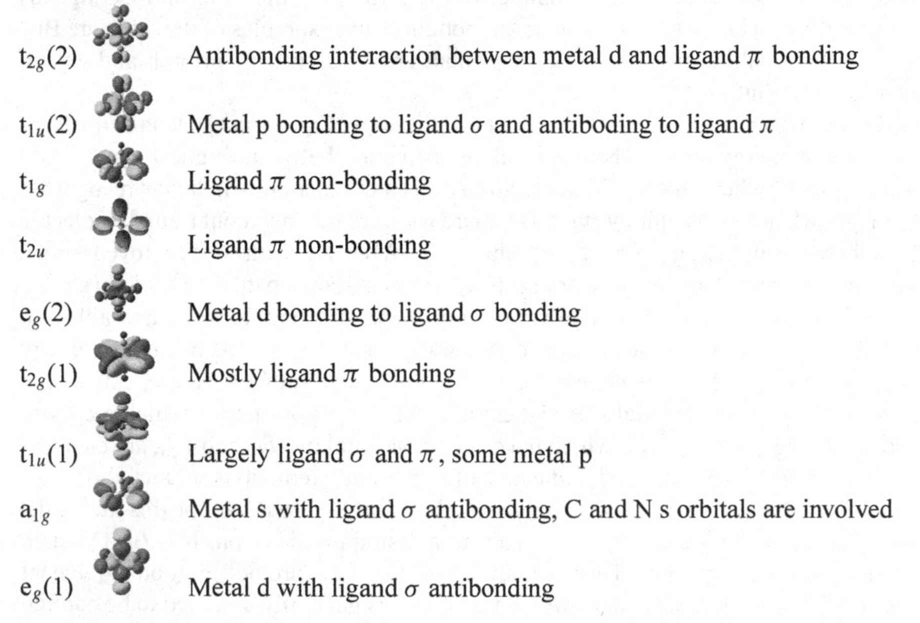

t2g(2) Antibonding interaction between metal d and ligand π bonding
t1u(2) Metal p bonding to ligand σ and antiboding to ligand π
t1g Ligand π non-bonding
t2u Ligand π non-bonding
eg(2) Metal d bonding to ligand σ bonding
t2g(1) Mostly ligand π bonding
t1u(1) Largely ligand σ and π, some metal p
a1g Metal s with ligand σ antibonding, C and N s orbitals are involved
eg(1) Metal d with ligand σ antibonding

Figure 8.39 Pictures of the molecular orbitals of Figure 8.38. The data for Figures 8.38 and 8.39 were kindly provided by Professor E. Diana

The second point concern the electronic spectra of transition metal ions. These spectra are important – they are responsible for the colour of transition metal complexes – and have been much studied. The discussion of their interpretation provides excellent examples in the application of group theory (and particularly of the O_h group) in chemistry. Typically, one is dealing with many (d) electron systems and this means that we have to form direct products between the irreducible representations of the O_h group in order to get the (single) symmetry species of wavefunctions from the symmetries of the many d electrons involved. Unfortunately, although not particularly difficult, this is a lengthy problem – so long that the space can only be justified in rather specialist texts and not generalist, like the present.

8.5 Summary

This chapter has been devoted to the important cubic point group O_h. The equivalence of the coordinate axes means the introduction of triply degenerate irreducible representations (p. 193), although it is the doubly degenerate that pose a greater problem (p. 202). The splitting of d orbitals in octahedral transition metal complexes into t_{2g} and e_g sets is of importance in inorganic chemistry (p. 226).

9 Point groups and their relationships

9.1 The determination of the point group of a molecule

The development of the subject in this book is now such that, at last, all of the different types of point group operations have been met. It is therefore a convenient point at which to briefly review the chemically important point groups and the allocation of a molecule to the correct one. That is, we shall tackle the question 'How do I decide what the symmetry of a particular molecule is?'. Unless a correct answer can be guaranteed, it would be only too easy to end up trying to use the incorrect character table. Even if this happens, it is seldom disastrous. Consequential problems usually arise quickly, which serves to place doubt on the original choice.

The way that most experienced workers identify the point group of a molecule is by a spontaneous knee-jerk type of reflex (most common) or to list as many symmetry operations as they can immediately see (much less common). Such a list is usually mental but the beginner may prefer to use pencil and paper. Even if incomplete, this list may at once identify the point group; if not, it will certainly reduce the number of possibilities to two or three. A glance at the list of operations across the head of the character tables of the possible groups (Appendix 3) will reveal the operations in which the possible groups differ. These operations (or, rather, the corresponding elements) are then explicitly looked for and thus the correct group selected. This procedure of scanning likely character tables is very strongly recommended to the beginner as the best way forward; it requires intelligent comparisons to be made of different point groups and this can be a very enlightening process. An alternative, the one recommended in most texts, is to mount a more systematic search for symmetry elements. Several schemes for such a search exist and one is given in Figure 9.1. One starts at the top and traces a path by answering the questions on the way, which ends with the correct point group (provided that no mistakes have been made!). Unfortunately, all such schemes (including that in Figure 9.1) tend to suffer from a basic defect – they have been compiled by someone who would never dream of using it! A more important defect is that they may not properly address the particular problem confronting the student at a particular moment. A fellow student may be someone worth talking to – they could have had the same problem. Alternatively, since what is a logical sequence for one person may not appear so for another, a search on the web will provide alternatives (search for 'flow + chart/diagram'

Symmetry and Structure: Readable Group Theory for Chemists Sidney F. A. Kettle

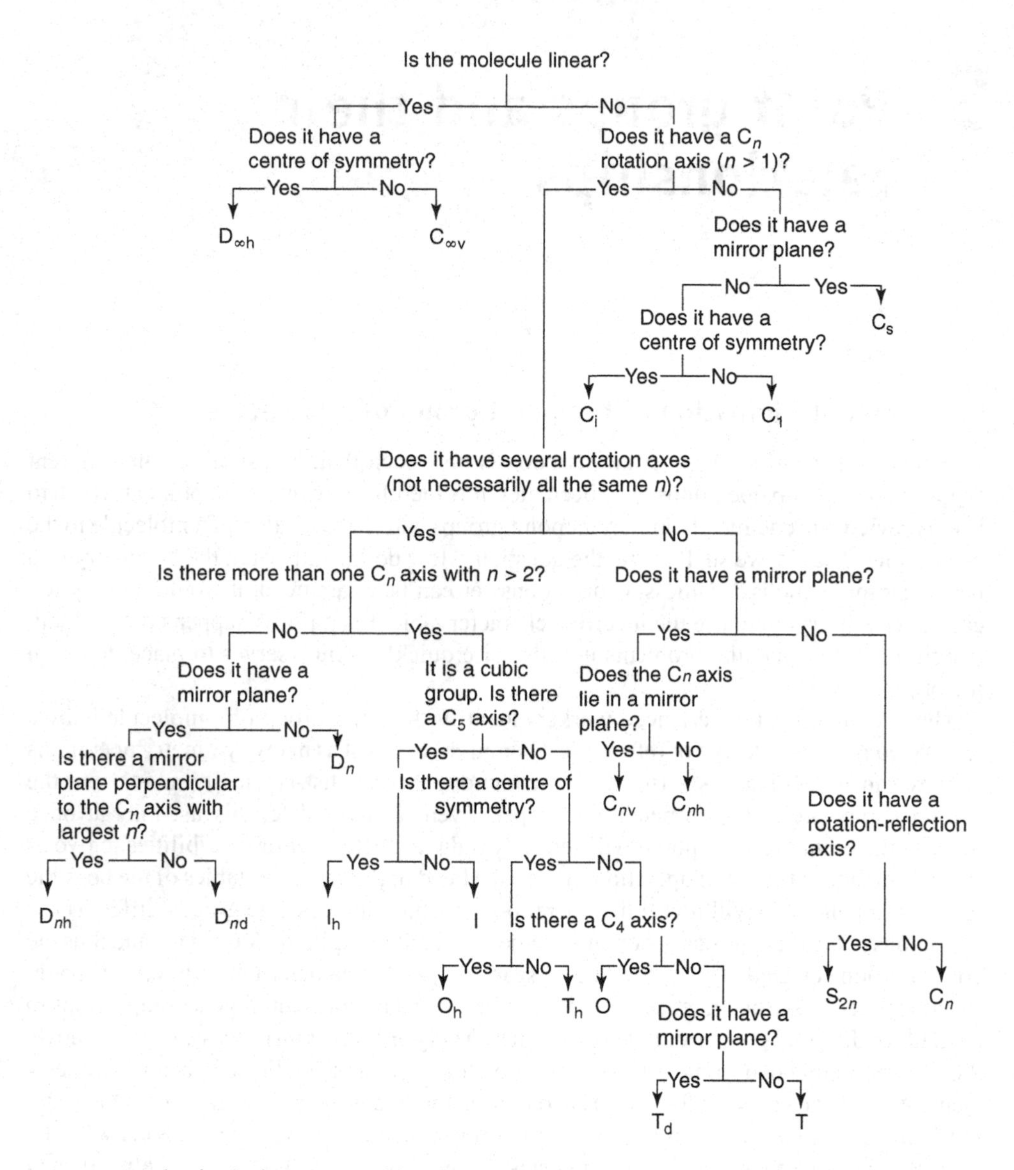

Figure 9.1 A 'yes' – 'no' response table. This is one of the many variants available which may be used to assign a molecule to the correct point group

and 'point group of a molecule'). What follows are some comments that should make the task of allocation of a point group an easier one.

When a molecule has a single rotational axis, C_n, this is usually quite evident (it becomes easier as n increases). It may be that this C_n axis is the only symmetry element, in which case the point group is C_n. Frequently, however, there will be other symmetry elements. If the only ones are mirror planes which contain the C_n axis (there must be n such σ_v planes because if there was only a single one, the existence of the C_n axis would create the other

$n-1$), then the point group is $C_{n\text{v}}$. Were there just a single mirror plane perpendicular to the C_n axis (a σ_h plane), it would be a $C_{n\text{h}}$ point group. The simultaneous existence of n σ_v mirror planes and a σ_h plane means that there must be further symmetry elements, in particular, n C_2 axes perpendicular to the original C_n axis, equally spaced around it (and symmetrically placed relative to the n σ_v mirror planes). Such point groups are $D_{n\text{h}}$ (D for Dihedral). Can these additional n C_2 axes exist without the n σ_v and σ_h? The answer is that they can; the point group produced is D_n. What of the simultaneous existence of the C_n, the n σ_v's and the n C_2's? These combinations also exist and lead to the point groups $D_{n\text{d}}$. The n σ_v axes now symmetrically interleave the n C_2 axes and so are, more correctly, referred to as σ_d mirror planes. In similar fashion the 'σ_v' mirror planes in the $D_{n\text{h}}$ point groups should be called σ_d. Unfortunately, many authors do not use this notation. In such groups with n even the vertical mirror plane reflection operations invariably fall into two classes. In many texts, one of these is usually – arbitrarily – denoted $(n/2)\sigma_\text{v}$ and the other $(n/2)\sigma_\text{d}$ (in Appendix 3 the notation $(n/2)\sigma_\text{d}$ and $(n/2)\sigma'_\text{d}$ has been used). The combination of just C_n, σ_h and n C_2 does not exist – the existence of these elements requires the co-existence of n σ_v and we are back to $D_{n\text{h}}$. In practice, most problems arise from C_2 axes – which can be somewhat hidden. As a rough test, if the reader can find all of the symmetry elements in spiropentane (Figure 9.2), a molecule of $D_{2\text{d}}$ symmetry (and, so, three C_2 axes, falling into two types, two mirror planes and an S_4 axis), they have little to fear.

In the previous chapter S_n operations were met for the first time. A set of such operations, together with the identity, can comprise a group, provided that n is even. Such groups are called S_n, although the case of $n = 2$, S_2, is usually called C_i because the operation S_2 is precisely equivalent to inversion in a centre of symmetry (Figure 9.3).

If it is clear that a molecule has non-coincident C_n and C'_n axes, where n and n' are both greater than two (n can be equal to n'), then the point group of the molecule is one of those for which the x, y and z axes are interchanged by some of the operations of the group. The Cartesian axes then transform together as a triply degenerate irreducible representation. If the molecule also contains a C_5 axis (it would actually have several) then the point

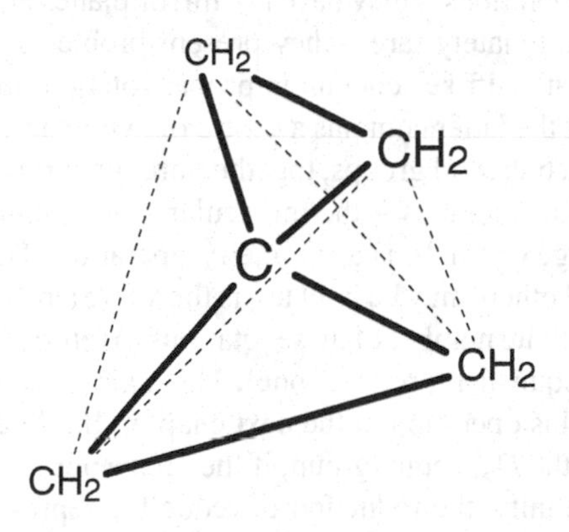

Figure 9.2 Spiropentane. Find three C_2 rotation axes, two mirror planes and an S_4 axis. The molecular symmetry, $D_{2\text{d}}$, is a subgroup of the tetrahedral and some symmetry elements may be easier to find there; the tetrahedron is shown dotted

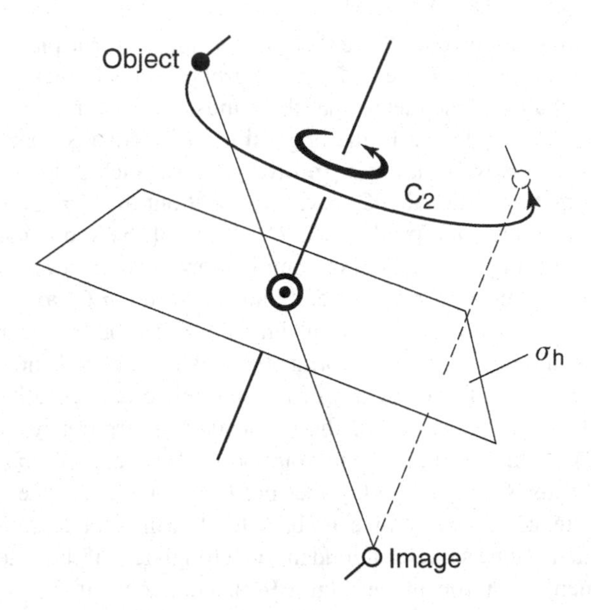

Figure 9.3 The operation i, indicated by a straight line through the centre of symmetry (itself shown as a black dot in a circle), is equivalent to 'rotate about an arbitrary C_2 axis (shown as an arc) containing the i and reflect in a σ_h containing the i (shown dotted)'. That is, $i \equiv S_2$

group would be an icosahedral one – I or I_h (pronounced 'eye aich'). An icosahedron of I_h symmetry is shown in Figure 9.4 and will be discussed in Chapter 10. If the molecule contains a C_4 axis (there would be three of them in all) then the point group would be cubic (or, equivalently, octahedral) – O or O_h. Finally, if its pure rotation axis of highest symmetry is a C_3 axis (there would be a total of four of these) then its symmetry would be that of a tetrahedral group – T, T_d or T_h. The distinction between each of the two icosahedral, the two octahedral or three tetrahedral groups is quite simple. Groups with no suffix are groups with only pure rotation operations – they have no mirror planes and no centre of symmetry, for example. They are fortunately rare – they present problems, which will be discussed in Chapter 11. Groups with suffixes contain improper rotation operations. The distinction between T_d and T_h is that the latter contains a centre of symmetry, whereas the former does not. The octahedral and tetrahedral groups, together, are often referred to as 'cubic' groups.

Linear molecules all have an axis – the molecular axis – about which rotation by any angle, no matter how large or small, is a symmetry operation. The 'fundamental' rotation operation, from which all others may be built up, is therefore a rotation by an infinitesimally small angle. It takes an infinite number of these rotations to return the molecule to its original position (rather than an equivalent, rotated, one). This axis is therefore a C_∞ axis. We will have more to say about this operation in the next chapter. If a linear molecule has a centre of symmetry they are of the $D_{\infty h}$ point group; if they have not, they are $C_{\infty v}$. Because they each have an order of infinity, the reduction of reducible representations in these groups has to be handled differently to the method developed in this text. The problem is one that is seldom encountered but is discussed at the end of the next chapter, after the relevant character tables have been given.

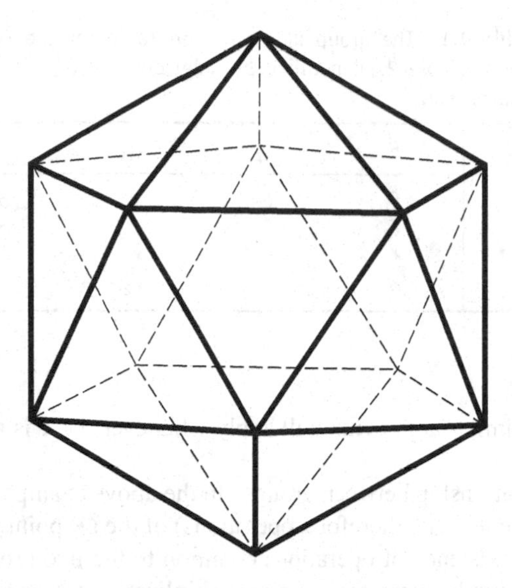

Figure 9.4 An icosahedron. A fivefold rotational axis passes through each pair of opposite corners and a threefold through the mid-points of each pair of opposite faces. It is discussed in detail in the next chapter

One point group remains. It is that which, in addition to the identity element, contains only the operation of reflection in a single mirror plane. It is denoted C_s ('cee ess').

9.2 The relationships between point groups

Now that we have, in principle at least, met all of the point groups which are of interest to the chemist, the next question to address is that of the relationships between different groups. Sometimes there is either little or none, but often relationships exist which can be used either to gain insights or to simplify work – or both.

As discussed at length in the previous chapter, a molecule in which a central atom M is surrounded by six identical atoms or groups L, ML_6, is of octahedral symmetry, O_h. Suppose one of the L is now replaced by a chemically similar, but different, atom or group X (for instance, both L and X could be halogen atoms). The molecule is then ML_5X and has, at most, C_{4v} symmetry (assuming that the change from L to X does not lead to a gross structural change in the molecule – no C_5 axis is introduced, for example). Strictly, then, a discussion of the ML_5X molecule should follow the general pattern developed in Chapter 6, where the electronic structure of BrF_5 was considered, due allowance being made for the presence of X. However, the difference between L and X could be negligible (if they are isotopic variants of the same element, for instance). In such a case the difference in conclusions between a discussion based on O_h symmetry and one based on C_{4v} should also be negligible. There must be a continuity between the C_{4v} and O_h cases because even if the difference between L and X is large, it could be broken down into a series of small,

Table 9.1 The group multiplication table for the C_{2v} group (Table 2.1) modified by deletion of the σ_v row and column

C_{2v}	E	C_2	~~σ_v~~	σ'_v
E	E	C_2	~~σ_v~~	σ'_v
C_2	C_2	E	~~σ'_v~~	σ_v
~~σ_v~~	~~σ_v~~	~~σ'_v~~	~~E~~	~~C_2~~
σ'_v	σ'_v	σ_v	~~C_2~~	E

hypothetical, steps. Similar arguments will apply whenever there is a similar relationship between two groups.

Just what is this relationship between groups? In the above example, it is clear that some of the symmetry elements (and, therefore, operations) of the O_h point group are not present in C_{4v}. However, the existence of operations common to the two groups means that there will be some relationship between their group multiplication tables and, very importantly, between their character tables. The group which has the smaller number of operations is referred to as a *subgroup* of the other; there are many fascinating relationships which exist between a group and its subgroups, some of which will be met in this section.

It is evident from group multiplication tables that the symmetry operations of a point group are not, in general, independent of one another – if one symmetry operation is removed then usually, as a consequence, others will be removed also. So, for example, if in the case of the C_{2v} point group one mirror plane reflection operation is to be deleted then this can only be done if a second operation is also deleted. This is shown in Table 9.1 where it is seen that deletion of the σ_v column and row still leaves a σ_v entry (as a product of C_2 and σ'_v). We can only totally remove σ_v entries from the table if we remove either σ_v *and* σ'_v or σ_v *and* C_2. Deletion of the former pair leaves the point group C_2 (operations E, C_2) as a subgroup of C_{2v} and deletion of the latter pair gives C_s (operations E, σ). Because there is only one possible way of producing the subgroup C_2 from C_{2v}, C_2 is said to be an *invariant subgroup* of C_{2v}. More rigorously, an invariant subgroup contains only complete classes of the parent group. It is thus understandable that the point group C_s is also an invariant subgroup of C_{2v}, despite the fact that it could be derived from either σ_v or σ'_v. Key is the fact that these two mirror plane reflection operations are not in the same class in C_{2v}. The existence of an invariant subgroup can be of key importance. In Chapter 14, for instance, the problem of the enormous size of space groups will be encountered. If the faces of a crystal are ignored then it is, effectively, infinite and there is an infinity of translation operations in the group. Fortunately, it will not prove necessary to work with a group of this size. This is because the (infinite) group of all translations is an invariant subgroup of the space group. In turn, this means that we can work with a point group; all this will be explained in detail in Chapter 14.

Not all subgroups are invariant. The C_{3v} point group, which was the subject of Chapter 7, provides an example. The multiplication table for this group is given in Table 9.2; the operations are those indicated in Figure 7.1. A clockwise rotation by 120° is denoted by C_3^+ and an anticlockwise rotation by C_3^-; the mirror plane reflections are $\sigma_v(1)$, $\sigma_v(2)$ and

Table 9.2

C_{3v}	E	C_3^+	C_3^-	$\sigma_v(1)$	$\sigma_v(2)$	$\sigma_v(3)$
E	E	C_3^+	C_3^-	$\sigma_v(1)$	$\sigma_v(2)$	$\sigma_v(3)$
C_3^+	C_3^+	C_3^-	E	$\sigma_v(2)$	$\sigma_v(3)$	$\sigma_v(1)$
C_3^-	C_3^-	E	C_3^+	$\sigma_v(3)$	$\sigma_v(1)$	$\sigma_v(2)$
$\sigma_v(1)$	$\sigma_v(1)$	$\sigma_v(3)$	$\sigma_v(2)$	E	C_3^+	C_3^-
$\sigma_v(2)$	$\sigma_v(2)$	$\sigma_v(1)$	$\sigma_v(3)$	C_3^-	E	C_3^+
$\sigma_v(3)$	$\sigma_v(3)$	$\sigma_v(2)$	$\sigma_v(1)$	C_3^+	C_3^-	E

The first operation is listed along the top and the second down the left-hand side

$\sigma_v(3)$. The multiplication table in Table 9.2 differs from all the other multiplication tables that have been explicitly given in this book. It is not symmetric about the leading diagonal (top left to bottom right). Put another way, for some combinations of operations the result depends on the order in which the operations are applied. Thus,

$$C_3^+\sigma_v(1) = \sigma_v(2)$$
$$\text{but}\quad \sigma_v(1)C_3^+ = \sigma_v(3)$$

Care therefore has to be taken to specify that the operations at the head of the columns in the multiplication table are on the *right* in expressions such as those above. Equivalently, they are the *first* operation. This may seem strange but, if so, it is only because we are accustomed to reading from left to right so that in the first example above we *read* C_3^+ before $\sigma_v(1)$. However, if the operations operate on some function, ψ say, then we have

$$C_3^+\sigma_v(1)\psi$$

and, clearly, here $\sigma_v(1)$ must operate *before* C_3^+.

Problem 9.1 Using Figure 7.1 check that the C_{3v} group multiplication table given in Table 9.2 is correct.

In the multiplication table (Table 9.2) the complete deletion of a single σ_v operation requires that the other two σ_v's are also deleted, to give the group C_3 as an invariant subgroup, a subgroup that can only be obtained in one way. However, deletion of the two C_3 operations causes the multiplication table to break up into three disconnected multiplication tables. This is because we can only remove the C_3^+ and C_3^- entries from Table 9.2 by both deleting the C_3^+ and C_3^- columns and rows and then deleting one of the pairs [$\sigma_v(1)$ and $\sigma_v(2)$] or [$\sigma_v(2)$ and $\sigma_v(3)$] or [$\sigma_v(3)$ and $\sigma_v(1)$]. For each of these three choices we arrive at C_s as a subgroup. That is, there are three different but equivalent ways that C_s can be obtained as a subgroup, so it is *not* an invariant subgroup of C_{3v} – one that can be obtained in only one way – (although, as has been seen, it is an invariant subgroup of C_{2v}).

Problem 9.2 Check the above assertions by deleting from Table 9.2:

(a) All C_3 operations. Is it possible to just delete C_3^+ but leave C_3^-?
(b) One σ_v operation.
(c) One σ_v operation and the C_3 operations.

The distinction between invariant and non-invariant subgroups may seem rather academic. In fact, it has quite a variety of consequences, as two examples will show. The first example concerns molecular dynamics. Suppose that a molecule of C_{3v} symmetry is momentarily distorted – by a molecular vibration, for instance – to give a molecule of C_s symmetry. Thus, in the ammonia molecule, one N—H bond might be momentarily longer (or shorter) than the other two. Because the symmetry of the molecule has been reduced to that of a non-invariant subgroup there exists other different but equivalent distortions (in the case of our ammonia molecule there are two such equivalent distortions, corresponding to distortion of one of the two other N—H bonds to give a different but equivalent arrangement of C_s symmetry). That is, because there are three different C_s subgroups of C_{3v} there will be three equivalent distortions; the molecule would be of the same energy in each of the three equivalent configurations. In this situation, the distortion can 'rotate' from one bond to the next with no nett cost in energy. That is, the presence of non-invariant subgroups means that a molecule may indulge in some unexpected gymnastics. It is clear that special care has to be taken in a detailed analysis of the vibrational and rotational properties of molecules with symmetries which have non-invariant subgroups.

The second example is concerned with the character tables of invariant subgroups. When the operations of a point group can be written as a product of the operations of two of its invariant subgroups, then its character table can also be derived from those of these subgroups. Consider the C_{2v} point group. We have seen that it has two invariant subgroups, C_2 and C_s. It follows that all of the operations of C_{2v} can be derived from those of these two subgroups. Take each of the operations of one invariant subgroup and combine it, in turn, with all of the operations of the other invariant subgroup. Thus, in our case, carry out the steps shown in Table 9.3. That is, the operations of C_{2v} are products of the operations of C_2 and C_s. Using the language of Section 5.3, the group C_{2v} is said to be the direct product of the groups C_2 and C_s, a relationship usually written as

$$C_{2v} = C_2 \times C_s$$

Table 9.3 The combination of operations of the invariant subgroups of C_{2v}

C_2 combines with C_s to give C_{2v}
E combines with E to give E
E combines with σ to give σ_v
C_2 combines with E to give C_2
C_2 combines with σ to give σ_v'

Table 9.4

C_2	E	C_2	E	C_2
A	1	1	1	1
B	1	−1	1	−1
A	1	1	1	1
B	1	−1	1	−1

X

C_s	E	σ
A'	1	1
A''	1	−1

=

C_{2v}	E	C_2	σ_v	σ_v'
A_1	1	1	1	1
B_1	1	−1	1	−1
A_2	1	1	−1	−1
B_2	1	−1	−1	1

(more strictly, the symbol $\otimes$ should be used in place of the multiplication sign). In Section 5.3 it was also seen that a similar property holds for the corresponding character tables. Thus, in the present case the character table for C_2 is taken and the whole of it multiplied by the characters of the C_s table, to give a table four times the size of that of C_2. This is shown in Table 9.4 where, for simplicity, the C_2 character table has been written out four times on the left. Each one is then multiplied by the corresponding C_s character to give the C_{2v} table.

The C_{2v} character table given in Table 9.4 is the same as that met in Chapter 2 (Table 2.4), with the A_2 and B_1 irreducible representations interchanged in position.

Problem 9.3 Check through the individual steps in Tables 9.3 and 9.4.

Examples of this relationship between character tables have already been met. In Chapter 5 the fact that D_2 and C_i are both invariant subgroups of D_{2h} was exploited (Tables 5.3 and 5.4). In the previous chapter the fact that O_h has invariant subgroups O and C_i was used in Table 8.2 and the preceding discussion.

Problem 9.4 Show that the operations of the group C_{3v} are the product of operations of the groups C_3 (E, C_3^+, C_3^-) and C_s (E, σ). However, because C_s is not an invariant subgroup of C_{3v}, the character table of C_{3v} is *not* the direct product of the character tables of C_3 and C_s. This is immediately seen when the character tables of C_{3v} and C_3 are compared (Appendix 3).

At the beginning of this chapter it was recognized that C_{4v} is a subgroup of O_h. It is not an invariant subgroup because there are eight C_4 operations in O_h but only two in C_{4v}. It is also evident that the character table of O_h is not a direct product of that of C_{4v} with any other group because that of O_h contains triply degenerate irreducible representations whereas C_{4v} does not. This is another illustration of the rule that the character table of a group is never the direct product of the character table of a non-invariant subgroup with that of another group.

Table 9.5

C_{3v}	E	~~$2C_3$~~	~~3~~σ_v
A_1	1	~~1~~	1
A_2	1	~~1~~	−1
E	2	−~~1~~	0

C_s	E	σ
A'	1	1
A''	1	−1

9.3 Correlation tables

Having discussed how the character table of a group may be related to that of its subgroups we now consider the opposite problem: how is the character table of a subgroup related to that of the parent group? Again, the general form of the relationship is best seen by considering an example. The example which we choose corresponds to the physical situation described earlier in this chapter, that in which a molecule of C_{3v} symmetry is distorted to give a structure with C_s symmetry (i.e. a distortion leading to the loss of the threefold axis). The character tables of the C_s and C_{3v} point groups are given in Table 9.5. Note that in the C_s character table a single prime as a superscript indicates something which is symmetric with respect to a mirror plane reflection and a double prime indicates antisymmetry. This use (and meaning) of primes reappears in other point groups – see Appendix 3. In the C_{3v} character table in Table 9.5 the loss of the C_3 axis has been indicated by deleting the column associated with the corresponding operations. Since loss of this axis also leads to the loss of two σ_v mirror planes (those generated by C_3 operations acting on the 'first' σ_v) the number 3 has also been deleted from the $3\sigma_v$ entry. It is clear from Table 9.5 that the remaining characters of the A_1 irreducible representation of the C_{3v} point group are those of the A′ irreducible representation of the C_s point group. One says that the 'A_1 irreducible representation of C_{3v} *correlates* with the A′ irreducible representation of C_s'. This means that any function or object which transforms as A_1 in C_{3v} *must* transform as A′ in C_s when the molecular symmetry changes. Similarly, Table 9.5 shows that the A_2 irreducible representation of C_{3v} correlates with A″ of C_s.

The E irreducible representation of C_{3v} is both interesting and important for it does not correlate uniquely with a single irreducible representation of C_s. Rather, it gives rise to a reducible representation, one which is readily seen to have A′ + A″ components. In summary, then, we have the correlations shown in Table 9.6.

This example illustrates the general theorem that each irreducible representation of a group gives rise to a representation, which may be either reducible or irreducible, of each of its subgroups. Tables showing these correlations – so-called *correlation tables* – are available

Table 9.6

C_{3v}		C_s
A_1	⟶	A′
A_2	⟶	A″
E	⟶	A′ + A″

but it is very easy to work them out using the example given above as a model. Working-out sometimes has an advantage over the use of tables. The D_{2h} group was described in Chapter 4 (its character table is given in Table 4.1). This group has C_{2v} as a subgroup and correlation of the irreducible representations of the two groups seems very easily.

Problem 9.5 Use either Tables 4.1 and 2.4 or Appendix 3 to correlate the irreducible representations of the D_{2h} and C_{2v} groups.

Hint: If you find this problem more difficult than expected, read the next part of this section.

As the reader may have discovered when tackling Problem 9.5, whilst the problem is not a difficult one, there is a catch in correlating from D_{2h} to C_{2v}. The D_{2h} group has three different C_2 axes. The precise correlation between the two groups depends on which of the three twofold axes is retained in going from D_{2h} to C_{2v}. This does not indicate any fundamental problem, rather that it may be necessary to relabel coordinate axes (and associated basis functions) in moving between the two groups. The twofold axis retained in C_{2v} may not be that labelled z in D_{2h}, although it would be called z in C_{2v}. In compilations of correlation tables it is usual to indicate all three possible $D_{2h} \rightarrow C_{2v}$ correlations but one still has to decide which correlation is appropriate before using the tables. In such cases even experienced workers may find that they are less likely to make a mistake by working out the correlation for themselves rather than by using the tables!

There is another way of showing correlations, and this is by use of a diagram. That for the C_{3v}–C_s correlation is given in Figure 9.5. Such diagrams emphasize another aspect of the consequences of a decrease in symmetry. Figure 9.5 shows, for example, that a function transforming as A_1 in C_{3v} and one of the two functions transforming as E have a common symmetry in C_s: that described by the A' irreducible representation. This means that in C_s symmetry these two functions can interact with each other, an interaction which is symmetry-forbidden in C_{3v} symmetry.

Another aspect of a reduction in symmetry, equally evident from either Table 9.6 or Figure 9.5 (since they describe the same data), is that a decrease in symmetry may lead to

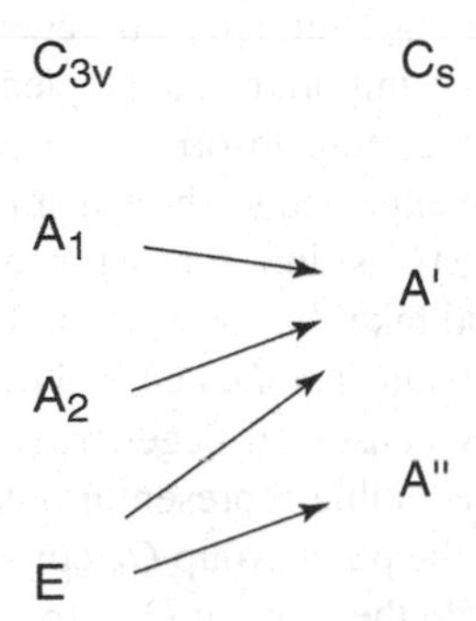

Figure 9.5 The correlation between the irreducible representations of the groups C_{3v} and C_s

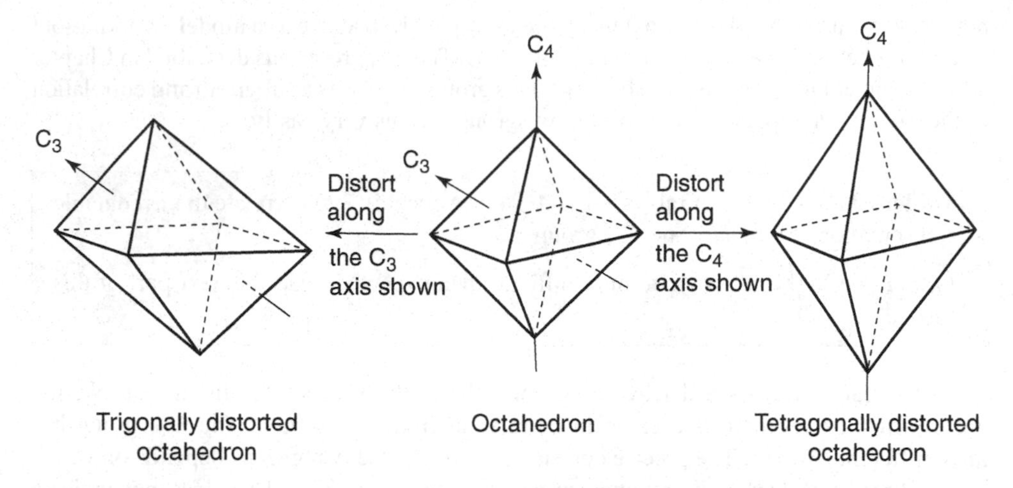

Figure 9.6 A symmetrical distortion of an octahedron (O_h) along a threefold axis gives a figure with D_{3d} symmetry whilst a symmetrical distortion along a fourfold axis gives a D_{4h} figure

a decrease in degeneracy. In the example above, the degeneracy of functions transforming as E in C_{3v} is lost in C_s. A particularly important case is that of octahedral transition metal coordination compounds, discussed in Chapter 8. Although much of the basic theory of such compounds is conveniently developed assuming full octahedral symmetry (O_h), real-life examples usually show some minor distortion. The most important cases are those in which such a distortion is either along a fourfold or a threefold axis (either distortion therefore destroying all other fourfold and threefold rotation axes), as shown in Figure 9.6. The appropriate correlation table is given in Table 9.7.

Problem 9.6 Use the character tables of the O_h, D_{4h} and D_{3d} point groups in Appendix 3 to check the correlations given in Table 9.7.

It is seen from Table 9.7 that in D_{4h} symmetry all degeneracies present in O_h symmetry are at least partially removed. One important consequence of this is that when a single spectral band is predicted in the electronic absorption spectrum of an octahedral transition metal complex this band would be expected to show a splitting if the real symmetry is D_{4h} and a degeneracy is involved in the transition (for instance, the excited state might be triply degenerate). Such a splitting could take the form of the observation of a separate peak, a shoulder or an asymmetry on the band. The D_{3d} case shows, however, that it is not always true that a reduction in symmetry causes all degeneracies to be relieved (i.e. a splitting to occur); thus the E_g and E_u irreducible representations of O_h persist in D_{3d}. There is, however, a trap for the unwary. In the point group O_h convention was followed in choosing a C_4 axis as the z axis; one would do the same in D_{4h}. In D_{3d}, the axis of highest symmetry is a C_3 axis and *this* is the z axis. It follows that, although E_g of O_h becomes E_g of D_{3d} it is NOT true that the basis functions for E_g in O_h, $x^2–y^2$ and $1/\sqrt{3}(2z^2 - x^2 - y^2)$, are

Table 9.7

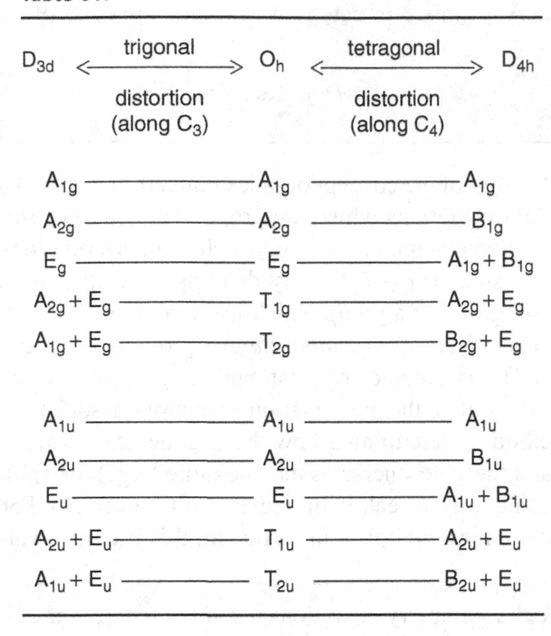

D_{3d}	← trigonal distortion (along C_3) →	O_h	← tetragonal distortion (along C_4) →	D_{4h}
A_{1g}		A_{1g}		A_{1g}
A_{2g}		A_{2g}		B_{1g}
E_g		E_g		$A_{1g} + B_{1g}$
$A_{2g} + E_g$		T_{1g}		$A_{2g} + E_g$
$A_{1g} + E_g$		T_{2g}		$B_{2g} + E_g$
A_{1u}		A_{1u}		A_{1u}
A_{2u}		A_{2u}		B_{1u}
E_u		E_u		$A_{1u} + B_{1u}$
$A_{2u} + E_u$		T_{1u}		$A_{2u} + E_u$
$A_{1u} + E_u$		T_{2u}		$B_{2u} + E_u$

basis functions for E_g in D_{3d}. A detailed analysis, using the methodology of Appendix 2, is needed to describe the correlations between basis functions in O_h and D_{3d}.

In practice, the correlations which exist between groups are quite important for two reasons. First, as indicated above, they enable the properties of low symmetry molecules to be related to those of high symmetry species. Another aspect of this occurs when a molecule is high symmetry but is trapped in a low symmetry environment – an octahedral molecule on a low symmetry lattice site in a crystal, for example. Any spectral splittings that occur as a result can give information on the degeneracies in the high symmetry situation. Second, some of the problems of degenerate representations – and some were met in the last chapter and more will be met in the next – can often be neatly side-stepped by pretending that a molecule has a lower symmetry than is in fact the case – so that the degeneracy is split (or 'relieved') – and, after working in the low symmetry group, using a correlation relationship to apply the result to the high symmetry case.[1]

Another interesting aspect of the relationship of a group to its subgroups is that the number of symmetry operations in a group (the order of the group) is a simple multiple of the number of symmetry operations of any of its subgroups. The multiplication factor – which is always an integer – is called the *index* of the subgroup (relative to the particular parent group). Thus, the C_3 group (of order 3) is subgroup of index 2 of the point-group C_{3v} (of order 6). However, the same C_3 group (of order 3) is a subgroup of index 40 of the point group I_h (of order 120).

[1] A particular attraction of a reduction in symmetry is that interactions can be caused to become 1:1 – ambiguities about which of the functions within a degenerate set interact with members of another degenerate set no longer exist.

Problem 9.7 Use Appendix 3 to determine the index of each of the following subgroups of O_h:

$$D_{4h},\ C_{4v},\ D_{3d},\ C_{3v},\ D_{2h},\ C_{2v}$$

An important application of the concept of index concerns rotational subgroups. A point group may only contain operations which are proper rotation operations (such as C_2, C_3 and so on) or it may contain some operations which are pure rotations and others which are improper rotations (such as σ_v, i, S_4). By deleting all of the improper rotations one can always obtain a subgroup of a group which itself contains both proper and improper rotations. What remains is the *pure rotational subgroup* of the parent group. This subgroup is always of index 2. The importance of rotational subgroups is their relationship to the (infinite) group consisting of all the pure rotation operations associated with a sphere. This group provides a method of determining how the degeneracies which may be associated with the free atom (and these degeneracies may be quite large) are split up when the atom is placed in the molecule (this is dealt with in detail in Chapter 10). For the metal atom at the centre of a transition metal complex, in particular, this is quite invaluable information.

Problem 9.8 Use the symmetry operations listed at the top of the character tables in Appendix 3 to show that deletion of improper rotation operations in the following point groups in each case leads to a pure rotational subgroup of index 2.

$$I_h,\ T_h,\ D_{5h},\ C_{2h},\ D_{3d}$$

Note: In several of these examples it is possible to obtain subgroups by deletion of all improper and some proper rotations. Such subgroups are not of index 2. The statements made in the text refer to the largest pure rotational subgroup of a given group.

9.4 Summary

There are relationships between a group and its subgroups. The operations of a group can immediately be obtained from the operations of its subgroups (p. 236), as can its character table (p. 239) provided that the subgroups are invariant (p. 236). Correlations exist between the irreducible representations of a group with its subgroups and are useful in discussions associated with molecules which approximate to high symmetry species (p. 240). Groups containing improper rotation operations always have a pure rotational subgroup of index 2 (p. 244).

10 Tetrahedral, icosahedral and spherical symmetries

10.1 An overview

This is a chapter largely devoted to three groups, tetrahedral, icosahedral and spherical, all of which are such that their coordinate axes, x, y and z, can be chosen to be symmetry-equivalent. Diagrams with each of these geometries are shown in Figure 10.1, where the axis sets about to be discussed for the tetrahedron and icosahedron are included and also extended to the spherical case. The fact that they can have equivalent x, y and z axes is a feature which they share with the O and O_h groups of Chapter 8. Indeed, much of the discussion in that chapter can be carried forward to this. In particular, transition metal coordination complexes with a tetrahedral geometry, the first geometry to be considered here, are sometimes conveniently discussed at the same time as octahedral complexes. This is because the character tables of the point groups T_d (the tetrahedral group almost invariably encountered) and O are isomorphous ('isomorphous' means that apart from the labels surrounding them, the tables are identical; for O and T_d the labels on the irreducible representations are also identical).

The first two geometries, each of which encompasses more than one point group, have a feature in common: they disobey the rule that the axis of highest rotational symmetry should be taken as the z axis. For tetrahedra the rule can be taken to apply if one adopts a rather broad definition of 'rotational' but for the icosahedra there is no such escape clause. In both cases, it is the C_2 axes which are taken to be x, y and z. The reason is simple – in both groups there is a set of C_2 rotational axes at 90° to each other, which is just what one needs for a coordinate set. For the tetrahedron, these axes are co-linear with three S_4 axis and so if one is prepared to regard S_4 as an axis of rotational symmetry (in other words, ignore the implicit 'pure' before 'rotational') then the normal rule is followed. However, if one insists on the 'pure', then the axes of highest rotational symmetry are C_3 axes, axes that are inclined relative to each other at the tetrahedral angle, 109°28″. If one C_3 is chosen as z then x and y, although perpendicular, point in rather aimless and unhelpful directions in space, whatever the details of the choice. They remain degenerate, of course, but this is not much help when working with such an apparently different set. Those that want to make life as easy as possible will opt for the C_2 choice of axes. For the icosahedral groups the axis of highest rotational symmetry, pure or not, is a C_5, of which there are six. The angle between C_5 axes is 63°26″ – again not much use in the search for 90°! For the icosahedron,

Symmetry and Structure: Readable Group Theory for Chemists Sidney F. A. Kettle

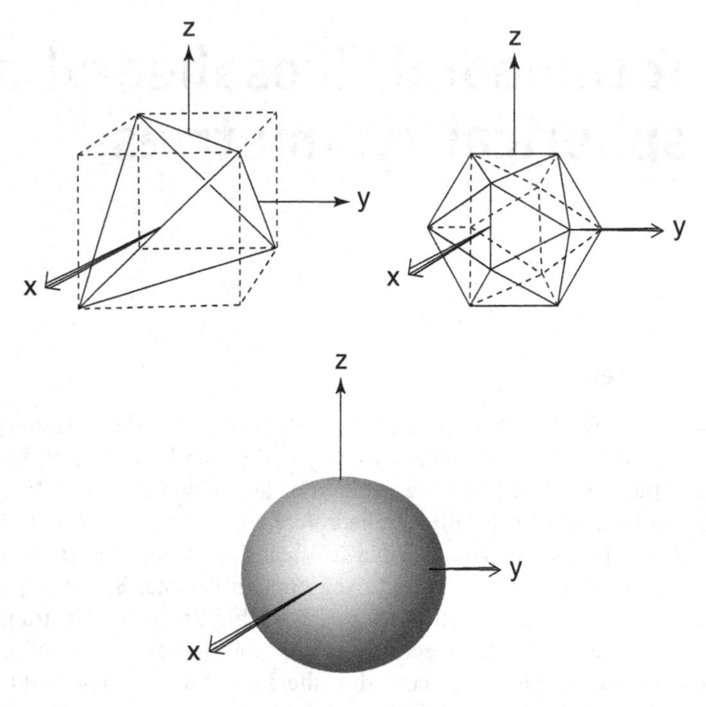

Figure 10.1

there are 15 C_2 axes (for the simple twelve-corner icosahedron – that of Figure 10.1 – and its 30 edges, one C_2 axis passes through each pair of opposite edges). From the 15 C_2, three that are mutually perpendicular are chosen as coordinate axes. Clearly, the choice of the trio to select is not unique.

For both the tetrahedron and the icosahedron the choice of axes has an important consequence for the use of the projection operator method in obtaining symmetry-adapted functions. When, as is likely for these two, there is no member of a basis set (atomic orbital, vibration, whatever) which lies on an axis, one has to take care. Basis sets have to be invented which respect the axes. So, take a tetrahedron of four atoms, which are labelled a, b, c and d (no diagram is given, one can be drawn any way the reader wishes). A twofold axis bisects the edge joining a and b (or any other pair). As basis functions $(a + b)$ and $(a - b)$ should be used (and their equivalents for two other edges). The function $(a + b)$ generates symmetry-adapted combinations which are symmetric under the C_2 whilst $(a - b)$ generates those that are antisymmetric. If either a or b were chosen instead, one would be imposing a C_3 axis as choice of coordinate axis on the problem. With such a choice, whilst the first function of any degenerate set would look reasonable, the other(s), corresponding to the other coordinate axes in rather general positions, would both be difficult to generate and aesthetically unpleasant when obtained. For the I_h group, the generation of all symmetry-adapted members of sets transforming as the G and H irreducible representations presents a problem. Again, the crafting of suitable basis sets can be a help, particularly if one has an idea of the likely answer(!). The latter can be obtained in one of two ways: from basis functions given in the character table or from the characteristics of the atomic orbital set that spans the irreducible representation (the way to discover a suitable set is described later in this chapter). What one is looking for is the sort of nodal pattern which is going to be present in the final functions. A useful trick is to pretend that the symmetry is lower than I_h; that is, to use the method described at the end of the previous chapter. It is also mentioned at the beginning of Appendix 4. If all else fails, there is a method called 'Schmidt orthogonalization' which always works, albeit at the expense of a lot of effort and, usually, inelegant final functions.

The final symmetry that we study, that of the sphere, has an even less unique choice of coordinate axes; any three mutually perpendicular axes are acceptable – and the highest rotational symmetry of each of these axes is infinite! Put another way, the basic rotational operation of a sphere, that of which all others are multiples, is an infinitesimally small rotation. Strange as this may seem, it is in fact very important. Much of quantum mechanics as we know it arises from the properties of these infinitesimal rotations. First, however, we start with something much more evidently understandable. The tetrahedron.

10.2 The tetrahedron

We have mentioned that it is often possible to discuss octahedral and tetrahedral transition metal complexes together; this is because their geometries are derived from a cube and the labels of the irreducible representations of the point groups O (octahedral, pure rotations only) and T_d (tetrahedral, as in methane) are identical. Tetrahedral complexes, of general formula ML_4, are of widespread occurrence but are not as common as octahedral. Together, species with geometries which approximate to either octahedral or tetrahedral account for at least 80% of all coordination compounds. In organic chemistry, of course, it is the tetrahedral geometry which is the important one. Clearly, it is appropriate that a discussion of tetrahedral molecules should be included in this text.

Because a tetrahedron is derived from a cube, the symmetry operations which turn a tetrahedron into itself are also symmetry operations of the cube (but the converse is not true). The corresponding symmetry operations are:

Cube (and octahedron)	E,	$8C_3$,	$6C_4$,	$3C_2$,	$6C_2'$,	i,	$8S_6$,	$6S_4$,	$3\sigma_h$,	$6\sigma_d$
Tetrahedron	E,	$8C_3$,		$3C_2$,				$6S_4$		$6\sigma_d$

The group of operations of the tetrahedron is given the shorthand label T_d (pronounced 'tee-dee').

Problem 10.1 Draw diagrams to show all of the symmetry operations of a tetrahedron. *Hint*: Figure 10.1 is helpful. The mid-point of each cube face in this figure corresponds to a corner of the octahedron shown in Figure 8.2.

Although there exists a group consisting solely of the pure rotations of the tetrahedron (E, $8C_3$, $3C_2$, a group called T), the group T_d is not a direct product group of T with any other group (if it were, there would be three, not two, additional classes of T_d compared

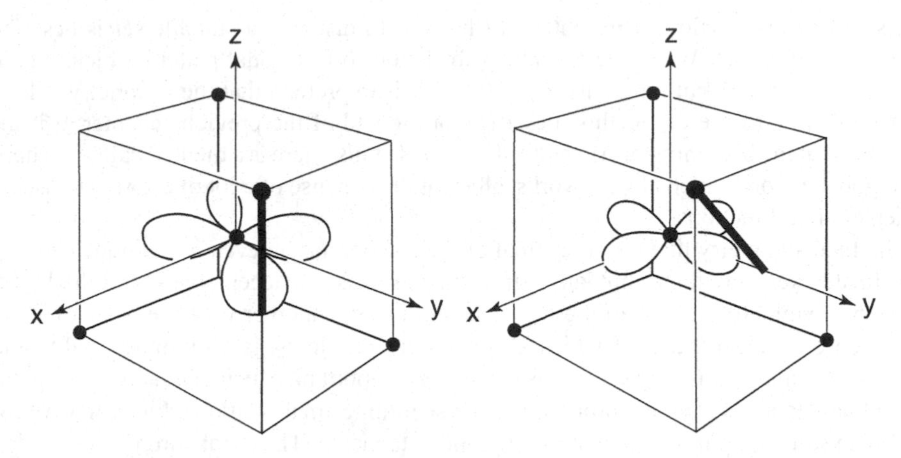

Figure 10.2 The lobes of the t_2 orbitals (left) point towards the mid-point of a cube edge whereas the lobes of the e orbitals (right) point towards the mid-points of cube faces. These distances are in a $1:\sqrt{2}$ relationship so it seems probable that the t_2 orbitals will experience the larger repulsion

with T and they would have 1, 8 and 3 operations in them). Remember, character tables are all square – they have as many irreducible representations as classes of operation.

Problem 10.2 Detail the arguments behind the assertion just made.

The character table of the T_d group is given in Table 10.1.

The bonding in tetrahedral molecules will not be discussed in detail but the essentials are given in Table 10.2, which summarizes the ways in which the various orbitals transform (Table 10.1 does not give them explicitly, only in coordinate axis form).

Problem 10.3 Check that the transformations given in Table 10.2 are correct. The generation of the correct reducible representation for the transformation of the apical atom π orbitals is not a trivial task and the reader who gets the correct answer is to be congratulated. Success depends on choosing the orientation of the π orbitals mindful of the symmetry operations under which they are to transform (this problem is best tackled by the techniques described in Appendix 4).

Problem 10.4 Use the projection operator method to derive explicit forms for the σ orbitals of four atoms arranged at the apices of a tetrahedron.

Problem 10.5 Use the data in Table 10.2 to describe the bonding in methane, CH_4. If the symmetry-adapted functions are to be generated, there are some helpful comments following the discussion of coordinate axes earlier in this chapter.

Table 10.1

T_d	E	$8C_3$	$3C_2$	$6S_4$	$6\sigma_d$	
A_1	1	1	1	1	1	$(x^2+y^2+z^2)$
A_2	1	1	1	−1	−1	
E	2	−1	2	0	0	(x^2-y^2), $1\sqrt{3}(2z^2-x^2-y^2)$
T_1	3	0	−1	1	−1	R_x, R_y, R_z
T_2	3	0	−1	−1	1	(x, y, z) (xy, yz, zx)

It will be noted that double and triple degeneracies exist in a tetrahedral environment and, rather important, that the p and three of the d orbitals of a central atom both transform as T_2. This means that the t_2 d orbitals in a tetrahedron will be mixed with a bit of p, and vice versa. In this lies, ultimately, the explanation of the fact that tetrahedral transition metal complexes tend to be more highly coloured than do octahedral. Because d and p orbitals mix, this mixing makes some electronic transitions more allowed in a tetrahedron than they are in an octahedron (pure d–d transitions are forbidden, but d–p are allowed). Just as for an octahedron, in a tetrahedral environment the d orbitals of a transition metal split into two sets; $d_{x^2-y^2}$ and $d_{(1/\sqrt{3})(2z^2-x^2-y^2)}$ are of E symmetry and, as has been commented, d_{xy}, d_{yz}, and d_{zx} of T_2. If a diagram analogous to Figure 8.35 is drawn for a tetrahedron then it is concluded that in this geometry splitting the T_2 set is of higher energy than the E – the inverse of the splitting found for an octahedron (Figure 10.2). This is the sort of problem for which one could be tempted to orient the z axis of the tetrahedron along a threefold axis. The orbital d_{z^2} would then point directly towards a ligand and one might hope for a more evident proof of the $E - T_2$ splitting than is apparent in Figure 10.2. Alas, not so. The 'new' d_{z^2} is not the same as the first and, indeed, it is not even an E orbital. The 'new' d_{z^2} is actually a mixture of orbitals from the 'old' E and T_2 sets. A rigorous proof that the d orbitals split into e and t_2 sets will be given later. The splitting of the d orbitals in a tetrahedron is only about one half of that for the corresponding ligands arranged octahedrally (more accurately, 4/9). This reduction in separation means that strong field tetrahedral complexes are virtually unknown; almost all are weak field. The 4/9 factor is actually the ratio of the

Table 10.2

		Symmetry
Orbitals of an atom at the centre of the tetrahedron	s	A_1
	(p_x, p_y, p_z)	T_2
	$(d_{x^2-y^2}, d_{(1/\sqrt{3})(2z^2-x^2-y^2)})$	E
	(d_{xy}, d_{yz}, d_{zx})	T_2
Orbitals of the four atoms at the apices of the tetrahedron		
	σ	$A_1 + T_2$
	π	$E + T_1 + T_2$

squares of the number of ligands in the two cases, 16/36, and so must be expected to be rather approximate. Identical metal–ligand bond lengths are assumed, for instance.

It is the fact that in both a tetrahedral and an octahedral environment the d orbitals of a transition metal split into a set of two (E in T_d, E_g in O_h) and a set of three (T_2 in T_d, T_{2g} in O_h) – and that the d orbitals of E symmetry in T_d are the same as those of E_g in O_h (and similarly for the T_2 and T_{2g} orbitals) – which enables a common discussion of the two symmetries in specialized texts. In this common discussion the orbital sets are referred to as E and T_2 (one can think of the discussion of octahedral molecules taking place in the group O – for there, these are the correct symmetry labels). The two geometries are then distinguished by the fact that the E–T_2 splittings are of opposite signs. One warning, however: a warning signalled in the discussion above. Although a set of three p orbitals of a central atom have T_{1u} symmetry in O_h, they have T_2 symmetry, *not* T_1, in T_d. The moral is clear – never transfer labels between the geometries without checking.

Problem 10.6 Show that the p orbitals of an atom at the centre of a tetrahedron have T_2 symmetry.

We have met two tetrahedral groups, T and T_d. There is a third, and although it is scarcely ever met it is not without interest. It is called T_h, and is the direct product of T with C_i (the group containing only the operation of inversion in a centre of symmetry in addition to the identity). Consider a molecule $[M(H_2O)_6]^{n+}$, where the M and the six surrounding oxygens comprise an octahedral complex typical of those discussed in the last chapter. Now add the two H's to each oxygen. They destroy the fourfold axes, at best leaving only C_2's (and even this requires that the hydrogens on *trans* oxygens be co-planar). If, finally, adjacent H_2O ligands are interrelated by a C_3 rotation, a T_h molecule results (Figure 10.3).

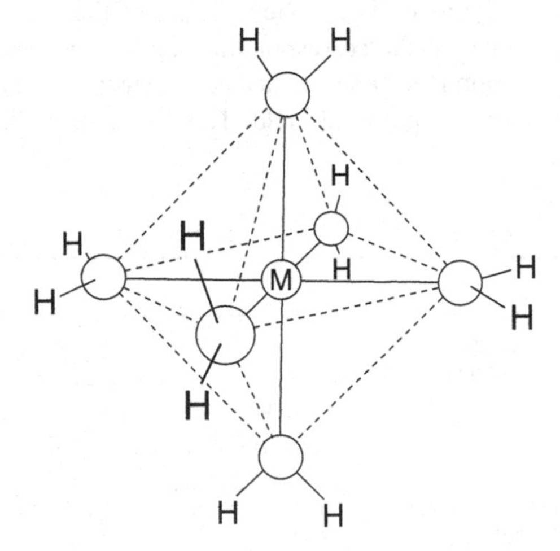

Figure 10.3 A $M(H_2O)_6$ complex of T_h symmetry. The oxygens are arranged in O_h fashion but the position of the hydrogens, whilst destroying the octahedral C_4 axes, retain the σ_h mirror planes. The arrangement of the hydrogens is key to the geometry and so has been emphasized

10.3 The icosahedron

Apart from spherical symmetry, the icosahedron has the highest symmetry possible for a three-dimensional object. It has a plethora of rotational axes, $6C_5$, $10C_3$ and $15C_2$. The group containing all the operations associated with these axes (together with E, of course), is labelled I. It has a deceptively simple character table, only five rows and columns. For each C_5 axis there are two C_5 rotation operations (clockwise and anticlockwise by 72°) and two $C_5{}^2$ rotation operations (clockwise and anticlockwise by 144°). For each C_3 axis there are two C_3 operations (clockwise and anticlockwise by 120°) and for each C_2 axis there is a single C_2 operation. The five classes of operations are E, $12C_5$, $12C_5{}^2$, $20C_3$, and $15C_2$, a total of 60. The theorems of Chapter 6 (see Section 6.4) indicate something strange. The sums of the squares of the characters under the identity operation, E, must sum to 60. And one of the numbers to be squared has to be a 1 (there has to be a totally symmetric irreducible representation). Subtracting this 1, we need four numbers which, when squared, sum to 59. Even if all were 3, the highest degeneracy that we have met so far, we would only get a total of 36 (4×3^2). We have to invoke higher degeneracies. In fact, degeneracies of 3, 3, 4 and 5 ($9 + 9 + 16 + 25 = 59$). Fourfold degenerate irreducible representations are labelled G and the fivefold H.

Whilst the group I is the simplest icosahedral group, it is not the one usually met. Almost invariably, it is the group I_h which is discussed. (The origin of the subscript h in the label is to be found in the early German literature; it indicates horizontal mirror planes perpendicular to the usual C_2 choice of coordinate axes. However, in the character table the mirror plane reflections are denoted σ, not σ_h, and it has always been so, even in the early German literature!.) The key distinction between the groups I and I_h is that the latter contains a centre of symmetry. I_h is the direct product of I and C_i (which contains just E and i). This means that the operations of I_h are the 60 of I together with another 60, in which each of the first 60 is individually combined with i. These second 60, like the first, fall into five classes. These are i, $12S_{10}$, $12S_{10}{}^3$, $20S_6$ and 15σ. Commonly, as we have met before (Chapter 8), the S_n ($n = 10, 6$) are referred to as 'rotation reflection' operations by chemists, whilst physicists and crystallographers tend to prefer 'rotation inversion'.

Although examples of molecules with an icosahedral geometry have long been known (the anion $B_{12}H_{12}{}^{2-}$ is the classic example and is shown in Figure 10.4a), and some viruses also have the symmetry, in recent years it has become much more important. The main reason for this is the discovery of the molecule C_{60}, shown in Figure 10.4b. It has spawned an entire chemistry, with the icosahedral group as a common reference point.[1] But there is another reason. In Chapter 12 we will see that it is not possible for a crystal to have C_5 axes. But icosahedral crystals exist! The whole subject of so-called quasicrystals, with apparent fivefold axes, is a fascinating one and we shall refer to it again in Chapter 13. Clearly, the icosahedral group I_h is worthy of some study; its character table is given in Table 10.3.

There are several other aspects of the I_h character table that are worthy of comment. First, the appearance of characters related to the angles of 72 and 144°, although the appearance is not the simple one that might have been expected. Characters involving cos 72° are associated with rotations of both 72 *and* 144°, and the same is true for characters involving

---

[1] A very readable, and pictorial, paper which uses carbon cages akin to C_{60} as exotic models with which to introduce point groups, particularly those of high symmetry, including I, is W.O.J. Boo, *J. Chem. Educ.* **69** (1992) 605.

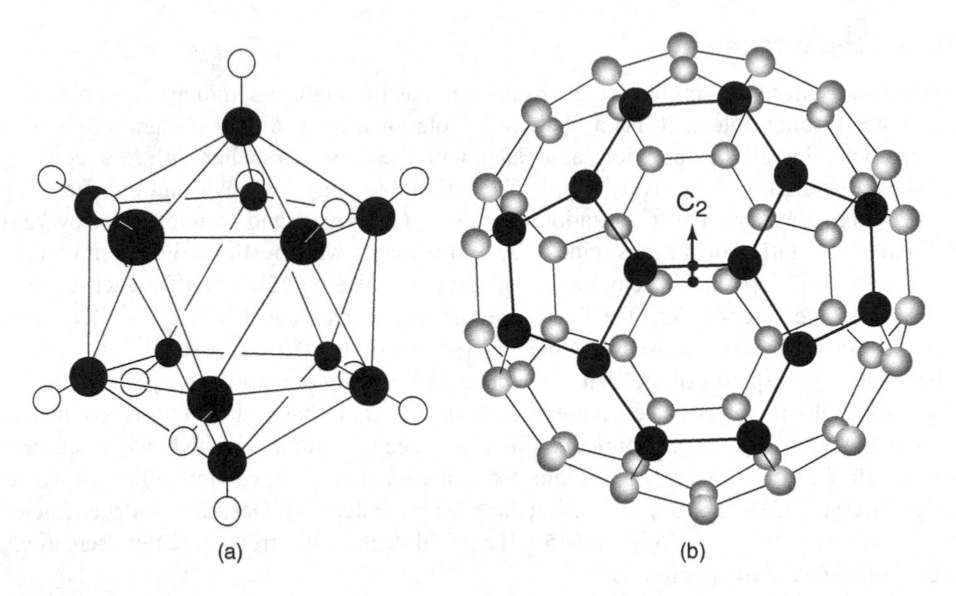

Figure 10.4 Two molecules with I_h geometry. (a) The $B_{12}H_{12}^{2-}$ anion. This view shows the C_3 axes passing through the centre of pairs of opposite faces. (b) Fullerene, C_{60}. This view is almost down a C_2 axis, which might well be chosen as a coordinate axis. Clearly visible are hexagonal rings, which are centred by C_3 axes, and pentagonal rings, which are centred by C_5 axes

cos 144°. Second, The use of the labels *G* and *H* to denote fourfold and fivefold degeneracies, respectively, is not as obscure as it seems. Although it is now relatively rare, one still encounters the use of *F*, rather than the *T* that we have used in this book, to denote triple degeneracy. So, one might meet the statement that 'the d orbitals of a metal in a tetrahedral complex split into e and f_2 sets', rather than the e and t_2 we used above. Following the alphabetical sequence, *E*(2), *F*(3), it is logical that *G* should be (4) and *H* (5).[2] Third, if there were a transition metal atom at the centre of a molecule of I_h symmetry (and molecules with a metal atom inside a C_{60} cage are well known, even if they are not usually at the centre), then there would be no splitting of the d orbitals such as occurs for octahedral and tetrahedral geometries. The d orbitals transform, as a set of five, under the H_g irreducible representation. This is shown in the character table above (Table 10.3) because the relevant parts of their explicit mathematical functions are given at the right-hand side of the table. Finally, the nodal patterns characteristic of the individual irreducible representations and their components exist, of course, but are very difficult to show in detail. For the O_h group we were just about able to indicate how to do it by projection onto the surface of a sphere (Figure 8.9). There the half-sphere that was, more or less, visible had to contain twenty-four segments (half the order of the group, because only half of the sphere was, more or less, visible). It would be possible to give a corresponding diagram for the icosahedron but sixty (half of 120) segments would have to be shown. The diagram would be impossibly complicated. We

[2] Even more logical would be to follow the sequence *T*(3), *U*(4) and *V*(5) – and, indeed, these labels can be found used in this way, *U* and *V* indicating quadruple and quintuple degeneracies, respectively. However, it is the labels *G* and *H* which are the more commonly used and so have been given in the text.

Table 10.3 The character table of the icosahedral group I_h; that of the group I is contained within the top left-hand quadrant, once the g subscripts on the irreducible representation labels are deleted

I_h	E	$12C_5$	$12C_5^2$	$20C_3$	$15C_2$	i	$12S_{10}$	$12S_{10}^3$	$20S_6$	15σ	
A_g	1	1	1	1	1	1	1	1	1	1	$x^2+y^2+z^2$
T_{1g}	3	−2cos144	−2cos72	0	−1	3	−2cos144	−2cos72	0	−1	(R_x, R_y, R_z)
T_{2g}	3	−2cos72	−2cos144	0	−1	3	−2cos72	−2cos144	0	−1	
G_g	4	−1	−1	1	0	4	−1	−1	1	0	
H_g	5	0	0	−1	1	5	0	0	−1	1	$(\frac{1}{\sqrt{6}}[2z^2-x^2-y^2], \frac{1}{\sqrt{2}}[x^2-y^2], xy, yz, zx)$
A_u	1	1	1	1	1	−1	−1	−1	−1	−1	
T_{1u}	3	−2cos144	−2cos72	0	−1	−3	2cos144	2cos72	0	1	(T_x, T_y, T_z) (x, y, z)
T_{2u}	3	−2cos72	−2cos144	0	−1	−3	2cos72	2cos144	0	1	
G_u	4	−1	−1	1	0	−4	1	1	−1	0	
H_u	5	0	0	−1	1	−5	0	0	1	−1	

shall not attempt to give it. An alternative would be to detail the atomic orbitals which form a basis for a particular irreducible representation. This is entirely feasible for some of the irreducible representations, those of low nodality – and we will give them later, although it is not difficult to anticipate them from the data at the right-hand side of Table 10.3. But help is at hand for the reader who is masochistic enough to want to see the complexities in full. In the next section we detail how to work out how any set of orbitals, s, p, d, f, g, . . ., no matter how complicated, transforms in any molecular symmetry. Applied to the present example, the masochist simply has to go further and further along the sequence, applying the method to I_h, ticking off the irreducible representations as they are met, until all have been encountered.

10.4 Spherical symmetry

The number of rotation operations in the spherical group is infinite; for a sphere, any angle of rotation about any axis turns the sphere into a sphere indistinguishable from the starting one. So, the operation of rotation by any angle, about any axis, is an acceptable symmetry operation. We cannot expect a simple character table! Indeed, the group character table contains an infinite number of classes of operations and an infinite number of irreducible representations. And this, of course, is without including any improper rotations – mirror plane reflections and the like. Of all the improper rotations the only one we need consider is that of inversion in a centre of symmetry. All of the other improper rotations are combinations of it with one of proper rotations. Recognizing this, it proves convenient for us to delay consideration of the operation of inversion in a centre of symmetry. Even so, given its infinite nature, all we can hope to do is to give a fragment of the character table of the spherical group. Such a fragment is given in Table 10.4, which is one that will be found in the literature, with a few trivial additional simplifications.

There is much that can be said about Table 10.4. To start with, it is evidently wrong. It shows a single C_2 operation, a single C_3 and so on – but we know that there has to be an infinite number of each. In each case, the one shown is representative; all the others of its infinity of partners have exactly the same properties. Next, despite the complexity of the group, the individual characters are very simple, ones that we recognize (the τ is a shorthand for the $-2\cos 144$ that we met in Table 10.3). This simplicity suggests that there

Table 10.4

K	E	C_2	C_3	C_4	C_5	C_6	. . .
S	1	1	1	1	1	1	. . .
P	3	−1	0	1	τ	2	. . .
D	5	1	−1	−1	0	1	. . .
F	7	−1	1	−1	$-\tau$	−1	. . .
G	9	1	0	1	−1	−2	. . .
H	11	−1	−1	1	1	−1	. . .
I	13	1	1	−1	τ	1	. . .
.	.	.	.	.	.	.	. . .

must be a relatively simple way of obtaining these characters, and indeed there is – we will meet it shortly. Thirdly, the labels given to the irreducible representations are labels that we recognize – *S*, *P*, *D*, *F*... Remembering that lower case symbols are used when irreducible representations are used to label orbitals, we realize that they are very familiar, s, p, d, f... Perhaps this is an appropriate point to remind the reader of a phrase earlier in this chapter; 'much of quantum mechanics as we know it arises from the properties of these infinitesimal rotations'. Well, something of the 'quantum mechanics as we know it' has just been met; the relevance of the infinitesimal rotations is to come! One final point about Table 10.4: the label given to the spherical group. We have used *K*. This has a tradition dating from the origins of the theory (it is the initial letter of the German word Kugelgruppe – 'spherical group'), although alternative labels will be found in the literature. A popular one is *R*(3), indicating the group of all rotations in three dimensions.

The way that the *K* character table has been given, incorrectly, is an indication that Table 10.4 is more for show than for use. Hopefully, however, it will have given an indication of the importance of the group. To investigate it further it is necessary to depart from the pattern established so far in the book. Up to this point character tables have been the focus of attention. Now, we turn to the fundamental operation of the group. The spherical group may be infinite, but it is built upon a single operation, that of an infinitesimal rotation (about any axis). Not surprisingly, to gain insight into the properties of this operation is to gain insight into the spherical group. In fact, that which follows applies to all rotations; infinitesimal rotations are of fundamental importance (in them lies the reason that the concept of angular momentum is so often invoked in textbooks on quantum theory). On the other hand, finite rotations are of practical importance, as we have seen many times – and shall see again. Because it is an operation, it is convenient to have a rotation (infinitesimal or not) operate on something; something which is subjected to the rotation. Of course, it could be almost anything, but it is easiest to work with a single object, one which behaves in a simple way under the operation. An obvious choice is a radial vector pointing out from the axis of rotation (Figure 10.5). Rotation turns it into a slightly different vector, but one of the same magnitude (length). We have met this situation before, in the discussion associated with Figure 7.4, and the reader may find it helpful to look back at this.

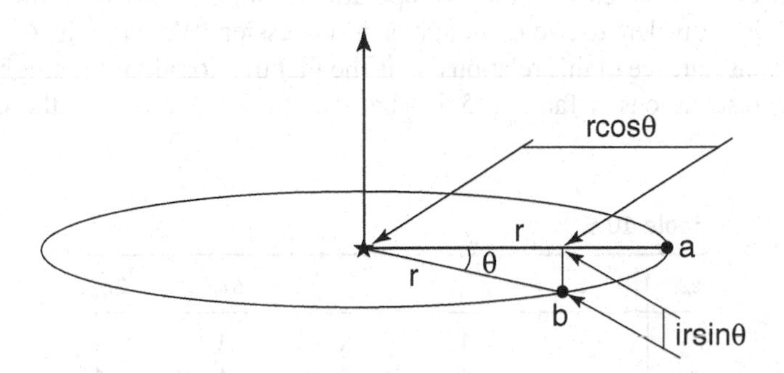

Figure 10.5 When a vector originally terminating at the point a is rotated by θ to the point b, the amount of the original vector contained within the final is rcos(θ). New is an additional component of rsin(θ); however, this is perpendicular to the original direction and this necessitates the inclusion of i, where $i^2 = -1$

Here, we focus on somewhat different aspects. First, the character under the operation and, related, the relationship between the 'start' and 'finish' vectors. The first of these, the character, is easy. We are looking for a function (of the angle) which means that the 'finish' more closely resembles the 'start' the smaller the angle (here, we are thinking of an angle which is of a significant magnitude; but the function found will apply to an infinitesimal rotation also). The obvious choice of function is $\cos(\theta)$. This has all the right properties. For instance, when $\theta = 180°$, and the 'finish' vector is the negative of the 'start', $\cos(\theta) = \cos 180° = -1$. But $\cos(\theta)$ clearly gives an incomplete description of the entire rotation. So, at $\theta = 90°$, $\cos 90° = 0$. But the vector does not vanish, as implied by the 0; rather, it is pointing in a perpendicular direction to that of the 'start'. One immediately thinks of $\sin(\theta)$, because this is 1 when θ is 90. Whilst this is true, it cannot be the entire story, because the vector remains the same length, irrespective of the angle of rotation. The function $\cos(\theta) + \sin(\theta)$ is not acceptable. To see this point, return to the $\theta = 0$ and 180° cases. The two vectors are of the same length but one is the negative of the other. The length of the vector is related to the square of the vector, not the vector itself. Actually, this is a point that we have met in a different guise many times in earlier chapters; for example, the application of the orthornormality theorems of Section 6.4 and in the normalization of functions throughout the book. If we want to maintain a constant length for the vector and yet discuss its variation with an angle θ, then we have to aim for a square which is $\cos^2(\theta) + \sin^2(\theta)$, because $\cos^2(\theta) + \sin^2(\theta) = 1$; the vector stays the same length, irrespective of angle. A function such as $\cos(\theta) + \sin(\theta)$ is no good, because its square is $\cos^2(\theta) + 2\cos(\theta)\sin(\theta) + \sin^2(\theta)$.

We can see another problem with $\cos\theta + \sin\theta$; it describes a harmonic motion but contains no explicit indication that something is rotating – it is the sort of expression that might be used to describe the vibrations of a violin string. How can we include the rotation? Here, an aspect of group theory comes to our aid, which can be seen in all of the character tables that have been met, although sometimes only in trivial fashion. Consider a singly degenerate irreducible representation (this avoids the possibility of mixing functions). Table 6.1, the character table of the C_{4v} group, with four singly degenerate irreducible representations, is ideal for this and is reproduced in Table 10.5. The key point is one that we have met several times in earlier chapters when talking of direct products. When two operations multiply (are applied in succession) the character of the (single) operation which is their product is the product of the characters of the two operations. So, in Table 10.5, the operation of rotation by C_2 is equivalent to two C_4 rotations in succession. We can write $C_2 = C_4 \cdot C_4$, or, $C_2 = {C_4}^2$. A consequence of this relationship is the fact that for all of the singly degenerate irreducible representations in Table 10.5 the character under C_2 is *always* the square of that

Table 10.5

C_{4v}	E	$2C_4$	C_2	$2\sigma_v$	$2\sigma_v'$
A_1	1	1	1	1	1
A_2	1	1	1	−1	−1
B_1	1	−1	1	1	−1
B_2	1	−1	1	−1	1
E	2	0	−2	0	0

under C_4. This result helps us in the problem of how to indicate that the rotational operations in spherical symmetry *are* rotations. We have been looking at a unit vector, one that has the property that rotation by 180° converts it into its negative. That is, its character under C_2 is -1. We still have the relationship $C_2 = C_4{}^2$; what then is the character under C_4, the thing that we are interested in? The answer is i, because $i^2 = -1$.[3] This suggests that a suitable expression to describe the rotation is $\cos(\theta) + i\sin(\theta)$. But what of its square, which gives us the length of the vector – and which has to equal 1? It would seem that it will involve i, a complex number. The answer is that with complex numbers one does not form a square as with normal numbers; instead, the square is formed using the complex conjugate of the original. Here, the complex conjugate is $\cos(\theta) - i\sin(\theta)$. If this seems strange, note that this expression is equally suitable as a description of the rotation since $-i$ behaves in the same way as i ($-i^2 = -1$). We have found two solutions, but they are partners, they go together. The product of complex conjugates:

$$[\cos\theta + i\sin(\theta)][\cos(\theta) - i\sin(\theta)] = \cos^2(\theta) + \sin^2(\theta) = 1$$

is just the result we wanted; the vector retains its length when rotated!

Let us look at the expression $\cos(\theta) + i\sin(\theta)$ in more detail. The first term, the $\cos(\theta)$, is the character associated with the rotation, in the sense that it has been used throughout this book. The second term, $i\sin(\theta)$, is something quite new. It does not contribute to the character; rather, its presence serves to ensure that the length of the vector does not change throughout the rotation. There is an identity, $\cos(\theta) + i\sin(\theta) = \exp(i\theta)$, and it is both common – and convenient – to use the latter expression rather than the former. However, in the present text we will persist in the expanded version, because it enables us to focus on the character associated with a rotation. Add a couple of lines of simple mathematics and we will have a very valuable result.

Consider a set of five d orbitals. Fortunately, we do not need to work with their full mathematical expressions; because we are discussing angles it is convenient to work with the angular forms of these orbitals rather than labels such as $d_{x^2-y^2}$. These angular forms are easy, and relate to a common theme in this book, nodal planes. Taking our axis of rotation as the z axis (as is conventional), then the $d_{x^2-y^2}$ and d_{xy} orbitals both have two nodal planes containing the z axis. We take this number 2 and use it to compile a list of the five d functions:[4]

$$\cos(2\theta),\ \cos(\theta),\ \cos(0),\ \cos(-\theta),\ \cos(-2\theta)$$

Here, the expressions involving 2θ relate to the d orbitals with two nodal planes along the z axis, $d_{x^2-y^2}$ and d_{xy}. Those involving θ relate to the d orbitals with one nodal plane along the z axis (d_{xz} and d_{yz}). Finally, that involving $\cos(0)$ ($= 1$) *is* d_{z^2}, with 0 nodal planes along the z axis. This approach works for all other orbitals. So, p orbitals have a maximum of 1 nodal planes and the expressions run from $\cos(\theta)$ to $\cos(-\theta)$; f orbitals have a maximum of 3 and so the expressions run from $\cos(3\theta)$ to $\cos(-3\theta)$; g orbitals have a maximum of 4

[3] The student who has problems with this argument may find it helpful to turn to the next section, where a similar – but different – discussion is applied in the C_4 group (Section 10.5).

[4] In what follows we use the same symbol to denote the angular form of orbitals, θ, as in the discussion of angular rotations. Strictly, this is bad practise, but it is adopted here to give a smooth continuity to the discussion. We also confine our discussion to the θ component of the angular function.

nodal planes and the expressions run from $\cos(4\theta)$ to $\cos(-4\theta)$, and so on for even higher orbitals.

Problem 10.7 In the above list there was no mention made of s orbitals. Does the generality include them?

We now ask the key question, although its importance is not immediately self-evident. What is the character generated by the complete set of d orbitals under the operation of rotation by the angle θ? Before answering this question, we have to recognize a limitation in the earlier derivation of the expression $\cos(\theta) + i\sin(\theta)$. We obtained it by looking at the vector of Figure 10.5, a vector which became the negative of itself by a rotation of 180°; that is, it behaved like $\cos(\theta)$. Had we, instead, considered something which behaved like $\cos(n\theta)$, then we would have obtained the expression $\cos(n\theta) + i\sin(n\theta)$, where $\cos(n\theta)$ is the character and $i\sin(n\theta)$ is the length-maintaining component. With this generalization, we can immediately answer the question of the character generated by the complete set of d orbitals under the operation of rotation by the angle θ. It is:

$$\text{Character}(\theta) = \cos(2\theta) + \cos(\theta) + \cos(0) + \cos(-\theta) + \cos(-2\theta)$$

This is an unwieldy expression; can we put it into a more compact form? Fortunately, the answer is 'yes' – with a little manipulation. Multiply each side of this expression by $\sin(\theta/2)$. We get:

$$\begin{aligned}\sin(\theta/2)\cdot\text{Character}(\theta) = {}& \sin(\theta/2)\cdot\cos(2\theta) + \sin(\theta/2)\cdot\cos(\theta) + \sin(\theta/2)\\ & + \sin(\theta/2)\cdot\cos(-\theta) + \sin(\theta/2)\cdot\cos(-2\theta)\end{aligned}$$

Now, expand the terms on the right-hand side using the identity:

$$\sin(a)\cos(b) = 1/2[\sin(a+b) + \sin(a-b)]$$

and we obtain:

$$\begin{aligned}\sin(\theta/2)\cdot\text{Character}(\theta) = {}& 1/2[\sin(5\theta/2) + \sin(-3\theta/2) + \sin(3\theta/2) + \sin(-\theta/2)\\ & + \sin(\theta/2) + \sin(\theta/2) + \sin(-\theta/2) + \sin(3\theta/2) + \sin(-3\theta/2) + \sin(5\theta/2)]\end{aligned}$$

This is not as bad as it seems, because $\sin(-x) = -\sin(x)$. When this expression is applied to the terms on the right-hand side of the above expression they largely cancel out, and we are left with the expression:

$$\sin(\theta/2)\cdot\text{Character}(\theta) = \sin(5\theta/2)$$

that is:

$$\text{Character}(\theta) = \frac{\sin\left(\frac{5\theta}{2}\right)}{\sin\left(\frac{\theta}{2}\right)}$$

We will apply this expression immediately, but first note its generalization, using the symbol L to denote the maximum number of nodal planes in a set:

$$\text{Character}(\theta) = \frac{\sin\left[(2L+1)\frac{\theta}{2}\right]}{\sin\left(\frac{\theta}{2}\right)}$$

It is not difficult to see that this generalization is valid because the cancellation of terms above is general; only the first and last terms in the right-hand expansion will survive, whatever the value of L.

To see why this is a useful expression, let us apply it first to the case of a set of d orbitals in an octahedral ligand field, one discussed in detail in Chapter 8, and then to a tetrahedral ligand field, discussed in this chapter. For the octahedral case we list the pure rotational operations of the O_h group (pure rotations because we are working with θ). These are listed below, with the appropriate values of θ:

E	$8C_3$	$6C_4$	$3C_2$	$6C_2'$
0	120	90	180	180

Next, we insert each value of θ, in turn, into the expression

$$\frac{\sin\frac{5\theta}{2}}{\sin\frac{\theta}{2}}$$

and obtain:[5]

E	$8C_3$	$6C_4$	$3C_2$	$6C_2'$
0	120	60	180	180
5	−1	−1	1	1

Turning to the improper rotations of the O_h group, we list them in the order corresponding to combining each of the proper rotations with the operation of inversion in a centre of symmetry. They are:

$$i \quad 8S_6 \quad 6S_4 \quad 3\sigma_h \quad 6\sigma_d$$

Because we are considering d orbitals, they are centrosymmetric and so we simply repeat the above set of characters (had the orbitals been centroantisymmetric we would also have repeated the set of characters, but with all signs changed):

i	$8S_6$	$6S_4$	$3\sigma_h$	$6\sigma_d$
5	−1	−1	1	1

Bringing all these together, we have:

E	$8C_3$	$6C_4$	$3C_2$	$6C_2'$	i	$8S_6$	$6S_4$	$3\sigma_h$	$6\sigma_d$
5	−1	−1	1	1	5	−1	−1	1	1

and this is a reducible representation with, surprise, surprise, $e_g + t_{2g}$ components (lower case symbols are used because it is an orbital set which is under discussion). With this example as a model we can now turn to the T_d case.

The T_d case is more difficult because it is not the direct product of a pure rotational group with C_i (which contains just E and i). But we can proceed by recognizing that although the group does not contain a centre of symmetry, individual improper rotation operations may be written as a product of a pure rotation with inversion in a centre of symmetry. So, the

[5] The equation cannot immediately be applied to the $\theta = 0°$ case. Here, one has to use the relationship that as $\theta \to 0$, $\sin\theta \to \theta$. Note that if one chooses to think of the operation E as a rotation by $\theta = 360°$, the equation works with multiples of 180°.

operations of the T_d group, E $8C_3$ $3C_2$ $6S_4$ $6\sigma_d$, may be rewritten as:

E	$8C_3$	$3C_2$	$6C_4i$	$6C_2i$

We now add the appropriate θ values:

E	$8C_3$	$3C_2$	$6C_4i$	$6C_2i$
0	120	180	90	180

For a set of d orbitals we can either again use the expression

$$\frac{\sin\left(\frac{5\theta}{2}\right)}{\sin\left(\frac{\theta}{2}\right)}$$

or, more simply, steal the answers from the calculations on O_h. Either way we obtain the reducible representation:

E	$8C_3$	$3C_2$	$6C_4i$	$6C_2i$
5	−1	1	−1	1

which, of course, has $e + t_2$ components.

Problem 10.8 What would have been the result had we been considering centroantisymmetric functions?

The technique that has just been introduced is a very valuable one, for it enables the determination of the way that atomic entities behave in lower symmetry environments. Although the examples given above refer to atomic orbitals, these were chosen because the results are known to the reader; they have already been met. When many-electron functions are under study then entries way beyond those in Table 10.3 have to be handled. So, for the Mn(II) ion, with five d electrons, one has to go to H and I for some excited states. For many-f-electron systems these are met as ground states, and one has to go even higher for the excited! But go as high as you wish, the equation

$$\text{Character}(\theta) = \frac{\sin(2L+1)\frac{\theta}{2}}{\sin\frac{\theta}{2}}$$

is always simple to use; just plug in the correct values of L and θ. In this way it is seen that 0-noded functions (S in Table 10.4) transform as A_g in I_h (Table 10.3); 1-noded functions (P in Table 10.4) transform as T_{1u}, 2-noded functions (D in Table 10.4) transform as H_g and 3-noded functions (F in Table 10.4) as $T_{2u} + G_u$. Actually, except for the last, all of these results are given in Table 10.3 by the basis functions at the right-hand side.

Problem 10.9 Those who identify with the masochist mentioned earlier are invited to risk an evening at the disco by using the above expression to discover how much further one has to go in the spherical harmonics (an alternative name for the irreducible representations of Table 10.4) to generate all of the irreducible representations of Table 10.3.

Table 10.6

$C_{\infty v}$	E	$2C_{\infty}^{\phi}$	....	$\infty\sigma_v$
$A_1 \equiv \Sigma^+$	1	1	....	1
$A_2 \equiv \Sigma^-$	1	1	....	−1
$E_1 \equiv \Pi$	2	$2\cos\phi$	....	0
$E_2 \equiv \Delta$	2	$2\cos2\phi$	....	0
$E_3 \equiv \Phi$	2	$2\cos3\phi$	....	0
....	.	.	....	.

10.5 Linear molecules

This is an appropriate point at which to include linear systems. There are only two relevant point groups, distinguished by whether or not they have the operation of a centre of symmetry: $C_{\infty v}$, without a centre of symmetry, and $D_{\infty h}$ which has one; N_2 is an example of a $D_{\infty h}$ molecule and CO an example of a $C_{\infty v}$. Both of these groups have a unique axis, a C_{∞}, in which the basic symmetry operation is that of an infinitesimal rotation, the one discussed in the previous section. That is, there is a C_{∞} axis and the infinity of associated operations. The character table gives the character for the operation of rotation by an arbitrary angle ϕ denoted C_{∞}^{ϕ}. Not only is there an infinite number of operations based on C_{∞}, but each group also has an infinite number of σ_v mirror planes. Fortunately, they all fall into a single class. The two groups are related: $D_{\infty h}$ is the direct product of $C_{\infty v}$ with C_i. We can therefore concentrate our discussion on $C_{\infty v}$; its character table is given in Table 10.6.

There are several interesting things about Table 10.6. First, the listing of alternative symbols for the irreducible representations.[6] Both sets are found in the chemical literature, which is why both have been given in Table 10.6. It is the Σ, Π, Δ system which is perhaps the more commonly used. Second, the operations. As for the spherical group, the basic rotation operation is that of an infinitesimal rotation, denoted C_{∞}, carried out sufficient times to give a real rotation of ϕ (although ϕ is the equivalent of θ used in the previous section, it is the symbol generally used in this character table and so is that adopted here). The two together, C_{∞} and ϕ, are written $C_{\infty}{}^{\phi}$. The prefix 2 on C_{∞}^{ϕ} in the character table arises because the rotations can be either clockwise or anticlockwise. Following the $2C_{\infty}^{\phi}$ entry a row of dots indicates the infinite number of values that ϕ may assume. So, this infinity is the sum of all of these rotation operation entries. In contrast, all of the infinite number of mirror planes containing the C_{∞} axis are grouped together as a single entry, $\infty\sigma_v$. Third, all but two of the irreducible representations are doubly degenerate, the characters of $2\cos n\theta$ being understandable in terms of the discussion of the spherical group above. Note that there is no $i \sin n\theta$ term. Length is preserved because the doubly degenerate functions may mix under the rotation (an explicit example of this is discussed in Chapter 7). Finally, there is an apparently major problem with Table 10.6. Because the group is infinite, the usual method of reducing a reducible representation will not work (it involves dividing by the order of

[6] Actually, alternative systems exist for all of the point groups. Those who delve into the physics literature may encounter the use of a notation based on the Γ symbol. The Γ is used in a general way to indicate an irreducible representation whilst different irreducible representations are denoted by subscripts, Γ_1, Γ_2 etc.

the group, the number of operations in the group). In practice there is seldom a problem; reduction by inspection is usually possible (the number of times that $2\cos n\theta$ appears in the reducible representation is the number of times that the E_n irreducible representation occurs in the irreducible sum). However, in the past the topic has been a popular one in the chemical education literature and some (usually!) easy-to-read references are given here.

L. Schäfer and S.J. Cyvin, *J. Chem. Educ.* **48** (1971) 295.

D.P. Strommen and E.P. Lippincott, *J. Chem. Educ.* **49** (1972) 341.

J.M. Alvariño, *J. Chem. Educ.* **55** (1978) 307.

R.L. Flurry Jr., *J. Chem. Educ.* **56** (1979) 638.

D.P. Strommen, *J. Chem. Educ.* **56** (1979) 640.

J.M. Alvariño and A. Chamorro, *J. Chem. Educ.* **57** (1980) 785.

10.6 Summary

For the tetrahedral and icosahedral groups it is C_2 axes which are taken as coordinate axes rather than axes of highest symmetry (p. 245). In icosahedral groups degeneracies of four and five can occur (p. 251). In the spherical group there is no limit to the possible degeneracy (p. 254). Fortunately, there exists a simple equation which enables the reduction in degeneracy in real-life situations to be determined (p. 258). The fundamental rotation operation of a sphere, that of rotation by an infinitesimally small angle, is also a characteristic of linear molecules and these are also discussed (p. 261).

11 π-Electron systems

11.1 Square cyclobutadiene and the C_4 point group

One of the areas of chemistry in which relatively simple quantum mechanical ideas have had a very important impact has been in the field of unsaturated organic molecules. When a molecule contains alternate single and double carbon–carbon bonds then it is found that as a first approximation those electrons involved in π-bonding can be considered on their own – that the σ electrons can be ignored. It seems that these π electrons largely determine the chemistry of such molecules, a recognition which has given an understanding of the chemical stability and reactions of these molecules and also of their spectroscopic properties. The distinction between σ and π orbitals was made in Section 5.5. It is important to recognize that when a molecule contains a series of atoms linked by alternate single and double bonds then on *each* atom in the series there is an orbital involved in the π bonding. It is usually the case that this orbital is a p orbital. The ready availability of detailed and accurate numerical calculations on simple organic molecules has shown that the idea of σ–π separability rests on less secure foundations than was once held to be the case. The orbital symmetry distinctions persist but configuration interaction, effectively electron–electron repulsion, serves to mix different electron configurations. Nonetheless, there is no doubt that the predictions made by the simple theory are rather good, even if a detailed and general justification for this is not available. It is when the results of the simple model are symmetry-determined that the most evident justification occurs and it is such applications which will be the concern of this chapter.

The symmetry aspects of Hückel theory, the best known π-electron model, are most readily seen from an example. A simple molecule, but one which serves to illustrate all of the main points of the theory, will be considered. The molecule is a very unstable and fugitive one, cyclobutadiene, C_4H_4, which will be taken to be a planar molecule with its four carbon atoms arranged at the corners of a square. The carbon atoms are known to have this arrangement when the molecule is stabilized by complexing with a transition metal atom, as in the molecule $C_4H_4Fe(CO)_3$.[1] Figure 11.1 shows square cyclobutadiene together with the four $2p_\pi$ orbitals that will be of interest (we suppose that the carbon 2s and the other carbon 2p orbitals are involved in the bonding of the σ framework). The molecular symmetry is D_{4h} and so this is the obvious group in which to work. However, we shall not.

[1] In fact, in the spin singlet ground state the molecule is rectangular with somewhat localized double bonds; in its spin triplet ground state it is square. For a simple discussion of this point see 'Why do some molecules have symmetry different from that expected?' by E. Heilbronner, *J. Chem. Educ.* **66** (1989) 471.

Symmetry and Structure: Readable Group Theory for Chemists Sidney F. A. Kettle

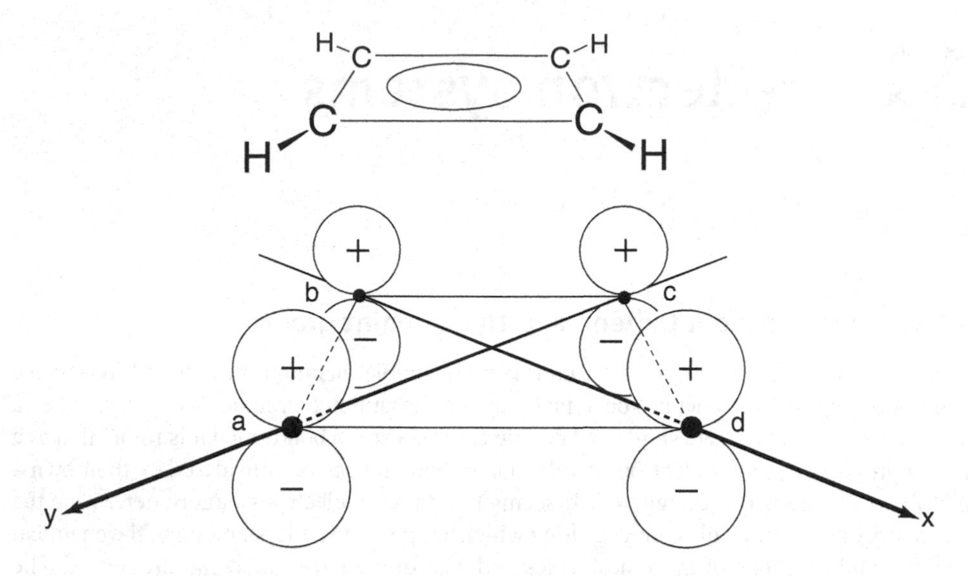

Figure 11.1 Cyclobutadiene, C_4H_4, and the four carbon $2p_\pi$ orbitals

Although it is not particularly obvious from the way that the D_{4h} character table is usually written (Appendix 3), the D_{4h} group is the direct product of $C_{4v} \times C_s$. That is, add a σ_h mirror plane to C_{4v} and other symmetry elements are at once generated so that the group becomes D_{4h}. Now, the problem that we are considering immediately defines the effect of this σ_h mirror plane. We are only interested in the p_π orbitals shown in Figure 11.1 and these, and anything derived from them, are antisymmetric with respect to reflection in the σ_h mirror plane. So, it might well be simpler to work in the C_{4v} point group and, at the end, move to D_{4h} by recognizing this σ_h antisymmetry. It is probable that most workers would be content to stop here and work in C_{4v}, but we shall press on!

Problem 11.1 (a) Using Appendix 3 and Figure 11.1 show that square planar cyclobutadiene has D_{4h} symmetry. (b) Using Appendix 3, show that the D_{4h} group is the direct product of C_{4v} and C_s.

The C_{4v} group possesses two sorts of σ_v mirror planes; $2\sigma_v$ and $2\sigma_v'$. Either the σ_v or σ_v' mirror planes (it does not matter which label we choose, although the distinction between the B_1 and B_2 irreducible representations depends on our choice) cut vertically through the carbon p_π orbitals of cyclobutadiene. They therefore relate one side of each lobe of this orbital to the other side (Figure 11.2). But these sides must be of the same phase. So, the operation of reflection in these mirror planes gives no new information. The operations are superfluous and perhaps can be discarded. But, if they are discarded we must also discard the other mirror planes – point groups exist with both $2\sigma_v$ and $2\sigma_v'$ but there are no, and can be no, point groups with just one set. The most sensible thing to do would be to play safe and keep them all – after all, not much additional work is involved. We shall be more

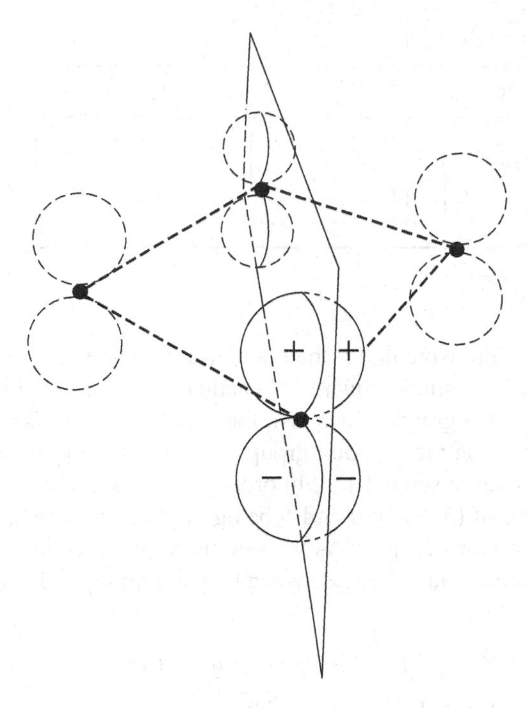

Figure 11.2 A σ_v (or σ'_v) plane in C_{4v} cuts each carbon $2p_\pi$ orbital in half; the relationship between the two halves is determined by the orbital, not by the mirror plane

daring, however, and eliminate them because this will give us the opportunity to work in what seems a rather odd group – the C_4 point group. It is unusual to discuss a group of pure rotations such as the C_4 group in a text at the level of the present one and as a consequence these groups tend to be regarded as rather strange and difficult. However, we have already covered the basic ideas in the last chapter and so it seems sensible to build on this. It must be admitted, however, that the discussion which results is a bit more difficult than would have been the case had we worked in an 'easier' group. As the reader may check for him or herself (Problem 11.10), we shall ultimately obtain the same answers as would have been obtained in C_{4v} (or D_{4h})!

The C_4 character table is given in Table 11.1. Note that it is an Abelian group – there is only one operation in each class. In particular, note that C_4 and C_4^3 (the C_4 operation carried out three times in the same sense) are in different classes; there is no operation that interconverts their effects. In Chapter 6 the importance of the definition of 'class' was mentioned and a formal definition of 'class' is given in Appendix 1. The proof that C_4 and C_4^3 are in different classes in the C_4 group is explicitly given in this appendix.

There are two apparently odd things about the C_4 character table: the appearance of $i(=\sqrt{-1})$,[2] and the failure of the number 2 to appear against the E irreducible representation

[2] It may be helpful to note that the complex conjugate of i, denoted i^*, is $-i(ii^* = 1)$. In some books i^* is found where in this text $-i$ has been used.

Table 11.1

C_4	E	C_4	C_2	C_4^3
A	1	1	1	1
B	1	−1	1	−1
E	$\begin{cases}1\\1\end{cases}$	i $-i$	−1 −1	$-i$ i

$i = \sqrt{-1}$.

under the identity column. Note that *if* the number 2 did appear, Theorem 2 of Chapter 6 would not be obeyed. The sum of squares of characters in the identity column would not be equal to the order of the group (which is 4, the number of operations in the group). The main reason for working in the C_4 point group is to give an opportunity to look at this E irreducible representation in some detail. In order to do this, it has to be remembered that the complex conjugate of $(a + ib)$, a and b being ordinary numbers, is $(a - ib)$, a point already used in the previous chapter. As we saw there, these complex conjugates have a special relationship to each other because when they are multiplied together a real number results:

$$(a + ib)(a - ib) = a(a - ib) + ib(a - ib) = a^2 - iab + iab - (i)^2b^2 = a^2 + b^2$$

because $-(i)^2 = -(-1) = 1$. In contrast, neither $(a + ib)^2$ nor $(a - ib)^2$ are free from i. Note that where, in the E irreducible representation, one component contains i, the other contains $-i$. These components are complex conjugates (as is easily shown if one sets $a = 0, b = 1$ in the expressions earlier in this paragraph).

In previous chapters in this book it sometimes happened that an irreducible representation was multiplied by itself (when forming direct products, for instance). It always happened when reducible representations were decomposed into their irreducible components; the method depends on it. The decomposition procedure used in the earlier chapters is followed when working with the C_4 point group for all irreducible representations except E. For such applications involving the E irreducible representation, it is complex conjugates that have to be multiplied. That is, one multiplies the first component of this doubly degenerate representation by the second and vice versa. In this way real, not complex, answers are obtained. Examples will follow!

The E irreducible representation of the C_4 point group is said to be a *separable degenerate* representation. Some purists object to this name – holding that it is self-contradictory – but it is the name commonly used. The word 'degenerate' is used because functions transforming as this representation have the same energy – an example will be met shortly. 'Separable' is used because it is possible to design an experiment on a molecule of C_4 symmetry which shows that all functions transforming as the E irreducible representations are not necessarily quite equivalent. In order to illustrate this we shall return to the topic of optical activity, first met in Section 4.7. A brief reminder; a molecule is optically active when, in an electronic transition, there is a helical movement of charge density. A characteristic of a helix is that movement along it corresponds to a simultaneous translation and rotation and so, as was outlined in the discussion of Section 4.7, optically active molecules are those in which a transition is simultaneously both electric dipole (charge translation) and magnetic dipole

Table 11.2

C_4		E	C_4	C_2	C_4^3		
A		1	1	1	1	$z; R_z; z^2; x^2+y^2$	
B		1	-1	1	-1	$x^2-y^2; xy$	
E	E_1	1	i	-1	$-i$	$x+iy; R_x+iR_y$	(yz, zx)
	E_2	1	$-i$	-1	i	$x-iy; R_x-iR_y$	

(charge rotation) allowed; these two have to transform as the same irreducible representation. The rule we arrived at was:

> Molecules[3] may be optically active when they have a symmetry such that T_α and R_α ($\alpha = x, y$ or z) transform as the same irreducible representation.

We saw that this condition is only satisfied when such molecules possess neither a centre of symmetry nor a mirror plane. They do not have any improper rotation operations. As an alternative general statement, one can say that optically active molecules do not have any S_n axis, where n can assume any value ($n = 1$ corresponds to a mirror plane and $n = 2$ to a centre of symmetry).

Problem 11.2 The separation of the cobalt complex ion $[Co(en)_3]^{3+}$ into optical isomers is a common undergraduate experiment. The complex is, essentially, octahedral and 'en' is the bidentate ligand ethylenediamine, $NH_2.CH_2.CH_2.NH_2$, which is bonded to the cobalt through the nitrogen atoms on adjacent (*cis*) coordination sites. Determine the symmetry of this molecule and thus show that it has no S_n axis.

In order to see that there is some anisotropy in the xy plane of a C_4 molecule, a defect in Table 11.1 has to be remedied; it contained no basis functions. This table is repeated as Table 11.2, but this time with basis functions. In Table 11.2 the individual components of the E irreducible representation are labelled E_1 and E_2. This labelling is for the convenience of the following discussion and only that (as a study of Appendix 3 will show, the labels E_1 and E_2 are often used to distinguish different sets of doubly degenerate irreducible representations – this is not the usage here!).

In the particular case of the C_4 point group, T_z and R_z both transform as A (although the former is not explicitly included in the table, the entry z implies it) and the complex combination $T_x + iT_y$ transforms in the same way as $R_x + iR_y$. Similarly, $T_x - iT_Y$ transforms isomorphically with $R_x - iR_y$. The complex form of these latter combinations is a bit off-putting, although such forms were met in the last chapter (to see the connection note that y is at 90° to x and that the problem of length retention reappears; more on this later).

[3] Note the word 'molecules' in this statement. It does not apply to crystals which, under some circumstances, can contain mirror planes of symmetry and yet be optically active.

Even so, ignoring this, it is clear that T_α and R_α transform isomorphically in the C_4 group so that a molecule of C_4 symmetry is potentially optically active. A beam of polarized light incident on such a molecule down the fourfold axis might suffer a rotation. Clearly, this is not compatible with the isotropy which one normally associates with degeneracy in the xy plane. The explanation lies, not surprisingly, in the appearance of complex coefficients in the character table.

Problem 11.3 Despite the discussion of optical activity in the context of cyclobutadiene in the text, it is believed that cyclobutadiene is not optically active. Why?

11.2 Working with complex characters[4]

All of the character tables met in all of the chapters of this book, except the last, contained simple quantities, usually integers, as characters. This is a pattern common to most texts at the level of the present. As a consequence, most people approach complex characters with some apprehension, expecting some strange twists. This apprehension is justified! One example of the different pattern is seen in the statement made towards the end of Section 4.6 that 'we will only get the totally symmetric irreducible representation as the direct product when the two irreducible representations that we are multiplying are the same'. This statement remains true for the C_4 point group but needs some elaboration. Consider the direct product of the first component of the E irreducible representation of Table 11.2 with itself:

	E	C_4	C_2	C_4^3
$E(1)$	1	i	-1	$-i$
$E(1)\otimes E(1)$	1	-1	1	-1

this direct product is the B irreducible representation, not the A. To obtain the A, the direct product has to be formed of $E(1)$ with its *complex conjugate*, $E(2)$:

$E(1)$	1	i	-1	$-i$
$E(2)$	1	$-i$	-1	i
$E(1)\otimes E(2)$	1	1	1	1

That is, when working with a separately degenerate representation, one has to elaborate on the statements made in Chapter 4 about direct products, after that, the way to proceed is reasonably straightforward, and helpful too because it sheds light on other areas of quantum chemistry. Thus, the general expression for an overlap integral given in texts on quantum mechanics is

$$S_{\mathrm{ab}} = \int \psi_{\mathrm{a}}^{*}\psi_{\mathrm{b}}\delta v$$

[4] A short, readable, paper which deals with the problems covered in this section is 'Representations with imaginary characters' by R.L. Carter, *J. Chem. Educ.* **70** (1993) 17.

the asterisk on ψ_a^* indicating the complex conjugate of ψ_a (the implication being that both ψ_a and ψ_b are complex). If ψ_a and ψ_b are not complex this reduces to the simple form:

$$S_{ab} = \int \psi_a \psi_b \delta v$$

However, when ψ_a and ψ_b are both complex the more general form must be used. We have seen that group theory can simplify discussions of overlap integrals (Section 4.4); when ψ_a and ψ_b are complex, the corresponding complex conjugate irreducible representations must be used when carrying out the associated group theory. An explicit example of this will be met in the next section. We can use the above result to deepen our understanding of the C_4 character table and, in particular, the appearance of i and $-i$ (although in doing so we will, to some extent, be repeating material already met). In the previous chapter, we met the general character of a rotation of a simple vector by θ as:

$$\cos\theta + i\sin\theta$$

Clearly, this applies to the C_4 character table because $\theta = 90$, so that $\cos\theta = 0$ and $\sin\theta = 1$, giving $0 + i \cdot 1 = i$. So, the entry i in the character table, although it would be called a character, is not the same as those met in earlier chapters, where a character was regarded as a multiplier. Here, the multiplier is 0. Nonetheless, in carrying out the group theory corresponding to forming the overlap integral between one of the E functions and itself, we must get the answer 1 for all of the operations. The function remains itself whatever the operation. In doing the group theory equivalent to such a normalization, for this is what it is, we have to multiply characters to obtain the totally symmetric direct product. The only way that we can obtain the number 1 from i is to multiply it by $-i$, its complex conjugate.

Problem 11.4 Modify the discussion of selection rules in Section 4.4 so that it covers the case where the wavefunctions are complex.

11.3 The π orbitals of cyclobutadiene

We now return to the problem of the π-electrons of cyclobutadiene. We know that these π electrons interact with each other – they form π bonds of some sort, and so the first problem is that of finding the π molecular orbitals which they occupy. This will be tackled in two stages. First, the irreducible representations generated by the transformations of the four carbon p_π orbitals is determined and their symmetry-adapted combinations generated. Second, the approximate relative energies of these symmetry-adapted combinations will be determined.

It is easy to show that the transformations of the four carbon p_π orbitals of cyclobutadiene in the C_4 point group (Figure 11.1) generate the reducible representation

E	C_4	C_2	C_4^3
4	0	0	0

and that this gives rise to $A + B + E$ irreducible components.[5] The zeros avoid any problems with complex characters (for the moment!).

The determination of the symmetry-adapted combinations is straightforward, and follows the projection operator procedure detailed in Chapter 6 very closely. Using the labels shown in Figure 11.1 for the four p_π orbitals and neglecting overlap between these orbitals, the two E components given in Table 11.2 are used separately to give the linear combinations (we work with the orbital labelled a):

$$\begin{aligned}\psi(A) &= \tfrac{1}{2}(a+b+c+d)\\ \psi(B) &= \tfrac{1}{2}(a-b+c-d)\\ \psi(E_1) &= \tfrac{1}{2}(a-ib-c+id)\\ \psi(E_2) &= \tfrac{1}{2}(a+ib-c-id)\end{aligned}$$

Problem 11.5 Use the projection operator technique to obtain the above linear combinations. The normalization of the E functions will be discussed in the text below.

Hint: The derivation is similar to that detailed in Section 6.6.

As indicated in the above problem, the only difficult point in this derivation concerns the two E functions. First, a hidden catch. In using the projection operator technique to generate a function transforming as a component of a separably degenerate representation one has to use its *complex conjugate* in the derivation. Thus, using the characters of the *second E* component in Table 11.2, the function listed above as $\psi(E_1)$ is obtained by the projection operator technique in un-normalized form: $a - ib - c + id = \psi$, say. It is easy to show that this procedure has given the correct answer – that $\psi(E_1)$ transforms as the *first* E component in Table 11.2. As Table 11.2 shows, the effect of a C_4 rotation on a function transforming as the first E component is to multiply it by i. Now this rotation permutes the p_π orbitals thus:

$$\begin{array}{ccc} a & \rightarrow & b\\ \uparrow & & \downarrow\\ d & \leftarrow & c\end{array}$$

so that it turns $\psi(E_1)$ into

$$b - ic - d + ia$$

which is $i(a - ib - c + id) = i\psi(E_1)$, as expected for the first E component. The next step is to normalize $\psi(E_1)$; that is, multiply it by a coefficient such that:

$$\int \varphi^* \varphi \delta v = 1$$

[5] Reducible representations like this one – in which the number which is the order of the group appears in the identity operation column with all other entries zero – are called 'the regular representation' (of the particular point group). They always span each and every irreducible representation, the number of times an irreducible representation is spanned being given by the number in the identity operation for that particular irreducible representation (i.e. the dimension of the irreducible representation). The regular representation plays a part in the proof of some theorems of group theory.

where φ^* is the complex conjugate of φ (and φ is the normalized $\psi(E_1)$). The complex conjugate of a function is obtained by replacing i by $-i$ within it, so the complex conjugate of $\psi(E_1)$ is:

$$a + ib - c - id = y(E_1)^*$$

This function has been met before, it is $\psi(E_2)$; $\psi(E_1)$ and $\psi(E_2)$ are complex conjugates of each other.

It follows that the overlap integral of $\psi(E_1)$ with itself has the value

$$\int \psi(E_1)\,\psi^*(E_1)\,\psi\delta v = \int (a + ib - c - id)(a - ib - c + id)\delta v = \int aa\delta v$$
$$+ \int bb\delta v + \int cc\delta v + \int dd\delta v = 4$$

where, as mentioned earlier, it has been assumed that the functions a, b, c and d do not overlap each other. The fact that a and b, for example, do not overlap each other means that the overlap integral $\int ab\delta v$ is equal to zero. Because a and b are separately normalized, $\int aa\delta v = \int bb\delta v = 1$. From the value of the overlap integral obtained above, 4, it follows that the normalization constant for $\psi(E_1)$ – and, equally, $\psi(E_2)$ – must be 1/2, the value used in the linear combinations above.

Problem 11.6 Show that $\int \psi(E_2)^* \psi(E_2)\delta v = 4$.

11.4 The energies of the π orbitals of cyclobutadiene in the Hückel approximation

The limit at which simple group theory can help the discussion has now been reached. To proceed, chemical knowledge has to be added or, failing that, chemical intuition! In practice, this means that the next step involves using some model which provides a recipe for obtaining relative orbital energies. Such a model was used in earlier chapters of this book when a nodal plane criterion was used to obtain orbital energies – the more nodes that an orbital contains, the higher its energy is expected to be. This model was augmented by an overlap criterion – the greater the overlap between two orbitals, the larger the energetic consequences of the interaction between them. The latter part of the discussion of Section 3.5 provides a good example of the augmentation of symmetry arguments by these models.

In the present section the nodal plane argument will be used in a more mathematical form (when the functions obtained in the previous section were obtained it was assumed the overlap between p_π orbitals on adjacent carbon atoms in cyclobutadiene is zero. It would therefore scarcely be convincing to use an overlap model at this point!). The mathematical form to be used is that contained in Hückel theory; this is the simplest of all mathematical models of chemical bonding and one that is particularly appropriate to unsaturated organic molecules.[6] Although they were not explicitly discussed in Section 4.4, the conclusions there are applicable to energy integrals. These are of the form

$$\int \psi_a \mathcal{H} \psi_b \delta v$$

[6] It was subsequently extended, in a purely numerical form, to inorganic molecules too.

where $\mathcal{H}$ is the so-called Hamiltonian operator for the system. The Hamiltonian operator represents all the energies (kinetic, potential, attractive, repulsive) of the system. Fortunately, a detailed expression is not needed for $\mathcal{H}$ in the present context; all one needs to note is that at equivalent points within a molecule the blend of attractive and repulsive forces must be identical. $\mathcal{H}$ has the symmetry of the molecule; it is totally symmetric. So, just like the corresponding overlap integral, the energy integral is only non-zero when ψ_a and ψ_b are of the same symmetry species. Such energy integrals are important in Hückel theory. In this application, the orbitals ψ_a and ψ_b are p_π orbitals and so, in cyclobutadiene, they are the orbitals a, b, c and d of Figure 11.1. The energy of each of these orbitals, before each is involved in any interaction with its partners, is the same. This energy is conventionally designated α. For the orbital a we have, then

$$\int a\mathcal{H}a\delta v = \alpha$$

with similar expressions for b, c and d. At this point in the development we are dealing with isolated atoms; the molecular symmetry is not yet relevant. The energy of interaction between adjacent p_π orbitals is called β. So, the interaction between a and b is

$$\int a\mathcal{H}\mathrm{b}\delta v = \int b\mathcal{H}a\delta v = \beta$$

with similar expressions for the pairs b/c, c/d and d/a. Those p_π orbitals which are not adjacent are assumed not to interact so that, for instance,

$$\int a\mathcal{H}c\delta v = \int c\mathcal{H}a\delta v = 0$$

and similarly for b/d. Strictly, up to this point we are working with a set of diatomic interactions; the next step is to combine them into the molecule, whereupon the C_4 symmetry requirements become applicable.

To obtain the energy, within the Hückel model, of the A combination

$$\psi(A) = \tfrac{1}{2}(a + b + c + d)$$

we simply have to evaluate

$$\int \psi(A)\mathcal{H}\psi(A)\delta v = \frac{1}{4}\int(a + b + c + d)\mathcal{H}(a + b + c + d)\delta v$$

Expansion of the right-hand side of this expression and substitution of α, β and 0 as appropriate for the resulting integrals gives the energy of $\psi(A)$ as

$$\mathcal{E}[\psi(A)] = \alpha + 2\beta$$

Problem 11.7 (i) Show that the energy of $\psi(A)$ is $\alpha + 2\beta$. (ii) Show that the energy of $\psi(B)$ is $\alpha - 2\beta$.

As requested in the above problem, it is a simple matter to show that the energy of the $\psi(B)$ orbital is

$$\mathcal{E}[\psi(B)] = \alpha - 2\beta$$

but that of $\psi(E_1)$ is a bit more difficult. This is because the form of the energy expression appropriate to complex functions has to be used. This is

$$\int \psi_a^* \mathcal{H} \psi_b \delta v$$

In our case if we take ψ_b to be $\psi(E_1)$ then ψ_a^* is its complex conjugate, that is, $\psi(E_2)$. It follows that we have to evaluate

$$\mathcal{E}[\psi(E_1)] = \tfrac{1}{4} \int (a + ib - c - id)\mathcal{H}(a - ib - c + id)\delta v$$

On expansion of this expression all of the complex quantities disappear.

Problem 11.8 Show that the energy of $\psi(E_1)$ is α.

The energy of $\psi(E_2)$, which is given by

$$\mathcal{E}[\psi(E_2)] = \tfrac{1}{4} \int (a - ib - c + id)\mathcal{H}(a + ib - c - id)\delta v$$

will be the same as that for $\psi(E_1)$ because the right-hand side of this expression on expansion is identical to that for $\psi(E_1)$. We have, then,

$$\mathcal{E}[\psi(E_1)] = \mathcal{E}[\psi(E_2)] = \alpha$$

Evidently, for the present problem at least, it is entirely reasonable that $\psi(E_1)$ and $\psi(E_2)$ should be called 'degenerate'. Actually, this degeneracy between them is general – the algebraic expressions obtained for their energies were identical and so the degeneracy did not result from the Hückel approximations, which came later.

Problem 11.9 Show that the order of energy levels just obtained is also the order of increasing nodality.

Because the interaction between two of the p_π orbitals is one that leads to a stabilization – it requires more energy to ionize an electron from a stabilized orbital than from an isolated $p\pi$ orbital – the energy β is negative (as too is α, but because its contribution to all of the energy levels is the same its value does not affect the relative order of orbital energies). It is concluded that the relative energies of the π molecular orbitals of cyclobutadiene are those given in Figure 11.3. There are four $p\pi$ electrons – one from each carbon atom – located in these orbitals and so we conclude that in the most stable arrangement they will be distributed as shown in Figure 11.3, the degenerate E orbitals containing one electron each; these two electrons will, in the ground state, have parallel spins (the maximum spin multiplicity principle). The total π electron stabilization, compared to four carbon $p\pi$ orbitals of energy α, is 4β (2β from each electron in the A orbital).

Suppose that, instead of a delocalized π system, we had two localized, non-interacting, π bonds – that is, suppose cyclobutadiene is rectangular, rather than square:

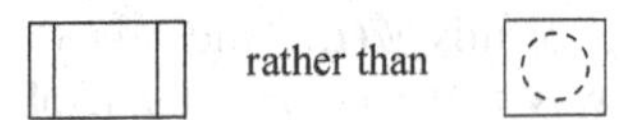

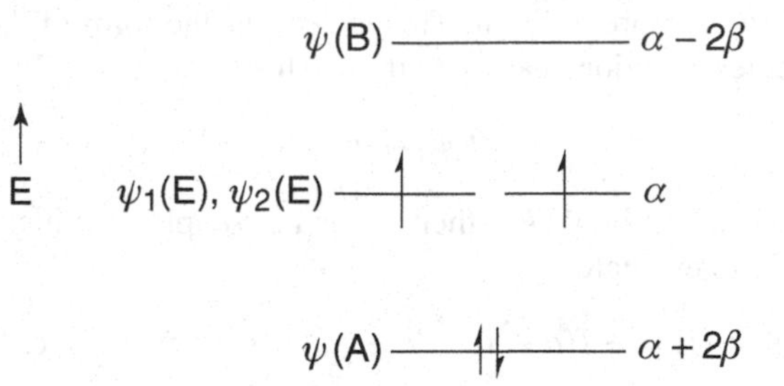

$\psi(B)$ ——— $\alpha - 2\beta$

E ↑ $\psi_1(E), \psi_2(E)$ ——— α

$\psi(A)$ ——— $\alpha + 2\beta$

Figure 11.3 Relative energies of the π molecular orbitals of cyclobutadiene in the Hückel approximation

Each of the two isolated π bonds will have the form derived in Section 5.5 for ethene

$$\frac{1}{\sqrt{2}}(e + f)$$

The energy of this function is given by

$$\tfrac{1}{2}\int(e + f)\mathcal{H}(e + f)\delta v$$

which, on expansion and substitution of the Hückel values for the integrals, leads to an energy of

$$\alpha + \beta$$

There would be two such π bonding orbitals, each doubly occupied so that the total π electron stabilization would, again, be 4β. As far as the π electrons are concerned, at this level of approximation, there is nothing to choose between rectangular and square cyclobutadiene. Cyclobutadiene is a very reactive compound – it readily dimerizes (a reaction that can be discussed by the methods of the next section) – but it has been prepared at 35 K in an argon matrix. In its ground state it seems almost certain that it is a planar molecule with no unpaired electrons. Whether it is rectangular or square is not known (for the latter the electron configuration a^2e^2 gives rise to both triplet and singlet spin states so, notwithstanding the predictions of Hückel theory, the possibility of a square molecule is not completely excluded by the observation of a singlet spin state). We shall have more to say about such spin functions in the next chapter.

Of the π-electron wavefunctions obtained working in the C_4 point group, two, $\psi(A)$ and $\psi(B)$, are identical in form to two that would have been obtained working in either C_{4v} or D_{4h} (the symmetry labels would have been different, of course). On the other hand, the $\psi_1(E)$ and $\psi_2(E)$ wavefunctions are different. The reason for this can be traced back to the existence of operations in C_{4v} and D_{4h} which have the effect of either mixing or interchanging the two degenerate functions. If the x and y axes through the carbon atoms as shown in Figure 11.1 had been taken, then the $2\sigma_v$ mirror planes discarded in C_{4v} would have had the effect of interchanging x and y (and so also any functions transforming like them). Hence, they would have transformed *as a pair*. In contrast, there is in C_4 no operation which will interchange or mix $\psi(E_1)$ and $\psi(E_2)$ – they are separate functions, although degenerate. The only way that we can, in C_4, obtain those E functions which would have been obtained in C_{4v} is to mix $\psi(E_1)$ and $\psi(E_2)$ although, of course, this is not permissible in the C_4 group itself. Taking the sum and difference of $\psi(E_1)$ and $\psi(E_2)$, the sum gives:

$$\psi(E_1) + \psi(E_2) = \frac{1}{2}(2a - 2c)$$

or, renormalizing

$$\psi_1'(E) = \frac{1}{\sqrt{2}}(a - c)$$

and the difference gives:

$$\psi(E_1) - \psi(E_2) = \tfrac{1}{2}(2ib - 2id)$$

or, renormalizing and remembering that the complex conjugate of $(ib - id)$ is $(-ib + id)$ one obtains

$$\psi_2'(E) = \frac{1}{\sqrt{2}}(b - d)$$

The functions $\psi_1'(E)$ and $\psi_2'(E)$ are those E functions which would have been obtained working in C_{4v} (or D_{4h}).

Problem 11.10 Work in either the point group D_{4h} or C_{4v} and: (a) obtain the explicit forms of the four p_π molecular orbitals of cyclobutadiene, (b) check that the doubly degenerate functions obtained have an energy of α within the Hückel approximation.

Before we finally leave the C_4 group there is one further point that should be made. In deriving the C_{2v} character table in Chapter 2 it was asserted that there is no other set of characters other than those considered there which, when substituted for the operations of the C_{2v} group in the group multiplication table, would give a table which is arithmetically correct. The possibility of complex characters such as those which occur in the C_4 character table was not explored. However, it is clear that substitution of a set of characters such as

$$1 \quad i \quad -1 \quad -i$$

in the C_{2v} character table would not lead to a multiplication table which is arithmetically correct (because, for instance, when i multiplies i it gives -1 on the leading diagonal rather than 1). It is evident from this discussion, and can be readily checked, that the multiplication tables of C_4 and C_{2v} are not isomorphous. Any operation in C_{2v} carried out twice leads to E, whereas in C_4 the C_4 and C_4 operations have to be carried out four times to give E. As we first discussed in the last chapter, this explains the appearance of i in the C_4 character table. Because

$$C_2 \times C_2 = E$$

only characters of either 1 or -1 for the C_2 operation are possible for a singly degenerate irreducible representation (because $1 \times 1 = -1 \times -1 = 1$). In the C_4 group we also have that

$$C_4 \times C_4 = C_2$$

In other words, the character for the C_4 operation, squared, must give the character of the C_2. This presents no problems when the character for C_2 is 1, because that for the C_4 can then be either $+1$ or -1 (leading to the A and B irreducible representations of C_4). When the character for C_2 is -1, however, the only possibilities are that the character for the C_4 operation is either i or $-i$ (either of these squared gives -1), leading to the two components of the E irreducible representation. We will meet a very similar pattern, but in a very different context, in the next chapter.

One example of the application of symmetry to the energy levels of π-electron systems has just been given. There are many others, but, having established the principles and procedures involved, the subject will not be pursued further in detail. Suffice to say that the concept of aromaticity in organic chemistry is closely related to the type of stabilization arguments used when comparing square and rectangular cyclobutadiene. Roughly speaking, aromatic systems are those for which the delocalized system is more stable than any corresponding localised one (so cyclobutadiene is not an aromatic system).

11.5 Symmetry and chemical reactions

There have been many attempts to apply symmetry concepts to molecular reactions. This is a difficult area; it is necessary to assume some geometry for the key step in the reaction and often the only reasonable symmetry is low and so of little help. Further, large molecular distortions are usually involved in chemical reactions; that is, the molecules involved are vibrationally very excited. This has two consequences. First, the analysis given of vibrations in Chapter 4 evidently needs modification for large amplitude vibrations – when there are several symmetry-related atoms in a molecule the evidence is that one bond breaks before the others, whereas the discussion of Chapter 2 would lead us to expect several bonds to break simultaneously. This is akin to the failure of simple molecular orbital theory at large internuclear distances (it predicts a mixture of dissociation products, not one), a failure which is also of relevance.

By far the most fruitful of the applications of symmetry to molecular reactivity has been the symmetry correlation method introduced by Woodward and Hoffmann and which is applicable to many organic reactions. A simple example of the application of their approach will be given, although the formalism which will be used was introduced by other workers. Consider the possible reaction of two ethene molecules to give cyclobutane, a molecule which, for simplicity, will be assumed to be planar

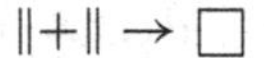

Written like this, it seems a perfectly feasible reaction, yet it is not one that readily occurs; the question then is 'why does it not occur?' The answer is not difficult to find. Pictorially, place two ethene molecules close together so that they are just about to react. The 'before' and 'after' reaction bonding arrangements are shown in Figure 11.4. The actual symmetry shown in Figure 11.4 is D_{2h}, but it is common to work in C_{2v}, so that the geometrical constraints on the molecular arrangement are not as rigid as required by D_{2h} symmetry. If symmetry constraints arise in C_{2v} (as they do) they are likely to be yet more severe in D_{2h}. Working in C_{2v} and choosing the C_2 axis as shown in Figure 11.4, it is easy to show that the symmetry species subtended by the two π bonding molecular orbitals in the two ethene

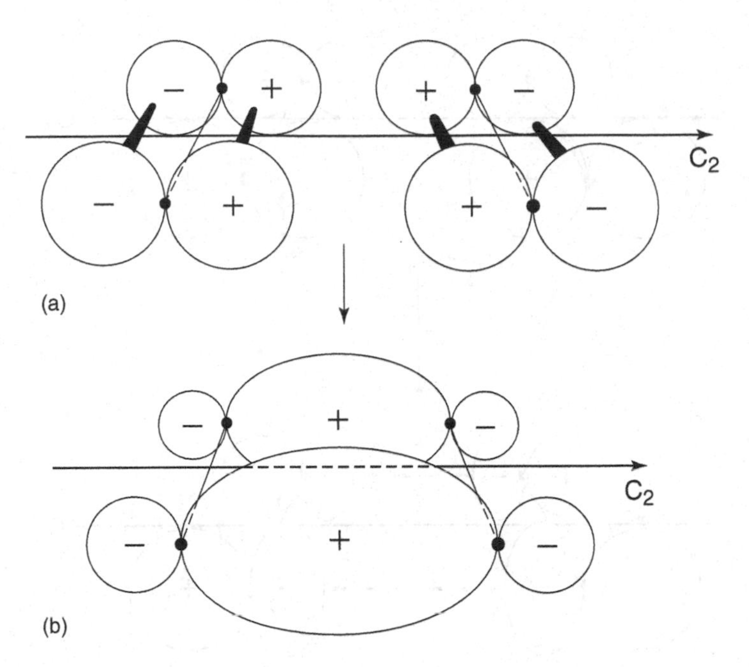

Figure 11.4 Each of these diagrams show two orbitals, not more. (a, 'before') Each π-bonding orbital of each C_2H_4 molecule individually transforms in C_{2v} as A_1. (b, 'after') The two new σ-bonding orbitals in C_4H_8 transform together in C_{2v} as $A_1 + B_1$

molecules shown in that figure is $2A_1$ (it is assumed that the reader is reasonably familiar with the C_{2v} character table!). It is these two π orbitals that are involved in the reaction and that are assumed to smoothly become the two new C—C σ bonds as the reaction takes place. These two new carbon σ bonds give rise to the symmetry species $A_1 + B_1$. This is not the same as those generated by the π orbitals with which the problem started. There is a discontinuity; the π bonds cannot smoothly become the new σ bonds and so a ready reaction is not to be expected.

Let us look at this further by asking whether there exist any B_1 orbitals in the two ethene molecules? The answer is 'yes'. There are two of them and they are derived from the two π antibonding orbitals of the two ethene molecules (Figure 11.5). Correspondingly, the σ antibonding orbitals corresponding to the two newly formed C—C σ bonds in cyclobutane have symmetries $A_1 + B_1$ (Figure 11.5). We are led to the orbital correlation diagram shown in Figure 11.6 which shows the correspondences between the 'before' and 'after' reaction orbital patterns. In this figure the detailed pattern of σ orbital energy levels in cyclobutane has been obtained using the nodal pattern method of determining relative related energy levels met in the early chapters of this book – the more nodal planes, the higher the energy.

Problem 11.11 Use the nodal criterion (used, for example, in Section 3.6) to show that it is reasonable to expect both a_1 levels in Figure 11.6 to be more stable then the corresponding b_1 levels.

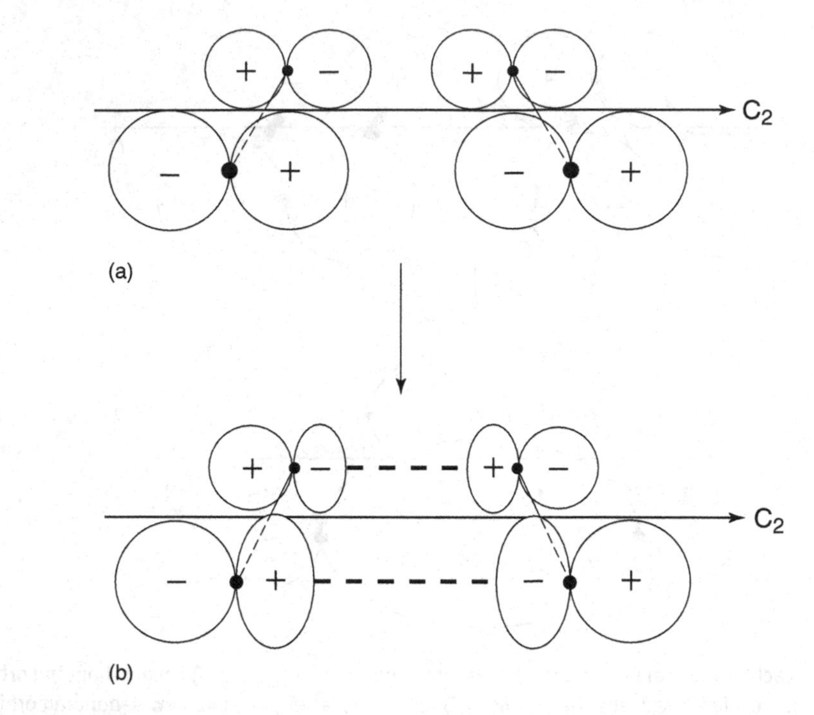

Figure 11.5 (a) The π-antibonding orbital of each individual C_2H_4 molecule transforms in C_{2v} as B_1. (b) The two new σ-antibonding orbitals in C_4H_8 (dashed) transform together in C_{2v} as $A_1 + B_1$. (The diagram shows *two* orbitals; if regarded, however, as a single symmetry-adapted orbital it is the B_1. The A_1 is obtained by changing the phases of all lobes of the σ-antibonding orbital at the 'front' of the diagram)

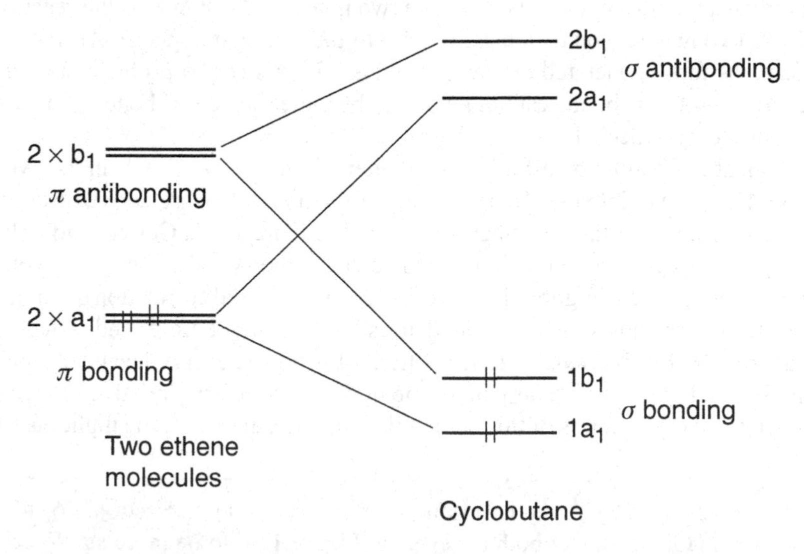

Figure 11.6 Correlation between the π-bonding and antibonding orbitals of two ethene molecules and the σ bonding and antibonding orbitals of cyclobutadiene

It is clear from Figure 11.6 that there will be no strong bonding driving force to form cyclobutane from two ethene molecules. As the reaction proceeds the lowest energy orbitals would be expected to be filled. Up to the energy level crossing-over point in the middle of Figure 11.6 this means that the two A_1 orbitals will be filled. But whilst the stability of one increases with decreasing separation between the two ethenes (as a π bond becomes a σ bond), that of the other will decrease rapidly (as a π bond becomes a σ antibonding orbital) so that no reaction is to be expected.

There is an alternative approach to this problem, an approach which is based on states and terms rather than orbitals. States and terms are two topics that will be covered in some detail in the next chapter; only a superficial knowledge is assumed in what follows. The ground state electronic configuration of two ethene molecules is $(a_1)^2(a_1)^2$, a configuration which gives rise to a $^1A_1$ term (we shall be concerned with spin singlet terms throughout the following discussion). A configuration such as $(a_1)^2(b_1)^2$ is an excited state configuration but also gives rise to a $^1A_1$ term (as is readily seen since the quadruple direct product $A_1 \otimes A_1 \otimes B_1 \otimes B_1 = A_1$). In cyclobutane the situation is reversed. The ground state configuration (considering only the newly formed σ orbitals) is $(1a_1)^2(1b_1)^2$. In contrast, $(1a_1)^2(2a_1)^2$ is an excited state configuration. The important thing is that both of these configurations give rise to $^1A_1$ terms. The appropriate *term correlation diagram* is shown in Figure 11.7, where the *non-crossing rule* has been invoked (terms of the same symmetry species only cross in very rare circumstances). Physically, this application of the non-crossing rule in the present example arises because electron repulsion favours electrons being as spatially separated as possible and the energy gained from this separation can contribute the energy apparently required to promote an electron to a higher orbital. Figure 11.7 demonstrates rather more

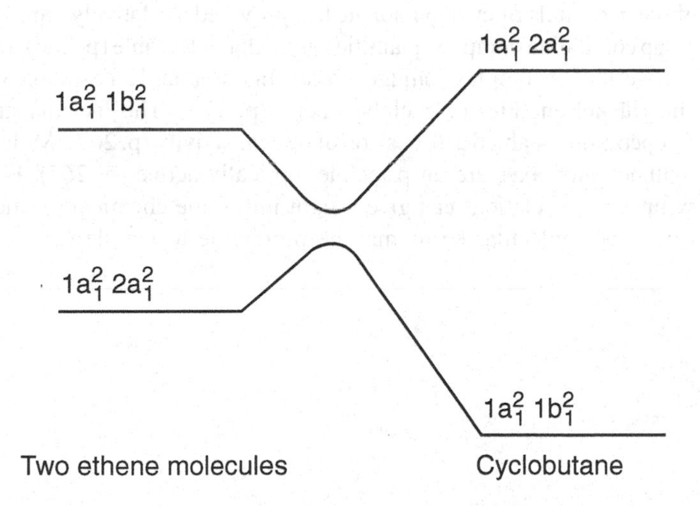

Figure 11.7 Correlation between $^1A_1$ terms in the dimerization of two ethene molecules to give cyclobutane. The two bends in the plots are an example of the non-crossing rule. If the bends were accidentally deleted from the diagram, the natural correction would be with straight lines. In fact, even a very small energy term connecting the two configurations in the cross-over region is sufficient to strongly mix the two wavefunctions. The sensitivity of this mixing to the position in the cross-over region leads to the result shown

clearly than does Figure 11.6 that ethene should not be expected to spontaneously dimerize to cyclobutane. Even here, our discussion is somewhat simplified but it does correctly indicate that one can sometimes be misled by a simple 'filling of the lowest orbitals' approach to chemical bonding. It might be appropriate to remind the reader that the complicating effects of repulsive forces on simple pictures of chemical bonding were also encountered in the first chapter of this book.

The discussion which has been presented above can readily be extended to photochemically induced reactions (that is, reactions involving electronically excited molecules). Many very readable accounts of the topic have been written but these tend to use symmetry arguments in a rather less formal manner than the present text. Commonly, particular symmetry operations are selected and orbital behaviour classified as either A (antisymmetric) or S (symmetric) under these operations; these labels are equivalent to the characters -1 and 1 used in this book.

One final cautionary note. In this discussion the concern has been with a single reaction mechanism. Other mechanisms may exist which provide an alternative, and more accessible, route to a particular product. Thus, although ethene does not dimerize to cyclobutane, reaction between the ethene derivatives $CH_3(H)C{=}C(H)OC_2H_5$ and $(CN)_2C{=}C(CN)_2$ proceeds smoothly at room temperature to give the corresponding cyclobutane derivatives. In this case there is evidence that a zwitterion intermediate, $(CN)_2C^-{-}C(CN)_2{-}CH(CH_3){-}^+CH(OC_2H_5)$ is formed. Symmetry arguments are powerful, but nature may be yet more cunning and have unexpected tricks!

11.6 Summary

Discussion of the π orbitals of cyclobutadiene has provided a relatively simple example of the use of a group containing complex quantities in its character table (p. 266). It is necessary in such cases to work with complex conjugate basis functions and an example was provided in deriving the Hückel energies of cyclobutadiene (p. 271). The fact that the C_4 group contains no S_n operations enabled a discussion of optical activity (p. 267). Molecules having symmetries without such axes are, in principle, optically active (p. 267). Finally, it was shown that symmetry correlations can give insight into some chemical reactions (p. 276). Correlations between molecular terms may be preferable to simple orbital correlations (p. 278).

12 The group theory of electron spin

12.1 The problem of electron spin

Although in earlier chapters electrons have been encountered in the context of molecular bonding, electron spin has not really been mentioned. The nearest we have approached the topic is in talk of 'an electron pair', the implication being that one electron has spin 'up' and the other 'down', whilst in Chapter 8 we counted unpaired electrons. A question such as 'what is the symmetry of an unpaired electron in the water molecule?' would have been met with, at best, a frown. Of course, one could reason that the electron would be associated with some orbital and that the thing to do would be to ask questions about this orbital. Alas, not so. The answer can be independent of orbital. Even worse, the question cannot be answered in the context of the C_{2v} point group! Clearly, there is some basic ground to be covered before we can answer such a deceptively simple question.

In Chapter 10 we met the equation:

$$\text{Character}(\theta) = \frac{\sin\left[(2L+1)\frac{\theta}{2}\right]}{\sin\left(\frac{\theta}{2}\right)}$$

which is used when moving from spherical symmetry to a point group geometry. Here, the angle θ is the angle associated with some pure rotation operation and L is the maximum number of nodal planes containing the z axis which go with an orbital set (so, $L = 2$ for a set of d orbitals). However, L is more commonly given another name; it is referred to as an angular momentum quantum number. The reason for this is seen in another aspect of Chapter 10; the fact that the basic rotational symmetry operation of a sphere is that of rotation by an infinitesimally small angle. All other, more evidently real, rotations are multiples of this basic operation. There is a whole algebra based on these infinitesimal rotation operations and one result, not really surprisingly, is that angular momentum is a quantized quantity. So, for a set of d orbitals the angular momentum quantum number is a maximum of 2, with components with the values 1 and 0 (and, of course, -1 and -2). The orbitals d_{xy} and $d_{x^2-y^2}$ (with two nodal planes containing the z axis) are linear combinations of the orbitals with angular momentum 2 and -2. The orbitals d_{yz} and d_{zx} (with one nodal plane) are linear combinations of the orbitals with angular momentum 1 and -1. The orbital d_{z^2} (with

Symmetry and Structure: Readable Group Theory for Chemists Sidney F. A. Kettle

zero nodal planes) *is* the orbital with angular momentum 0. When talking of d orbitals, it does not matter which definition of L one uses, one gets the same answer in the above equation. But there is a difference when one is talking of electron spin. Expressions such as 'the electron behaves as if it were spinning' (which is subtly different from the statement that 'the electron is spinning') are met. What this means is that the electron behaves as if it has an angular momentum, an angular momentum which can have the quantum numbers $^1/_2$ and $-^1/_2$ (these are generally referred to as the 'spin quantum numbers'). But what of the other interpretation of these quantum numbers, where the positive quantum number of greatest value gives the number of nodal planes containing the z axis? How can one have $^1/_2$ a nodal plane containing the z axis? The only sensible way forward is to double the $^1/_2$ and so obtain a single nodal plane, and to do this we will have to double everything – and this means, for example, that the identity operation will have to be associated with a rotation of 720°, not 360°. This sounds strange but it is not, and shortly we will give it a physical basis. But first it may be helpful to arrive at the same conclusion by a different route.

The equation given above was obtained in Chapter 10 after considering the character obtained for a set of d orbitals subject to a rotation of $\theta°$; this character was:

$$\text{Character}(\theta) = \cos(2\theta) + \cos(\theta) + \cos(0) + \cos(-\theta) + \cos(-2\theta)$$

We can write a similar equation for the two spin functions and this is:

$$\text{Character}(\theta) = \cos(^1/_2\theta) + \cos(-^1/_2\theta).$$

This equation is much easier to simplify than that for the d orbitals because $\cos(-^1/_2\theta) = \cos(^1/_2\theta)$; we obtain

$$\text{Character}(\theta) = 2\cos(^1/_2\theta)$$

Let $\theta = 360°$. Character (360) $= 2\cos(^1/_2\theta) = 2\cos(180) = -2$. But a rotation of 360° is normally associated with the identity and then can only give positive characters, never negative, as here. To compensate for the $^1/_2$ in the above expression we have to work with 2θ, so that a rotation of 720° has to be associated with the identity and, consequently, a positive character is always obtained. What, then, is the operation of rotation by 360°, in this new world? It does not seem to have been given a simple, one word, name but it is commonly denoted I (some authors use R - and this does not exhaust the list). To add some flesh to the bones of this argument, let us look at a specific case – that of the C_{2v} point group, or, rather, its equivalent when we are talking of spin, proves convenient. What are the operations of this group? We can start by listing those of C_{2v}; E, σ_v, C_2, and σ_v'. This list is deceptive; we are living in a 720° world. E, then, is either rotation by 0° or rotation by 720°. C_2 is still rotation by 180°, but in a 720° world. The mirror plane reflections are similar because, as has been mentioned more than once (see Figure 4.18, for instance), a σ_v reflection operation is equivalent to a C_2 combined with an i inversion. This C_2 is also in the 720° world. In Figure 12.1 is shown a diagram of the relationships of these symmetry operations, along with that of rotation by 360°, I, all in a 720° world. It is very evident that there are some operations missing and that these can be obtained by combining I with the operations of the C_{2v} point group (except E, which has already been used to get the I of Figure 12.1). A diagram of all of the operations is shown in Figure 12.2. Because it has twice the number of operations of C_{2v}, it is called 'the C_{2v} double group'. Again, there is no established

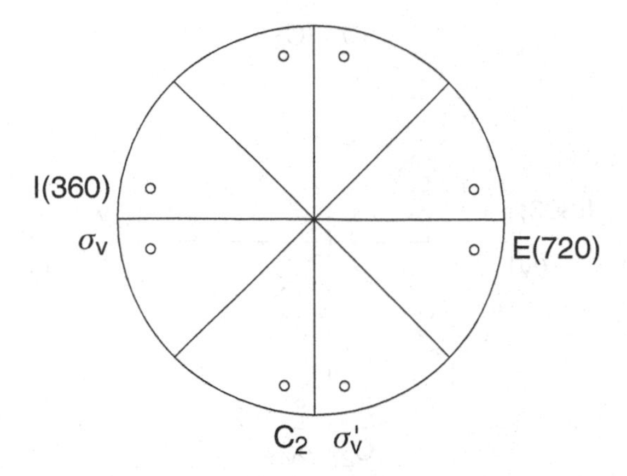

Figure 12.1 The operations of Figure 2.14 re-drawn in a 720° world, with the addition of the new operation of rotation by 360°, I

notation for such double groups but that preferred by the author[1] is to label them with a pre-superscript 2, thus; ${}^2C_{2v}$. Whenever one is dealing with an odd number of electrons, strictly, it is a double group which should be used, whatever the point group. This comment will become more understandable in a moment. For clarity and conformity, in the present chapter we will usually refer to the 'ordinary' C_{2v} point group as ${}^1C_{2v}$. Interestingly, Figure 12.2 solves the problem of 'how can one have $^1/_2$ a nodal plane containing the z axis?'. In Figure 12.2 it can be seen that one half of the (line representing the) nodal plane runs from the centre point, horizontally, to the left-hand side of the circle and the other half of the line (continuous with the first half) runs from the centre point to the right-hand side of the circle. Two halves which, because they are one line, represent a single nodal plane. In a 360° world we were only aware of the first half of the line.

The discussion of the previous paragraph, with its involvement of an apparently somewhat artificial 720° world, seems rather abstract, remote even. Can it be given a more evident physical reality? Fortunately, the answer is in the affirmative. The way forward lies in the Möbius strip. Take a long narrow piece of paper, such as those used in Supermarket checkouts, and join the ends together. There are many ways in which the ends can be joined. Simplest is to join them to form a hoop (Figure 12.3, middle). More complicated is to introduce a twist before they are joined. Hold one end fixed and introduce the twist into the other as it is brought towards the fixed end. There are two ways of introducing this twist; either it can be made in a clockwise or an anticlockwise sense (Figures 12.3, upper and lower). This far from exhausts the ways that the two ends of the strip of paper can be joined and we will meet some of the others later. For the moment, the two single-twisted joins, the two Möbius strips, are our concern. Now, on one side of the piece of paper, before it is joined, list the operations of the C_{2v} group: E, σ_v, C_2, and σ_v'. On the other side list the additional operations of the ${}^2C_{2v}$ group: I, $I\sigma_v$, IC_2, and $I\sigma_v'$. The order in which they

[1] The reason for this preference is that one can envisage triple groups (rotation by 3 × 360 being the identity), quadruple groups (rotation by 4 × 360 being the identity) and so on. There is no known application of such higher groups but perhaps it is as well to use a notation which recognizes their existence.

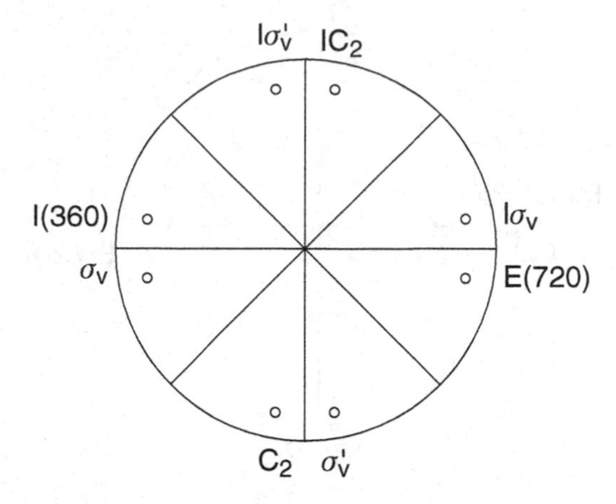

Figure 12.2 Figure 12.1 completed, with the addition of all the operations unique to a 720° world. The complete set comprises the operations of the C_{2v} double group, labelled $^2C_{2v}$

are listed is important; it should follow the sequence of Figure 12.2. Figure 12.4 shows the result. Now take two such labelled strips and use them to form Möbius strips, one with a clockwise and the other with an anticlockwise rotation when joining. For either, starting at the point labelled E it takes two complete rotations, one over each surface, to regain this point, a rotation of 720°. The two alternative Möbius strips can be taken as representing the spin $^1/_2$ and $-^1/_2$ situations; we will take the anticlockwise rotation as spin $^1/_2$.

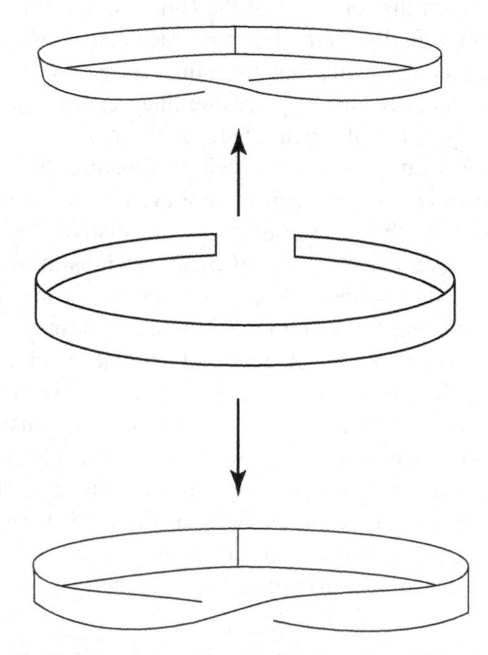

Figure 12.3 Möbius strips. It is possible to twist the central hoop half a turn in either of two directions before the ends are joined. This gives two different Möbius strips which correspond to the two different spin quantum numbers

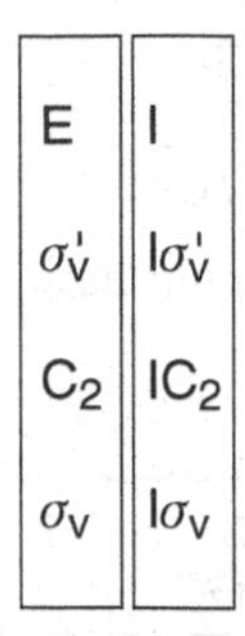

Figure 12.4 The symmetry operation labels to be written on either side of the strips of paper used to form the Möbius strips of Figure 12.3. It is important that the sequence is that shown, so that the sequence in Figure 12.2 is reproduced

We now go back to the problem of the additional ways of rotating the paper ends before joining them. We do so using the pattern just developed; a spin of $^1/_2$ is represented by an anticlockwise twist and a spin of $-^1/_2$ by a clockwise twist. What if we have one electron with spin $^1/_2$ and one with spin $-^1/_2$? We twist first in an anticlockwise sense and then clockwise. The second twist 'undoes' the first. We stay on the 'C_{2v}' side of the paper and never venture into the 720° world. What if we have two electrons with parallel spins, both with spin $^1/_2$ say? The first anticlockwise twist takes us into the 720° world but the second immediately brings us back again into the 360°. Again, starting with 'E', we stay on the 'C_{2v}' side of the paper (except for the infinitesimally small piece of paper between the two twists). It is only for an odd number of electrons, no matter their spin, that we are forced into the 720°, the double group, world. So, as we add individual electrons to a system (the conclusion is general and does not just apply to C_{2v}) we alternate between the 'ordinary' and corresponding double group as each additional electron is added.

Problem 12.1 Make a Möbius strip as shown in Figures 12.3 and 12.4 (the longer and narrower the better). Use it to describe the situations described in the text, involving varying numbers of electrons of α and β spin. Some imagination will have to be used if a pair such as $\alpha\alpha$ are to have an infinitesimally small space between their twists.

Whilst we have gained some insight into the $^2C_{2v}$ group, we have yet to meet its character table (although we have met the fact that the two spin functions of $^1/_2$ and $-^1/_2$ give rise to a character of -2 under the I operation). To generate this character table, several methods are available. A figure such as Figure 12.2 enables a group multiplication table to be compiled, if need be, and a discussion such as that in Chapter 2 would enable some, if not all, of the irreducible representations to be obtained from it. Alternatively, a simple algebraic method, such as that used for the C_{4v} group in Chapter 6, would be applicable and the development would, almost word for word, be the same. The groups C_{4v} and $^2C_{2v}$ are isomorphous. Not surprisingly, the operation of C_2, in a 720° world, parallels that of the operation of C_4 in a 360° world, and similarly for the other operations. The class structure of the C_{4v} group is now given with that of $^2C_{2v}$ beneath it. We have not proved that this is the correct class structure

$$\begin{array}{llllll} C_{4v} & E & 2C_4 & C_2 & 2\sigma_v & 2\sigma_v' \\ {}^2C_{2v} & E & C_2 & I & \sigma_v & \sigma_v' \\ & & IC_2 & & I\sigma_v & I\sigma_v' \end{array}$$

for the latter group but this can be done quite easily, in case of doubt, using the methods of Appendix 1.

Table 12.1

${}^2C_{2v}$	E	C_2 IC_2	I	σ_v $I\sigma_v$	σ_v' $I\sigma_v'$
A_1	1	1	1	1	1
A_2	1	1	1	−1	−1
B_1	1	−1	1	1	−1
B_2	1	−1	1	−1	1
$E_{1/2}$	2	0	−2	0	0

It will be immediately apparent that, in C_{4v} there are two operations grouped together in a class, but in ${}^2C_{2v}$ they are listed separately (although still in the same class, shown by them being given in the same column). The reason for this will become apparent shortly; use some shorthand for the pair at your own risk! The ${}^2C_{2v}$ character table is given in Table 12.1.

We will give nodal diagrams of the irreducible representations of ${}^2C_{2v}$ shortly, and these should resolve any residual problems that the reader may have with the singly degenerate ones. First, however, the doubly degenerate irreducible representation, $E_{1/2}$. As this label hints, this is the irreducible representation under which the two spin functions, $^1/_2$ and $-^1/_2$, together, transform. We have seen that the character that they generate under the operation of rotation by θ is:

$$\text{Character}(\theta) = 2\cos(^1/_2\theta).$$

When θ is 180° (C_2) the character is $2\cos(90) = 0$. Because $\sigma_v = C_2 \cdot i$, Character $(\sigma_v) =$ Character$(C_2) \cdot$ Character(i). Again, Character$(C_2) = 0$ and so Character $(\sigma_v) = 0$. These are sufficient to establish that, indeed, $^1/_2$ and $-^1/_2$, together, transform as $E_{1/2}$. But this is not the end; $E_{1/2}$ has some tricks in store. They will become apparent in the nodal diagrams which we will shortly draw of it. It may be recalled that in Chapter 6, when first we met two operations falling into the same class, we commented (at the beginning of Section 6.2) 'the correct procedure is to consider the transformation of each object under each of the individual operations in the class and to take the average of the characters generated'. As the diagrams will show, for $E_{1/2}$ one can have a character of 1 under the C_2 operation and a character of −1 for the IC_2. It is the average of the two, 0, which is the $E_{1/2}$ character for the class. The reason for retaining a separate listing of C_2 and IC_2 at the head of the character table is now evident – one may need to work with each, separately from the other. Analogous arguments apply to σ_v and $I\sigma_v$ and also to σ_v' and $I\sigma_v'$. At last, we have – almost – answered the question 'what is the symmetry of an unpaired electron in the water molecule?' It is $E_{1/2}$, irrespective of the symmetry of the orbital in which it finds itself. Except that the answer, unlike the question, recognizes that there are two possible spin functions. What are the nodal properties of the irreducible representations of ${}^2C_{2v}$? In a sense the answer is old and in a sense it is new. Because ${}^2C_{2v}$ and C_{4v} are isomorphous, so

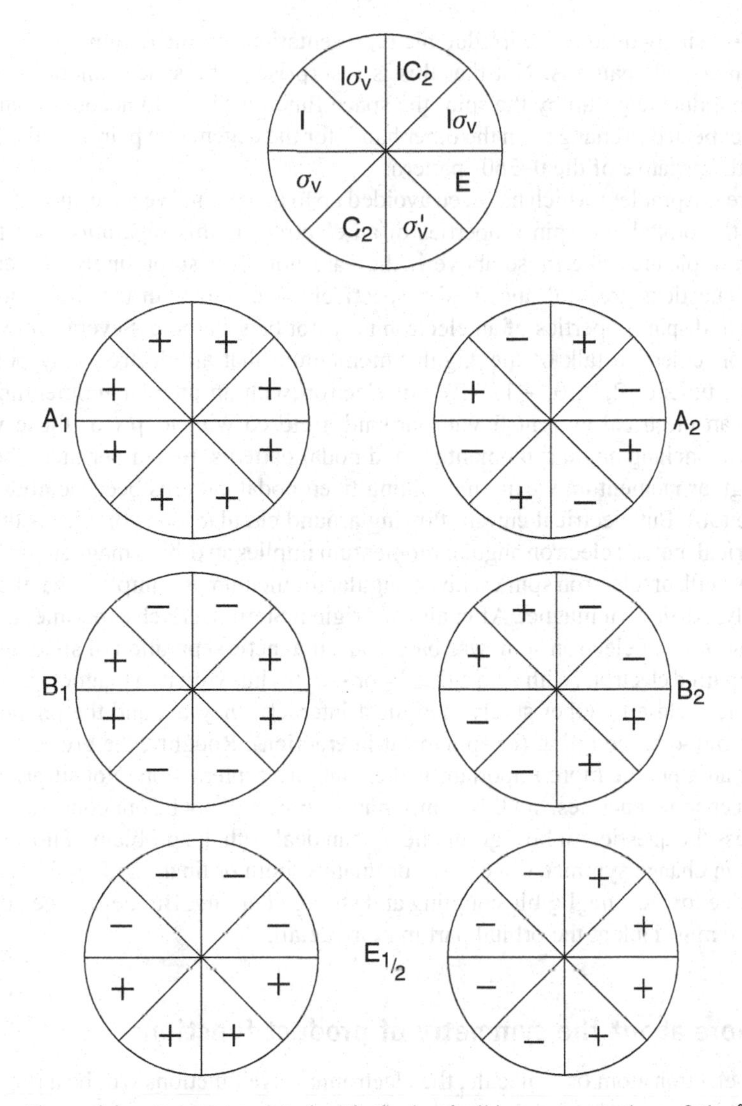

Figure 12.5 The nodal patterns associated with the irreducible representations of the $^{2}C_{2v}$ group. For the singly degenerate functions, the patterns are the same as those in Figure 3.1 from 0–360° and this pattern is repeated from 360–720°. As a consequence all have an even number of nodal planes. For the $E_{1/2}$ functions the 360–720° pattern is a repeat of the 0–360° but with all phases changed. As a consequence it has an odd number of nodal planes. Similar even-odd nodal plane patterns are found in all double groups

too are the nodal patterns associated with their irreducible representations (provided that we show the whole of 0–720° space for the former). We have already met those of C_{4v} in Figure 6.9, but new is the excursion into the 720° world. The diagrams for $^{2}C_{2v}$ are given in Figure 12.5. Notice the general pattern; those irreducible representations which carry the same label as irreducible representations of $^{1}C_{2v}$ have identical 0–360 nodal patterns as the latter. These, singly degenerate irreducible representations, in the region 360–720° simply repeat their 0–360° patterns. Not that this is a surprise – the space functions were only forced into a double group by the spin; the space functions have to accommodate this but cannot be expected to change. On the other hand, for the degenerate pair, $E_{1/2}$, the 360–720° pattern is the negative of the 0–360° pattern.

But there is a problem which has been avoided up to this point. We have always separately discussed the orbital and spin properties of an electron. Is this separation justified? And can we draw pictures like those above if they are not? Not surprisingly, the answers to these two questions are 'no' and 'yes' respectively. We start with the 'no': the fact that the orbital and spin properties of an electron may not be separable. Several times we have found it convenient to talk of the angular momentum that an electron may possess (for d electrons, units of 2, 1, 0, −1, −2). An electron with an angular momentum may be likened to an electrical current flowing around a closed wire loop (for those who have problems reconciling angular momentum and nodal patterns, remember that when we are talking angular momentum we are not talking fixed nodal patterns because nodal patterns may rotate too). But electrical current flowing around circular loops of wire is the basis of most electrical motors; electron angular momentum implies an orbital magnetism. Similarly, although we talk of 'electron spin', with an angular momentum it is implied that the electron, intrinsically, is a tiny bar magnet. After all, the original Stern–Gerlach experiment, which led to the recognition of electron spin, was based on a magnetic separation of silver atoms with a single unpaired electron (with spin either $^1/_2$ or $-^1/_2$). Thus with two magnets – one orbital, one intrinsic – close together surely they must interact;[2] they do, and the phenomenon is known as spin–orbit coupling (or spin–orbit interaction). Roughly, the greater the atomic number of an atom the more important is the spin–orbit interaction. Not surprisingly, this interaction changes energies, but this – important – topic will not be our concern. Rather, we shall address the question of how group theory can deal with the problem of how spin–orbit coupling can change symmetry species. Fortunately, both ordinary and double groups can handle both extremes: negligible coupling and strong coupling. But before we address this problem we must look at the orbital part in more detail.

12.2 More about the symmetry of product functions

In a many-electron atom or molecule, the electronic wave functions will be a product wave function which, at the simplest level, takes the form

$$\psi = \phi_1\phi_2\phi_3 \cdots \phi_n$$

where $\phi_1 \cdots \phi_n$ are individual one-electron orbitals and ψ is the single wave function which describes all n electrons. We have already met something similar; products of one-electron wavefunctions had to be discussed as soon as we became interested in overlap between orbitals in Section 3.5. Even earlier, we met products in Section 3.1 when normalization was the topic. At first no name was given to the process but in Section 4.3 the name 'direct product' was introduced. Direct products involving two or more degenerate irreducible

[2] There are those who will argue that the description just given of the spin–orbit interaction is inadequate (relativistic effects are involved). Nonetheless, it correctly establishes grounds for the existence of the phenomenon.

Table 12.2

C_{4v}	E	$2C_4$	C_2	$2\sigma_v$	$2\sigma_v'$
A_1	1	1	1	1	1
A_2	1	1	1	−1	−1
B_1	1	−1	1	1	−1
B_2	1	−1	1	−1	1
E	2	0	−2	0	0

representations invariably give rise to a reducible representation as a product. Consider the C_{4v} point group. The C_{4v} character table is given in Table 12.2 (it was first met as Table 5.6) and it is evident that the direct product $E \otimes E$ must be reducible because the number that will appear in the identity column (4) is larger than any character in the table. The direct product $E \otimes E$ is:

	E	$2C_4$	C_2	$2\sigma_v$	$2\sigma_v'$
E ⊗ E	4	0	4	0	0

which is readily seen to be a representation with components

$$A_1 + A_2 + B_1 + B_2$$

Problem 12.2 Show that the direct product table for the C_{4v} group is:

C_{4v}	A_1	A_2	B_1	B_2	E
A_1	A_1	A_2	B_1	B_2	E
A_2	A_2	A_1	B_2	B_1	E
B_1	B_1	B_2	A_1	A_2	E
B_2	B_2	B_1	A_2	A_1	E
E	E	E	E	E	$(A_1 + A_2 + B_1 + B_2)$

12.3 Configurations and terms

It is instructive to consider the meaning of the $E \otimes E$ direct product in the C_{4v} point group in more detail. Suppose that there are two electrons, one of which is to be placed in the degenerate pair of orbitals of E symmetry denoted individually e_1 and e_2. The second electron is to be placed in a different degenerate pair of orbitals of E symmetry which we individually denote by E_1 and E_2. The possible two-electron functions are

$$e_1E_1 \quad e_1E_2 \quad e_2E_1 \quad e_2E_2$$

That is, they are four in number (in agreement with the number 4 which appears in the identity column when the $E \otimes E$ direct product is formed). Group theory tells us that it is possible to take linear combinations of these four functions such that one combination

Table 12.3

Operation	E	C_4^+	C_4^-	C_2	$\sigma_v(1)$	$\sigma_v(2)$	$\sigma_v'(1)$	$\sigma_v'(2)$
	e_1	$-e_2$	e_2	$-e_1$	$-e_1$	e_1	e_2	$-e_2$
	E_1	$-E_2$	E_2	$-E_1$	$-E_1$	E_1	E_2	$-E_2$
	e_1E_1	e_2E_2	e_2E_2	e_1E_1	e_1E_1	e_1E_1	e_2E_2	e_2E_2

has A_1 symmetry, one has A_2 symmetry, one has B_1 symmetry and one has B_2. These symmetry-adapted functions may be obtained by the projection operator method described in Chapter 5 and, more particularly – because it deals with a non-Abelian group – Chapter 6. We first simply choose one function – e_1E_1 for instance – and work out how it transforms under the operations of the group. For this, we need to know how the individual functions e_1 and E_1 transform. This information is detailed in Table 6.7 (where it is necessary to replace p_x by e_1 or E_1 and p_y by e_2 or E_2). In this way Table 12.3 is obtained. Multiplication by the A_1 characters and adding, in the usual projection operator method, leads to the conclusion that

$$\psi(A_1) = \frac{1}{\sqrt{2}}(e_1E_1 + e_2E_2)$$

Problem 12.3 Use Table 12.3 to show that

$$\psi(B_1) = \frac{1}{\sqrt{2}}(e_1E_1 - e_2E_2)$$

Problem 12.4 Derive a table similar to Table 12.3 but appropriate to the function e_1E_2. Use it to show that

$$\psi(A_2) = \frac{1}{\sqrt{2}}(e_1E_2 - e_2E_1)$$

$$\psi(B_2) = \frac{1}{\sqrt{2}}(e_1E_2 + e_2E_1)$$

In the belief that a specific example would help the reader, the above discussion was concerned with electronic wavefunctions in an 'ordinary' group. The method, however, is not limited to such wavefunctions. Thus, the pairs (e_1, e_2) and (E_1, E_2) could equally have been vibrational wavefunctions, in which case the product wavefunctions would have been the ones relevant to a discussion of combination bands in a vibrational spectrum (vibrational excitations in which two different vibrations are excited by a single quantum of energy). Indeed, the discussion is of general applicability, with one exception. Note that the members of the two pairs of E functions are different – (e_1, e_2) and (E_1, E_2), not (e_1, e_2) and (e_1, e_2). This is not to say that the case in which the members of the pairs are identical is not important. In vibrational spectroscopy, for instance, overtone bands arise from double excitations – where (e_1, e_2) is combined with (e_1, e_2). Such cases are not immediately covered by the discussion because when products are formed within the members of a doubly degenerate set only *three* product functions can be distinguished. Two quanta of vibrational energy can be excited in e_1 or e_2 or one quantum can be excited in each so that the distinguishable excited states are of the form:

$$e_1e_1 \quad e_2e_2 \quad e_1e_2$$

What has happened to the fourth function e_2e_1? Why should it have been discarded and yet e_1e_2 retained? There is no reason; we have made a mistake – these three functions do not transform as irreducible representations. This is seen from the functions above involving e and E – they have to translate into the present problem but with E_1 and E_2 replaced by e_1 and e_2 respectively. With this change the functions become:

$$\psi(A_1) = \frac{1}{\sqrt{2}}(e_1e_1 + e_2e_2)$$

$$\psi(A_2) = \frac{1}{\sqrt{2}}(e_1e_2 - e_2e_1)$$

$$\psi(B_1) = \frac{1}{\sqrt{2}}(e_1e_1 - e_2e_2)$$

$$\psi(B_2) = \frac{1}{\sqrt{2}}(e_1e_2 + e_2e_1)$$

Look at the $\psi(A_2)$ function. The two terms on the right-hand side mutually cancel, at least when we are talking about vibrations. We are left with three overtone vibrational functions, as we expected. Why should $\psi(A_2)$ be so different? The fundamental difference can be seen if we interchange the suffixes 1 and 2 on the e's in the above expressions. $\psi(A_1)$, $\psi(B_1)$ and $\psi(B_2)$ are unchanged. They are all symmetric with respect to the interchange. In contrast, $\psi(A_2)$ becomes -$\psi(A_2)$ when the suffixes 1 and 2 are interchanged. It is antisymmetric with respect to the interchange. That is, the direct product $E \otimes E$ can be divided into a symmetric direct product $A_1 + B_1 + B_2$ and an antisymmetric direct product A_2. This is a general situation; when a direct product is formed between a degenerate irreducible representation and itself, the product is always a sum of symmetric and antisymmetric components. In the vibrational example considered above, the antisymmetric component vanished, but this is not always the case. What is general is the fact that the two components behave differently. In particular, when they refer to orbital functions the symmetric and antisymmetric direct products are associated with spin functions of different multiplicities, a point to which we will return. The interested reader will find the electronic case developed in Ballhausen[3] and more about the vibrational in Wilson, Decius and Cross.[4] Both give detailed derivations of expressions for the symmetric and antisymmetric direct products; the two derivations are closely related.

There is another simple application of direct products that is important to consider. As has been seen, if, in ${}^1C_{2v}$ symmetry, we have a molecule with the electron configuration $a_2^1b_2^1$ (or, as it was previously expressed, the product wavefunction is $\varphi_1(A_2)\varphi_2(B_2)$), then it is said that this configuration gives rise to a *term* of B_1 symmetry (or, in the form used earlier, there is a product wavefunction $\psi(B_1)$), the direct product $A_2 \otimes B_2$ being B_1. Such a term would normally have a pre-superscript which indicates the (electron) spin multiplicity. Thus a triplet spin term would be ${}^3B_1$, a singlet ${}^1B_1$, and so on. For the moment we neglect

---

[3] C.J. Ballhausen, *Introduction to Ligand Field Theory*, McGraw Hill, New York, 1962, p.48.

[4] E.B. Wilson, J.C. Decius and P.C. Cross, *Molecular Vibrations*, McGraw Hill, New York, 1955, p.152.

spin, something which is formally expressed by saying that we are only concerned with orbital terms. Thus, in $^1C_{2v}$ we say that the orbital configuration $a_2^1b_2^1$ gives rise to the orbital term B_1. Similarly, for the C_{4v} example considered above, the electron configuration e^1E^1 gives rise to the terms A_1, A_2, B_1 and B_2. More correctly, and following the notation used in earlier diagrams in this book, one says that the electron configuration $1e^12e^1$ gives rise to the terms A_1, A_2, B_1 and B_2.

When a singly degenerate orbital is occupied by two electrons the product wavefunction describing this situation is totally symmetric – because the direct product is totally symmetric (see, for example, Table 4.4). It is not so easy to see that the same result follows when a set of degenerate orbitals is completely occupied by electrons because simply forming direct products leads, apparently, to a large number of terms. However, the Pauli exclusion principle eliminates all but one of these terms. There is only one way of filling all orbitals of a degenerate set and that is by putting two electrons into each orbital. There is, then, only a single wavefunction and so there must be a singly degenerate term. This simple argument does not tell us whether or not this term is totally symmetric. It seems intuitively likely that it will be, and this is confirmed by following the transformations of the product wavefunction under the operations of the point group in the way that was done for e_1E_1 above. This leads to a general and very valuable conclusion:

Closed shells of electrons are invariably totally symmetric.

Here, by 'closed shell' is meant configurations like a_1^2, b_{1u}^2, e^4, t_{2g}^6 and so on. This conclusion means that for a many-electron molecule the possible terms arising from a configuration can be obtained simply by considering those orbitals which are partially filled. Those that are totally filled are ignored – unless these are the only ones present, in which case the orbital term is totally symmetric.

Problem 12.5 Show that in the C_{4v} group the electron configuration e^4 gives rise to a term of A_1 symmetry.

Hint: It may be helpful to write this configuration, using the notation adopted earlier in this chapter, as $e_1^1e_1^1e_2^1e_2^1$. The table constructed as part of Problem 12.4 can then be modified to be used in the present problem.

It might be thought that, having determined that a totally symmetric term results from a closed shell and that the many-electron wavefunction is totally symmetric, this would be the end of the matter. This is not the case. Consider the situation shown in Figure 12.6a, in which for a C_{2v} molecule in addition to a filled A_2 orbital there is an empty B_2 orbital at higher energy. Both the configuration a_2^2 and the configuration in which both electrons are promoted into the b_2 orbital, b_2^2, have orbital symmetry A_1. In general, it is found by detailed calculation that although the ground term wavefunction is well represented as one derived solely from a configuration such as a_2^2, this wavefunction is improved if there is mixed in with it a contribution from the excited term configuration b_2^2, which also gives rise to a term of orbital symmetry A_1. Such *configuration interaction* is an important step in

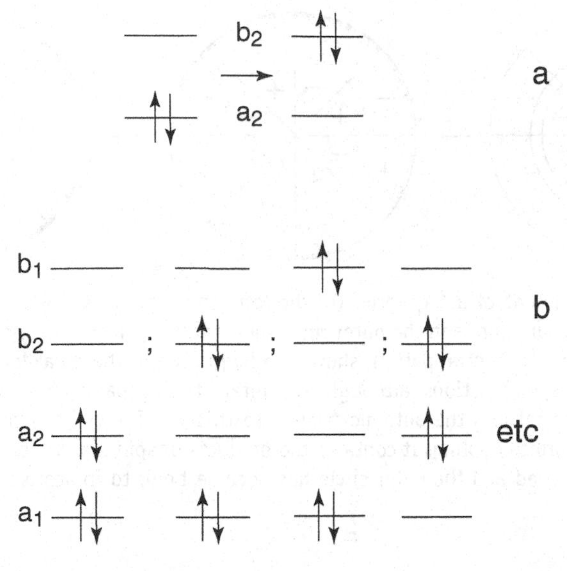

Figure 12.6 (a) A ground state a_2^2 configuration and a b_2^2, both of the same symmetry ($^1A_1$). (b) Examples of states, all of $^1A_1$ symmetry, between which interaction may occur

most detailed calculations of molecular properties, although more than one excited term is usually involved in mixing with the term arising from the ground term configuration (and there may be reasons for excluding some others). So, if, in the present example, the ground term configuration were one in which a doubly occupied orbital of A_1 symmetry had above it a doubly occupied orbital of A_2 symmetry followed by empty orbitals of B_2 and B_1 symmetries (Figure 12.6b), then configuration interaction would be expected between the A_1 terms $a_1^2a_2^2, a_1^2b_2^2, a_1^2b_1^2, a_2^2b_2^2, a_2^2b_1^2, b_2^2b_1^2$ and $a_1^1a_2^1b_2^1b_1^1$ and also configuration interaction between the excited B_1 terms arising from the configurations $a_1^2a_2^1b_2^1$, $a_1^1a_2^2b_1^1$, $a_1^1b_2^2b_1^1$ and $a_2^1b_2^1b_1^2$.

Problem 12.6 Check that the (first) set of seven configurations given above all give rise to terms of A_1 symmetry and that the (second) set of four all give rise to B_1 terms.

As has just been mentioned, the inclusion of configuration interaction is usually an important step in accurate calculations on the electronic structure of molecules – for instance, in obtaining those results which have been used at several points in this book. Two points should be made. First, for configuration interactions, just as for orbital interactions, only terms of the same symmetry species interact (they also have to be of the same spin multiplicity). Second, it is evident that as the number of orbitals included in a molecular problem increases – also and the number of configurations that arise – so the number of terms of a given symmetry species which may interact under configuration interaction increases. The improvement that results in the description of the ground term (and, usually, the lowest excited terms) is often considerable, so an upper limit on the improvement is usually set by the capacity of the computer available: its ability to handle the enormous number of integrals that have to be calculated.

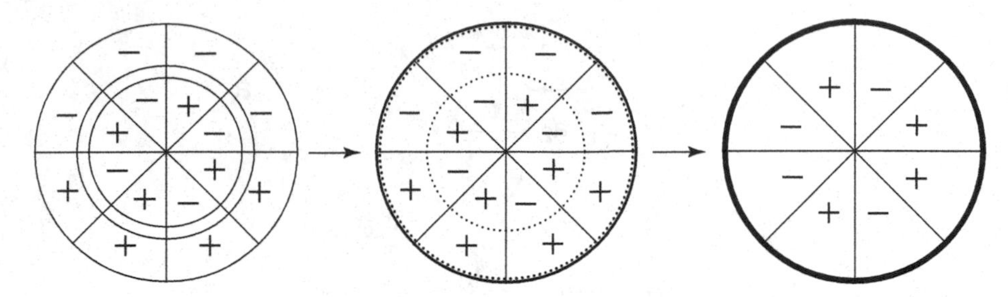

Figure 12.7 A component of a $^2A_2$ term. On the left, the orbital, A_2, function is in the centre and an $E_{1/2}$, doublet, function is in the outer ring. Their total separation is symbolized by the open ring separating them. The central pattern shows the beginning of the breakdown of the separation between orbital and spin functions, although they retain their separate phase patterns. The inner circles begin to fragment and the outermost circle to thicken. The right-hand diagram shows the limit of strong spin–orbit coupling. It contains the products of spin and orbital phase patterns; the central cirle has vanished and the outer circle has become bold, to indicate the strong spin–orbit coupling limit.

12.4 The inclusion of electron spin

We now return to a problem left incomplete in the previous two sections, that of spin and spin–orbit coupling. For simplicity, we shall introduce it in the context of single-electron functions although the discussion applies equally well to the many-electron functions met in the last section. Again for simplicity, we will work with the C_{2v} point group, either $^1C_{2v}$ or $^2C_{2v}$. Consider the latter and the case of an unpaired electron in an orbital of A_2 symmetry. The latter label applies to both $^1C_{2v}$ and $^2C_{2v}$ but because of the single unpaired electron we must be working in the $^2C_{2v}$ point group, where the electron spin functions (spin 'up' and 'down') transform as $E_{1/2}$. In this description, the spin and orbital functions are totally independent and so they can have different symmetry labels. But what of the transition to the other limit, strong spin–orbit coupling, where the spin and orbital functions are completely coupled and it is meaningless to talk of them as independent? The transition is shown diagrammatically in Figure 12.7. On the left-hand side the orbital and spin functions are shown in a single nodal diagram, the orbital component being in the centre and the (outer, concentric) spin diagram separated from it by an empty circle (empty to emphasize the separation). In the middle diagram this empty circle is shown both fragmented and thinner, indicating a decreasing independence of the orbital and spin components. The fragments of the circle that are removed may be regarded as adding to the outermost circle, which thickens. In the right-hand diagram the inner circle has vanished and the outer circle has reached its thickest, indicating the strong spin–orbit coupling limit. The nodal pattern in this diagram is the product of the two independent patterns in the zero-coupling limit (the left-hand diagram). That is, it is the direct product of the spin and orbital functions. The final, strong coupling, function has $E_{1/2}$ symmetry. True, it is not one of the $E_{1/2}$ functions of Figure 12.5, it is more noded, but it has the essential characteristic that the signs in the 360–720° region are the negative of those in the 0–360° region. The example of Figure 12.7 was one component of a $^2A_2$ term; for completeness, Figure 12.7 is repeated in Figure 12.8 along with the corresponding diagrams for the second component of the doublet.

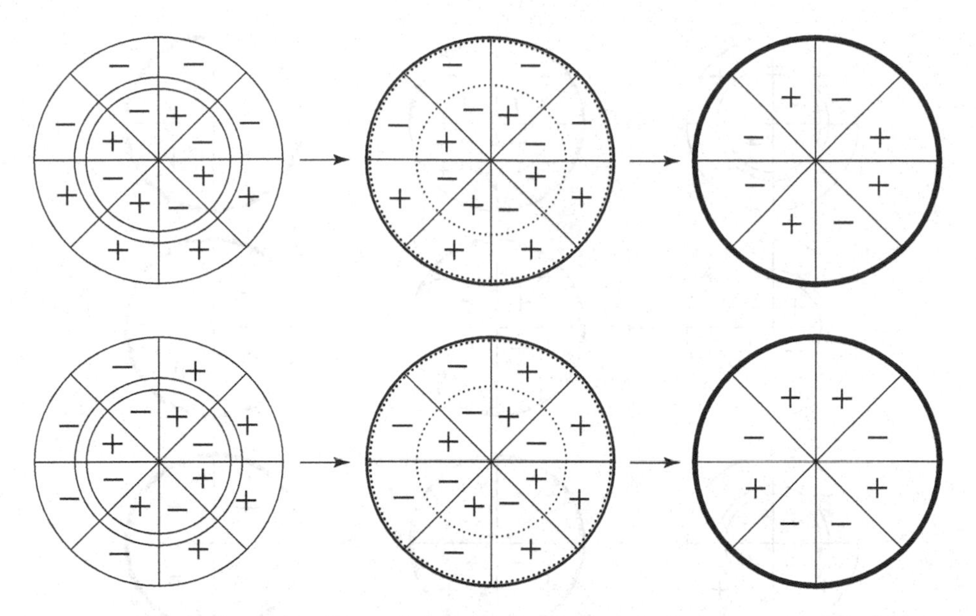

Figure 12.8 This is identical to Figure 12.7 except that it shows both components of the $^2A_2$ term. Comparison with Figure 12.5 may be helpful

Problem 12.7 By studying its transformations under the operations of the $^2C_{2v}$ point group show that the functions at the right-hand side of Figure 12.8 have $E_{1/2}$ symmetry.

Problem 12.8 By drawing diagrams akin to that of Figure 12.8:

(a) show that the terms $^2A_1$, $^2B_1$ and $^2B_2$ all become $E_{1/2}$ in the limit of strong spin–orbit coupling, and

(b) by consideration of the $E_{1/2}$ diagrams, show that although all four terms $^2A_1$, $^2A_2$, $^2B_1$ and $^2B_2$ become $E_{1/2}$ in the strong spin–orbit coupling limit, these four $E_{1/2}$ functions are all orthogonal (i.e. have zero overlap integrals with each other). If they were not orthogonal we would have to talk about mixing between them.

The above discussion involved the use of the $^2C_{2v}$ group; it was chosen because there is a clear symmetry separation between space and (doublet) spin functions. The former carried the same symmetry labels that they did in $^1C_{2v}$ whilst the latter were $E_{1/2}$. When we consider triplet spin functions this separation is lost – as we saw earlier in this chapter, we have to work in $^1C_{2v}$ with both. The immediate question is that of how the spin functions transform. This is less of a problem than it seems at first sight. The real problem is that we are accustomed to thinking of the triplet spin functions being something like $\alpha\alpha$, $\alpha\beta$ and $\beta\beta$. But these, respectively, correspond to angular momenta of 1, 0 and -1. This is reminiscent of the p orbitals which can be chosen so that they have just these angular momenta. But in $^1C_{2v}$ these are not the p orbital functions used; we use their standing wave counterparts, p_x, p_y and p_z – orbitals which are linear combinations of those with angular momenta 1, 0

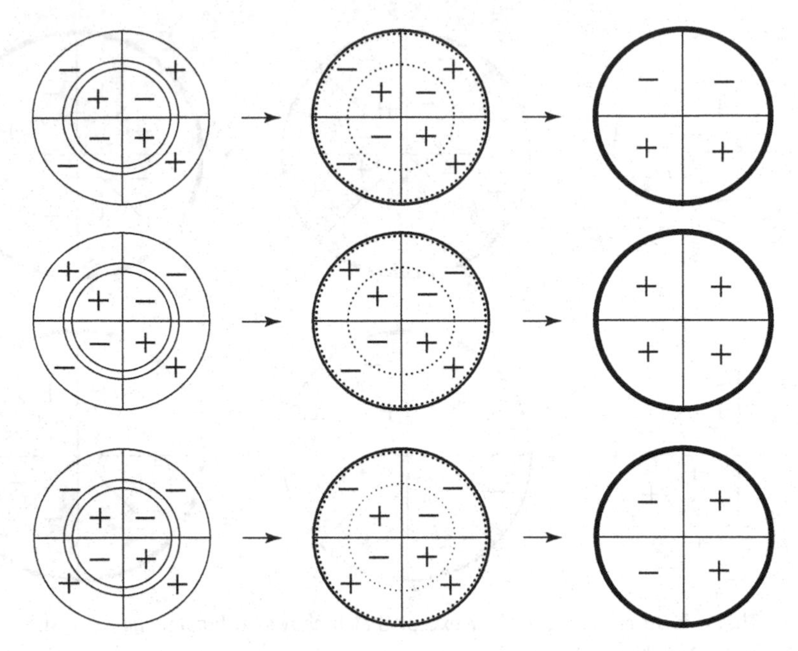

Figure 12.9 Diagrams analogous to those of Figure 12.8 for the three components of a $^3A_2$ term using the $^1C_{2v}$ group. The phase patterns are derived from those of Figure 3.1

and −1. And so it is with the spin functions; we need spin functions akin to p_x, p_y and p_z. For simplicity, but at the risk of confusion with s orbitals, we will call them s_x, s_y and s_z (s for 'spin'). However, care must be taken in giving them symmetry labels. We know that there are differences between orbital and spin. So, those orbital transitions which are spectroscopically allowed are electric dipole transitions but the spin are magnetic dipole (Table 4.5). The former correspond to T_x, T_y and T_z, the latter to R_x, R_y and R_z (Table 4.5). Why should this be important? Look back at Figure 4.19. Rotations and translations behave differently under inversion in a centre of symmetry. The same equation (that given at the beginning of this chapter) applies to the proper rotations of both. However, the fact that improper rotations are a combination of a pure rotation with inversion in a centre of symmetry means that, in $^1C_{2v}$, for the σ_v operations, the spin functions have the negative of the characters of the corresponding orbital functions. So, s_x has B_1, s_y has B_2 and s_z has A_2 symmetry.[5] Armed with this information, we can now construct diagrams akin to that of Figure 12.8 for $^3A_1$, $^3A_2$, $^3B_1$ and $^3B_2$ but now using, of course, the $^1C_{2v}$ group. These are given for $^3A_2$ in Figure 12.9 where, as before, increasing coupling between spin and orbital is indicated by the narrowing, fragmentation and transfer of density from the inner to the outer circle. Also as before, the strong coupling limit is given by the direct product of the spin and space symmetries and shown as such by the use of a thick outer circle in the diagram. There is another interesting aspect of the above discussion; it indicates something

[5] This discussion reveals something hidden by the use of the phrase 'spin triplet'. The three components of the spin triplet can have different energies.

in common between spin-orbit coupling and optical activity (but in very different fields, of course); both involve the linking of linear and rotatory charge displacements.

Another aspect of the two-unpaired electron case which is worthy of study is that we could tackle it starting from ${}^{2}C_{2v}$. We would use it for the single-electron case, where the electron would have the symmetry $E_{1/2}$. What if we added a second electron, which of course has the same symmetry? We would have to form the direct product $E_{1/2} \otimes E_{1/2}$ and this, with components $A_1 + A_2 + B_1 + B_2$, would contain a symmetric part and an antisymmetric part, A_2. Of course, the two-unpaired electron problem would be handled in ${}^{1}C_{2v}$, not ${}^{2}C_{2v}$, but what happens to the four components found in ${}^{2}C_{2v}$ when we move to ${}^{1}C_{2v}$? We have found the components of the symmetric direct product, $A_1 + B_1 + B_2$; they are the triplet spin functions (and symmetric with respect to interchange, since they originate in functions like $\alpha_1\alpha_2$, where we have added subscripts in an obvious way). The other spin function is A_2, the antisymmetric direct product, and this is the singlet spin combination. Such a spin combination must exist for the two-electron case but it was ignored in the above discussion.

There is a communality between the orbital and spin discussions above. When talking of degeneracies, we could for each distinguish between symmetric and antisymmetric direct products. What of their combination, where we specify both orbital and spin functions (as part of the description of a wavefunction, for instance)? Not all combinations are allowed. To obtain results which are in conformity with observation, the combination has itself to be antisymmetric. That is, a symmetric space part combines with an antisymmetric spin part, and vice versa. Why? This is a topic of debate, not least amongst some writers of science fiction, who envisage another universe where the allowed spin and space combinations are symmetric with respect to interchange.

12.5 Summary

The quantum number $1/2$ associated with electron spin has to be compensated by doubling angles, so that the identity becomes 720° (p. 282). A double group has twice as many operations as the corresponding 'normal' group (p. 284); the class structure is also different (p. 286). Nodal patterns may be drawn of the irreducible representations of double groups (p. 287). By combining spin and space functions in a single diagram it is possible to follow the evolution of nodal patterns with increasing spin–orbit coupling from the Russell–Saunders to the *jj* limit (p. 294).

13 Space groups

13.1 The crystal systems

Crystals and the determination of their structures at the atomic level represent important parts of chemistry and so in this chapter it will be assumed that the reader has some basic familiarity with associated terms such as 'monoclinic'. Although the solid state is an active part of chemistry, solid state symmetry is a much neglected area, leading to conceptual gaps which make the subject much more difficult than need be the case. In an attempt to meet this problem, in the present chapter some subject areas will be explored which are normally given a very brief treatment: the relationships between the 7 crystal systems, the 14 Bravais lattices, the 32 crystallographic point groups and the 230 crystallographic space groups. An attempt will be made to answer the fundamental question 'why are there 230 space groups and not some other number?'. This will force us into a deeper understanding of the space groups themselves, which is the object of the exercise. This quest will mean that a rather unusual attitude to the topic is adopted, one that is closer to the conventional physics approach than to that of chemistry.

A convenient starting point for the discussion is the concept of an empty lattice. In talking about crystal structures one expects to be concerned with the arrangement of atoms in space and the way that they fill unit cells. Despite this very reasonable expectation, our starting point will be quite different: no atoms, no unit cells, just the fiction of an empty lattice. The reason for this approach is that it will enable the introduction of atoms into the lattice as a step which is quite distinct from anything to do with the lattice itself. As will be seen, the step of introducing atoms adds possibilities which do not exist for the bare lattices. What, then, is a 'bare lattice' (and why is it a fiction)? For our purposes it is convenient to regard a lattice as arising from a three-dimensional network of vectors, vectors that connect equivalent points in an empty space (Figure 13.1a). These vectors are easy to picture; think of an arrow drawn from one point to another (Figure 13.1b). Really, of course, the equivalent points will only be equivalent after we have introduced atoms – in empty space all points are equivalent and form a continuum. So, in selecting a regular array of points from within a continuum, our model is definitely fictional but it will prove helpful to persist with it. Strictly, it is the network of points that form the lattice in empty space, although, group theoretically, the vector set which connects any one point with all of the others is more important. The reason is that the operations of the lattice which parallel the C_2, σ_v and σ_v' of C_{2v} are translations, those translations which turn the lattice into a lattice indistinguishable from the starting lattice (we assume that the lattice is infinite, so no

Symmetry and Structure: Readable Group Theory for Chemists Sidney F. A. Kettle

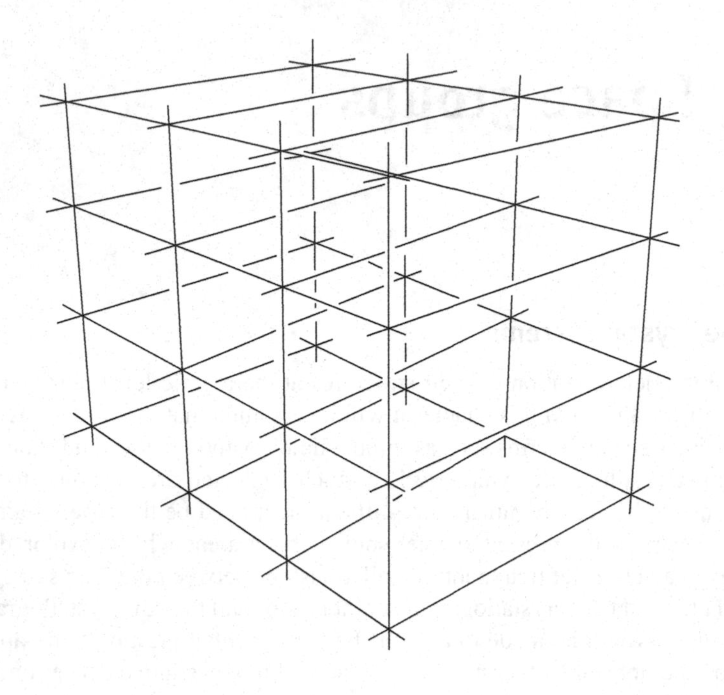

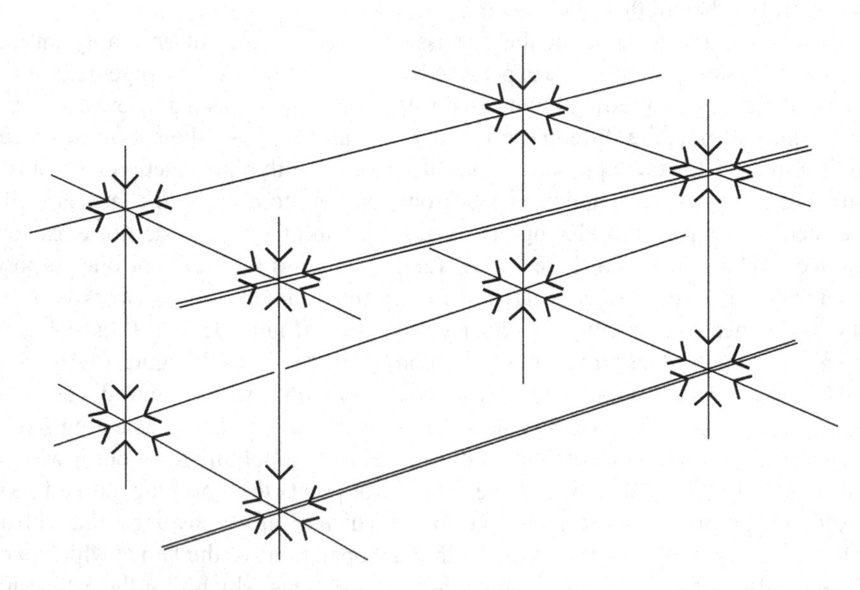

Figure 13.1 (a) A lattice typical of a crystal. Equivalent points in space are linked by sets of equally spaced parallel lines extending to infinity. For simplicity of construction, in this figure all lines are drawn as intersecting at right angles but this is not a general requirement. (b) A single segment of the lattice of (a). Each line in (a) is really two superimposed vectors, one the negative of the other, represented here by an arrowhead at either end of each line

problems arise by reaching the edges of a crystal). And these operations are the members of the set of all the translation vectors. The vectors connecting the separated and equivalent points of the lattice fill all of three-dimensional space in the manner indicated in Figure 13.1a; they look rather like sets of parallel fishing nets, the corners of the mesh (and it is these corners that are the points of the lattice) in any one net being linked by additional pieces of string sideways to the corresponding adjacent corners of the nets on either side of it. It is as if there were a three-dimensional array of string-edged boxes filling all space. It is tempting to call these boxes 'unit cells', and, indeed, this is what they would normally be called. However, this temptation will be resisted – as has been said earlier, no atoms, no unit cells. Rather, as we have seen, an arrowhead will be attached at each end of every piece of string to demonstrate that sets of translation vectors are under consideration; this has been done in Figure 13.1b. Each vector has a negative (if a is connected to b, b is connected to a), which is why arrowheads are needed at each end of each segment of string, at each equivalent lattice point. Our concern is to work out all the possible different symmetries that can be spanned by such sets of lattice points, and, of course, the translation vectors. This will lead us to the seven crystal systems. The argument that follows is directed at, and limited to, finding the symmetries of all those lattices conventionally called 'primitive', for there is one for each crystal system. We shall have more to say about non-primitive (centred) lattices later.

Problem 13.1 The subject matter of the following paragraphs is unlikely to be new to the reader, although it is equally unlikely that it will have been well understood. Write down, briefly, what you know about the seven crystal systems. This should help focus attention on any problem areas that exist.

It is natural to start with a set of translation vectors in which the vectors are all of the same length and all mutually perpendicular; this is the most symmetrical arrangement possible. A cubic lattice results (Figure 13.2). [1] The fact that we have chosen to work with an infinite set of vectors means that it is sensible to pause before passing from the statement that 'it is a cubic lattice' to the statement that 'the lattice has O_h symmetry'. It is true that in this lattice each lattice point has O_h symmetry. That is, the operations of this O_h point group turn a given vector either into itself or into an equivalent vector. However, if one were to choose another point in space other than a lattice point,[2] the symmetry would usually be different, although the lattice would still be turned into itself by the relevant symmetry operations. For example, points 'along' the vectors of Figure 13.2 have C_{4v} symmetry, except the mid-points of such vectors, which have D_{4h} symmetry. The lattice is turned into itself by the operations of C_{4v} and D_{4h} applied at the relevant points. This is not new. In an O_h molecule, for instance, there will be points within it where the local symmetry is lower than O_h. The molecular symmetry is that of the highest symmetry. Recognizing this, one can say that 'the lattice of Figure 13.2 has O_h symmetry'. At the end of the present

[1] The reader can be forgiven for relating this arrangement to the properties of the primitive cubic unit cell; note, however, one danger in this association – it leaves unanswered the question of just how the body-centred and face-centred cubic unit cells arise.

[2] Here we *really* ignore the fact that empty space is a continuum. We are saying that the points that we have selected are somehow different; they are different in that we have selected them! Points external to the chosen network have their symmetry defined relative to that network.

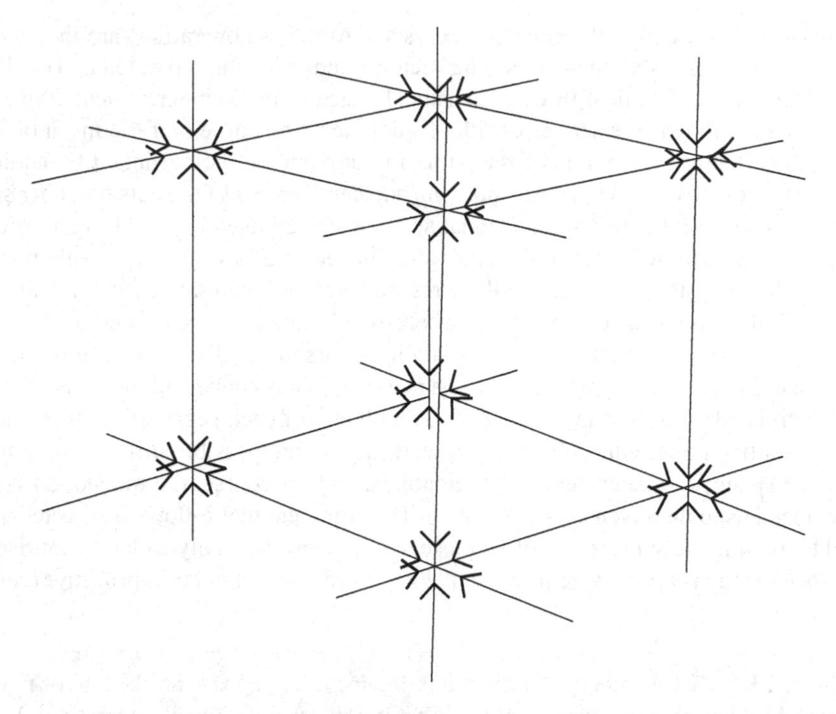

Figure 13.2 A segment of a cubic lattice drawn in the manner of Figure 13.1b. This segment has O_h symmetry; this is the symmetry of the point at the centre of the cube shown (although this point is not itself indicated) and also the symmetry of each point at which the vectors meet

chapter a particular choice of unit cell will be described which makes the lattice symmetry unambiguously clear.

Having established the symmetry of the most symmetrical of lattices, it seems reasonable to hope to obtain all the other possible lattices by reducing the symmetry. Unlike their string counterparts, vectors can be continuously deformed: stretched, contracted and re-orientated. The most natural way of proceeding is to focus on the O_h symmetry of the cubic lattice and to lower that symmetry by considering all the subgroups of O_h. This is the path that will be followed (even though it does not give all possible lattice symmetries) but first a basic and important fact. In Figure 13.1b each vector was drawn as two-headed because both a vector and its negative interrelate equivalent lattice points. Such a statement holds for each and every vector and so also for each and every sum of vectors, no matter how twisted and convoluted the path traced in three-dimensional space by such a sum – the resultant vector and its negative interrelate equivalent lattice points. There is a much more succinct way of stating this. It is that the lattice is centrosymmetric: contains centres of symmetry (all related to each other by members of the translational vector set). This statement is true of all the translation vector sets, no matter how low their symmetry otherwise. *All lattices are centrosymmetric*. It is certainly not true to say that all unit cells are centrosymmetric and so a good reason for excluding the phrase 'unit cell' from the present discussion is evident. The packing of atoms within a lattice can destroy the inherent centrosymmetry of that lattice. This is the reason that consideration of atoms is temporarily excluded from the discussion. The

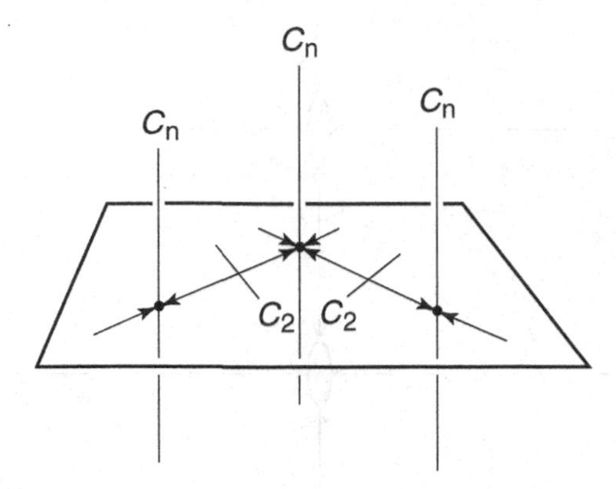

Figure 13.3 Perpendicular to a set of C_n axes there will be sets of translation vectors, as shown in this figure. In a plane perpendicular to the C_n axes and bisecting the translation vectors shown, will be sets of C_2 axes. For simplicity, and to emphasize the fact that translation vectors do not *have* to coincide with rotation axes, arrowheads are only shown on the translation vectors, not on the rotation axes

fact that all lattices are centrosymmetric simplifies the problem of finding the lattices which can be obtained from the cubic by a reduction in symmetry. Each acceptable symmetry will be that of a point group, a subgroup of O_h. The inherent lattice centrosymmetry means that all subgroups of O_h which lack a centre of symmetry can immediately be excluded. Further, of the groups which remain, only those which are consistent with a lattice are possibilities. What does this mean?

Suppose the lattice has a C_n rotation axis, with $n > 2$ (C_2 are special and we have to look at them separately). There cannot be just a single C_n axis; there must be an infinite set of them, regularly arranged in space – and all parallel. The set of translation operations ensures this. But this is not all. The double-headed nature of these arrows along the C_n directions means that perpendicular to the C_n axes there will be C_2 axes and these C_2 axes are interrelated by the C_n. This is best shown with a diagram, and one is given in Figure 13.3. The angle between the three C_n axes shown depends, of course, on the value of n. Looking at the double-arrowed translation vectors interrelating the C_n: the C_2 axes shown interchange them and also the C_n axes. The conclusion is that only centrosymmetric subgroups of O_h with C_2 axes perpendicular to the C_n can give rise to lattices. For O_h, C_n with $n > 2$ means C_4 and C_3. Which centrosymmetric subgroups of O_h are left when we apply the 'C_2 axes perpendicular' requirement to these cases? Relatively few. D_{4h} is the only one containing a fourfold rotation and D_{3d} is the only one with a threefold rotation. What of the C_n, $n = 2$ case; why is it different? Here, we will be dealing with at least two (sets of) C_2 axes, the 'original' (that corresponding to a C_n set with $n = 2$, in Figure 13.3) and any sets which are perpendicular to it (such as the C_2 shown in Figure 13.3). The key difference is that a perpendicular C_2 axis is now rotated onto *itself* (or, rather, one equivalent to itself, a member of the same set) by the 'original' C_2, whereas for C_n with $n > 2$ it is rotated into a different C_2 axis. In the latter cases, therefore, no problem could arise because a C_2 axis failed to rotate onto itself. However, strange as may seem, this possibility exists for the case where $n = 2$. An example of a C_2 group with no such problems is D_{2h}, which

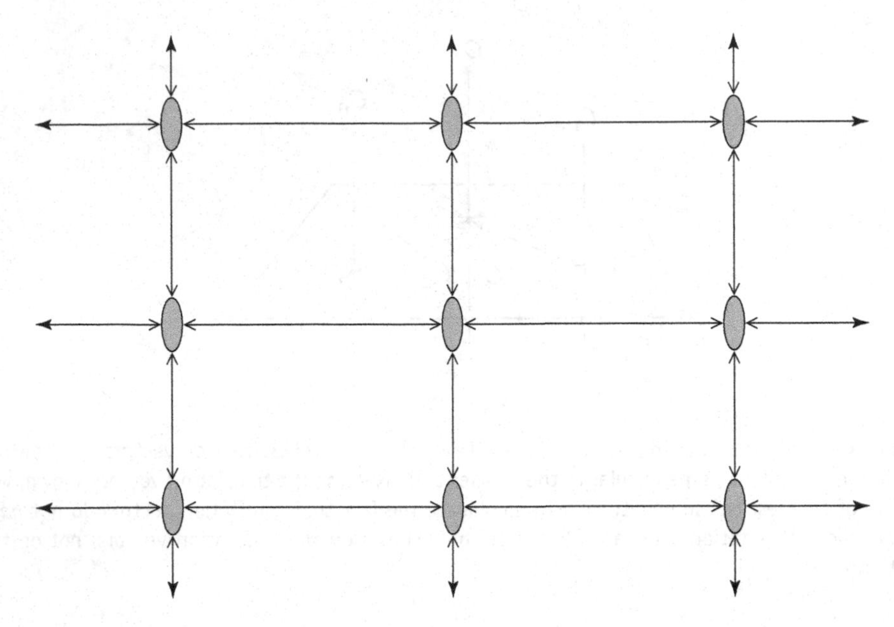

Figure 13.4 A layer of mutually perpendicular – and mutually compatible – twofold rotation axes. To extend the structure into three dimensions, a third set of C_2 axes is added, perpendicular to the two 'original' C_2 axis and through their points of intersection. These 'new' C_2 axes are shown as shaded ellipses. These 'new' axes must be perpendicular to both 'original' sets. If they are twisted at an angle (and so do not come straight out of the paper in this figure) then rotation of the 'old' axis set about the (oblique) 'new' C_2 leads to a non-existent axis set. The lattice cannot exist. All three C_2 sets must be mutually perpendicular.

contains three mutually perpendicular twofold rotation axes. Rotate about any one C_2 axis and the others interchange (Figure 13.4). But if the C_2 axes are not mutually perpendicular, problems arise. Consider Figure 13.5, which shows two sets of mutually perpendicular C_2 axes together with another set, coming out of the paper and slightly displaced relative to their position in Figure 13.4. The displaced set is incompatible with the in-plane sets. Matters are not improved if members of the out-of-paper set are inclined at an angle other than 90° to the others. In all cases an incompatible pattern is obtained. The only group that is compatible in this situation is C_{2h}, which is centrosymmetric and contains a single twofold rotation operation.

Finally, destruction of the twofold axis of C_{2h} by relaxing the requirement that additional axes be perpendicular to the first (an inescapable requirement for C_n axes with $n > 2$) leads to the last subgroup of O_h that has to be considered, C_i, which contains no operation except inversion in a centre of symmetry, the lowest symmetry that a lattice can have. Provided that all lattices can be obtained by reduction in symmetry from O_h, all have been generated. Alas, not all lattices can be obtained by such a reduction in symmetry; since O_h contains no sixfold rotation operation, a point group containing one can never be obtained by starting with it. Yet there is a perfectly good lattice that is based on a point group that contains such operations, D_{6h}. If this is so, it at once prompts the question 'how many others have been missed?'. Really, this is a question about whether there are any other point groups

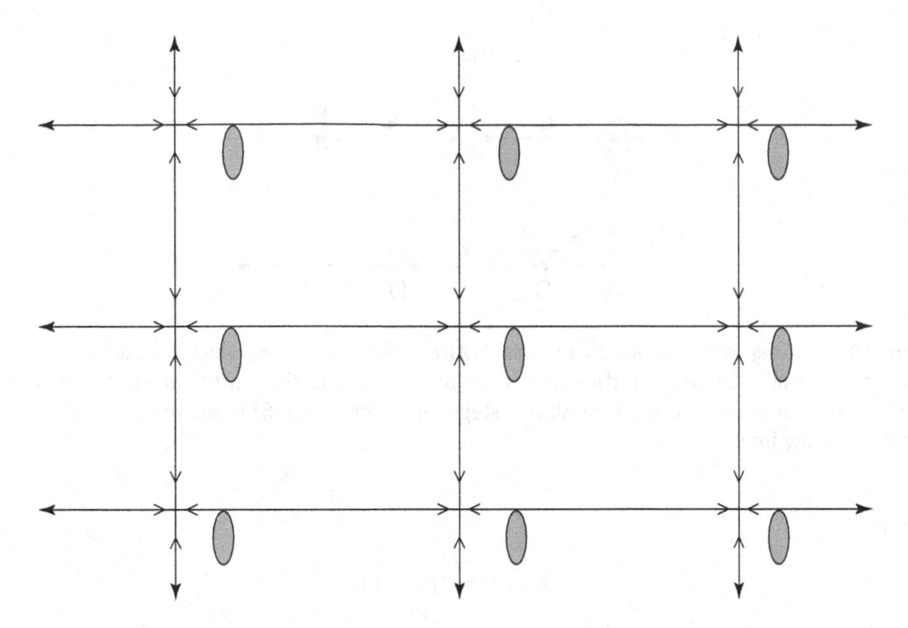

Figure 13.5 Being perpendicular (Figure 13.4) is a necessary but not sufficient condition for a lattice. In this figure, the C_2 axes perpendicular to the plane of the paper are not compatible with the in-plane sets

with symmetry operations not present in O_h which can give rise to acceptable lattices. Clear candidates are groups with fivefold and sevenfold rotation operations. Can they be excluded (and any others that may well occur to the reader)? The answer is 'yes', as is now demonstrated.

Consider Figure 13.6, which shows a two-dimensional lattice. Clearly, any constraints that apply to this lattice must apply to a three-dimensional lattice also, because any three-dimensional lattice may be regarded as linked networks of two-dimensional lattices, just as the fishing nets were linked at the beginning of this section (remember: the head-to-head linking of vectors in the manner of Figure 13.1 leads to straight-line arrays, never to kinks or bends). As we are looking for all rotation axes that are compatible with translation symmetry, we simply require the lattice points of Figure 13.6 to be interrelated both by pure translations and pure rotations and enquire into the compatibility of these two requirements. With no loss of generality for $n > 2$, assume that the C_n axes are perpendicular to the page and that one passes through each of the points A, B, C and D. The points A and B in the top row are separated, in general, by an integral number, m, of translation steps, a, and so by a distance of ma (the points C and D are separated by a single translation step, a, and so m will be a small number; for simplicity, in Figure 13.6 the case with $m = 2$ is shown). For the purposes of the argument, the lower row of points (that containing C and D) may be regarded as free to slide around, subject only to the requirement that the points are interrelated by the C_n axes. The angle of rotation associated with the C_n axis ($= 360/n$) is labelled θ so that, from Figure 13.6:

$$\mathrm{AB} = \mathrm{A'B'} = a + 2a\cos(\theta) = ma$$

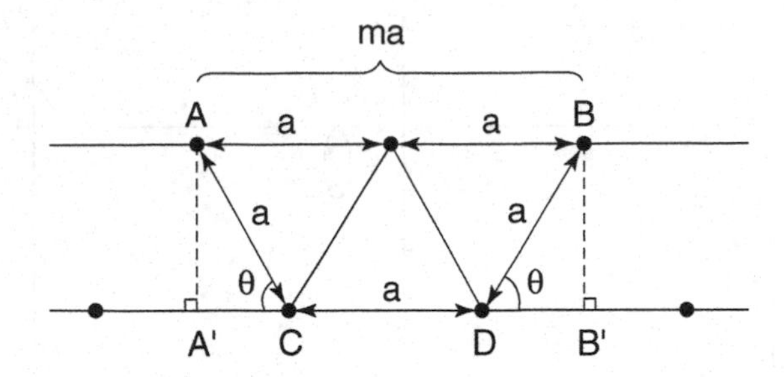

Figure 13.6 A fragment of a two-dimensional lattice, used in deducing which C_n axes through the dots and perpendicular to the lattice are compatible with the lattice. In this diagram the case of $m = 2$ (where m is the number of translation steps separating A and B) is shown but in principle m can assume any integer value

that is,

$$2a\cos(\theta) = (m - 1)a$$

and

$$2\cos(\theta) = (m - 1)$$

or

$$\cos(\theta) = \frac{(m - 1)}{2}$$

Now, m is an integer and $\cos(\theta)$ can only have values from 1 to -1. The possible solutions are therefore rather limited and are detailed in Table 13.1

Table 13.1

Value of m	Value of $\cos(\theta)$ (this equals $\frac{(m-1)}{2}$)	Comments
4	1.5	No solution possible
3	1	$\theta = 0$ or 360° (corresponding to a C_1 axis)
2	0.5	$\theta = 60^\circ$ (corresponding to a C_6 axis)
1	0	$\theta = 90^\circ$ (corresponding to a C_4 axis)
0	−0.5	$\theta = 120^\circ$ (corresponding to a C_3 axis)
−1	−1	$\theta = 180^\circ$ (corresponding to a C_2 axis)
−2	−1.5	No solution possible

Problem 13.2 Work through the above argument but instead of choosing $m = 2$, as in Figure 13.5, take $m = 3$.

Clearly, the only non-trivial rotation operations that are consistent with a lattice are C_2, C_3, C_4 and C_6. This means that the only case not covered by a reduction in symmetry from O_h is that of the C_6 rotations; C_5 and C_7, for instance, are not acceptable. The highest centrosymmetric point group containing a C_6 axis which also has the required perpendicular C_2 axes (remember Figure 13.3) is D_{6h}. It completes the list of acceptable lattices because there is no lattice that can be obtained from it which satisfies all the requirements and which has not already been generated.[3]

The seven point groups that have been obtained, and the seven lattices to which they give rise, characterize the seven *crystal systems*. For convenience, these are listed in Table 13.2. One point about this table is to be noted: that the characteristics of each *lattice* have been detailed. Compilations like Table 13.2 are commonly found in texts under a different heading, that of 'unit cells'. In such texts the unit cell is chosen as the smallest parallelepiped (box with three pairs of parallel edges) defined by the translation vectors. There are two reasons that this precedent is not followed here, quite apart from the fact that the name 'unit cell' has been avoided. First, the translation vectors are the fundamental quantities and this is better recognized by the name 'lattice'. Second, as will be seen, there is no unique choice of unit cell for any crystal structure – just some choices that are more convenient for some purposes than others (in particular, the choice that is convenient for x-ray crystallography may be inconvenient for spectroscopy, as will become apparent in the next chapter).

All of the seven three-dimensional lattices (or, rather, remembering the context in which they were sought, the seven crystal systems) which arise from translationally related repeat units have now been obtained. The relevance to crystal structures is clear. However, we should not close our eyes to other possibilities. For instance, suppose the repeat layers were not related one to another by simple translation operations. What if they were arranged in a spherical layer fashion, much as the layers of an onion? Perhaps such crystals could be formed if a nucleating unit led to crystal growth by successive spherical shells being added. As the shape of the popular form of a soccer ball, or C_{60}, shows, such a crystal could have fivefold rotation axes (six of them, all passing through the point which was the original crystal nucleus, giving a crystal with icosahedral symmetry), in complete contrast to the discussion above, where it was found that C_5 axes are not possible for regular arrays of translationally related units. In fact, crystals with C_5 axes are known (although it took time to convince the scientific community that they are real) and have been the subject of much debate. They are known as 'quasicrystals'. Some even seem to show the characteristics expected of C_5 axis perpendicular to a plane; however, whilst there is no problem about local C_5 axes (a molecule such as $Fe(C_5H_5)_2$ has one), these local C_5 axes cannot be inter related by simple translations. And C_5 axes are not the end; examples with C_8, C_{10} or C_{12} have also been described.

[3] Although it comes equally well from O_h, it is often convenient to think of D_{3d} as a subgroup of D_{6h}.

Table 13.2

Crystal (lattice) System	Characteristic point group	Lattice vector characteristics
Cubic	O_h	$a = b = c; \alpha = \beta = \gamma = 90°$
Tetragonal	D_{4h}	$a = b \neq c; \alpha = \beta = \gamma = 90°$
Orthorhombic	D_{2h}	$a \neq b \neq c; \alpha = \beta = \gamma = 90°$
Monoclinic	C_{2h}	$a \neq b \neq c; \alpha = \gamma = 90°, \beta > 90°$
Triclinic	C_i	$a \neq b \neq c; \alpha \neq \beta \neq \gamma$
Hexagonal	D_{6h}	$a = b \neq c; \alpha = \beta = 90°, \gamma = 120°$
Trigonal	D_{3d}	$a = b = c; \alpha = \beta = \gamma \neq 90°$

The quantities a, b and c are the absolute magnitudes of the three primitive translation vectors that define the lattice. The angles complement the axes; thus α is the angle between the vectors associated with b and c, β that between the vectors associated with c and a (it helps avoid hidden problems if one is consistent in the order in which axes are listed by being cyclic: $a \rightarrow b \rightarrow c \rightarrow a$). Note that in the monoclinic system there is conventionally a departure from the system followed for point groups where the axis of highest symmetry is chosen as z (the choice of β as the unique angle means that the y axis is chosen as the C_2 in monoclinic systems). The sixfold axis is not immediately evident in the vectors defining the hexagonal system but when they are used to define a lattice – and, so, an extended pattern generated – the sixfold axis becomes evident (it is along the vector associated with c). However, the threefold axis is not along any of the vectors used to define the trigonal lattice – the vectors associated with a, b and c are interrelated by the threefold axis. As will be discussed in the text, a better name for the trigonal case above is 'rhombohedral'. Trigonal is used here because it is the name normally encountered in a table such as this.

13.2 The Bravais lattices

In the preceding section the concept of a lattice was explored. It was found that all three-dimensional lattices have to conform to one of seven symmetry types, each characterized by a unique centrosymmetric point group, and that these are normally spoken of as 'the seven crystal systems'. However, there is more to say on the topic of lattices, even the topic of empty lattices. This can be seen by asking what, strictly, is an inadmissible question. The question is this: 'is it possible, for any of the seven lattices, that a second, identical, lattice be taken and interpenetrated into the first to give an arrangement which retains the symmetry of the first lattice?'. The question may be phrased in a rather less accurate but more colourful and understandable way. Suppose the first lattice were made of red string. Is it possible to construct within it a displaced but otherwise identical lattice made of blue string which does not destroy the symmetry pattern of the red? Such questions are inadmissible because any given crystal structure can only have a single lattice[4] (and so talk about 'interpenetrating lattices', whilst sometimes useful for teaching purposes, cannot accurately describe reality). However, having recognized the error, let us continue to sin – we will find that such sinning is a popular pastime! The question is illustrated in Figure 13.7 for the cubic lattice. The lattice generated in Section 13.1 (essentially, Figure 13.2) is shown in Figure 13.7a; it is shown

[4] This statement originates in the fact that there is a single translational subgroup of the space group. In discussions it is often very convenient to talk of the lattice composed of one set of atoms and to relate it to the lattice composed of another set. Such language is convenient rather than strictly accurate.

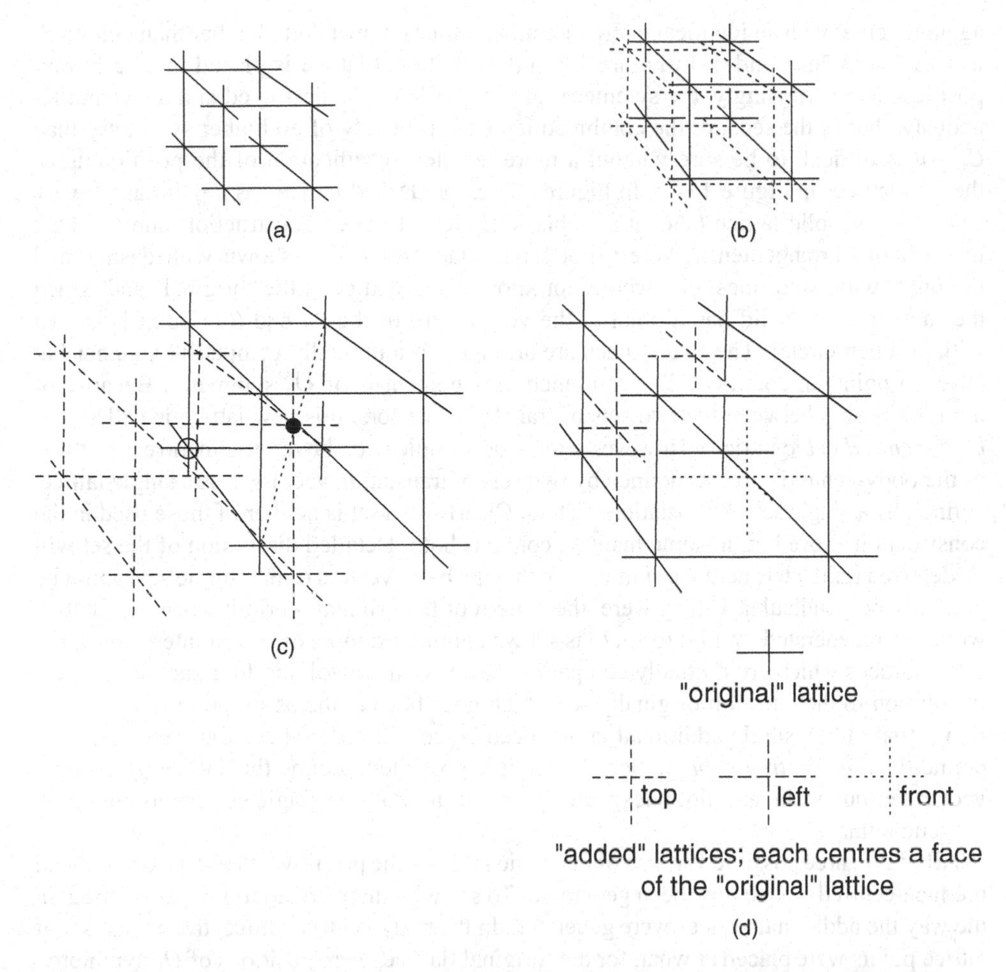

Figure 13.7 (a) The cubic lattice of Figure 13.2, shown without the arrowheads of that figure. (b) The lattice of (a) with an identical lattice (shown dashed) displaced from the 'first' by arbitrary translations (which means that the vectors associated with the two vector sets remain parallel). Except in very special cases such a 'second' lattice will destroy much of the symmetry of the 'first'. So, in this figure, the fourfold rotation axes of the 'first' lattice are destroyed by the presence of the 'second' and vice versa. (c) One of the special cases of (b) occurs when the 'corners' of the 'second' (dashed) lattice coincide with the centre of the cube defined by the first, because both have O_h symmetry. A diagonal of the 'first' lattice is shown dotted; the second lattice intersects the mid-point of this diagonal. The open circle shows that the converse pattern is also true; the 'corners' of the solid-line lattice fall at the centre of the cube defined by the dotted line lattice. (d) A second special case occurs when *three* 'additional' equivalent lattices are added to the 'first'. In this figure the 'additional' lattices are shown small-dashed, large-dashed and dotted

again together with an identical interpenetrating lattice (shown dotted rather than coloured) in Figures 13.7b, c and d. In Figure 13.7b the additional lattice is placed in an arbitrary position. Not surprisingly, the symmetry of the first lattice is destroyed and also simultaneously, that of the second; the combined lattice is probably of no higher symmetry than C_i – it is difficult to be sure without a more detailed specification of the positioning of the two lattices in Figure 13.7b. In Figure 13.7c, the dashed lattice has a point at the very centre of the solid lattice (shown as a black circle – a dotted construction line is added to show this arrangement). Were it not for the fact that one is shown with dashed and the other with solid lines, one would not know which lattice is the 'original' and which the 'added' – the solid has a point at the very centre of the dashed (the point is shown with an open circle). The two lattices are arranged in a mutually compatible manner but have no points in common. The combined arrangement is of O_h symmetry. Because of the relationship between the two sets of translation vectors, this new lattice is called 'the *body-centred cubic* lattice'. However, as has been indicated above, it is incorrect to think of the body-centred cubic as defined by two sets of translation vectors. It is a single lattice, defined by a single set of translation vectors. Clearly, this set is neither of those used in the construction above but, in some manner, contains both. Detailed discussion of the set will be deferred until later; here we simply note that the basic vectors of this single set cannot be mutually perpendicular. If they were, the pattern of the original – primitive cubic – lattice would be regenerated. In Figure 13.7d is shown another example of sets of interpenetrating cubic lattices which are mutually compatible but this time involving four such lattice sets (a solution to the question originally set which goes beyond the assumption in that question – that only a single additional lattice need be considered). These four sets, together, define the *face-centred cubic* lattice. Again, it is possible to define this lattice by a single vector set, but again, and not unexpectedly, the vectors of this single set are not mutually perpendicular.

All of the three possible variants of the cubic lattice – the primitive, the body-centred and the face-centred – have now been generated. To see why there are no more, consider again the way the additional lattices were generated. In the body-centred lattice, the second set of lattice points were placed at what, for the 'original' lattice, were positions of O_h symmetry. Clearly, the 'second' lattice was compatible with the 'first'. The case of the face-centred cubic was different. The lattice points of the 'first' lattice that were occupied by the 'second' were only of D_{4h} symmetry. Had just a single 'second' lattice been interpenetrated with the 'first' then the symmetry of the 'first' would have been destroyed and lattice points that were originally of O_h symmetry would have been reduced to D_{4h}. This problem was overcome by, effectively, adding a C_3 axis to this reduced D_{4h} symmetry to bring it back to O_h. This was achieved by adding three points around each 'original' lattice point, a step which required that three additional lattices, not just one, be added. It is less immediately obvious that this same step serves to turn each of the 'second' lattice points, originally of D_{4h} symmetry, into points of O_h symmetry. However, the fact that the combined lattice 'looks the same' whichever lattice is called 'first' establishes the point. It is not too difficult to see that the 'original' – solid line – lattice face centres each of the added lattices. The question then arises as to whether the same sort of trick can be played a different way. Is it possible, for example, to add several lattices at points of D_{3d} symmetry of the 'first' and, by adding four of them in total, thus regenerate an O_h lattice? The answer is 'no'; the reason is that although there are unique points of D_{4h} symmetry in the 'original' lattice there are none of D_{3d}. The same is true of all other subgroups of O_h. All of the possible cubic lattice patterns have been generated.[5]

The type of argument developed in the preceding paragraphs can also be applied to the six other different possible lattice symmetries. When all of the lattice patterns consistent with each various lattice symmetry type have been obtained, there are fourteen in all. They are known as the fourteen *Bravais lattices*, named after the Frenchman who first recognized their existence. They are shown in Figure 13.8; Figure 13.8a shows the 'primitive' lattices and Figure 13.8b the 'centred'. But beware, as has already been noted, it is misleading to call one set 'centred'; they too are actually primitive. This primitive lattice is shown for each in Figure 13.8b, where the primitive translation vectors are shown and labelled 'P'. In contrast, the non-primitive translation vectors correspond to the edges of the conventional centred lattices shown. Note three things. First that our lattices are still empty. Second, that each of the fourteen Bravais lattices is associated with a unique pattern of vectors – it is this vector pattern that distinguishes the various lattices. So, there are three qualitatively different vector sets that define cubic lattices. As has been mentioned above and can be seen from the P sets in Figure 13.8b, the vectors defining a cubic lattice do not *have* to be mutually perpendicular. This point is an important one; despite it, all the Bravais lattices have the symmetry of the crystal system and so all three cubic Bravais lattices *are* cubic, and of O_h symmetry. However, if we work with primitive translation vectors for the centred lattices, the vectors define the edges of a cell which does not have O_h symmetry! For completeness it is convenient at this point to list all fourteen Bravais lattices and this is done in Table 13.3.

Problem 13.3 Unconventionally, the name 'unit cell' has been (largely) avoided so far in this text. Make a short(!) list of the arguments given so far in support of this avoidance (weightier arguments will be given later).

One of the most evident things about Table 13.3 is its rather patchwork character – some crystal systems have body-centred lattices, others do not; some have one-face-centred lattices, others do not. It has already been seen why there cannot be a one-face-centred cubic – it would not be cubic (actually, it would be a primitive tetragonal lattice). The reason for the non-listing of the other apparently 'missing' lattices is similar – they are each equivalent to one already listed in Table 13.3. In Table 13.4 are given a list of the 'missing' lattices together with their equivalents – which are present.

13.3 The crystallographic point groups

The point has now been reached at which atoms must be introduced into the lattices although it will continue to be convenient to adopt the fiction that a lattice exists before atoms are introduced into it. Some consequences of the introduction of atoms are quite evident. Thus, although up to this point all lattices have been centrosymmetric, the introduction of atoms

[5] The way that these have been generated demonstrates a limitation in the way that the seven crystal systems were derived earlier in the text. *All* of the cubic lattices have O_h symmetry – this is why the derivation of the seven crystal systems had to be confined to 'those lattices conventionally called primitive'. Strictly, the derivation should have been concerned only with groups and subgroups but it was felt that a less abstract discussion would be more easily followed.

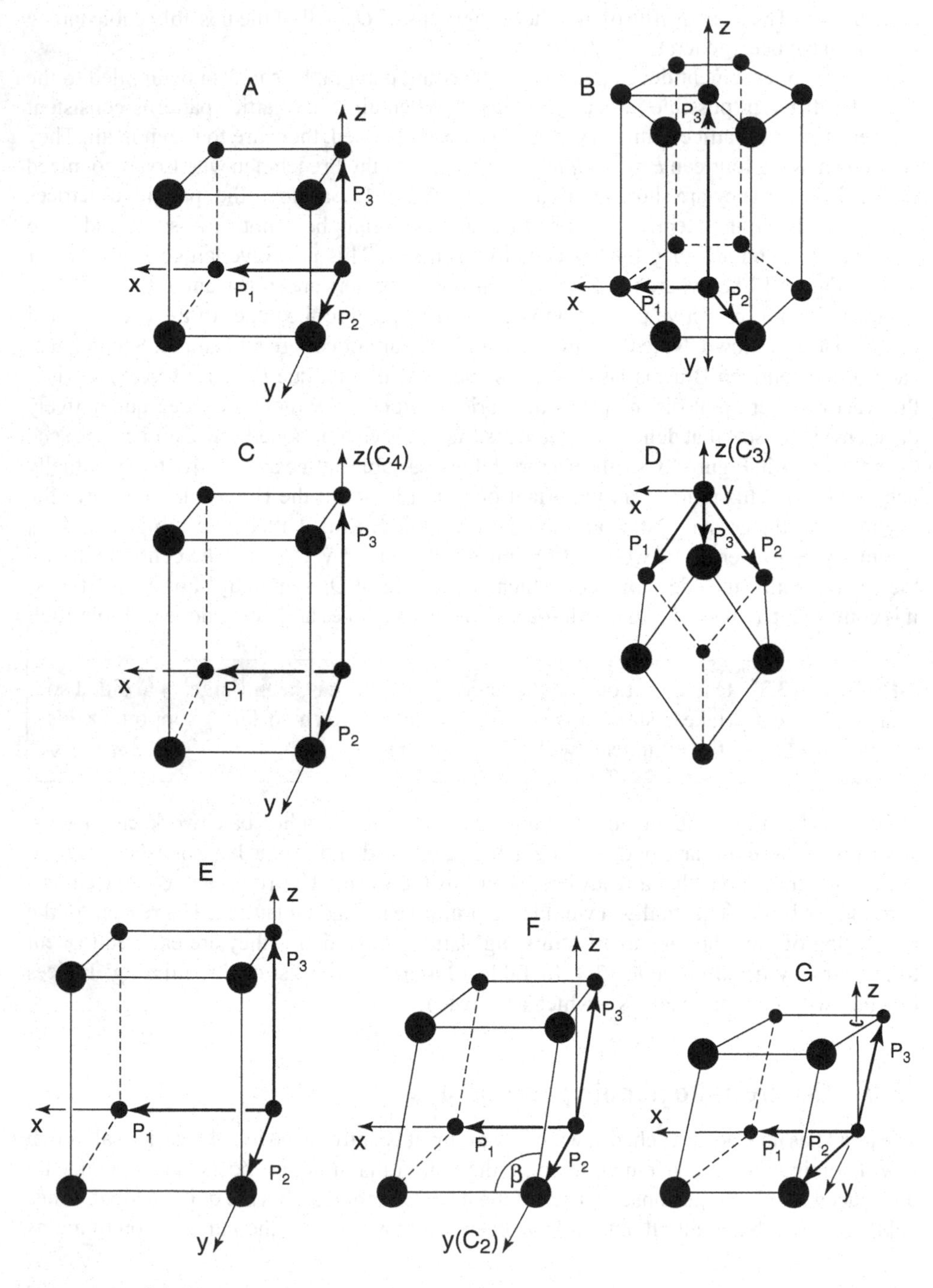

Figure 13.8 (a) The seven Bravais lattices conventionally called 'primitive', together with the associated primitive translation vectors (shown bold). Note that only in three cases are all the translation vectors in directions parallel to the Cartesian axes of the crystal. A, cubic; B, hexagonal; C, tetragonal; D, trigonal(rhombohedral); E, orthorhombic; F, monoclinic; G, trigonal. (*Continued*)

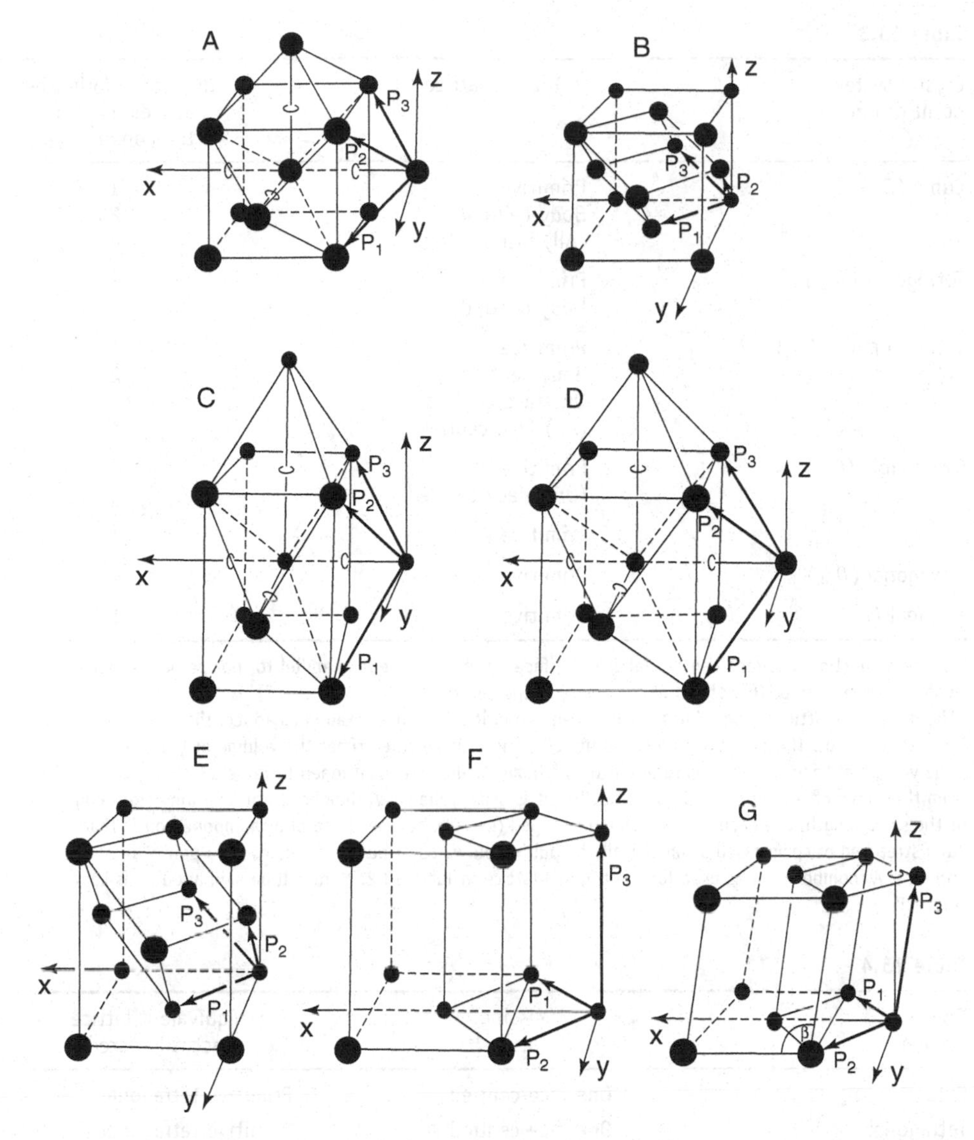

Figure 13.8 (*Continued*) (b) The seven Bravais lattices conventionally called 'centred', together with the associated primitive translation vectors (shown bold). Note that the translation vectors are generally in directions rather different from those of the Cartesian axes of the crystal. A, body-centred cubic; B, face-centred cubic; C body-centred tetragonal; D, body-centred orthorhombic; E, all-face-centred orthorhombic; F, one-face-centred orthorhombic; G face-centred monoclinic. Reproduced with permission from S.F.A. Kettle and L.J. Norrby, *J. Chem. Educ.* **70** (1993) 959, 963. ©1993, Division of Chemical Education, Inc.

Table 13.3

Crystal system point group	Bravais lattice	Number of 'primitive' lattices needed in the construction
Cubic (O_h)	Primitive	1
	Body-centred	2
	(All)-face-centred	4
Tetragonal (D_{4h})	Primitive	1
	Body-centred	2
Orthorhombic (D_{2h})	Primitive	1
	Body-centred	2
	One-face-centred	2
	(All)-face-centred	4
Monoclinic (C_{2h})[a]	Primitive	1
	(One)-face-centred	2
Triclinic (C_i)	Primitive	1
Hexagonal (D_{6h})[b]	Primitive	1
Trigonal (D_{3d}) [c]	Primitive	1

[a] In the monoclinic one-face-centred lattice, the face which is centred is parallel to, not perpendicular to, the twofold axis; it may be thought of as a σ_v plane being centred.
[b] The hexagonal lattice is sometimes drawn showing a unit cell with a hexagon as a face, this face being centred by a lattice point. This is not a primitive unit cell (it is actually three times the volume of the primitive).
[c] One variety of trigonal lattice is referred to as "rhombohedral", as mentioned in Table 13.2. This name arises from the shape of the corresponding unit cell as it is usually drawn. A rhombohedron contains eight edges, all of the same length, and each face is diamond-shaped (essentially, the shape of a rhombohedron is that of a cube stretched or compressed symmetrically by pulling outwards or pushing inwards on a pair of opposite corners). A rhombohedron is used for the trigonal lattice in Table 13.2. More will be said about this topic in the text.

Table 13.4

Crystal system	'Missing' lattice	Equivalent lattice actually listed
Cubic	One-face-centred	Primitive tetragonal
Tetragonal	One-face-centred	Primitive tetragonal
	All-face-centred	Body-centred tetragonal
Monoclinic	Body-centred	One-face-centred monoclinic
	All-face-centred	One-face-centred monoclinic
Triclinic	Any centring	Primitive triclinic
Hexagonal	Body-centred	One-face-centred monoclinic
	Unique-face-centred	Primitive orthorhombic
	All-face-centred	one-face-centred monoclinic
Trigonal (rhombohedral)	Body-centred	Primitive trigonal
	Unique-face-centred	Triclinic
	All-face-centred	Primitive trigonal

means crystal structures, which may or may not retain this centrosymmetry. Clearly, the actual arrangement of atoms, molecules or ions in the lattice is of key importance. Equally clearly there are considerable limitations on the admissible arrangements. For instance, to place atoms randomly in a cubic lattice would immediately destroy the multitude of rotation axes and other symmetry elements essential to a cubic lattice. The arrangement of atoms (the word 'atoms' will be used for simplicity; ions and molecules are not excluded) in a lattice must be consistent with the symmetry of that lattice for the lattice symmetry to be evident in the space group. This prompts the question 'what are the acceptable symmetries for a given crystal system?'. As the argument develops it will be seen that this question is correctly put – the different Bravais lattices falling into one crystal system do not have to be distinguished.

Because the high symmetry makes it particularly easy to visualize, the cubic case will be detailed. Rather than deal with the full O_h symmetry of the lattice it is easiest to focus on one aspect of this symmetry. This is that in cubic symmetry the x, y and z axes are all equivalent. All acceptable ways of introducing atoms into a cubic lattice must respect this equivalence. Whilst, no doubt, the number of acceptable ways of introducing atoms into a cubic lattice is infinite – it is this fact that serves to distinguish one cubic crystal structure from another – the number of different *symmetries* that these arrangements can have is rather limited. The symmetry of any acceptable arrangement has to be one in which x, y and z equivalence is retained. Relatively few point groups satisfy this condition, they are:

$$I_h,\ I,\ O_h,\ O,\ T_h,\ T_d \text{ and } T$$

Of these, the first two, of icosahedral symmetry, can be excluded because they require the existence of fivefold axes, and these are inconsistent with a cubic lattice, something that has already been mentioned (Figure 13.6 and the associated discussion). The other four are different; they are either O_h or subgroups of O_h. Clearly, all are consistent with a lattice of O_h symmetry. What would happen if an atomic arrangement with I_h symmetry were to be put into a cubic lattice? The lattice (that is, the arrangement of other I_h groupings in space) would destroy all the fivefold axes of I_h; the highest possible effective symmetry of the atomic arrangement would be T, this being the only subgroup of I_h (and I) which is contained in the above list. It is concluded, then, that the only symmetries of atomic arrangements consistent with a cubic lattice are:

$$O_h,\ O,\ T_h,\ T_d \text{ and } T$$

Three things are to be noted. First, only two of these groups (O_h and T_h) contain inversion in a centre of symmetry as an operation (this can be checked by reference to Appendix 3). The absence of this operation in the other groups means that the corresponding atomic arrangements will be such as to destroy the centrosymmetry originating in the lattice (remember, the translation operations of the lattice will only move the atomic arrangements, not turn them round in the way needed if the lattice centrosymmetry were to be preserved). Second, groupings of atoms with these symmetries can only form cubic lattices if the symmetry axes of the atomic grouping coincide with the corresponding axes of the lattice. So, an alkali metal salt, $K[MX_6]$, say, has a spherical cation and, probably, an octahedral anion. Both cation and anion, separately, are consistent with a cubic lattice. However, it does not automatically follow that the salt will crystallize in a cubic space group. Third, there has been nothing in the above discussion which confines it to the case of the primitive cubic lattice. It must therefore be concluded that it applies equally to all the cubic Bravais lattices. This is an important point, the relevance of which will become apparent when all possible space groups are counted.

Problem 13.4 Sketch two possible crystal structures for the salt $K[MX_6]$, one of which is cubic and the other which is not.

The pattern for the other crystal symmetries will follow that set by the discussion of the cubic case. Consider the tetragonal lattices, for instance. Suppose we have a molecule which, itself, is of a cubic symmetry – an octahedral molecule, ML_6, for instance – but which crystallizes with the molecules so arranged relative to one another that they form a tetragonal lattice. The crystal can be of no higher symmetry than tetragonal. Conversely, if a molecule is of low symmetry, for it to crystallize in a tetragonal lattice then the molecules have to be arranged in groups of four, symmetry related. This is because a symmetry consistent with a tetragonal lattice must have an atomic arrangement satisfying the basic requirement of a fourfold axis of some sort or other (the reason for the strange end of this sentence will become evident immediately). Point groups satisfying this condition (in addition to the cubic, which have been covered) are:

$$D_{4h},\ D_4,\ C_{4h},\ C_{4v},\ C_4,\ S_4 \text{ and } D_{2d}$$

The last two of these may cause some surprise because they do not contain C_4 rotation operations – but they do have S_4 rotation–reflection operations and so an essential tetragonality – and this is what is needed for compatibility with a tetragonal lattice. Having made this point, what *is* surprising is that the group D_{4d} is missing – this, at first sight, seems to have the essential requirement of C_4 rotation operations. Indeed it has, but it also has S_8 rotation–reflection operations and this is something that no lattice, cubic included, possesses. It follows that if a group of atoms of D_{4d} symmetry were to be placed in a tetragonal lattice its symmetry would be reduced (D_4 or C_{4v} are the highest symmetries that could result). Finally, note that of the seven point groups just listed, only two, D_{4h} and C_{4h}, contain the operation of inversion in a centre of symmetry. It would not be difficult to apply similar arguments to all the other crystal systems, but no new principle would emerge. We therefore content ourselves with listing, in Table 13.5, the correspondences between the crystal systems and acceptable point groups spanned by the atomic arrangements that may be described by them. These point groups number thirty-two in total and are usually referred to as 'the thirty-two crystallographic point groups'. In the next two sections it will be seen that they play a key role in determining the number of distinct crystallographic space groups.

Problem 13.5 Explain in detail (use of Appendix 3 may be needed) why a molecular arrangement of D_{4d} symmetry could, at best, lead to an arrangement of either D_4 or C_{4v} symmetry in a crystal.

Table 13.5

Crystal system	Acceptable point groups					
Cubic	O_h	O	T_h	T_d	T		
Tetragonal	D_{4h}	D_4	C_{4h}	C_{4v}	C_4	D_{2d}	S_4
Orthorhombic	D_{2h}	D_2	C_{2v}				
Monoclinic	C_{2h}	C_2	C_s				
Triclinic	C_i	C_1					
Hexagonal	D_{6h}	D_6	D_{3h}	C_{6h}	C_{6v}	C_6	C_{3h}
Trigonal (rhombohedral)	D_{3d}	D_3	C_{3v}	C_3	S_6		

Note the absence of D_{6d} in the hexagonal listing – it has S_{12} rotation-reflection operations, operations not possessed by D_{6h}, the parent of the hexagonal system. In contrast, note the presence of D_{3h}; this seems 'wrong' but is readily explained. The important point is that D_{6h}, the parent group of the hexagonal system, has a σ_h mirror plane reflection operation whereas D_{3d}, the parent of the trigonal system, has no such mirror plane reflection. D_{3h} has this mirror plane and so cannot be associated with a trigonal lattice. The reason why S_6, which looks as if it should be in the hexagonal system, is in the trigonal is given in the text.

Problem 13.6 The following symmetries, although acceptable as those of atomic arrangements, cannot persist in a crystal. In each case give the highest symmetry crystallographic point group arrangement that could result.

$$D_{5h},\ D_{5d},\ C_{5v},\ D_{7h},\ D_{7d},\ C_{7v}$$

In retrospect, it is possible to see a very simple way of relating a crystal system with the acceptable crystallographic point groups associated with it, a method that has been hinted at more than once in the above section. This relationship depends on the simultaneous satisfaction of two conditions. The first is that acceptable point groups are always subgroups of the symmetry of the crystal system. This condition is very important. For instance, it immediately shows that D_{3h} is associated with the hexagonal system and not the trigonal because D_{3h} is a subgroup of D_{6h} but not of D_{3d}. The second condition is that the subgroup is not also the subgroup of the parent group of a lower (= fewer symmetry operations) crystal system. It is the crystal system of lowest symmetry that is relevant. So, D_2 is a subgroup of O_h, D_{4h}, D_{6h} and D_{2h}. Of these, D_{2h} has the smallest number of symmetry operations and so D_2 is associated with the orthorhombic, D_{2h}, crystal system. It is also for this reason that S_6 is a trigonal crystallographic point group and not a hexagonal.

13.4 The symmorphic space groups

The discussion so far is sufficient to enable the first set of space groups to be obtained (about one-third of the total) but before doing so it is convenient first to review the present position. The seven crystal systems were first obtained as the seven different symmetries of translation vectors that can exist in three-dimensional space. We then found that in several cases there exist more than one distinct set of such vectors, all of the same symmetry. These give rise to the fourteen Bravais lattices. It was at this point that atoms were introduced into the discussion. It was found that for each Bravais lattice there exist several

Table 13.6

Crystal system	Number of Bravais lattices (B)	Number of crystallographic point groups (N)	The product (BN)
Cubic	3	5	15 (15)
Tetragonal	2	7	14 (16)
Orthorhombic	4	3	12 (13)
Monoclinic	2	3	6 (6)
Triclinic	1	2	2 (2)
Hexagonal	1	7	7 (8)
Trigonal	1	5	5 (13)
Total	14	32	61 (73)

symmetry-distinct ways of introducing atoms which are compatible with the symmetry of the Bravais lattice. Distinct space groups will differ *either* in their lattices *or* in the symmetry of their atom arrangement in space – or both. Space groups can be generated by combining each Bravais lattice of Table 13.3 with each of the crystallographic point groups corresponding to it in Table 13.5. Each space group that results will be a unique combination of lattice and point group. Effectively, from this point on the following approximate equation will be used to obtain space groups (the question of the points at which this equation is not quite correct will be at the heart of the following discussion):

(Bravais lattice) + (Corresponding point group) = (Space group)

How many space groups can be obtained in this way? The answer to this question is detailed in Table 13.6. This table summarizes the data in Tables 13.3 and 13.5 in a numerical format and then combines them.

In the extreme right-hand column of Table 13.6 is given in parentheses the actual number of space groups that exist of the sort that have been under discussion. In some cases the correct number has been obtained, but not in all. So, although perhaps not much is missing, something has to be added to the approach. In particular, the answer to the trigonal case is seriously wrong – and this will necessitate a serious discussion! The other errors are readily dealt with. In Table 13.7a are detailed the cases for which they occur. In this table the individual crystallographic point groups and Bravais lattices are given. The table shows the number of space groups that arise from each combination. The argument developed above leads to the expectation that the answer will be '1' in each and every case. It is where the number '2' appears that there is a problem!

The obvious thing about Table 13.7a is that most numbers *are* 1. Those that are 2 do not occur for the highest symmetry crystallographic point groups of a crystal system. This is a relevant point, as study of the C_{2v}, one-face-centred orthorhombic, example shows. In this example, although the lattice is D_{2h}, the crystallographic point group (the filling of the lattice with atoms) destroys all but a single set of parallel twofold axes (but the primitive translation vectors remain mutually perpendicular, which is why C_{2v} is an orthorhombic

Table 13.7a

System	Bravais lattice	Crystallographic point group				
Tetragonal	p; b.c.	D_{4h}	1	1		
		D_4	1	1		
		C_{4h}	1	1		
		C_{4v}	1	1		
		C_4	1	1		
		D_{2d}	2	2		
		S_4	1	1		
Orthorhombic	p; b.c; o.f.c; a.f.c	D_{2h}	1	1	1	1
		D_2	1	1	1	1
		C_{2v}	1	1	2	1
Hexagonal	p.	D_{6h}	1			
		D_6	1			
		D_{3h}	1			
		C_{6h}	1			
		C_{6v}	1			
		C_6	1			
		C_{3h}	1			

p = primitive, b.c = body-centred, o.f.c = one-face-centred, a.f.c = all-face centred.

and not a monoclinic crystallographic point group). Is the lattice face that is centred (in the one-face-centred case) parallel to or perpendicular to the twofold axes? The answer is that *both* are possible and so two space groups are obtained but not the expected one. The duality arises from a degree of freedom between the lattice and its relationship to the crystallographic point group that was ignored in the analysis of the previous sections. Similarly, in the D_{3h}, hexagonal, case, in the group defining the parent lattice, D_{6h}, there are two distinct sets of mirror plane reflections perpendicular to the sixfold axis. In D_{3h}, only one of the sets is retained. In the parent D_{6h} lattice one of the associated sets of symmetry elements contains the translation vectors and the other bisects the angle between them. Which set is retained in D_{3h}? The answer is that either is possible – but two different space groups are obtained as a result.

Problem 13.7 By sketching a one-face – centred orthorhombic lattice and placing atoms in it in two different ways, illustrate the two different space groups which were the subject of the above discussion.

Having thus seen how the relatively small errors in Table 13.6 arise, what of the problem of the apparent gross error in prediction for the trigonal case? The extension of Table 13.7a to cover this crystal system is given in Table 13.7b.

This extension is most strange when compared with Table 13.7a because again the number 1 is expected, if not everywhere, at least to be predominant – but it does not appear! Matters

Table 13.7b

System	Bravais lattice	Crystallographic point group	
Trigonal	p.	D_{3d}	3
		D_3	3
		C_{3v}	3
		C_3	2
		S_6	2

would perhaps be improved if there were two primitive trigonal Bravais lattices, not just one (for then the number 2 would be expected rather than 1), but even then there would be a problem – the 'additional' errors occur for the higher symmetry point groups, not the lower, which is where they were found in Table 13.7a. To deal with the former problem first. In fact, there has been a long-standing argument about the number of trigonal Bravais lattices. There have been those who have argued that there are fifteen, not fourteen, Bravais lattices, and that two of them are trigonal. Indeed, the very first listing ever given of these lattices gave a total of fifteen. The number was reduced to fourteen by Bravais who showed mathematically that two of the fifteen could be similarly described. Were there to be two trigonal lattices, then they would both have to be primitive – and, surprise surprise, this is where the argument has arisen! The primitive trigonal lattice was introduced as one of those obtained when the symmetry of a primitive O_h lattice is reduced but later, in tables, this was associated with the word 'rhombohedral'. A rhombohedral unit cell[6] can be pictured as obtained when opposite corners of a cube are either symmetrically compressed or stretched as shown in Figure 13.9. In a sense, this is the 'true' trigonal lattice. The second has already been met as the hexagonal. As Table 13.2 shows, this lattice is characterized by two equivalent vectors at 120° to each other and a third at 90° to the other pair. Nothing else is specified about this third vector. But, from Figure 13.6 and the associated discussion, the angle of 120° (a value that requires that there also be angles of 60°) could mean that the third vector is either a C_6 or a C_3. The former value gives rise to the hexagonal lattice, and the latter to a trigonal. But since the lattice has already been listed under the heading 'hexagonal' we cannot include it a second time as a trigonal. In that a sixfold axis implies a coincident threefold – but not vice versa – its listing as a hexagonal lattice is clearly correct.

One problem has now been solved; each and every entry in Table 13.7b should be '2' – but this still leaves unanswered the problem posed by the fact that three are '3'. Why? In fact, the answer has already been given. It was met when discussing the fact that in the hexagonal crystal system there were two different ways of introducing a D_{3h} arrangement of atoms. Either the vertical mirror planes of this group were *coincident* with the directions of two of the primitive translation vectors defining the hexagonal lattice or the mirror planes *interleaved* the vector directions. It is essentially these two possibilities which give rise to the additional trigonal space groups, except that for the D_3 case there are no mirror planes

[6] The following discussion becomes easier to visualize if unit cells are regarded as building blocks, rather than focusing on the associated lattices.

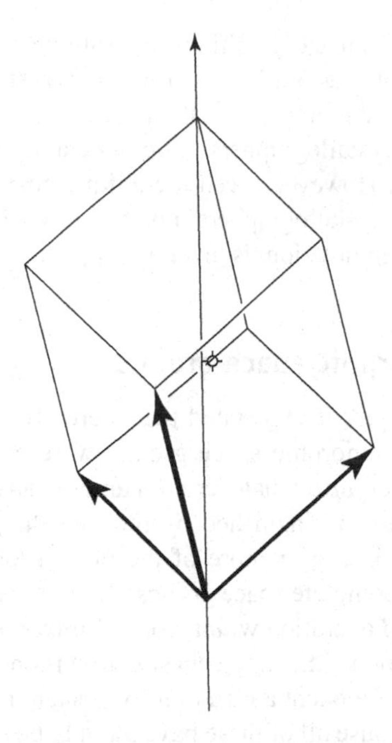

Figure 13.9 The three primitive translation vectors of the rhombohedral lattice are interrelated by threefold rotation operations. There are no restrictions on the angles between the vectors, although at certain angles special lattices are generated (90° gives the primitive cubic, for instance)

and it is the corresponding alternative orientations of the twofold axes which is relevant. For groups without these symmetry elements this ambiguity does not arise and so these have only the now-expected '2' in Table 13.7b.

Problem 13.8 Explain why the number 1 does not appear in Table 13.7b.

The end of this section has almost been reached but before concluding it there are two questions demanding answers. First, the section was headed 'The symmorphic space groups' – yet the word 'symmorphic' not not been explained. What is it all about? In this section our concern has been with those space groups that arise from the combination of translation operations with point group operations. In making these combinations, life was made simple by the fact that these two types of operations were quite distinct. The space groups that result are called the *symmorphic* space groups. The name itself is of little significance until we work with the other space groups, the *non-symmorphic* space groups. These will be the concern of the next section. The final question arises because Table 13.7a gives the number of space groups of a particular type that exist. This implies that there is some source book containing all such information. Indeed there is, and much study in the field is impossible without a copy to hand. The book is called *International Tables for Crystallography* and it is revised from time to time; with each revision it gets bigger, which is the reason why some prefer the early editions. An Internet version is now available.[7] As its name makes clear, the book was originally written for crystallographers, although a real attempt has been made to make it more generally accessible. A problem is that it is written using the nomenclature of crystallographers, a nomenclature that is rather different from that used so far in this book. However, given the unique position of the book, there is no alternative to working in the crystallographers' notation when using it. A brief introduction to this, the Hermann–Mauguin notation, is given in Appendix 5.

13.5 The non-symmorphic space groups

At the beginning of this chapter it was stated that there are 230 space groups but in the previous section only 73 symmorphic space groups were met. It follows that there are 157 *non-symmorphic space groups*, whatever the name means. The vast majority of non-symmorphic space groups are distinguished by the fact that, whilst the lattices are the Bravais lattices of Table 13.3, one or more of the point group symmetry elements that combine with them (to give complete space groups) contain a translation component. This combination of a point group operation with a non-primitive translation is a characteristic of non-symmorphic space groups.[8] So, a typical situation is one in which a twofold rotation operation has a translation component added to it. Any such translation component cannot be a primitive translation because all of these have already been included in the translation vector set. Double counting is not allowed! A C_2 rotation carried out twice is equal to the identity, leave alone, operation. If a translation is to be included along with the C_2 then the composite operation carried out twice must also give the identity. It follows that any translation associated with a C_2 has to be of one-half of a primitive translation in the direction of the twofold axis. Carrying out the operation twice would then give a single translation step. But as already has been said, this is in the translation group, not the point group. So, our identity remains the identity as far as the point group is concerned. The way that operations originally associated with a point group can apparently be transferred to a translation group clearly merits detailed discussion.

At the present point it is sufficient to recognize that only well-defined non-primitive translations can be associated with point group operations. The operation in which a non-primitive translation is associated with a rotation operation is called a *screw rotation* and the axis is a *screw axis*. In Hermann–Mauguin notation the screw axis just discussed is denoted a 2_1 axis (pronounced 'two one axis'). Here, the 2 is the Hermann–Mauguin equivalent of that which we have so far in this book called C_2. The subscript 1 means that associated with the 2 is a non-primitive translation of an amount equal to one of the two steps needed to give a pure translation. It therefore corresponds to one-half of a primitive translation in the direction of the 2 axis. In a similar way, a 3_1 axis involves one-third of a primitive translation in the direction of the threefold axis and 4_1 one-quarter of a primitive translation. These last two examples show more clearly than the first why the axes are called 'screw' axes. The act of putting a screw into a piece of wood involves a simultaneous rotation and translation

[7] Relevant to the present chapter is Volume A, details of which are available on-line at http://www.iucr.org/a.

[8] But it does not completely define them. As will be seen, it is possible to get the same effect by moving the position of a set of axes in space and the complete definition has to take account of this.

of the screw. So here, we have a combination of a rotation with a translation. However, these last two examples also point to a problem. Most screws are right-hand but some are left-hand (many a would-be mechanic has ruined a mechanism because of an unexpected left-handed screw!). Which do we have here? The answer is met by a convention: 3_1 and 4_1 refer to right-handed screws but 3_2 and 4_3 refer to left-handed (these latter two might equally well be written as 3_{-1} and 4_{-1} but this is never done). Right-handed screws go into the wood when rotated clockwise, viewed from the screwdriver end.

It is not just rotation operations that can be combined with non-primitive translations. So, too, can mirror plane reflection operations. Mirror plane reflections combined with non-primitive translations (and these are always halves of primitive translations because two reflections in a mirror plane give the identity) are called *glide planes* and are denoted by the direction in which the translation associated with the glide occurs. So, in an '*a* glide' the translation is one-half of a primitive translation in the x direction, in a '*b* glide' the translation is in the y direction, and so on. The requirement on the non-primitive translation associated with a glide plane is that it lies in a plane parallel to that of the mirror plane. It can, therefore, be composed of half-primitive translations along more than one axis. Such glides involving two half-primitive translations are denoted by the letter n (for '<u>n</u>et', because they span the diagonals of a two-dimensional net), and those involving three by the letter d (for '<u>d</u>iamond', because they occur in the diamond lattice). The identity cannot be associated with a non-primitive translation (if it did, the identity would lose its meaning in the point group) and the operation of inversion in a centre of symmetry is not associated with a non-primitive translation. For inversion in a centre of symmetry there is a choice. Either it could be combined with a non-primitive translation (which would have to be one-half of a primitive) or, because inversion is an operation which operates about a unique point, the point can simply be moved to a new position which is displaced from the 'original' by one-quarter of the corresponding primitive translation – the overall effect is the same (Figure 13.10). By adopting the latter choice the need to formally specify the translation involved is avoided and this makes life easier. So, the diagrams in *International Tables*, which show centres of symmetry as points, have this latter choice built-in. A similar (sideways) translation of rotation axes occurs when a non-primitive translation concerned is perpendicular to the rotation axis. However, these 'hidden' non-primitive translations seldom occur on their own. They normally occur as a consequence of the presence of one or more of the formally defined non-primitive translations.

At first sight, all this notation appears rather complicated and one fears that the detailed discussion of the non-symmorphic space groups will be too. In fact, these fears are not really justified. Given the plethora of possible ways of combining translations with rotations and reflections, and the multitude of ways in which axes and centres of symmetry can be displaced, an enormous number of non-symmorphic space groups would be expected. But, as we have already seen, whilst the number of non-symmorphic space groups is greater than the number of symmorphic, there are only about twice as many. The reason is to be found in the group algebra. The combination of any two operations must be equivalent to a single operation. This requirement drastically limits the possibilities. Table 13.8 is the non-symmorphic space group equivalent of an enlarged version of Table 13.7, listing the number of non-symmorphic space groups associated with each point group. It is evident that the non-symmorphic space groups are rather unevenly distributed. There can be more, the same or fewer (even zero) non-symmorphic space groups associated with the point

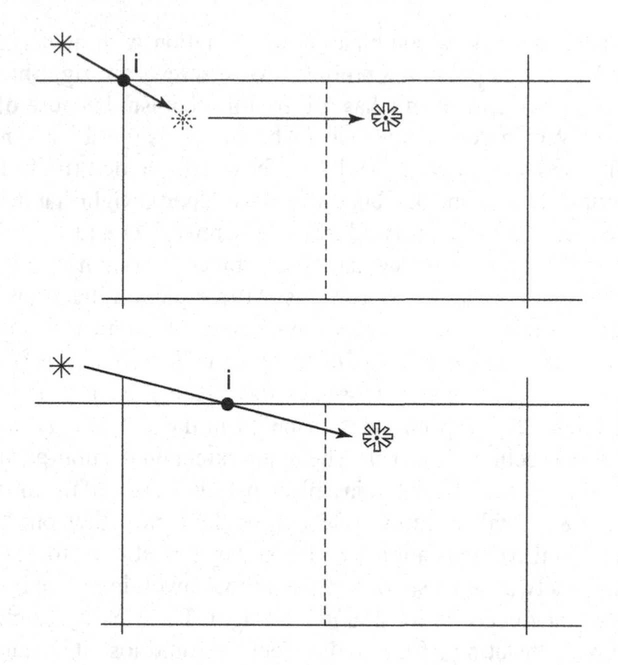

Figure 13.10 The action of inversion in a centre of symmetry followed by a translation of one half of a primitive translation (upper) is equivalent to inverting in a centre of symmetry which has been moved by one quarter of the primitive translation

group. For comparison, Table 13.8 gives, in parentheses, the corresponding data for the symmorphic space groups.

> **Problem 13.9** Draw separate diagrams to illustrate 2_1, 3_1, 4_1, 3_2 and 4_2 screw axes and others to illustrate *a*, *b*, *c*, *n* and *d* glides.

It is not possible to give a simple discussion of the number of non-symmorphic space groups to parallel that for the symmorphic. However, some general comments on Table 13.8 may be helpful. Generally, the number of non-symmorphic space groups associated with a crystallographic point group is comparable with the number of symmorphic. When the number of symmetry operations that can be combined with a non-primitive translation is small, so too is the number of non-symmorphic space groups. So, when there is a large number it tends to be with that point group which has the full lattice symmetry – although when there are alternative ways of matching a point group with a lattice an enhanced number also results. The non-primitive translations associated with a C_2 (2) axis and a mirror plane σ (*m*) are both one-half of a primitive translation. This communality means that they can interplay in space groups – and this increases the number possible. So, the groups in Table 13.8 that give rise to the largest number of space groups are those which contain both 2 and *m* operations. Beyond these generalities it is best to proceed with specific examples and a selection is given in Appendix 6.

Table 13.8

System	Crystallographic point group	Number of non-symmorphic space groups	(Number of symmorphic space groups)
Cubic	O_h	7	(3)
	T_d	3	(3)
	O	5	(3)
	T_h	4	(3)
	T	2	(3)
Tetragonal	D_{4h}	18	(2)
	D_4	8	(4)
	C_{4h}	4	(2)
	C_{4v}	10	(2)
	C_4	4	(2)
	D_{2d}	8	(4)
	S_4	0	(2)
Orthorhombic	D_{2h}	24	(4)
	D_2	5	(4)
	C_{2v}	17	(5)
Monoclinic	C_{2h}	4	(2)
	C_s	2	(2)
	C_2	2	(1)
Triclinic	C_i	0	(1)
	C_1	0	(1)
Hexagonal	D_{6h}	3	(1)
	D_6	5	(1)
	D_{3h}	2	(2)
	C_{6h}	3	(1)
	C_{6v}	3	(1)
	C_6	5	(1)
	C_{3h}	0	(1)
Trigonal	D_{3d}	3	(3)
	D_3	4	(3)
	C_{3v}	3	(3)
	C_3	2	(2)
	S_6	0	(2)

13.6 Unit cells

In the early part of this chapter care was taken to avoid use of the term 'unit cell'. Later, and particularly when space groups were discussed, it crept in, although its use was kept to a minimum. The name is so simple and useful that it cannot long be avoided. Why then should it be so assiduously avoided in the present text? The reason is that the concept of a unit cell is more complicated than one might suppose and it is preferable to avoid basing arguments on an unknowingly simplified concept. Where, then, lies the problem? The answer is that for no crystal structure is there a unique unit cell. Indeed, quite the opposite. For every crystal structure there is an infinite number of acceptable unit cells, *all* of them primitive (of course, crystallographers will prefer non-primitive unit cells for some structures). This infinite choice is important – it is the reason for including this section – but it should be

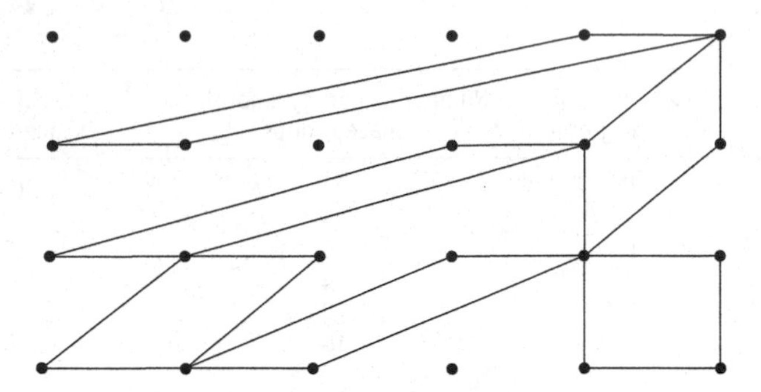

Figure 13.11 Six different equally acceptable (although not necessarily equally convenient) choices of two-dimensional unit cell for the two-dimensional lattice shown. All of the choices have identical areas. Clearly, there is an infinite number of acceptable choices. Similar considerations apply to the three-dimensional case

contrasted with the common use of the expression 'the unit cell is...' in research papers and textbooks when referring to individual crystal structures, a statement which by the use of the 'the' implies that only a single choice exists for a unit cell. Of course, the use is justified in that for one reason or another there is often a single *convenient* choice, but it is important not to overlook the possibility that for a different purpose a different choice might be preferable.

In Figure 13.11 is shown a two-dimensional grid that might be one layer in an orthorhombic lattice. The figure shows several alternative choices of two-dimensional unit cells, all of the same area but differing in shape. Clearly, a similar set of constructions is possible in the third dimension of an orthorhombic (or any other) lattice. For each of these constructions an infinite number of variants exist, at least for an infinite lattice (and on the atomic length scale the lattice of a real crystal is, effectively, infinite). The most obvious simplification, that almost invariably adopted, is to choose those translational vectors (or unit cell edges) which are the smallest in magnitude. These lead to the chunkiest possible unit cell, one which makes diagrams of the atomic arrangement within it easiest to draw and understand. Nonetheless, an infinity of alternative choices exist. But this discussion has been too restrictive! Yet more choice exists. First, it has unquestioningly been accepted that unit cells should be bounded by plane faces. Again, this is a choice of convention and convenience; it is not a requirement. The faces of a unit cell can be curved, dimpled, re-entrant, whatever. There is one requirement of a unit cell: that when repeatedly operated on by the primitive translation vectors it generates the entire crystal (or crystal lattice, depending on the context) completely.

Problem 13.10 Draw a diagram similar to that in Figure 13.11 but with all unit cell edges curved.

A second reason why our statement 'that there is an infinite variety of choice of unit cell' was too restrictive lies in the fact that all the unit cells considered had one thing in common; they were each bounded by three pairs of parallel faces. Even if we restrict ourselves to plane faces, a unit cell can have many more than three sets of parallel plane faces. Indeed, if one were to select one choice of unit cell as being preferred to all others then it would generally be one with many facets, many faces. After all that has just been said about unit cells this is a strange message – that one choice of unit cell is to be preferred. It is made more strange by the fact that the actual preferred choice of cell is very different from that which is familiar to chemists – the unit cells of crystallographers. The final section of this chapter describes these strange (to the chemist, but not the physicist) unit cells.

Problem 13.11 Repeat Problem 13.3. – the answer should be longer this time!

13.7 Wigner–Seitz unit cells

Wigner–Seitz unit cells are essential to a full understanding of the solid state. This is because Brillouin zones, which are at the heart of solid state physics, are a sort of Wigner–Seitz unit cell.[9] The construction of Wigner–Seitz unit cells is perhaps best explained in a somewhat unreal, anthropomorphic, way. Suppose the reader is reduced to the dimensions of an atom and is standing at a lattice site (alternatively, that the lattice is so enlarged that a person can stand inside it). From the chosen lattice site draw lines to all other (equivalent, of course) lattice sites. The chosen lattice site bristles with lines, rather like a curled up hedgehog/porcupine. Now, exactly halfway along each line, construct a plane perpendicular to the line; it is perhaps helpful to think of these planes as being rather solid. Standing at the chosen lattice site, and forgetting the lines used in their construction, the reader will find themself surrounded by the planes originating from the shortest lines, those running to the nearest lattice points. These planes will intersect and, although the planes themselves run to infinity, all that will be seen is the box formed by their intersection immediately around the chosen lattice point. On the real, atomic, scale, this box is the Wigner–Seitz unit cell of the lattice. In Figure 13.12 are given examples of Wigner–Seitz unit cells. Note several things. First, a Wigner–Seitz unit cell has a lattice point at its *centre* and nowhere else; the unit cells encountered in most textbooks have lattice points at their corners. Second, a Wigner–Seitz unit cell, by its very construction, contains only a single lattice point; all Wigner–Seitz unit cells are primitive. Third, the number of faces of a Wigner–Seitz unit cell is determined by the number of neighbours and their disposition in space. As will be seen from Figure 13.12, it is not unusual for a Wigner–Seitz unit cell to have a dozen or so faces – this is not unreasonable, because each sphere in an array of close-packed spheres has twelve nearest neighbours. Fourth, each Wigner–Seitz unit cell has the symmetry of its crystal system. That is, whereas the primitive unit cells of centred lattices shown in Figure 13.8b all had symmetries lower than those of their crystal systems – those listed in Table 13.2 – all of the cubic Wigner–Seitz unit cells are of O_h symmetry, for example.

What makes the Wigner–Seitz unit cell unique? Two things. First, it is the only choice of primitive cell which invariably has the point group symmetry of its crystal system. Second, it has a property shared by no other choice of unit cell, a property that is evident

[9] The development of Brillouin zone theory is not given in the present text; the author has written about it elsewhere in a manner which is entirely compatible with the present discussion (see S.F.A. Kettle, *Physical Inorganic Chemistry* Chapter 17, Oxford University Press, Oxford, 1998).

from its construction: it contains all (general, not lattice) points in space that are closer to the chosen lattice point than to any other lattice point. This is an important property. Suppose, for instance, that some spectroscopic property of a crystal is studied; the particular form of spectroscopy is unimportant. It is possible that the spectrum obtained would show evidence of interaction between individual atoms/bonds/molecules (depending on the particular spectroscopy) and their environment. In order better to understand the spectrum, one might attempt to calculate the interaction between a given atom/bond/molecule and every other in the crystal. Even with the fastest and most powerful of modern computers, this is a near-impossible task, one that would exhaust any research budget. As a compromise, it might be decided to carry out calculations of the interactions between the given atom/bond/molecule and all those others with which it interacts more strongly than does

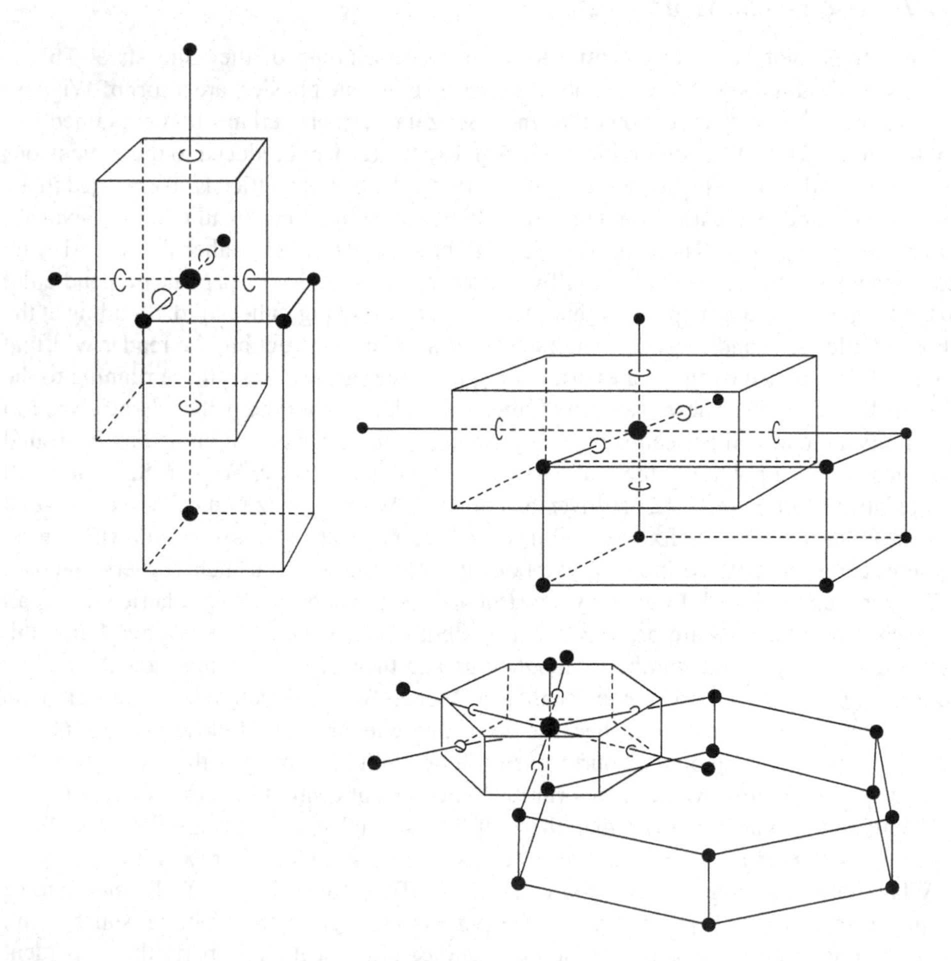

Figure 13.12 (a) The Wigner–Seitz unit cells of three primitive lattices. The conventional unit cell has lattice points at its corners; the Wigner–Seitz unit cell has a single lattice point at its centre. The Wigner–Seitz cells are drawn showing the proximate lattice points which give rise to faces of the cell (see the text for a discussion).(*Continued*)

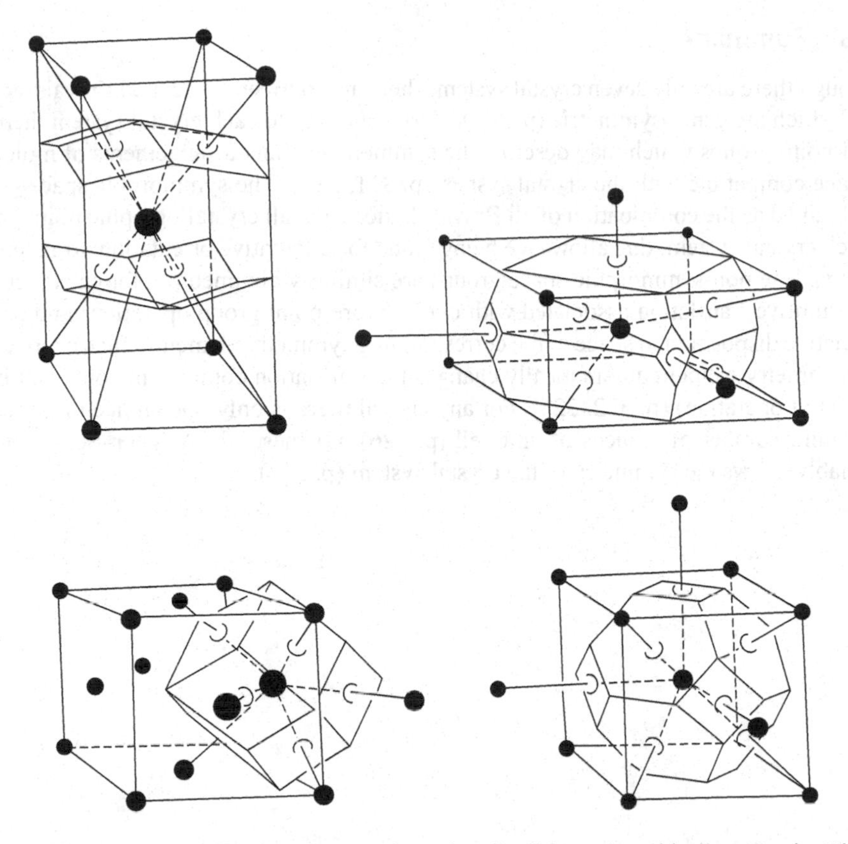

Figure 13.12 (*Continued*) (b) The Wigner–Seitz unit cells of four centred lattices. As the Wigner–Seitz unit cell is always primitive it is smaller (in three examples a factor of $^1/_2$ and in one example a factor of $^1/_4$) than its conventional counterpart. The Wigner–Seitz cells are drawn showing the proximate lattice points which give rise to faces of the cell (see the text for a discussion)

any other equivalent atom/bond/molecule (because these are likely to be the most important interactions). Which, then, are the atoms that have to be considered? The answer is simple: all those contained within the Wigner–Seitz unit cell which has the atom/bond/molecule at its centre. The only assumption contained within this statement is that the magnitude of the relevant individual interactions decreases with increase in separation between the interacting centres, as do all interactions of recognized chemical importance. There are more than fourteen Wigner–Seitz unit cells (only a selection is shown in Figure 13.12) because, although in principle there is one for each different Bravais lattice, the actual shape of a Wigner–Seitz unit cell depends on an axial ratio. In a primitive tetragonal lattice, for instance, does the unique translational vector have a magnitude which is greater or less than the magnitude of the other two translational vectors? The symmetry of the Wigner–Seitz unit cell is D_{4h} in both cases but the cells look qualitatively different. In total, there are twenty-four different-looking Wigner–Seitz unit cells.

13.8 Summary

Although there are only seven crystal systems there are fourteen associated Bravais lattices, all of which are centrosymmetric (p. 302). Corresponding to each crystal system there is a set of point groups which may describe the symmetry-distinct arrangements of molecules in space compatible with the crystal system (p. 311, 317). The symmorphic space groups are obtained as the combination of all Bravais lattices and all crystallographic point groups of each crystal system, due allowance being made for alternative orientation arrangements (p. 318). The non-symmorphic space groups are similarly obtained but for each there is a non-primitive translation associated with one or more point group operations and/or with the relative disposition in space of the corresponding symmetry elements (such a movement of a symmetry element automatically changes the translation component contained in the associated operation) (p. 322, 325). For any crystal there is only one choice of lattice but an infinite number of choices of unit cell (p. 326). Of these, the Wigner–Seitz unit cell invariably shows the symmetry of the crystal system (p. 327).

14 Spectroscopic studies of crystals

14.1 Translational invariance

The discussion of the previous chapter contained two contributory components – that coming from the lattice and that coming from the crystallographic point group. The allowed combinations of these parts led to the 230 space groups. However, the discussion was rather different from that in other chapters of this book in that there was no mention of a character table, only of symmetry elements and operations. In this chapter we will begin to introduce space group character tables. Again, there are two distinct approaches which may be adopted, that through the lattice and that through the crystallographic point group. The former is that appropriate to solid state physics – it leads to the development of band theory, relevant to such topics as the electronic energy level patterns in solids. The latter is more appropriate to spectroscopic studies of solids and is the one which will be the subject of this chapter. A major distinction between the two approaches arises from the fact that there is, effectively, an infinite number of translation operations in the translation group of the lattice but the crystallographic point group is finite. The full space group combines, in some way, both the translation group and the point group. Whichever one chooses to work with, one should really be using the full space group and its character table – and so the relationship between this full group and the one being used is very relevant. There are two aspects to this relationship – the mathematical and the physical. Although both are relevant it is arguable that the latter has the greater importance. Unless a mathematical relationship has some physical significance it remains nothing more than an elegant irrelevance. First then, it is necessary to look at spectroscopic measurements on crystals in the large, in order to discover those aspects which enable mathematical simplifications.

The most general and relevant aspect of spectroscopic measurements on crystals is that of scale. A typical translation vector in a crystal relating adjacent equivalent points has a magnitude of a few Ångstroms (this is a quantity which would normally be quoted as the length of a unit cell edge, although it could equally well be called a translation vector). In contrast, the wavelength of visible radiation is of the order of a few thousand Ångstroms. In the infrared the wavelength is much longer and even in the vacuum ultraviolet it is a few hundred Ångstroms. Only for x-rays does the wavelength of the radiation become comparable to or smaller than the length of a typical primitive translation vector in a crystal. This is relevant to the classical explanation of the interaction of light with matter given in

Symmetry and Structure: Readable Group Theory for Chemists Sidney F. A. Kettle

Section 4.5. There, typically, the electric vector of an incident light wave was seen as inducing a transient charge displacement in a molecule. This charge displacement changes sign with the oscillations of the electric vector; when these oscillations coincide with a natural frequency of the molecule then resonance occurs, typically with transfer of energy to the molecule. This picture carries over, unmodified, into the spectroscopy of crystals. However, it can be enlarged by recognition of the enormous difference between the wavelength of the incident radiation and the magnitude of a typical primitive translation vector. In a crystal, molecules which are close to each other and related by pure translation operations will experience essentially the same incident electric vector originating in the light wave. Indeed, they can be thousands of Ångstroms apart and still experience, essentially, the same electric vector. The fact that different molecules are related by a pure translation means that they have precisely identical orientations with respect to the incident light wave (ignoring any small imperfections in the crystal). To a first approximation, then, for all common spectroscopies the interaction of light with a crystal is a translation-independent process.

Problem 14.1 Neutrons can have wavelengths which are comparable to interatomic separations. What main extensions would have to be made to the discussion in the text to cover their use to probe the vibrational properties of crystals (they can be inelastically scattered, in a sort of Raman process)? Diffraction effects can be ignored.

Translationally related molecules behave in the same way; get the answer for one molecule and you have it for them all. The same cannot be said of molecules that are related by point group operations (or composite operations with a point group component). A 2 (C_2) rotation, for example, 'turns a molecule over' so that if there two molecules interrelated by a 2 (C_2) they would experience precisely opposite transient dipoles induced by the electric vector of a light wave. For two molecules interrelated by a 2_1 screw operation, the 2 would ensure that the induced dipoles are out of phase; the non-primitive translation component of the operation would be irrelevant for the reasons given above. The conclusion is that 'the only important aspect of solid state symmetry which is relevant to spectroscopy is that contained in the crystallographic point group'. Even if the actual 'point' group is a derivative of a real point group (if the actual 'point' group contains non-primitive translation operations, as discussed in Chapter 13), then it still will be possible to work with one of the thirty-two crystallographic point groups. This, then, is the physical picture. Can it be given a more formal, mathematical, justification? The answer is 'yes'; indeed, more than one such formal justification exists. The different justifications are those associated with different models – two, in particular, are important – the unit cell group model and the factor group model. Fortunately, these two models invariably lead to identical predictions although the latter is perhaps the closer related to the detailed discussion above. The next section is devoted to these models.

One final word about the physical picture. The content of the present chapter is based on the assumption that there is an interaction of some sort between the molecules in the crystal under study. If they behave as isolated individuals then there are no spectroscopic complications arising from the fact that the solid state is involved. Indeed, quite the opposite. The molecules may behave as if they were in the gas phase (where there certainly would be no molecule–molecule interactions) but they are, in fact, fixed in a crystal and that means that, in contrast to the gas phase, they are fixed in their orientations. This is a topic discussed at the end of Chapter 4, where it was pointed out that this can mean a change in the spectral bands excited as the orientation of the crystal is changed (provided that oriented, polarized, radiation is used). This, so-called *oriented gas model* will not be discussed in detail (in truth, there is little that could be added to what has already been said) although it should be emphasized that it can well happen that it is applicable to some spectral bands arising from a crystal – but that one of the models to be covered in the next section has to be used for others. This is because, for instance, some vibrational modes of a molecule may be well insulated from those of the surrounding molecules but other vibrational modes of the same molecule are not. A vibration which changes the dipole of a molecule is more likely to be coupled with the same vibrations of other molecules than is a vibration which is, say, quadrupole active because dipole–dipole coupling attenuates less rapidly with distance than does quadrupole–quadrupole. Another possibility is that a molecule is sensitive to its general environment but, nonetheless, is insulated from specific interactions with other molecules. It is sensitive only to the symmetry of the site in the crystal at which it is situated. Almost always, this site is of lower symmetry than that of the isolated molecule and so the *site symmetry* model is characterized by the splitting of degeneracies and by the increased strength (and, perhaps, the appearance) of transitions forbidden in the isolated molecule. The first task of any analysis is to determine which model is appropriate for each spectral feature. An example of the application of the oriented gas and site symmetry models will be given later in this chapter.

Problem 14.2 $AsBr_3$ has a C_{3v} structure, like NH_3 (Chapter 7). Some is trapped within a plastic film matrix, in the hope that when the film is stretched oriented molecules of $AsBr_3$ will be obtained. First, however, it is necessary to determine whether the oriented gas or site symmetry model best describes the molecular vibrations in the unstretched film. Suggest an experimental distinction, assuming that all molecules occupy identical sites in the film.

14.2 The factor group and unit cell group models

In this section, as in the previous, it is convenient to talk in terms of ‘molecules’, although in the appropriate context the discussion could equally well apply to atoms or to ions. The first approach to be considered is the *factor group* model. Clearly, the first task is to define what is meant by ‘factor group’. In principle any group could have one or more associated factor groups. The character tables of factor groups are invariably simpler than those of the parent group to which they relate[1] – an attractive feature; for the case of space groups, the character tables of the corresponding factor groups are enormously simpler. The concept of

[1] This statement is true whenever a factor group is non-trivial. A few groups have only factor groups which are trivial, being identical to the parent group itself. In such cases the parent group is itself very simple, having no non-trivial invariant subgroup.

a factor group is closely linked to that of an invariant subgroup, a topic which is covered in some detail in Section 8.6. There it was shown that when a group is the direct product of two invariant subgroups then its character table is the direct product of the character tables of those of the two invariant subgroups. The particular case considered was a demonstration that the point group C_{2v} is the direct product of the groups C_2 and C_s. Let us consider this case again; Table 8.11 is particularly relevant. We have that:

$$C_2 \otimes C_s = C_{2v}$$

(remember, the symbol $\otimes$ is used to indicate a direct product). Writing each group in full:

$$\{ E \ C_2 \} \otimes \{ E \ \sigma \} = \{ E \ C_2 \ \sigma_v \ \sigma_v' \}$$

The left-hand side of this expression can be written differently:

$$[E\{ E \ \sigma \} C_2\{ E \ \sigma \}]$$

Written in this form, one sees a generality; the 'inner' group $\{ E \ \sigma \}$ could be varied without changing the general form of the expression (although a change would mean that it was no longer applicable to C_{2v}). Alternatively, in this particular form of the expression, it can be regarded as a constant which multiplies both the E and the C_2. As a constant, it is playing the role of an identity element. The group $[E\{ E \ \sigma \} C_2\{ E \ \sigma \}]$ is said to be the *factor group* of C_{2v} with respect to the group $\{ E \ \sigma \}$, the group which plays the part of the identity element. Most important is the fact that the character table of the factor group is that (really, is isomorphic to that) of C_2. This is a simple, almost trivial, example of a factor group. Some meaning would be attached to it if it were possible to make some measurement on a molecule of C_{2v} symmetry, the result of which was independent of the σ_v and σ_v' operations. The result would depend only on the E and the C_2. In such a case, the factor group above, a group isomorphic to C_2, contains all of the relevant information. One could work in the full group but to do so would be to add nothing new.

The situation is quite different in the solid state. As has been explained in the previous section, to a very good approximation, spectroscopic phenomena are independent of the translation operations. Effectively, all measurements concern transitions which transform as the totally symmetric irreducible representation of the translation group. A detailed study of the way that these phenomena transform under the translation operations is therefore not of any value. Further, the translation group is always an invariant subgroup of the full space group. It is therefore possible to form a factor group of the space group with respect to the relevant translation group (the detailed operations of which are therefore not of concern). The translational group plays the role of the identity; can there be a better way of getting rid of it and all of its problems? So important are these factor groups that they are simply referred to as 'the factor group' (of a particular space group). Just as each and every space group is different so, too, are the corresponding factor groups. Sometimes, the difference will lie in the details of the translation group and so not be evident. More evident will be the relationship between the group of the point-group-derived operations and the corresponding crystallographic point group. In practice, one works with the character table of the relevant crystallographic point group. It is necessary to make the correct substitutions (of 2_1 for 2, for example) and it is here that the *International Tables of Crystallography* prove invaluable,[2] although they do not contain the character tables themselves. A word of consolation. Those accustomed to working with point groups and not with factor groups may well find the prospect of handling screw rotations and glide planes somewhat daunting – these operations may appear in the 'corrected' crystallographic point group. In contrast, those accustomed to working with factor groups welcome the appearance of screws and glides! The reason is that, almost invariably, the character generated under such operations is zero, whatever the problem under discussion.[3] An example of the use of a factor group in a vibrational analysis of a solid is given in the next section.

Apparently quite different from the factor group approach is the *unit cell* model. In this, the problem of the translation operations is dealt with by the simple expedient of ignoring them! The justification for this simplification is that given in the previous section – that the spectroscopic phenomena under consideration are translation-independent. The method consists of considering a unit cell of the crystal and the operations which interrelate the molecules that it contains. These operations are taken to be moduli primitive translation; any operation that takes an object out of the unit cell is held to bring it back again through the opposite face (so, if it disappears through the top face, it reappears through the bottom – where the operation is completed). The unit cell method ends by using the same mapping of crystallographic point group operations onto their space group derivatives as does the factor group and so the two methods lead to identical results. Of the two, possibly because of its more evident connection with the results of crystal structure determinations, the unit cell method is perhaps the more popular. It has to be recognized, however, that it suffers from two potential weaknesses. The first is that it gives undue prominence to a particular choice of unit cell. As has been emphasized, there is an infinite number of choices of acceptable unit cell for any crystal. A particular choice of cell invites statements such as 'because in the unit cell they are well separated...', which are strictly unacceptable. Acceptable alternatives are along the lines 'because in the crystal they are well separated . . .' Second, it is usual for the crystallographically determined unit cell to be used in the unit cell model. This poses a problem when the crystallographic unit cell is centred because it is a primitive unit cell which has to be used in the unit cell model – and the method lays down no rules for moving from the centred to the primitive. Errors have appeared in the literature as a result. Authors have been known to work with the centred cell rather than the primitive (and so predict too many spectral features). Others, aware of the problem, have worked with the centred cell and simply divided the predictions by the relevant factor (those at the right-hand side of Table 13.3). Unfortunately, this procedure does not guarantee the correct answer either. An example of the use of the unit cell method is given in the next section.

---

[2] But beware the problem of centred unit cells in the *International Tables*. Because the size of the unit cell is increased so, too, is the number of point-group-derived operations which interrelate points within the unit cell – a doubled unit cell means a doubling in the number of operations and so on. The 'extras' are really translation operations masquerading as point-group-derived operations, appearing as glides and screws. In looking for the correct set of point-group-derived operations (strictly, a set which multiply correctly, as described in Appendix 6) pure point group operations should always be retained in preference to derivatives containing a non-primitive translation component. So, if there are 2 and 2_1 parallel to each other in a centred unit cell, the 2_1 should be ignored.

[3] The exceptions to this statement are few. A long polymer chain, aligned along a screw axis, in which one monomer unit is related to the next by the screw operation is one such exception. In such a molecule a vibration could map onto itself under the screw operation.

14.3 Examples of use of the factor and unit cell group models

As indicated above, in practice the factor group and unit cell group models lead to identical predictions. Indeed, despite the fact that they were developed rather differently in the previous section, someone looking over the shoulder of a spectroscopist might well have some difficulty in deciding which of the two was being used. The reason is that the development of the factor group model given above concentrated on how the translation group could be factored out of the problem. This having been agreed, the next step is to turn to the character table of the relevant crystallographic point group – and this is the first step in the unit cell model also. The two methods differ in subtle ways, which relate to the nuances of their different models. In the following account, these differences will be slightly exaggerated. Further, since it was presented as a problem area above, the unit cell model will be applied to a crystal structure which, crystallographically, is treated as having a centred unit cell.

14.3.1 The $\nu(CO)$ spectra of crystalline $(C_6H_6)Cr(CO)_3$

Whilst the effects which are the subject of this chapter may be found in all forms of spectroscopic measurements on crystals they are more important in some forms than in others. Roughly, the more local the phenomena observed, the less important are the effects. So, in Mössbauer spectroscopy, where the excited states of suitable nuclei are probed, the phenomena are so local that, essentially, only the atoms bonded to the atom under study have any influence. On the other hand, if in a particular form of spectroscopy the spectral bands are very broad, the effects can be masked within the bandwidth. Many measurements made in the visible and ultraviolet regions of the spectrum, where electronic transitions are studied, fall in this category. The solid state effects can be measured and studied but rather special conditions are often needed – low temperatures, single crystals together with polarized radiation and, perhaps, doping of the crystal with an isomorphous diluent. One of the spectroscopic areas in which the phenomena are easily studied is in that of vibrational spectroscopy, an area which has the advantage that the reader may well encounter the relevant phenomena in the laboratory. Particularly attractive for study are transition metal carbonyl species. For these, the $\nu(CO)$ modes are the particular concern. They fall in a region of the spectrum which is almost free from other modes, making assignment easy. They are associated with strong spectral bands, making measurement easy. They couple together rather strongly giving symmetry-determined modes, making interpretation easy[4] The species which is the subject of this section, $(C_6H_6)Cr(CO)_3$, crystallizes in a relatively simple space group – the space group $P2_1/m$, number 11, $C_{2h}{}^2$. There are two molecules in the primitive unit ('in the unit cell' would be the way that this is commonly put). As the m in the $P2_1/m$ indicates, the space group contains mirror planes and the $(C_6H_6)Cr(CO)_3$ molecules lie on these, as shown in Figure 14.1. This means that the site symmetry is C_s, in contrast to the molecular symmetry – which is C_{3v}. The factor group is isomorphic to the crystallographic point group, which is C_{2h} (in $P2_1/m$ the C_2 of C_{2h} is 'replaced' by the 2_1). In Figure 14.1 and in the following discussion, only the C≡O stretching modes will be considered.

[4] Easy it may be for simple species but, inevitably, research exploits this to enable the study of species which are complicated and the spectral interpretation no longer easy!

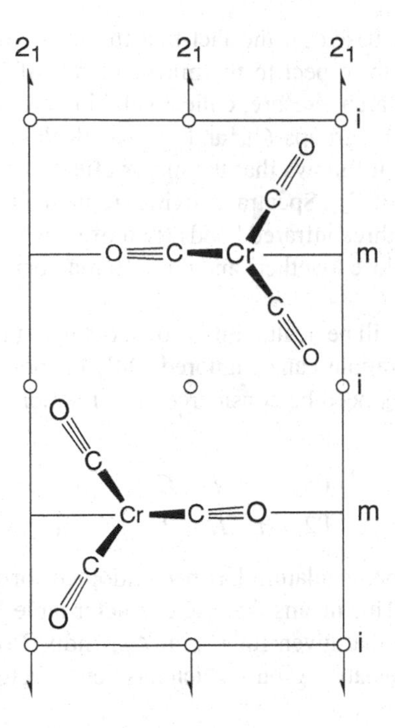

Figure 14.1 The crystal structure of the species $(C_6H_6)Cr(CO)_3$; only the $Cr(CO)_3$ groups are pictured, with their perspective being exaggerated. A primitive unit is shown and consists of two molecules. The space group is $P2_1/m$ ($C_{2h}{}^2$) and the molecular site symmetry is C_s.

Because the molecular symmetry is C_{3v}, the prediction of the symmetries of the ν(C≡O) stretching modes follows the discussion of Section 7.3. This latter dealt with the 1s orbitals of the hydrogen atoms in ammonia, but if the three (C≡O) stretches of $(C_6H_6)Cr(CO)_3$ are considered instead, identical results are obtained; $A_1 + E$ are the symmetry species. Application of the criteria described in Section 4.5 shows that these modes are both infrared and Raman active. This is a single-molecule model, which in a solid state context would be called the 'isolated molecule model'. It is also that of the oriented gas model, leading to identical predictions. The oriented gas model differs from that of the isolated molecule because the former would recognize that the molecular C_3 axes are almost exactly aligned along the crystal c (z) axis. This means that if a single crystal were studied and if the incident infrared radiation were polarized along z then only the A_1 mode would appear with any great intensity in the infrared spectrum (the A_1 mode is polarized along the molecular C_3 axis). When polarized perpendicular to the z axis, only the E modes would be seen. With suitable experimental arrangements, a similar separation could also be achieved in the Raman. Without these experimental distinctions, the isolated molecule and oriented gas models both simply predict two bands, coincident in infrared and Raman.[5]

[5] A further distinction between isolated and oriented gas models is that there is an environment-induced frequency shift of spectral features from the isolated molecule to the oriented gas model. However, it is rare for isolated molecule data to be available (as opposed to data from species in solution – which is not really that of isolated molecules) so this is seldom a useful distinction.

The site group model is based on the fact that the molecules of $(C_6H_6)\ Cr(CO)_3$ are symmetrically arranged with respect to the mirror planes of $P2_1/m$. The three $\nu(C{\equiv}O)$ vibrators of each molecule are therefore, collectively, in an environment of C_s symmetry. The relationship between the groups C_{3v} and C_s was dealt with in Chapter 9 (Table 9.6 or Figure 9.5), from which it follows that the major effect of site symmetry is to split the degeneracy of the E modes of C_{3v}. Spectral activities remain unchanged so the prediction of the site group model is for three infrared bands (two of which, the components of the split E mode, will probably be close together) and three Raman bands, coincident with those in the infrared.

The factor group model will be dealt with in some detail. At the heart of the model is the fact that the translation operations can be ignored. Only the point-group-derived operations isomorphous to those of C_{2h} need be considered. The relationship between the operations of $P2_1/m$ and C_{2h} is:

C_{2h}	E	C_2	i	σ_h
$P2_1/m$	E	2_1	i	m

where a somewhat mixed nomenclature has been adopted for the operations of the crystallographic point group.[6] This means that the character table for the $C_{2h}{}^2$ $(P2_1/m)$ factor group can be derived from that given for C_{2h} in Appendix 3 (conventionally, there is no explicit reference to the translation group which has been 'factored out') and is:

Table 14.1

C_{2h}^2	E	2_1	i	m		
A_g	1	1	1	1	R_z	$x^2;y^2;z^2;\ xy$
B_g	1	−1	1	−1	$R_x;R_y$	$yz;zx$
A_u	1	1	−1	−1	T_z	z
B_u	1	−1	−1	1	$T_x;T_y$	$x;y$

The next task is to use the six $\nu(C{\equiv}O)$ vibrators of two $(C_6H_6)Cr(CO)_3$ molecules (Z, the number of molecules in the primitive unit, is 2) as a basis to generate a reducible representation. In doing this it has to be remembered that, for example, both of the mirror planes in Figure 14.1 are, effectively, equivalent. The act of reflection of an object in two different mirror planes in this figure will lead to results which differ only in primitive translations – and these have been taken out of the problem by the use of the factor group. As far as the point group component is concerned, the final results are identical. So, the fact that the two molecules shown in Figure 14.1 lie on two apparently different mirror planes is no problem; the mirror planes are treated as one. Alternatively, each $M(C{\equiv}O)_3$ unit is reflected in the mirror plane on which it lies. From Figure 14.1 the reducible representation

[6] In the Hermann–Mauguin notation the identity element is denoted by 1; the centre of symmetry is $\bar{1}$. However, these are symmetry elements and in group theory it is the corresponding operations that are relevant. Further, characters such as 1 and −1 will be generated in the application of these operations. To avoid possible confusion, the Schönflies symbols E and i are therefore used to denote the operations.

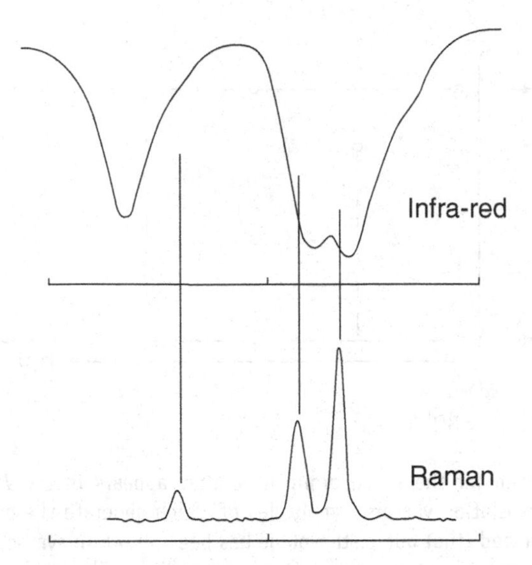

Figure 14.2 A comparison of the infrared and Raman spectra of crystalline $(C_6H_6)Cr(CO)_3$ in the $\nu(C{\equiv}O)$ region. Either the infrared or Raman on its own could be interpreted as originating in the $A_1 + E$ modes of the isolated molecule (the E mode of the C_{3v} molecule being split by the lowered site symmetry, C_s). However, comparison of the two indicates a general non-coincidence, explicable only in terms of the factor group model

generated by the transformation of the $\nu(C{\equiv}O)$ vibrators is easily shown to be:

E	2_1	i	m
6	0	0	2

which has components $2A_g + B_g + A_u + 2B_u$.[7] As the character table above shows, all of the modes with a g suffix are Raman active and all of those with a u are infrared active. The factor group predictions are therefore for three infrared bands and three Raman bands, non-coincident with the infrared. The observed spectra are shown in Figure 14.2 and are entirely in accord with these predictions. In both infrared and Raman, two bands are close together and identified as derived from the split E mode discussed under the site symmetry model above. The sequence of increasing complexity:

Oriented gas model → Site symmetry model → Factor group model

should not be taken as meaning that the applicability of a more sophisticated model automatically invalidates all of the conclusions derived from a simpler model. So, as here, the site symmetry model can help in the application of the factor group, because it correctly predicts a split E mode.

[7] Note, as mentioned earlier, the character of 0 under the operation which contains a non-primitive translation component. This is because such operations almost invariably interrelate different molecules (and so, here, different vibrators).

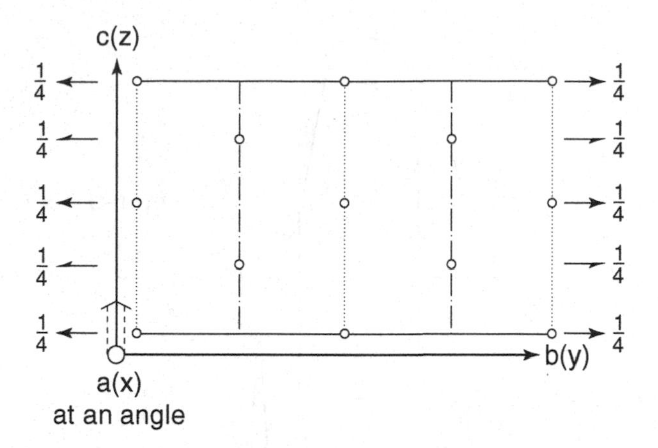

Figure 14.3 The diagram for the space group C2/*c* that appears in the *International Tables of Crystallography*. The translation vectors – multiples of which generate the entire crystal from this unit cell – have been added (that out-of-the-plane has been shown in symbolic fashion; actually, it is not perpendicular to the other two – the lattice is monoclinic). The meaning of the two different sorts of dotted lines is given in the text

14.3.2 The vibrational spectrum of a $M(C{\equiv}O)_3$ species crystallizing in the C2/c ($C_{2h}{}^6$) space group using the unit cell model

One of the simplest of the centred space groups is C2/*c* and this simplicity is why it has been chosen as an example; it provides sufficient generality for more complicated cases subsequently to be treated with some confidence. A modified version of the diagram that appears in the *International Tables for Crystallography* for this space group is given in Figure 14.3. The C in C2/*c* indicates that it is the face perpendicular to the *c* (*z*) axis which is centred, the unique (twofold) axis being *b* (*y*) (because this is the crystallographers' convention for monoclinic systems). As the alternative name for the space group, $C_{2h}{}^6$, shows, the relevant unit cell group is C_{2h} (the superscript 6 indicates that it is the sixth C_{2h} space group listed in the *International Tables*, nothing more). The first question that arises is that of the relationship between the operations of the two groups. Figure 14.3 shows an immediate problem, one that has been mentioned previously - but largely in a different context and with a different explanation: it contains too many symmetry elements. For instance, not only are there twofold rotation axes (shown as the arrows pointing along *y*) but, interleaving them, 2_1 axes (shown as half-headed arrows pointing in the *y* direction). As befits C_{2h}, the glide planes are perpendicular to the twofold axes but, again, there are two sorts. Those shown dotted are glides in which the translation component is along *c* (out of the plane of the paper), as required by the /*c* in the name of the space group. Shown dot-dashed are glides in which the translation contains both *c* and *a* components (one-half of a unit cell edge in each case). There is no mention of these latter glides in the name of the space group. Finally, although less obvious, there are twice as many centres of symmetry as are expected. These doublings result from the fact that Figure 14.3 contains two primitive units, units which are interrelated by a pure translation operation. If, as is convenient for crystallographers, Figure 14.3 is regarded as containing a *single* unit cell then this pure translation has to be combined with point group operations if the symmetry

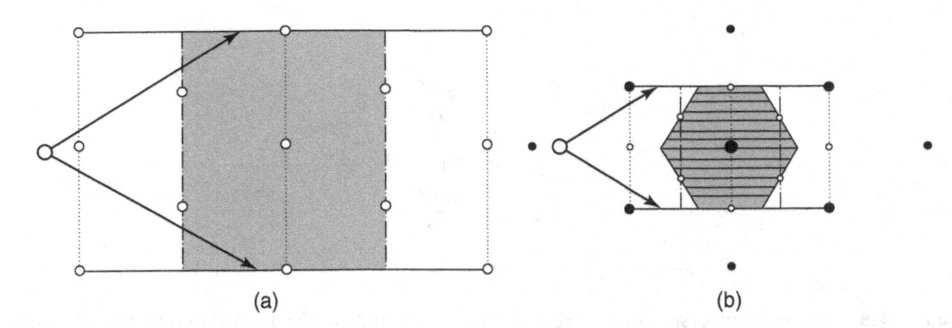

Figure 14.4 (a) The grey area represents a primitive unit cell obtained from that in Figure 14.3 by chopping it in half. The (genuine) primitive translation vectors, multiples of which generate the entire crystal from this unit cell, have been added (that out-of-the-plane has been shown as in Figure 14.3). The points connected by the vectors are general and have no special properties. (b) The Wigner–Seitz unit cell corresponding to that in Figure 14.3. The (genuine) primitive translation vectors, multiples of which generate the entire crystal from this unit cell, have been added (that out-of-the-plane has been shown as in Figure 14.3). As required, these are the same as those shown in (a). So that the construction of this unit cell can be followed, equivalent points are shown as black balls and the scale has been reduced compared to that in (a). The edges of the Wigner–Seitz cell bisect the lines drawn from the central point to the surrounding points. Each edge is perpendicular to the line that it bisects

relationship between all points is to be recognized. Hence the doubling. The 'extras' are indicated by the fact that they contain extra translation components compared with their genuine counterparts. So, in a unit cell analysis the 2_1 and glide containing a and c translation components are discarded. It is the position in space of the 'extra' centres of symmetry which contains their translation component (this aspect was discussed in Chapter 13; the reason that the twofold rotation axes – the full arrowheads – are at $1/4c$ was also covered there).

A unit cell model requires a unit cell. As has been emphasized many times, there is no unique choice. Two, both shaded, are given in Figures 14.4a and 14.4b (which should be thought of as cross-sections of three-dimensional unit cells). That in Figure 14.4a is perhaps the more obvious, being a rectangular block from the crystallographic unit cell of the *International Tables*. That shown in Figure 14.4b is the Wigner–Seitz unit cell, which for some purposes has advantages (to show that it *is* a Wigner–Seitz unit cell it is drawn with a smaller scale so that adjacent equivalent points can be shown). In both parts of Figure 14.4 the primitive translation vectors in the plane of the paper are shown; they are identical in the two parts, as they have to be, and are non-orthogonal (not at 90°) – in contrast to the crystallographically preferred choice of axes (shown in Figure 14.3).

In Figure 14.5 are shown sets of $M(C{\equiv}O)_3$ groups in the unit cell of Figure 14.4a (there seems to be no actual species with data that enable the discussion to be of a real-life example). The four sets of $M(C{\equiv}O)_3$ groups, labelled α, β, γ and δ, are interrelated by the operations of the unit cell group; the ability to carry out these conversions is at the heart of the method. The results of the operations (applied to α) are:

$C2/c$	E	2	i	c
	α	β	γ	δ

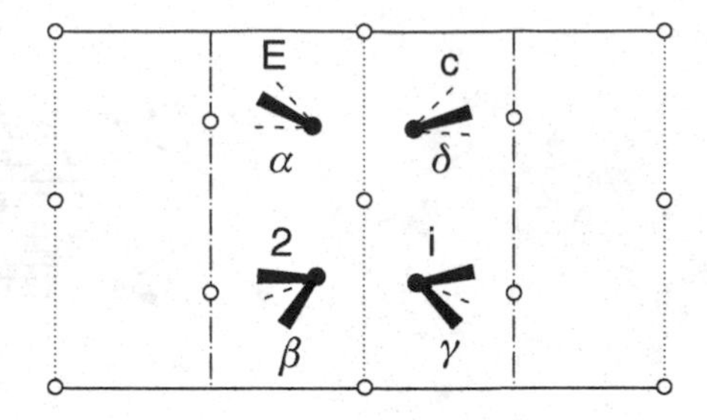

Figure 14.5 The interconversion of four sets of $M(C{\equiv}O)_3$ groups in the primitive unit cell of Figure 14.4a. The relevant operations of Figure 14.3 are used (here, *i* is used to denote inversion in a centre of symmetry and *c* to indicate the *c* glide).

Of these, only the operation of the *c* glide presents any difficulty. The *c* axis is perpendicular to the plane of the paper, so the *c* glide operation involves reflection in a mirror plane (corresponding to the mirror plane reflection of C_{2h}) followed by a translation of *c*/2 perpendicular to the plane of the paper. The result of this *c*/2 translation depends on the choice of direction of translation, down or up. One of these will lead to the generation of a $M(C{\equiv}O)_3$ group within the unit cell, and the other to the generation of a $M(C{\equiv}O)_3$ group in the adjacent unit cell. In the unit cell group, the closure requirement (Appendix 1) is achieved by the $M(C{\equiv}O)_3$ group which 'goes out' of the unit cell and 'comes back in' through the opposite face (Figure 14.6, where *c* is in the plane of the paper). That is, the unit cell group is defined so that it does not matter whether the translation of *c*/2 in the c glide is 'up' or 'down', they lead to the same result. A similar situation holds for 2_1 screw axes (for 3_1 and similar screw axes the situation is a little more complicated[8]). The character table for the C_{2h}^6 unit cell group is obtained from that for C_{2h} using the correspondences:

C_{2h}	E	C_2	i	σ_h
$C2/c(C_{2h}^6)$	E	2	i	c

and is:

Table 14.2

C_{2h}^6	E	2	i	c		
A_g	1	1	1	1	R_z	$x^2;y^2;z^2;xy$
B_g	1	−1	1	−1	$R_x;R_y$	$yz;zx$
A_u	1	1	−1	−1	T_z	z
B_u	1	−1	−1	1	$T_x;T_y$	$x;y$

[8] If a 3_1 operation 'takes a point out' of the unit cell, then its reappearance through the opposite face means that it is equivalent to $(3_{-1})^2$. This is perhaps most readily seen by analogy with the point group relationship $C_3^+ \equiv (C_3^-)^2$, contained in Table 9.2.

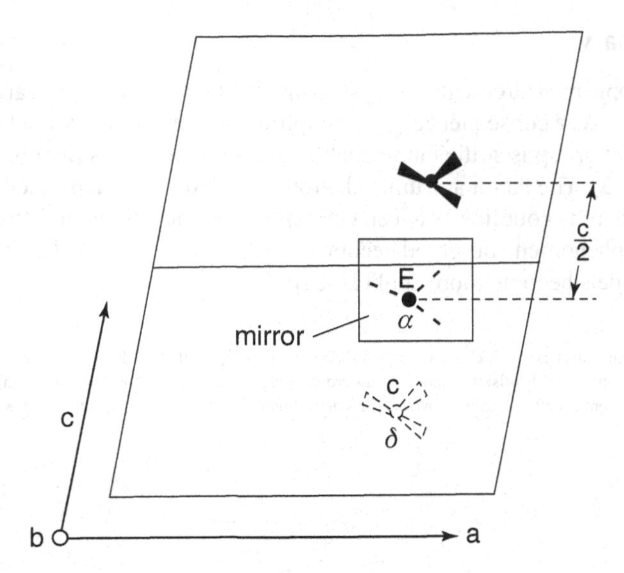

Figure 14.6 The c glide operation in the unit cell group. The 'starting' molecule (centre) is first reflected in a mirror plane (shown as a square around the molecule; this mirror plane is perpendicular to the 2 (C_2) axis, as required in the point group C_{2h}). This reflection is followed by a c/2 translation to complete the c glide operation. If the $c/2$ translation is upwards (to give the black molecule) then this molecule 'reappears' in the original unit cell (to give the molecule shown dotted). Had the $c/2$ translation component been taken in the downwards direction, the dotted molecule would have been that generated without the need to 'come back into' the unit cell

The transformations of the four $M(C{\equiv}O)_3$ groups of Figure 14.5 (and so the corresponding $\nu(C{\equiv}O)$ vibrations) are straightforward and give rise to the reducible representation:

E	2	i	c
12	0	0	0

which has $3A_g + 3B_g + 3A_u + 3B_u$ components, leading to a prediction of six infrared active modes and six Raman active, with no coincidences. Several comments are relevant. Because all of the characters (except that for the identity operation) are 0 no error would have resulted had we, in error, worked with the crystallographic (doubled) unit cell and divided the resulting reducible representation by two. However, if the problem had been one in which there were $M(C{\equiv}O)$ groups lying on a 2 (C_2) axis, then these would have given a non-zero character under this operation. On the other hand, they would have given a zero character under the 2_1 operations which, apparently, exist in the (doubled) unit cell. The reducible representations generated would have depended on just how this (unreal) dilemma was handled. It is scarcely likely that the correct prediction would have been obtained by dividing any of the possible answers by two!

14.4 Summary

Most spectroscopic measurements on crystals involve phenomena which are translationally invariant (p. 321). As a consequence, great simplification results and knowledge of the crystallographic point group is sufficient to enable spectral predictions (and/or interpretations) to be made (p. 333). The factor and unit cell groups lead to identical predictions but are only relevant when there is coupling between corresponding transitions in different molecules[9] (p. 333). As the phenomena observed become increasingly localized, the site symmetry and oriented gas models become more applicable (p. 332).

---

[9] Several alternative names exist which are used to describe the resulting splittings; different names tend to be the province of different areas of spectroscopy. Whilst the names 'factor group' and 'unit cell group' splittings are readily understood because of the content of the present chapter, the alternative names 'correlation field' or 'Davydov' splitting may also be encountered.

Appendix 1 Groups and classes: definitions and examples

A1.1 Groups

In Chapter 2 a definition was given of a group which was adequate for the purposes at that point in the text but which was incomplete; all of the requirements were not detailed. The first object of this appendix is to remedy this deficiency and to accurately define the word 'group'. At some points in this appendix it will implicitly be assumed that the group under discussion does not contain an infinite number of elements. This excludes $C_{\infty v}$ and $D_{\infty h}$ – but all of the general statements made can be shown to apply to these two groups also.

Suppose we have a collection of elements (some examples will be given shortly which will help to indicate the breadth of the term 'element'). The set of these elements, A, B, $C, \ldots$ form a group G, written as

$$G = \{A, B, C, \ldots\} \text{ if}:$$

1 There is some *law of combination* which relates the elements one to another. No matter what the precise nature of the operation of combination, it is called *multiplication*. So, the fact that A combines with B to give C would be written:

$$AB = C \tag{A1.1}$$

At the end of this section several different laws of multiplication will be given to illustrate equation (A1.1).

Note: (a) Whenever a group is specified it is, formally, necessary to also specify the law of combination. (b) The order in which elements multiply is important. There is NO general requirement that, for instance,

$$AB = BA$$

so it must be assumed that, in general,

$$AB \neq BA \tag{A1.2}$$

Symmetry and Structure: Readable Group Theory for Chemists Sidney F. A. Kettle

More detailed consideration of this inequality will lead to the concept of *class* later in this appendix.

2 Multiplication is closed (the *closure requirement*). That is, the product of any two elements within a group is an element within the group.

Note: 'An element' here means a *single* element. Multiplication is single valued; there can never be any ambiguity about the outcome of a multiplication. Thus, it can happen that $AB = C$ and $AB = D$ if and only if

$$C = D$$

3 Multiplication is *associative*. One might think that once the multiplication of two elements is defined there would be no problem about multiplying any number together. This is not the case. Consider the triple product

$$ABC$$

Because we have only defined how to multiply pairs of elements we have to select a pair from this trio and multiply them first. There is a choice between

$$(AB)C \text{ and } A(BC)$$

But suppose $AB = C$ (as above) and $BC = D$. Then our products are

$$CC \text{ and } AD$$

It is by no means evident that these are equal unless this equality is introduced as a requirement. This is just what the statement 'multiplication is associative' does. It means that it must be true that

$$(AB)C = A(BC)$$

for the elements to form a group.

Note: This means that a string of elements can now be multiplied together. Thus,

$$(AB)CD = A(BC)D = AB(CD)$$

4 The group contains a *unit element* (often denoted E or I). This unit element plays a role which in some ways resembles that of the number 1 in ordinary arithmetic. Thus, when it multiplies any other element of the group, A, say, the product is A, i.e.

$$EA = AE = A \tag{A1.3}$$

5 For each element in the group there is a unique element which is its *inverse*. Loosely speaking, the inverse of an element 'undoes' the effect of that element. Thus, C_3^- is the inverse of C_3^+.

The inverse of the element A is usually written A^{-1} (in ordinary arithmetic think of multiplying by, say, the number 7; this multiplication can be cancelled out by multiplying again, this time by the number $7^{-1} = 1/7$), that is,

$$AA^{-1} = A^{-1}A = E \tag{A1.4}$$

Note: The element which has here been called A^{-1} would normally have another label within the group; it could be B, for instance, or it could be A itself if A were self-inverse. The label A^{-1} is here used in preference to, say, B, because the label B does not reveal the special relationship between A and A^{-1} given by the equation above.

Problem A1.1 Apply the relationships given above to the elements of the C_{2v} group (E, C_2, σ_v, σ_v') and thus, formally, show that they comprise a group.

A1.2 Some examples of groups

The multiplication table for the C_{3v} group has already been met – in Table 9.2 and the associated discussion. It is reproduced below because the examples chosen in this section have a similarity with it. The reader could be well advised to check out these similarities and thus to illustrate the meaning of the term 'isomorphous groups'.

Table A1.1

C_{3v}	E	C_3^+	C_3^-	$\sigma_v(1)$	$\sigma_v(2)$	$\sigma_v(3)$
E	E	C_3^+	C_3^-	$\sigma_v(1)$	$\sigma_v(2)$	$\sigma_v(3)$
C_3^+	C_3^+	C_3^-	E	$\sigma_v(2)$	$\sigma_v(3)$	$\sigma_v(1)$
C_3^-	C_3^-	E	C_3^+	$\sigma_v(3)$	$\sigma_v(1)$	$\sigma_v(2)$
$\sigma_v(1)$	$\sigma_v(1)$	$\sigma_v(3)$	$\sigma_v(2)$	E	C_3^+	C_3^-
$\sigma_v(2)$	$\sigma_v(2)$	$\sigma_v(1)$	$\sigma_v(3)$	C_3^+	E	C_3^-
$\sigma_v(3)$	$\sigma_v(3)$	$\sigma_v(2)$	$\sigma_v(1)$	C_3^-	C_3^+	E

1 *Permutation groups.* The groups formed by the operations permuting n objects form a fascinating subject for study. The character table for the so-called 'symmetric group' (the permutation group) with $n = 2$ is isomorphic to that of C_2, that for $n = 3$ is isomorphic to C_{3v} and that for $n = 4$ is isomorphic to T_d. The groups with $n \geq 5$ are not isomorphic to any point group. The symmetric groups are of potential importance when identical particles are of interest. In chemistry these particles could be identical nuclei but more frequently they are electrons.

The symmetric group with $n = 3$ has six operations; to describe them we label the three particles a, b and c. If (a) indicates that a is not permuted, (ab) means 'interchange a and b' and (abc) means cyclically permute a, b and c, then the six operations (and, beneath each, the commonly used shorthand label) are:

$$\begin{array}{cccccc} (a)(b)(c) & (abc) & (acb) & (ab)(c) & (ac)(b) & (a)(bc) \\ E & P_1 & P_2 & X_1 & X_2 & X_3 \end{array}$$

Using the shorthand symbols indicated (P – cyclic Permutation; X = eXchange) the following group multiplication table is obtained.

Problem A1.2 Check that Table A1.2 is correct.

Table A1.2

SECOND OPERATION	FIRST OPERATION					
	E	P_1	P_2	X_1	X_2	X_3
E	E	P_1	P_2	X_1	X_2	X_3
P_1	P_1	P_2	E	X_2	X_3	X_1
P_2	P_2	E	P_1	X_3	X_1	X_2
X_1	X_1	X_3	X_2	E	P_1	P_2
X_2	X_2	X_1	X_3	P_2	E	P_1
X_3	X_3	X_2	X_1	P_1	P_2	E

2 *Substitution groups.* These are fun – and have played an important part in the development of group theory – but do not seem to have any general application. Consider the six functions (which, strictly, should be written $E(x)$, $P(x)$ etc):

$$E = x \qquad P = 1/(1-x) \qquad Q = (x-1)/x$$
$$R = 1/x \qquad S = 1-x \qquad T = x/(x-1)$$

These form a group when the law of combination is substitution as function of a function. Thus,

$$SR = S(1/x) = 1 - (1/x) = (x-1)/x = Q$$

and

$$PT = P[x/(x-1)] = \frac{1}{1-\left(\dfrac{x}{x-1}\right)} 1 = 1 - x = S$$

The multiplication table, given below, is isomorphic to that of the permutation group given above (and to C_{3v}). But the reader should be warned. The isomorphism is not self-evident; work will be required to demonstrate it.

Problem A1.3 Check that Table A1.3 is correct.

Table A1.3

SECOND OPERATION	FIRST OPERATION					
	E	P	Q	R	S	T
E	E	P	Q	R	S	T
P	P	Q	E	T	R	S
0	Q	E	P	S	T	R
R	R	S	T	E	P	Q
S	S	T	R	Q	E	P
T	T	R	S	P	Q	E

3 An example of a two-colour group has been given in the discussion associated with Figure 2.5. There, the changing of a colour was introduced as a component of a symmetry operation. Colour groups are of some importance in chemistry in the context of space groups, although beyond that discussed in Chapters 13 and 14. The operations of space groups have the effect of relating molecules in crystal lattices to one another. But what if the molecules are not quite identical? For instance, the molecules could be atomically identical but have opposite magnetic properties (because the electron spins are arranged in opposite ways, for example). In this case the operation – put colloquially – of 'turn the magnet over' is similar to the 'change the colour' operation; it forms a composite with another symmetry operation to relate not-quite identical objects. *Two-colour* space groups are also known as *black and white* groups or *Shubnikov* groups and can be used to describe such magnetic structures. This is not the end; grey groups (random arrangements of two types of magnetic units over a lattice) and polychromatic groups also exist.

A1.3 The classes of a group

When in the previous section the definition of a group was detailed it was found necessary to recognize that the multiplication of any two elements, A and B, of a group could not be assumed to be *commutative*. That is, it is not generally true that

$$AB = BA$$

(when either A or B is the identity, E, the equation is always true – it is equation (A1.3)). This equation may hold for some pairs of operations within a group but not others (for example, it is true for all pairs of σ_v operations in the C_{3v} point group, but is untrue when a C_3 is combined with a σ_v, see Table A1.1). Groups for which it is true for all pairs of elements are *Abelian* point groups; C_2 (Chapters 2 – 4), D_{2h} (Chapter 5) and C_4 (Chapter 11) are examples of Abelian point groups. In Abelian point groups there are never two elements in the same class. Non-Abelian point groups may have more than one element in each class and so, in giving a more precise meaning to the word 'class', equation A1.2 is a good starting point since it applies to at least some of the operations of non-Abelian groups:

$$AB \neq BA$$

Multiply each side of this equation, on the right, by the operation A^{-1}. This gives:

$$ABA^{-1} \neq BAA^{-1}$$

But $AA^{-1} = E$ (equation (A1.4)) and so $BAA^{-1} = BE = B$ (by equation (A1.3)). That is,

$$ABA^{-1} \neq B$$

The product ABA^{-1} must be equivalent to a single operation in the group. To be general, let us call this single operation D. That is,

$$ABA^{-1} = D \qquad (A1.5)$$

There is a hidden symmetry in equation A1.5. To see this, multiply on the left of each side of the equation by A^{-1} and on the right of each side by A. The result is:

$$A^{-1}(ABA^{-1})A = A^{-1}(D)A$$

Because multiplication is associative, this can be written:

$$(A^{-1}A)B(A^{-1}A) = A^{-1}DA$$

Which, by equation (A1.4), becomes

$$B = A^{-1}DA \qquad \text{(A1.6)}$$

which is to be compared with equation (A1.5). Because of this relationship between B and D they are said to be *conjugate* elements of the group. But A was picked at random in the above development – no restrictions were placed on it. Suppose a different element, C, say, had been chosen in its place? There is no theorem which would require that because

$$ABA^{-1} = D$$

then

$$CBC^{-1} = D$$

Rather, it must be assumed that CBC^{-1} gives yet another element (even if, sometimes, it does not). Consider the case where it does not give D but another element, F, say. So,

$$CBC^{-1} = F \qquad \text{(A1.7)}$$

But the arguments leading up to equation (A1.6) above can be paralleled with a similar development to show from equation (A1.7) that

$$B = C^{-1}FC \qquad \text{(A1.8)}$$

That is, B is conjugate with F as well as with D. Not surprisingly, this sequence requires that F and D are also conjugate elements, as may be shown by combining equations (A1.6) and (A1.8).

$$A^{-1}DA = B = C^{-1}FC$$

Consider the two outer expressions and multiply each on the left by A and on the right by A^{-1}.

$$(AA^{-1})D(AA^{-1}) = (AC^{-1})F(CA^{-1})$$

That is,

$$D = (AC^{-1})F(CA^{-1}) \qquad \text{(A1.9)}$$

Equation (A1.9) is of a form analogous to equations (A1.5), (A1.6), (A1.7) and (A1.8) provided that it can be shown that (AC^{-1}) and (CA^{-1}) are inverses of each other. If they are inverses then they satisfy equation (A1.4) and so they should multiply together to give E. We have:

$$\begin{aligned}(AC^{-1})(CA^{-1}) &= AC^{-1}CA^{-1}\\ &= A(C^{-1}C)\ \ A^{-1}\\ &= A(E)A^{-1}\end{aligned}$$

$$= AA^{-1}$$
$$= E.$$

That is, (AC^{-1}) and (CA^{-1}) are, indeed, inverses. Now AC^{-1} must be equal to a single element of the group; call it H. CA^{-1} must then be H^{-1} so that equation (A1.9) becomes

$$D = HFH^{-1} \tag{A1.10}$$

which is of the form required. We conclude that B, D and F are all conjugate elements and comprise a subset of the set of all the group operations. Each set of conjugate elements in a group forms a *class* of the group. Of course, in a particular group the elements just met may not all be distinct; we could have $C = D$, for instance (and, indeed, even in non-Abelian groups, some classes will contain only a single element).

Formally, then, in order to find all members of a group which are of the same class as B, each element of the group in turn (including B) is taken as A in the expression

$$ABA^{-1}$$

(see equation (A1.5)) and all of the products are collected together. They comprise all the elements which fall in the same class as B.

As an example consider the substitution group given in the previous section and use its multiplication table (Table A1.3). First, from the table the inverse of each element is identified:

Element	Inverse
E	E
P	Q
Q	P
R	R
S	S
T	T

To obtain all elements in the same class as P, work down this list forming the products of the form APA^{-1}, where A and its inverse are obtained from the listing above. The results are given below

$$\begin{aligned} EPE &= P \\ PPQ &= P \\ QPP &= P \\ RPR &= Q \\ SPS &= Q \\ TPT &= Q \end{aligned}$$

It is concluded that P and Q are in the same class (a result which could have been anticipated because they are isomorphous with the C_3^+ and C_3^- operations of C_{3v}).

Problem A1.4 Check the above argument.

As a second example we consider the problem encountered in Chapter 11, that C_4 and C_4^3 are in different classes in the C_4 group. The group multiplication table for the C_4 group is (note its diagonal symmetry):

Table A1.a

C_4	E	C_4	C_2	C_4^3
E	E	C_4	C_2	C_{43}
C_4	C_4	C_2	C_4^3	E
C_2	C_2	C_4^3	E	C_4
C_4^3	C_{43}	E	C_4	C_2

from which it is evident that the inverses are:

Element	Inverse
E	E
C_4	C_4^3
C_2	C_2
C_4^3	C_4

In the class containing C_4 there will be

$$\begin{aligned} EC_4E &= C_4 \\ C_4C_4C_4^3 &= C_4 \\ C_2C_4C_2 &= C_4 \\ C_4^3C_4C_4 &= C_4 \end{aligned}$$

That is, the operation C_4 is in a class of its own. It is easy to similarly show that C_4^3 is in a class of its own, as too is C_2. This shows that C_4 is an Abelian group.

Problem A1.5 Demonstrate that C_4^3 is in a class of its own.

Problem A1.6 Show that the C_{2v} group is an Abelian group.

A1.4 Class algebra

When the C_{3v} character table was introduced in Chapter 7 it was done so in the form

Table A1.b

C_{3v}	E	$2C_3$	$3\sigma_v$
A_1	1	1	1
A_2	1	1	−1
E	2	−1	0

and this and the character tables of all other non-Abelian groups are given in this form in Appendix 3. Why? Why put elements which fall in the same class, such as C_3^+ and C_3^-, together as $2C_3$? Why not write this character table as:

Table A1.c

C_{3v}	E	C_3^+	C_3^-	$\sigma_v(1)$	$\sigma_v(2)$	$\sigma_v(3)$
A_1	1	1	1	1	1	1
A_2	1	1	1	−1	−1	−1
E	2	−1	−1	0	0	0

After all, this is the form in which, effectively, it was used in the projection operator method (Section 7.3). First, we note that not all of the character table orthonormality relationships (Section 6.3) would remain true if this form of character table were used (some columns in the extended character table are identical). There is, however, another and fundamental reason: that there exists a *class algebra*. Take the C_{3v} group as an example. It contains three classes with elements

Class 1		E	
Class 2	C_3^+		C_3^-
Class 3	$\sigma_v(1)$	$\sigma_v(2)$	$\sigma_v(3)$

Express this mathematically, thus:

$$\begin{aligned}\mathbb{C}_1 &= E\\ \mathbb{C}_2 &= \tfrac{1}{2}(C_3^+ + C_3^-)\\ \mathbb{C}_3 &= \tfrac{1}{3}[\sigma_v(1) + \sigma_v(2) + \sigma_v(3)]\end{aligned}$$

These classes can be multiplied together. Thus,

$$\begin{aligned}\mathbb{C}_2\mathbb{C}_2 &= \tfrac{1}{4}(C_3^+ + C_3^-)(C_3^+ + C_3^-)\\ &= \tfrac{1}{4}[C_3^+C_3^+ + C_3^+C_3^- + C_3^-C_3^+ + C_3^-C_3^-]\end{aligned}$$

which, from Table A1.1, is

$$\begin{aligned}&= \tfrac{1}{4}[C_3^- + E + E + C_3^+]\\ &= \tfrac{1}{2}E + \tfrac{1}{4}(C_3^+ + C_3^-)\\ &= \tfrac{1}{2}(C_1 + C_2)\end{aligned}$$

A class multiplication table can thus be compiled and which is easily shown to be:

Table 1.4

C_{3v}	$\mathbb{C}_1$	$\mathbb{C}_2$	$\mathbb{C}_3$
$\mathbb{C}_1$	$\mathbb{C}_1$	$\mathbb{C}_2$	$\mathbb{C}_3$
$\mathbb{C}_2$	$\mathbb{C}_2$	$\tfrac{1}{2}(\mathbb{C}_1 + \mathbb{C}_2)$	$\mathbb{C}_3$
$\mathbb{C}_3$	$\mathbb{C}_3$	$\mathbb{C}_3$	$\tfrac{1}{3}(\mathbb{C}_1 + 2\mathbb{C}_2)$

Problem A1.7 Check that Table A1.4 is correct.

Problem A1.8 Show that the above classes do *not* form a group under the operation of class multiplication. (*Hint*: Refer to the relationships used to define a group at the beginning of this appendix.)

The classes of Abelian groups form groups under class multiplication but this is trivial because the classes are isomorphic to the elements of the Abelian group itself.

Problem A1.9 Check the truth of the above assertion by reference to the C_{2v} point group.

From the class multiplication table given above it is seen that, in general, the product of multiplying two classes together is of the form

$$\mathbb{C}_j\mathbb{C}_i = \sum_k c_k\mathbb{C}_k$$

where the sum k is over all classes and c_k is a coefficient. We now ask what may appear a rather strange question. Is it possible to obtain a linear sum of the classes of the form

$$\mathcal{E} = \sum_j a_j\mathbb{C}_j$$

which has the property that when multiplied by any class, $\mathbb{C}_i$ say, it satisfies an equation of the form

$$\mathbb{C}_i\mathcal{E} = \lambda\mathcal{E}$$

where λ is a number (possibly complex)?

Those with some knowledge of quantum mechanics will recognize this as an eigenvalue equation. The eigenvalues, λ, when determined, lead directly to the characters in the character table (these characters are not the λ's but are related to them by simple, well defined, numerical coefficients). That is, the characters in a character table are intimately related to the classes. This is the reason why character tables are given in the way that they are. Clearly, the mathematics given above can be developed to provide a method for the calculation of character tables. This development will not be given here but the interested reader will find a very readable account in a book by G.G. Hall, *Applied Group Theory*, Longman, Harlow, U.K., 1967.

Appendix 2 Matrix algebra and group theory

This book contains a non-mathematical treatment of what, in fact, is a mathematical subject. The present appendix goes some way towards reinstating the mathematics. However, it cannot claim to be comprehensive – if it were, its length would be very much greater.

A2.1 Matrix algebra and symmetry operations

An array of quantities – often numbers – such as those given below is called a *matrix*

$$\begin{bmatrix} 3 & 2 \\ 4 & -1 \\ 0 & 2 \end{bmatrix} \quad \text{and} \quad \begin{bmatrix} 3 & 2 & -2 \\ 4 & -1 & 0 \\ 0 & 2 & 3 \end{bmatrix}$$

Clearly, matrices can be square – contain the same number of rows as they have columns – or they may be rectangular – the number of rows may be greater or less than the number of columns. Each number or other quantity appearing in a matrix is referred to as a *matrix element.* If represented by an algebraic symbol a matrix element is often given suffixes to indicate in which row and which column it lies in the matrix.

Matrices of the same size may be added; this is done by adding together the corresponding entries (elements). We illustrate this by adding two matrices; as an aid to clarity the elements of one matrix are given as letters

$$\begin{bmatrix} 3 & 2 & -2 \\ 4 & -1 & 0 \\ 0 & 2 & 3 \end{bmatrix} + \begin{bmatrix} a & b & c \\ d & e & f \\ g & h & i \end{bmatrix} = \begin{bmatrix} (3+a) & (2+b) & (-2+c) \\ (4+d) & (-1+e) & f \\ g & (2+h) & (3+i) \end{bmatrix}$$

Problem A2.1 Fill in the missing quantities in the following matrix equation

$$\begin{bmatrix} \sin^2\theta & \frac{1}{\sqrt{2}} & 3 \\ \cdot & \cdot & \cdot \\ 3 & 1 & -\sin^2\theta \end{bmatrix} + \begin{bmatrix} \cdot & \cdot & 0 \\ 1 & \cos^2\phi & -1 \\ -1 & 5 & \cdot \end{bmatrix} = \begin{bmatrix} 1 & \sqrt{2} & \cdot \\ 1 & 0 & -4 \\ \cdot & \cdot & \cos 2\theta \end{bmatrix}$$

Symmetry and Structure: Readable Group Theory for Chemists Sidney F. A. Kettle

The application of matrix algebra to the theory of groups is relatively limited and we shall have no occasion to add or subtract matrices. Key to our use of them, however, is the multiplication of matrices. Matrix multiplication does NOT parallel matrix addition; one does NOT simply multiply corresponding pairs of elements together. Although pairs of elements are, indeed, multiplied, each element in a complete row is multiplied by the corresponding element in a complete column – so that the row and column have to be of equal length, to contain the same number of elements – and the products are added together. It is this sum of products that is an element in the product matrix. To obtain the entry in the mth row and the nth column of the product matrix, the elements in the mth row of the first matrix are multiplied by those in the nth column of the second.

Consider the two matrices which were added above. Now, let us multiply them. The entry at the top left-hand corner of the product matrix, the one in the first row ($m = 1$) and first column ($n = 1$), is given by:

$$\text{first row} \rightarrow \begin{bmatrix} 3 & 2 & -2 \\ \cdot & \cdot & \cdot \\ \cdot & \cdot & \cdot \end{bmatrix} \times \begin{bmatrix} a & \cdot & \cdot \\ d & \cdot & \cdot \\ g & \cdot & \cdot \end{bmatrix} = \begin{bmatrix} (3a + 2d - 2g) & \cdot & \cdot \\ \cdot & \cdot & \cdot \\ \cdot & \cdot & \cdot \end{bmatrix}$$

↑
first column

and the reader who is unfamiliar with matrix multiplication should check several of the elements of this product.

Problem A2.2 Fill in the blanks in the following matrix equation:

$$\begin{bmatrix} 3 & -1 \\ 2 & \cdot \end{bmatrix} \times \begin{bmatrix} 2 & -1 \\ \cdot & 3 \end{bmatrix} = \begin{bmatrix} 0 & \cdot \\ 10 & \cdot \end{bmatrix}$$

The multiplication of two matrices may be expressed algebraically. If the product of the matrices A and B (A being on the left) is denoted AB, then

$$(AB)_{mn} = \sum_t A_{mt} B_{tn} \tag{A2.1}$$

where m and n carry the meanings given above and t is simply a convenient running label which enables us to distinguish the individual matrix element products which have to be added together to give the element in the mth row and nth column of the product matrix AB.

In the example above, the two matrices which were multiplied together were square but this is not a requirement; the sole restriction is the obvious one stated above – that the number of elements in each row of the matrix on the left of the multiplication sign equals the number of elements in each column of the matrix on the right.

The relevance of this to molecular symmetry can be seen by reference to Figure 3.2. This shows the transformation of the hydrogen 1s orbitals, h_1 and h_2, under the symmetry operations of the C_{2v} point group. It was discussed in Section 3.2. Figure 3.2 shows that under the identity operation E, h_1 and h_2 remain unchanged. Let us look at this apparently trivial result in some detail. The fact that h_1 and h_2 each remain unchanged under the operation E can be expressed by the matrix product

$$\begin{bmatrix} 1 & 0 \\ 0 & 1 \end{bmatrix} \begin{bmatrix} h_1 \\ h_2 \end{bmatrix} = \begin{bmatrix} h_1 \\ h_2 \end{bmatrix} \tag{A2.2}$$

where, following convention, the multiplication sign between the two matrices multiplied together has been omitted. Writing them side by side in this way is taken as meaning that they are to be multiplied. The reader should check that, arithmetically, equation (A2.2) is correct. It may be correct, but what does it mean? On the left-hand side the hydrogen 1s orbitals are written as the elements of a column matrix. The order in which they are written is, ultimately, unimportant but that used is clearly the more natural. When the matrix multiplication is carried out this column matrix is regenerated, unchanged, on thre right-hand side of the equation. That is, multiplication by the matrix $\left[\begin{smallmatrix} 1 & 0 \\ 0 & 1 \end{smallmatrix}\right]$ has the same effect on h_1 and h_2 as the identity operation. However, had a different matrix been used to multiply the h_1, h_2 column matrix a different result would have been obtained. It is therefore reasonable to say that the matrix represents the operation E.

Figure 3.2 shows that the C_2 rotation interchanges h_1 and h_2. The reader can readily show that this is expressed by the matrix product

$$\begin{bmatrix} 0 & 1 \\ 1 & 0 \end{bmatrix} \begin{bmatrix} h_1 \\ h_2 \end{bmatrix} = \begin{bmatrix} h_2 \\ h_1 \end{bmatrix}$$

Here, the matrix $\left[\begin{smallmatrix} 0 & 1 \\ 1 & 0 \end{smallmatrix}\right]$ has a similar effect on $\left[\begin{smallmatrix} h_1 \\ h_2 \end{smallmatrix}\right]$ as the C_2 operation has on the h_1 and h_2; h_1 and h_2 are interchanged.

It is left as an exercise for the reader to show that the effects of the σ_v and σ_v' operations on h_1 and h_2 are paralleled in the matrix products:

$$\sigma_v : \begin{bmatrix} 1 & 0 \\ 0 & 1 \end{bmatrix} \begin{bmatrix} h_1 \\ h_2 \end{bmatrix} = \begin{bmatrix} h_1 \\ h_2 \end{bmatrix} \tag{A2.4}$$

$$\sigma_{v'} : \begin{bmatrix} 0 & 1 \\ 1 & 0 \end{bmatrix} \begin{bmatrix} h_1 \\ h_2 \end{bmatrix} = \begin{bmatrix} h_2 \\ h_1 \end{bmatrix} \tag{A2.5}$$

Problem A2.3 Show, by expansion and comparison with Chapter 3, that equations (A2.4) and (A2.5) correctly describe the action of σ_v and σ_v' on h_1 and h_2.

In Chapter 2 it was shown that sets of numbers such as 1, 1, −1, −1 multiply in a manner which is isomorphic to the multiplication of the operations of the C_2 point group (see Table 2.3 and the associated discussion, for example). The important thing about the square matrices in equations (A2.2)–(A2.5) is that when multiplied under the rules of matrix multiplication they, too, multiply isomorphically to the C_2 operations. The multiplication of these 2×2 matrices is given in Table A2.1.

Table A2.1

		right hand matrix in the product			
		E	C_2	σ_v	σ_v'
		$\begin{bmatrix}1&0\\0&1\end{bmatrix}$	$\begin{bmatrix}0&1\\1&0\end{bmatrix}$	$\begin{bmatrix}1&0\\0&1\end{bmatrix}$	$\begin{bmatrix}0&1\\1&0\end{bmatrix}$
left hand matrix in the product	$\begin{bmatrix}1&0\\0&1\end{bmatrix}$	$\begin{bmatrix}1&0\\0&1\end{bmatrix}$	$\begin{bmatrix}0&1\\1&0\end{bmatrix}$	$\begin{bmatrix}1&0\\0&1\end{bmatrix}$	$\begin{bmatrix}0&1\\1&0\end{bmatrix}$
	$\begin{bmatrix}0&1\\1&0\end{bmatrix}$	$\begin{bmatrix}0&1\\1&0\end{bmatrix}$	$\begin{bmatrix}1&0\\0&1\end{bmatrix}$	$\begin{bmatrix}0&1\\1&0\end{bmatrix}$	$\begin{bmatrix}1&0\\0&1\end{bmatrix}$
	$\begin{bmatrix}1&0\\0&1\end{bmatrix}$	$\begin{bmatrix}1&0\\0&1\end{bmatrix}$	$\begin{bmatrix}0&1\\1&0\end{bmatrix}$	$\begin{bmatrix}1&0\\0&1\end{bmatrix}$	$\begin{bmatrix}0&1\\1&0\end{bmatrix}$
	$\begin{bmatrix}0&1\\1&0\end{bmatrix}$	$\begin{bmatrix}0&1\\1&0\end{bmatrix}$	$\begin{bmatrix}1&0\\0&1\end{bmatrix}$	$\begin{bmatrix}0&1\\1&0\end{bmatrix}$	$\begin{bmatrix}1&0\\0&1\end{bmatrix}$

Problem A2.4 Check that Table A2.1 is correct.

Table A2.1 should be compared with Table 2.1. Each matrix in Table A2.1 will be found to transform isomorphically to the operation associated with it. Is this property limited to 2×2 matrices? No, provided that they are square matrices, matrices of any order can be found which multiply isomorphically to the operations of the C_{2v} point group. Indeed, the numbers which behaved like this in Chapter 2 may be regarded as 1×1 matrices! As an example of this, the following four matrices describe the transformations of the hydrogen atoms

$$\begin{bmatrix}H_a\\H_b\\H_c\\H_d\end{bmatrix}$$

of Figure 2.27 (and shown in Figures 2.28 and 2.29).

$$E: \begin{bmatrix}1&0&0&0\\0&1&0&0\\0&0&1&0\\0&0&0&1\end{bmatrix}$$

$$C_2: \begin{bmatrix}0&0&1&0\\0&0&0&1\\1&0&0&0\\0&1&0&0\end{bmatrix}$$

$$\sigma_v : \begin{bmatrix} 0 & 1 & 0 & 0 \\ 1 & 0 & 0 & 0 \\ 0 & 0 & 0 & 1 \\ 0 & 0 & 1 & 0 \end{bmatrix}$$

$$\sigma_v' : \begin{bmatrix} 0 & 0 & 0 & 1 \\ 0 & 0 & 1 & 0 \\ 0 & 1 & 0 & 0 \\ 1 & 0 & 0 & 0 \end{bmatrix}$$

Further, the multiplication of these matrices is isomorphic to that of the corresponding operations of the C_{2v} point group.

Problem A2.5 Show that the above matrices do, indeed, describe the transformations of the hydrogen atoms of Figure 2.27.

Problem A2.6 Show that the multiplication of the above matrices is isomorphous to that of the operations of the C_{2v} point group (it may be helpful to use Figure 2.28 as a check).

Matrix multiplication, then, provides a method of describing in detail the transformation of several objects under the operations of a point group. But in the text – in Section 3.2 – something similar has been described. It was in Section 3.2 that we first used the transformation of several objects under the operations of a point group to obtain reducible representations. Not surprisingly, the two methods – the transformations of several objects and the matrix – are connected. In Section 3.2 what was described was a method of obtaining the characters of reducible representations. The bridge between this and the matrix formalism appears when it is recognized that 'character' is the name given to the arithmetic sum of all of the elements on the leading diagonal (top left to bottom right) of a square matrix. So, application of this definition to the four matrices given immediately above gives their characters as:

	Matrix associated with			
	E	C_2	σ_v	σ_v'
Character of the matrix	4	0	0	0

This set of characters is just that obtained for the reducible representation generated by the transformations of the four hydrogen atoms in Figure 2.27.

Problem A2.7 Check that the transformations of the hydrogen atoms of Figure 2.27 lead to the above representation.

Note: The representation which has the number which is equal to the order of the group (here, 4) under the E operation and zeros elsewhere is called *the regular representation*. It is of importance because it is used in the proof of some group theoretical theorems (but none which are included in this book). It is generated by a basis set which is not associated with any symmetry element. Thus, here, the hydrogen atoms are in general positions – they do not lie on a mirror plane or symmetry axis and so the regular representation is generated.

The rules for the generation of characters given in boxes in Section 3.2 are now seen as arising from the definition of the character of a matrix and the fact that it is only when its transformation is described by an entry on the leading diagonal that an object remains unmoved under a symmetry operation. (A word of caution: this last statement will need some modification shortly when fractions will appear on the diagonal.)

Just as one distinguishes between reducible and irreducible representations, so one may distinguish reducible from irreducible matrix representations. Irreducible matrix representations will be met later in this section and the connection between reducible and irreducible is covered in Section A2.4. Both the 2×2 and 4×4 matrices given above are sets of reducible matrices.

Those functions whose transformations are described by matrices in the way just described are called *basis functions*. Those basis functions given at the right-hand side of character tables (Appendix 3) are ultimately related to the transformation of the x, y and z coordinate axes. It is therefore important to consider the transformation of a set of coordinate axes under typical group symmetry operations. This is not a difficult problem. For example, the inversion operation, *i*, is described by

$$i: \begin{bmatrix} -1 & 0 & 0 \\ 0 & -1 & 0 \\ 0 & 0 & -1 \end{bmatrix} \begin{bmatrix} x \\ y \\ z \end{bmatrix} = \begin{bmatrix} -x \\ -y \\ -z \end{bmatrix}$$

Reflection in a mirror plane (let us choose the yz plane as the mirror plane) is:

$$\sigma(yz): \begin{bmatrix} -1 & 0 & 0 \\ 0 & 1 & 0 \\ 0 & 0 & 1 \end{bmatrix} \begin{bmatrix} x \\ y \\ z \end{bmatrix} = \begin{bmatrix} -x \\ y \\ z \end{bmatrix}$$

The problem of the rotations was first met in Chapter 4 (Figure 4.16) and then again in Chapter 7, where the rotation of axes was discussed (see Figure 7.4 and the related discussion). When the axes x and y are rotated by an angle α around the z axis and are then relabelled x' and y', then, as Figure 7.4 shows,

$$x' = x \cos\alpha + y \sin\alpha$$

Similarly,

$$y' = -x \sin\alpha + y \cos\alpha.$$

We can add, trivially,

$$z' = z.$$

It follows that the matrix describing the effect of a rotation, $R_z(\alpha)$, of an angle α around the z axis is the 3×3 matrix in the middle of equation (A2.6).

$$R_z(\alpha)\begin{bmatrix} x \\ y \\ z \end{bmatrix} = \begin{bmatrix} \cos\alpha & \sin\alpha & 0 \\ -\sin\alpha & \cos\alpha & 0 \\ 0 & 0 & 1 \end{bmatrix}\begin{bmatrix} x \\ y \\ z \end{bmatrix} = \begin{bmatrix} x' \\ y' \\ z' \end{bmatrix} \tag{A2.6}$$

A study of the elements on the leading diagonal of this matrix – those that contribute to the character – will show the basis for the rule given at the end of Section 7.1:

> When an axis is rotated by an angle α its contribution to the character for that operation is $\cos\alpha$.

This relationship enables a more detailed study of the rotation of x and y axes by 45° shown in moving from Figure 6.6 to Figure 6.7 as well as the rotation to give the more general set in Figure 6.8; the following discussion is based on these figures and the reader will have to refer back to them.

It is evident that the character generated by the x and y axes under a C_4 operation is identical for either choice of x and y axes shown in Figures 6.6 and 6.7 (the character is 0 because x and y directions are transposed by the operation). It should also be evident that the same character under this operation is obtained for the more general x and y axes of Figure 6.8 (if it is not evident, use equation (A2.6) suitably adapted to the problem). Less evident is the fact that the general axis set gives the same character as the other sets under improper rotations. Consider the operation of reflection in the $\sigma_v(2)$ mirror plane of Figure 6.3, a mirror plane which is the xz plane in Figure 6.6. All three axis sets give a character of 1 for the z axis. For the (x, y) axis sets of Figure 6.6 and 6.7 characters of 0 are obtained for this reflection operation, but what of the axis set of Figure 6.8? If the angle between y' and the adjacent Br—F bond axis contained in the $\sigma_v(2)$ mirror plane is denoted θ then the relationship between x', y' and their images x'', y'' is found to be (Figure A2.1).

$$x'' = -x'\cos 2\theta - y'\sin 2\theta$$

$$y'' = -x'\sin 2\theta + y'\cos 2\theta$$

That is,

$$\begin{bmatrix} -\cos 2\theta & -\sin 2\theta \\ -\sin 2\theta & \cos 2\theta \end{bmatrix}\begin{bmatrix} x' \\ y' \end{bmatrix} = \begin{bmatrix} x'' \\ y'' \end{bmatrix}$$

Clearly, the character of the 2×2 transformation matrix is 0 ($-\cos 2\theta + \cos 2\theta$), just as was the case for the axis choice of Figures 6.6 and 6.7.

So far, in all of the axis transformations that have been considered the z axis has remained unmoved. If it, too, varies then the problem becomes that of describing the relationship between two generally orientated sets of axes. It is easy to see that rotation by three independent angles about coordinate axes is necessary to describe the relationship between two sets of arbitrarily orientated Cartesian axes. If the first rotation is about z, then x' and y' must remain in the original xy plane. If the second rotation is about x' then this axis will remain in

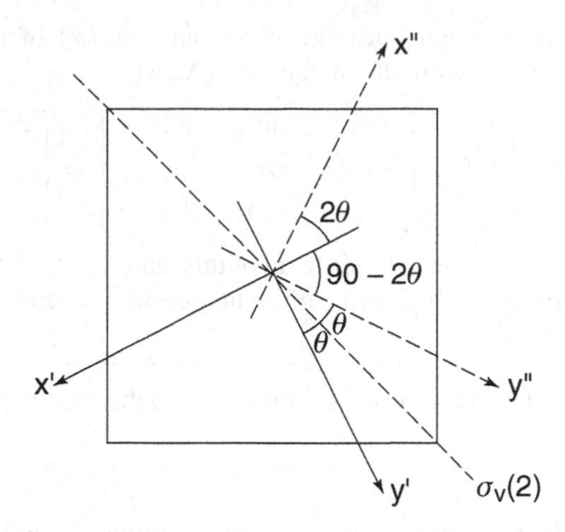

Figure A2.1 The effect of a mirror plane reflection ($\sigma_v(2)$) on x' and y' of Figure 5.6

the original xy plane; it does not assume a general position. Two rotations are not sufficient to place all three original axes in general positions; a third is needed.

The general transformation is shown in Figure A2.2. A rotation by ϕ about z is followed by one of θ about x'. Under this latter rotation z becomes z' and y' becomes y''. The final rotation is one of ψ about z', whereupon x' becomes x'' and y'' becomes y'''. Mathematically, equation (A2.6) is applied to each of these transformations in succession. Just for the record, the final, rather ugly, result is given in equation (A2.7).

$$\begin{bmatrix} (\cos\psi\cos\phi - \cos\theta\sin\phi\sin\psi) & (\cos\psi\sin\phi + \cos\theta\cos\phi\sin\psi) & (\sin\psi\sin\theta) \\ (-\sin\psi\cos\phi - \cos\theta\sin\phi\cos\psi) & (-\sin\psi\sin\phi + \cos\theta\cos\phi\cos\psi) & (\cos\psi\sin\theta) \\ (\sin\theta\sin\phi) & (-\sin\theta\cos\phi) & \cos\theta \end{bmatrix} \times \begin{bmatrix} x' \\ y' \\ z' \end{bmatrix} = \begin{bmatrix} x'' \\ y''' \\ z' \end{bmatrix} \quad \text{(A2.7)}$$

Because a set of p orbitals transforms in the same way as the coordinate axes, this relationship is needed to answer the problem left unresolved in Section 3.2, the transformation of a complete set of p orbitals referred to arbitrary axes under the operations of the C_{2v} point group.

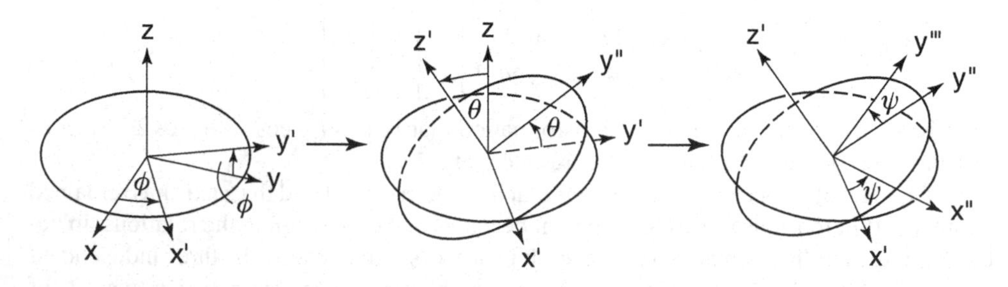

Figure A2.2 The interconversion of two sets of axes (x, y, z) and (x'', y''', z') which are in a general relationship of one set to the other

Table A2.2

C_{3v}	E	C_3^+	C_3^-	$\sigma_v(1)$	$\sigma_v(2)$	$\sigma_v(3)$
A_1	(1)	(1)	(1)	(1)	(1)	(1)
A_2	(1)	(1)	(1)	(−1)	(−1)	(−1)
E	$\begin{bmatrix} 1 & 0 \\ 0 & 1 \end{bmatrix}$	$\begin{bmatrix} -\frac{1}{2} & -\frac{\sqrt{3}}{2} \\ \frac{\sqrt{3}}{2} & -\frac{1}{2} \end{bmatrix}$	$\begin{bmatrix} -\frac{1}{2} & \frac{\sqrt{3}}{2} \\ -\frac{\sqrt{3}}{2} & -\frac{1}{2} \end{bmatrix}$	$\begin{bmatrix} -1 & 0 \\ 0 & 1 \end{bmatrix}$	$\begin{bmatrix} \frac{1}{2} & -\frac{\sqrt{3}}{2} \\ -\frac{\sqrt{3}}{2} & -\frac{1}{2} \end{bmatrix}$	$\begin{bmatrix} \frac{1}{2} & \frac{\sqrt{3}}{2} \\ \frac{\sqrt{3}}{2} & -\frac{1}{2} \end{bmatrix}$

In the section above, the particular concern has been with sets of matrices which usually are reducible representations. However, similar considerations apply to irreducible representations. That is, there exist sets of irreducible matrices for each group. As will be seen in Section A2.4, it is always possible to manipulate a set of matrices which form a basis for a reducible representation in such a way that they can be re-written as a sum of the irreducible matrices.

As an example of the irreducible matrix representations of a group, those for the C_{3v} point group are given in Table A2.2. This table should be compared with the C_{3v} character table given in Table 7.1 (and also in Appendix 3). Comparison of Table A2.2 with the C_{3v} character table reveals two important things. First, whereas individual operations are listed separately in Table A2.2, in the character table they are grouped into classes. Second, for a given irreducible representation, the irreducible matrices of all operations in any one class have the same character and this is the character listed for the class in the character table. The *rapprochement* between the 'individual operation' and 'classes' presentations is provided by the class algebra which was introduced in Appendix A1.4. There are some applications of group theory where it is necessary to use the complete matrix representations of groups.

A2.2 Direct products

In the main text three different uses of the phrase 'direct product' were met. First, the operations of some groups were said to be the direct product of the operations of two other groups. An example is the D_{2h} group, discussed in Section 5.3, which should be briefly reviewed before proceeding. Each individual operation of the D_{2h} point group may be regarded as a product of an individual operation of the D_2 group with an individual operation of the C_i group. For such cases the character table of the product group was also said to be the direct product of those of the other two groups. The phrase 'direct product' was also used to describe the multiplication together of two representations of a group, a topic which was discussed at some length in Chapter 4 (where, in Section 4.3, the symbol $\otimes$ was used to denote this particular application of the direct product). Clearly, the concept of a direct product is one of wide applicability in group theory; it also is an important one.

At the end of the previous section it was noted that there exists a close connection between the characters in a character table and sets of irreducible matrices. Just like the case of the (reducible) matrix representations which were discussed in that section in some detail, the multiplication of irreducible matrices is isomorphic to the multiplication of the group operations. Because of this isomorphism and because direct products of group operations can be formed, we would expect there to be, correspondingly, a direct product of matrices. At the beginning of the previous section one way of multiplying two matrices was described; but this was such that the size of the product matrix is often the same as the size of the matrices from which it was formed (e.g. when square matrices are multiplied together). A characteristic of direct products is that an *increase* in size is the norm – the D_{2h} group is larger than D_2 and C_2. This suggests that the direct product of the matrices involves a second form of matrix multiplication. This is not a unique situation. For instance, there are two different ways of combining – multiplying – vectors together.

The *direct product of two matrices* is obtained by individually and separately multiplying every element of each of the two matrices together. Thus, the direct product of the matrices

$$\begin{bmatrix} 3 & 2 \\ 4 & -1 \end{bmatrix} \quad \text{and} \quad \begin{bmatrix} a & b \\ c & d \end{bmatrix} \quad \text{is}$$

$$\begin{bmatrix} 3a & 3b & 2a & 2b \\ 3d & 3e & 2d & 2e \\ 4a & 4b & -a & -b \\ 4d & 4e & -d & -e \end{bmatrix}$$

If a general element of the matrix A is a_{ij} (i labelling the row and j the column in which a_{ij} occurs in A) and a typical element of B is b_{km} (k'th row, m'th column), then the general element of the matrix C which is the direct product of A and B is:

$$a_{ij} \cdot b_{km} = c_{ij,km} \tag{A2.8}$$

Of course, the general element $c_{ij,km}$ could simply be labelled according to the row and column in which it occurs in C. To do this, however, would be to lose sight of its origins; the more explicit, although apparently more unwieldy, expression in equation (A2.8) is therefore preferred. An example of the reason for this follows.

For the matrices themselves,

$$A \otimes B = C \tag{A2.9}$$

where $\otimes$ again indicates that a direct product is being formed.

There are several ways in which the matrix C may be written; a convenient one is

$$C = \begin{bmatrix} a_{11}B & a_{12}B & \cdot & \cdot & \cdot \\ a_{21}B & a_{22}B & \cdot & \cdot & \cdot \\ a_{31}B & a_{32}B & \cdot & \cdot & \cdot \\ \cdot & \cdot & \cdot & \cdot & \cdot \\ \cdot & \cdot & \cdot & \cdot & \cdot \end{bmatrix} \tag{A2.10}$$

where $a_{11}B$ means that each element of the matrix B is multiplied, in order, by a_{11}.

Table A2.3

	Operations of C_{3v}
Operations of C_i	Operations of D_{3d}

	E	C_3^+	C_3^-	$\sigma_v(1)$	$\sigma_v(2)$	$\sigma_v(3)$
E	E	C_3^+	C_3^-	$\sigma_v(1)$	$\sigma_v(2)$	$\sigma_v(3)$
i	i	S_6^-	S_6^+	$C_2(1)$	$C_2(2)$	$C_2(3)$

Problem A2.10 Fill in the blanks in the following matrix equation (it may be helpful to regard the 4 × 4 matrix as consisting of four 2 × 2's, corresponding to B above.

$$\begin{bmatrix} 1 & 2 \\ 0 & \cdot \end{bmatrix} \otimes \begin{bmatrix} \cdot & \cdot \\ -1 & \cdot \end{bmatrix} = \begin{bmatrix} 3 & 6 & 0 & 0 \\ \cdot & -3 & 0 & 0 \\ \cdot & \cdot & -4 & -8 \\ \cdot & \cdot & \cdot & \cdot \end{bmatrix}$$

The elements of the group D_{3d} are formed as the direct product of the elements of the C_{3v} group with the elements of the C_i group. The relationship between the operations of D_{3d}, C_{3v} and C_i is indicated in Table A2.3.

A precisely parallel relationship exists between the irreducible matrix representations of the D_{3d}, C_{3v} and C_i groups, a relationship detailed below. First, however, note that the group D_{3d} is also the direct product of D_3 with C_i, and it is this latter product which is conventionally taken to determine the labels of the irreducible representations of D_{3d}.

Problem A2.11 Show that $D_{3d} = D_3 \otimes C_i$.

The matrix representations of the C_i group are:

Table A2.4

C_i	E	i
A_g	(1)	(1)
A_u	(1)	(−1)

so that the direct product with the C_{3v} matrix representations given in Table A2.2 leads to the irreducible matrix representations for the D_{3d} group given in Table A2.5. It is entirely reasonable that this isomorphism should exist between multiplication of operations and

Table A2.5

D_{3d}	E	C_3^+	C_3^-	$\sigma_v(1)$	$\sigma_v(2)$	$\sigma_v(3)$	i	S_6^-	S_6^+	$C_2(1)$	$C_2(2)$	$C_2(3)$
A_{1g}	(1)	(1)	(1)	(1)	(1)	(1)	(1)	(1)	(1)	(1)	(1)	(1)
A_{2g}	(1)	(1)	(1)	(−1)	(−1)	(−1)	(1)	(1)	(1)	(−1)	(−1)	(−1)
E_g	$\begin{bmatrix} 1 & 0 \\ 0 & 1 \end{bmatrix}$	$\begin{bmatrix} -\frac{1}{2} & -\frac{\sqrt{3}}{2} \\ \frac{\sqrt{3}}{2} & -\frac{1}{2} \end{bmatrix}$	$\begin{bmatrix} -\frac{1}{2} & \frac{\sqrt{3}}{2} \\ -\frac{\sqrt{3}}{2} & -\frac{1}{2} \end{bmatrix}$	$\begin{bmatrix} -1 & 0 \\ 0 & 1 \end{bmatrix}$	$\begin{bmatrix} \frac{1}{2} & -\frac{\sqrt{3}}{2} \\ -\frac{\sqrt{3}}{2} & -\frac{1}{2} \end{bmatrix}$	$\begin{bmatrix} \frac{1}{2} & \frac{\sqrt{3}}{2} \\ \frac{\sqrt{3}}{2} & -\frac{1}{2} \end{bmatrix}$	$\begin{bmatrix} 1 & 0 \\ 0 & 1 \end{bmatrix}$	$\begin{bmatrix} -\frac{1}{2} & -\frac{\sqrt{3}}{2} \\ \frac{\sqrt{3}}{2} & -\frac{1}{2} \end{bmatrix}$	$\begin{bmatrix} -\frac{1}{2} & \frac{\sqrt{3}}{2} \\ -\frac{\sqrt{3}}{2} & -\frac{1}{2} \end{bmatrix}$	$\begin{bmatrix} -1 & 0 \\ 0 & 1 \end{bmatrix}$	$\begin{bmatrix} \frac{1}{2} & -\frac{\sqrt{3}}{2} \\ -\frac{\sqrt{3}}{2} & -\frac{1}{2} \end{bmatrix}$	$\begin{bmatrix} \frac{1}{2} & \frac{\sqrt{3}}{2} \\ \frac{\sqrt{3}}{2} & -\frac{1}{2} \end{bmatrix}$
A_{1u}	(1)	(1)	(1)	(1)	(1)	(1)	(−1)	(−1)	(−1)	(−1)	(−1)	(−1)
A_{2u}	(1)	(1)	(1)	(−1)	(−1)	(−1)	(−1)	(−1)	(−1)	(1)	(1)	(1)
E_u	$\begin{bmatrix} 1 & 0 \\ 0 & 1 \end{bmatrix}$	$\begin{bmatrix} -\frac{1}{2} & -\frac{\sqrt{3}}{2} \\ \frac{\sqrt{3}}{2} & -\frac{1}{2} \end{bmatrix}$	$\begin{bmatrix} -\frac{1}{2} & \frac{\sqrt{3}}{2} \\ -\frac{\sqrt{3}}{2} & -\frac{1}{2} \end{bmatrix}$	$\begin{bmatrix} -1 & 0 \\ 0 & 1 \end{bmatrix}$	$\begin{bmatrix} \frac{1}{2} & -\frac{\sqrt{3}}{2} \\ -\frac{\sqrt{3}}{2} & -\frac{1}{2} \end{bmatrix}$	$\begin{bmatrix} \frac{1}{2} & \frac{\sqrt{3}}{2} \\ \frac{\sqrt{3}}{2} & -\frac{1}{2} \end{bmatrix}$	$\begin{bmatrix} -1 & 0 \\ 0 & -1 \end{bmatrix}$	$\begin{bmatrix} \frac{1}{2} & \frac{\sqrt{3}}{2} \\ -\frac{\sqrt{3}}{2} & \frac{1}{2} \end{bmatrix}$	$\begin{bmatrix} \frac{1}{2} & -\frac{\sqrt{3}}{2} \\ \frac{\sqrt{3}}{2} & \frac{1}{2} \end{bmatrix}$	$\begin{bmatrix} 1 & 0 \\ 0 & -1 \end{bmatrix}$	$\begin{bmatrix} -\frac{1}{2} & \frac{\sqrt{3}}{2} \\ \frac{\sqrt{3}}{2} & \frac{1}{2} \end{bmatrix}$	$\begin{bmatrix} -\frac{1}{2} & -\frac{\sqrt{3}}{2} \\ -\frac{\sqrt{3}}{2} & \frac{1}{2} \end{bmatrix}$

multiplication of matrices in this application of the direct product. The isomorphism exists in each of the individual groups involved in the direct product.

Problem A2.12 Show that $D_{3d} = C_{3v} \otimes C_i$ (see above); then check Table 2.3.

The definition of a direct product given by equations (A2.8) and (A2.9) and the convention given by equation (A2.10) is, of course, also applicable to the direct products formed between two representations of the same group. Thus the direct product matrices $A_2 \otimes E$ of the C_{3v} point group are (from Table A2.2)

$$A_2 \otimes E =$$

$$\begin{matrix} E & C_3^+ & C_3^- & \sigma_v(1) & \sigma_v(2) & \sigma_v(3) \\ \begin{bmatrix} 1 & 0 \\ 0 & 1 \end{bmatrix} & \begin{bmatrix} -\frac{1}{2} & -\frac{\sqrt{3}}{2} \\ \frac{\sqrt{3}}{2} & -\frac{1}{2} \end{bmatrix} & \begin{bmatrix} -\frac{1}{2} & \frac{\sqrt{3}}{2} \\ -\frac{\sqrt{3}}{2} & \frac{-1}{2} \end{bmatrix} & \begin{bmatrix} 1 & 0 \\ 0 & -1 \end{bmatrix} & \begin{bmatrix} -\frac{1}{2} & \frac{\sqrt{3}}{2} \\ \frac{\sqrt{3}}{2} & \frac{1}{2} \end{bmatrix} & \begin{bmatrix} \frac{1}{2} & -\frac{\sqrt{3}}{2} \\ -\frac{\sqrt{3}}{2} & \frac{1}{2} \end{bmatrix} \end{matrix}$$

Examination of these matrices shows that the characters of the direct product matrices are the same as those obtained by the method described in the text – using a character table and multiplying the characters of the two irreducible representations together. The technique of multiplying characters to obtain direct product characters, although simple, ignores the subtle changes that have taken place in the corresponding matrices, particularly in those representing the σ_v operations (for which multiplying by the character 0 might well have appeared trivial).

Problem A2.13 Using the C_{3v} character table, form the direct product of the A_2 and E irreducible representations; compare the answer with the matrix form given above.

As a final example we consider the direct product $E \otimes E$ in C_{3v} but confine the discussion to just two of the product matrices. The two direct products which will be evaluated are those direct product matrices corresponding to $\sigma_v(1)$ and C_3^+ which arise from the $E \otimes E$ direct product. For the first of these, expression in the form given by equation (A2.10) leads to

$$\left\{ \begin{matrix} -1\begin{bmatrix} -1 & 0 \\ 0 & 1 \end{bmatrix} & 0\begin{bmatrix} -1 & 0 \\ 0 & 1 \end{bmatrix} \\ 0\begin{bmatrix} -1 & 0 \\ 0 & 1 \end{bmatrix} & 1\begin{bmatrix} -1 & 0 \\ 0 & 1 \end{bmatrix} \end{matrix} \right\}$$

which, on expansion gives

$$\begin{bmatrix} 1 & 0 & 0 & 0 \\ 0 & -1 & 0 & 0 \\ 0 & 0 & -1 & 0 \\ 0 & 0 & 0 & 1 \end{bmatrix}$$

a matrix with a character of 0, the same character as obtained working with the C_{3v} character table.

The second direct product, that corresponding to C_3^+, involves more work. In the form of equation (A2.10) the product is:

$$\begin{bmatrix} -\frac{1}{2}\begin{bmatrix} -\frac{1}{2} & -\frac{\sqrt{3}}{2} \\ \frac{\sqrt{3}}{2} & -\frac{1}{2} \end{bmatrix} & -\frac{\sqrt{3}}{2}\begin{bmatrix} -\frac{1}{2} & -\frac{\sqrt{3}}{2} \\ \frac{\sqrt{3}}{2} & -\frac{1}{2} \end{bmatrix} \\ \frac{\sqrt{3}}{2}\begin{bmatrix} -\frac{1}{2} & -\frac{\sqrt{3}}{2} \\ \frac{\sqrt{3}}{2} & -\frac{1}{2} \end{bmatrix} & -\frac{1}{2}\begin{bmatrix} -\frac{1}{2} & -\frac{\sqrt{3}}{2} \\ \frac{\sqrt{3}}{2} & -\frac{1}{2} \end{bmatrix} \end{bmatrix}$$

leading to

$$\begin{bmatrix} \frac{1}{4} & \frac{\sqrt{3}}{4} & \frac{\sqrt{3}}{4} & \frac{3}{4} \\ -\frac{\sqrt{3}}{4} & \frac{1}{4} & -\frac{3}{4} & \frac{\sqrt{3}}{4} \\ -\frac{\sqrt{3}}{4} & -\frac{3}{4} & \frac{1}{4} & \frac{\sqrt{3}}{4} \\ \frac{3}{4} & -\frac{\sqrt{3}}{4} & -\frac{\sqrt{3}}{4} & \frac{1}{4} \end{bmatrix}$$

and the expected character of 1.

Because these direct product matrices are 4×4 it is clear that they must describe the transformation of four quantities – basis functions – which must themselves be related to the basis functions for the E irreducible representation. The exploration and exploitation of such relationships is an important aspect of advanced group theory but the full development of this is beyond the scope of the present text, although a start has been made in Section 12.2.

Problem A2.14 Evaluate the direct product matrices of $E \otimes E$ in C_{3v} for the operations C_3^- and $\sigma_v(2)$.

Appendix 3 Character tables of the more important point groups

At the right of each character table in this compilation are given two columns of bases for irreducible representations; Rotations and Translations in the first column are needed for vibrational analyses (see Chapter 4) and for some other forms of spectroscopy; the second column is useful for some spectroscopies and for discussions of molecular bonding. Invariably, x^2, y^2 and z^2 in some combination, or independently, transform as the totally symmetric irreducible representation of a group. It follows that any linear combination of these functions and, in particular, $x^2 + y^2 + z^2 = r^2$ transform under this irreducible representation. The function r^2 is spherically symmetrical and so is associated with the s orbital of an atom.

'*Note*' comments have usually either been repeated where relevant or cross-references given. However, the reader encountering problems should scan the notes for related character tables, where he or she may well find related suggestions.

Whenever possible the direct product nature of a character table has been indicated by divisions within the character table itself. It is often possible to simplify a problem by working in a subgroup instead of the full group and the divisions within the character table in this appendix are intended to facilitate this. However, these divisions can often only be brought about by a rearrangement of the character table found in many books.

1 THE ICOSAHEDRAL GROUPS

I_h: The icosahedral group.

I_h	E	$12C_5$	$12C_5^2$	$20C_3$	$15C_2$	i	$12S_{10}$	$12S_{10}^3$	$20S_6$	15σ		
A_g	1	1	1	1	1	1	1	1	1	1		$x^2+y^2+z^2$
T_{1g}	3	$-2\cos 144$	$-2\cos 72$	0	-1	3	$-2\cos 144$	$-2\cos 72$	0	-1	(R_x, R_y, R_z)	
T_{2g}	3	$-2\cos 72$	$-2\cos 144$	0	-1	3	$-2\cos 72$	$-2\cos 144$	0	-1		
G_g	4	-1	-1	1	0	4	-1	-1	1	0		
H_g	5	0	0	-1	1	5	0	0	-1	1	$\left(\frac{1}{\sqrt{6}}[2z^2-x^2-y^2], \frac{1}{\sqrt{2}}[x^2-y^2], xy, yz, zx\right)$	
A_u	1	1	1	1	1	-1	-1	-1	-1	-1		
T_{1u}	3	$-2\cos 144$	$-2\cos 72$	0	-1	-3	$2\cos 144$	$2\cos 72$	0	1	(T_x, T_y, T_z)	(x, y, z)
T_{2u}	3	$-2\cos 72$	$-2\cos 144$	0	-1	-3	$2\cos 72$	$2\cos 144$	0	1		
G_u	4	-1	-1	1	0	-4	1	1	-1	0		
H_u	5	0	0	-1	1	-5	0	0	1	-1		

Notes: (1) This character table is the direct product of I with C_i, indicated by lines.

(2) When working with groups containing a C_5 axis the following relationships will be found useful.

$$-2\cos 72^\circ = -0.61803 = \frac{1}{2}(1-5^{1/2}) = \alpha$$

$$-2\cos 144^\circ = 1.61803 = \frac{1}{2}(1+5^{1/2}) = \beta$$

$$\alpha^2 = 1+\alpha; \quad \beta^2 = 1+\beta; \quad \alpha\beta = -1; \quad \alpha+\beta = 1$$

(3) The icosahedral groups are the only ones which contain fourfold (G) and fivefold (H) degenerate representations. In some compilations these are given the alternative labels of U (fourfold) and V (fivefold). Both the icosahedral and cubic groups contain triply degenerate representations; in this compilation they are denoted by T; in some others, and more seldom, the symbol F is used.

Example: The icosahedron shown in Figure 7.31.

I: The group of pure rotations of an icosahedron.

I	E	$12C_5$	$12C_5^2$	$20C_3$	$15C_2$		
A	1	1	1	1	1		$x^2+y^2+z^2$
T_1	3	−2 cos 144	−2 cos 72	0	−1	$(R_x, R_y, R_z),(T_x, T_y, T_z)$	(x, y, z)
T_2	3	−2 cos 72	−2 cos 144	0	−1		
G	4	−1	−1	1	0		
H	5	0	0	−1	1		$\left(\frac{1}{\sqrt{6}}[2z^2-x^2-y^2], \frac{1}{\sqrt{2}}[x^2-y^2], xy, yz, zx\right)$

Note: see comments under the I_h group.

Example: The easiest way of obtaining a figure of I symmetry is to take one edge of the icosahedron shown in Figure 7.31 and make it a zigzag (the C_2 axis passing through the mid-point of this edge must be preserved). The pure rotation operations are then used to generate this zigzag in each of the other edges. A figure of I symmetry results. Faces that are planar in the icosahedron become fluted under the above procedure.

2 CUBIC POINT GROUPS

The two most important cubic point groups are O_h and T_d

O_h	E	$8C_3$	$6C_4$	$3C_2$	$6C_2'$	i	$8S_6$	$6S_4$	$3\sigma_h$	$6\sigma_d$		
A_{1g}	1	1	1	1	1	1	1	1	1	1		$x^2+y^2+z^2$
A_{2g}	1	1	−1	1	−1	1	1	−1	1	−1		
E_g	2	−1	0	2	0	2	−1	0	2	0		$\left(\frac{1}{\sqrt{6}}[2z^2-x^2-y^2], \frac{1}{\sqrt{2}}[x^2-y^2]\right)$
T_{1g}	3	0	1	−1	−1	3	0	1	−1	−1	(R_x, R_y, R_z)	
T_{2g}	3	0	−1	−1	1	3	0	−1	−1	1		(xy, yz, zx)
A_{1u}	1	1	1	1	1	−1	−1	−1	−1	−1		
A_{2u}	1	1	−1	1	1	−1	−1	1	−1	1		
E_u	2	−1	0	2	0	−2	1	0	−2	0		
T_{1u}	3	0	1	−1	−1	−3	0	−1	1	1	(T_x, T_y, T_z)	(x, y, z)
T_{2u}	3	0	−1	−1	1	−3	0	1	1	−1		

Notes: (1) In some texts inorganic chemists refer to one set of the d orbitals on a transition metal atom at the centre of an octahedral complex as d_γ (which are e_g above) and d_ϵ (which are t_{2g}).

(2) The O_h group is a direct product of O and C_i. This is indicated by the lines in the character table.

(3) Although the σ_h mirror planes also satisfy the definition for σ_d here and later (for the D_{nh} groups) the h subscript is given precedence.

(4) The $1/\sqrt{6}$ and $1/\sqrt{2}$ in the expressions for e_g basis functions normalize each function to unity. In the text the $1/\sqrt{2}$ on x^2-y^2 was not used (to keep the discussion as simple as possible) and $1/\sqrt{3}$ therefore appeared in front of $2z^2-x^2-y^2$.

Examples: (a) A cube, an octahedron (Figure 7.1). (b) SF_6 (Figure 7.2 when $M = S, L = F$).

T_d	E	$8C_3$	$3C_2$	$6S_4$	$6\sigma_d$		
A_1	1	1	1	1	1		$x^2+y^2+z^2$
A_2	1	1	1	−1	−1		
E	2	−1	2	0	0		$\left(\frac{1}{\sqrt{6}}[2z^2-x^2-y^2], \frac{1}{\sqrt{2}}[x^2-y^2]\right)$
T_1	3	0	−1	1	−1	(R_x, R_y, R_z)	
T_2	3	0	−1	−1	1	(T_x, T_y, T_z)	(x, y, z); (xy, yz, zx)

Examples: (a) A tetrahedron (Figure 7.1). (b) CH_4, P_4, $Ni(CO)_4$, $C(CH_3)_4$ in its most symmetrical configuration.

Other cubic point groups

O : The group of the pure rotation operations of the octahedron.

O	E	$8C_3$	$3C_2$	$6C_4$	$6C_2'$		
A_1	1	1	1	1	1		$x^2+y^2+z^2$
A_2	1	1	1	−1	−1		
E	2	−1	2	0	0		$\left(\frac{1}{\sqrt{6}}[2z^2-x^2-y^2], \frac{1}{\sqrt{2}}[x^2-y^2]\right)$
T_1	3	0	−1	1	−1	$(T_x, T_y, T_z)(R_x, R_y, R_z)$	(x, y, z)
T_2	3	0	−1	−1	1		(xy, yz, zx)

Example: To obtain a figure of O symmetry follow the instructions given under the I character table, replacing 'Figure 7.30' by 'Figure 7.1', 'icosahedron' by 'octahedron' and 'I' by 'O'.

T : The group of the pure rotation operations of the tetrahedron.

T	E	$4C_3$	$4C_3^2$	$3C_2$		
A	1	1	1	1		$x^2+y^2+z^2$
E	$\begin{cases}1\\1\end{cases}$	$\begin{matrix}\varepsilon\\\varepsilon^2\end{matrix}$	$\begin{matrix}\varepsilon^2\\\varepsilon\end{matrix}$	$\begin{matrix}1\\1\end{matrix}$		$\left(\frac{1}{\sqrt{6}}[2z^2-x^2-y^2], \frac{1}{\sqrt{2}}[x^2-y^2]\right)$
T	3	0	0	−1	$(T_x, T_y, T_z)(R_x, R_y, R_z)$	(x, y, z); (xy, yz, zx)

Note: ε $\left(\varepsilon = \exp\left[\frac{2\pi i}{3}\right]\right)$ and ε^2 $\left(\varepsilon^2 = \exp\frac{4\pi i}{3} = \exp\frac{-2\pi i}{3}\right)$ are complex conjugates. See also Chapter 11 and the notes under the C_3 group below.
Example: To obtain a Figure of T symmetry follow the instructions given under the I character table, replacing 'Figure 7.3' by 'Figure 7.1', 'icosahedron' by 'tetrahedron' and 'I' by 'T'.

T_h	E	$4C_3$	$4C_3^2$	$3C_2$	i	$4S_6^5$	$4S_6$	$3\sigma_h$		
A_g	1	1	1	1	1	1	1	1		$x^2+y^2+z^2$
E_g	$\begin{cases}1\\1\end{cases}$	ε ε^2	ε^2 ε	1 1	$\begin{cases}1\\1\end{cases}$	ε ε^2	ε^2 ε	1 1		$\left(\frac{1}{\sqrt{6}}[2z^2-x^2-y^2], \frac{1}{\sqrt{2}}[x^2-y^2]\right)$
T_g	3	0	0	−1	3	0	0	−1	(R_x, R_y, R_z)	(xy, yz, zx)
A_u	1	1	1	1	−1	−1	−1	−1		
E_u	$\begin{cases}1\\1\end{cases}$	ε ε^2	ε^2 ε	1 1	$\begin{cases}-1\\-1\end{cases}$	$-\varepsilon$ $-\varepsilon^2$	$-\varepsilon^2$ $-\varepsilon$	−1 −1		
T_u	3	0	0	−1	−3	0	0	1	(T_x, T_y, T_z)	(x, y, z)

Notes: (1) The T_h character table is the direct product of T with C_i. This is indicated by the lines in the character table.
(2) For the meaning of ε see under the T character table above.

Example: An 'octahedral' complex $[M(OH_2)_6]$ in which all the atoms of *trans* H_2O ligands lie in a σ_h plane. (Figure 10.3)

3 THE GROUPS D_{nh}

A regular, planar polygon with n sides has D_{nh} symmetry. So, an equilateral triangle has D_{3h} symmetry, a square has D_{4h} symmetry. The label D arises because of the presence of twofold axes (Dihedral axes) perpendicular to a C_n axis. There are n of these twofold axes. The subscript h means that all of the groups have a unique mirror plane perpendicular to the C_n axis (if this axis is vertical then the mirror plane is horizontal). Although these groups all have σ_d operations, the σ_h takes precedence in the labelling. To avoid possible confusion with the D_{nd} groups many authors label as σ_v some or all of the σ_d mirror planes.

D_{2h}	E	$C_2(z)$	$C_2(y)$	$C_2(x)$	i	$\sigma(xy)$	$\sigma(zx)$	$\sigma(yz)$		
A_g	1	1	1	1	1	1	1	1		$z^2; x^2; y^2$
B_{1g}	1	1	−1	−1	1	1	−1	−1	R_z	xy
B_{2g}	1	−1	1	−1	1	−1	1	−1	R_y	zx
B_{3g}	1	−1	−1	1	1	−1	−1	1	R_x	yz
A_u	1	1	1	1	−1	−1	−1	−1		xyz
B_{1u}	1	1	−1	−1	−1	−1	1	1	T_x	z
B_{2u}	1	−1	1	−1	−1	1	1	1	T_y	y
B_{3u}	1	−1	−1	1	−1	1	−1	−1	T_z	x

Notes: (1) Because there are three mutually perpendicular C_2 axes the choice of x, y and z is arbitrary. A relabelling of these axes will lead to an interchange of the labels B_1, B_2 and B_3. Similarly, the h, v subscript notation on the mirror planes is unhelpful and so the mirror planes and the corresponding operations are defined by the Cartesian axes that lie in them.

(2) The D_{2h} group is a direct product of D_2 and C_i. This is indicated by the lines in the character table.
Examples: C_2H_4; B_2H_6 (see Chapter 4).

D_{3h}	E	$2C_3$	$3\sigma_v$	σ_h	$2S_3$	$3\sigma_d$		
A_1'	1	1	1	1	1	1		$z^2; x^2+y^2$
A_2'	1	1	−1	1	1	−1	R_z	
E'	2	−1	0	2	−1	0	(T_x, T_y)	$(x, y); \left(\frac{1}{\sqrt{2}}[x^2-y^2], xy\right)$
A_1''	1	1	1	−1	−1	−1		
A_2''	1	1	−1	−1	−1	1	T_z	z
E''	2	−1	0	−2	1	0	(R_y, R_x)	(zx, yz)

Notes: (1) The D_{3h} group is a direct product of D_3 and C_s. This is indicated by the lines in the character table. Irreducible representations symmetric with respect to reflection in the σ_h mirror plane are denoted by ′ while antisymmetry is denoted by ″.

(2) The $1/\sqrt{2}$ factor on (x^2-y^2) as an E' basis function means that, like xy, it is normalized to unity.
Example: A triangular prism (Figure A3.1)

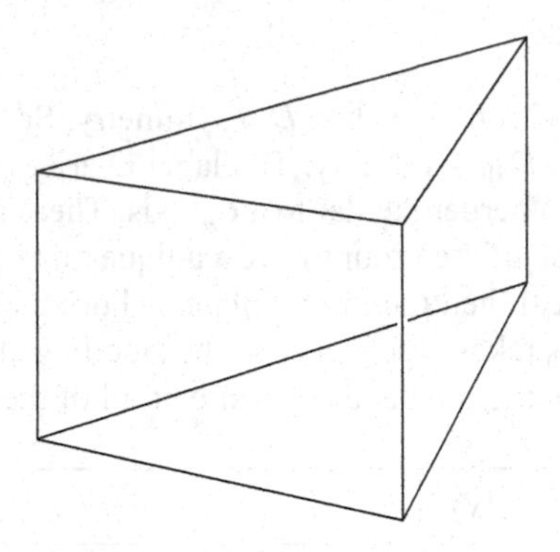

Figure A3.1 A triangular prism, of D_{3h} symmetry.

D_{4h}	E	$2C_4$	C_2	$2C'_2$	$2C''_2$	i	$2S_4$	σ_h	$2\sigma_d$	$2\sigma'_d$		
A_{1g}	1	1	1	1	1	1	1	1	1	1		z^2, x^2+y^2
A_{2g}	1	1	1	−1	−1	1	1	1	−1	−1	R_z	
B_{1g}	1	−1	1	1	−1	1	−1	1	1	−1		x^2-y^2
B_{2g}	1	−1	1	−1	1	1	−1	1	−1	1		xy
E_g	2	0	−2	0	0	2	0	−2	0	0	(R_x, R_y)	(zx, yz)
A_{1u}	1	1	1	1	1	−1	−1	−1	−1	−1		
A_{2u}	1	1	1	−1	−1	−1	−1	−1	1	1	T_z	z
B_{1u}	1	−1	1	1	−1	−1	1	−1	−1	1		
B_{2u}	1	−1	1	−1	1	−1	1	−1	1	−1		
E_u	2	0	−2	0	0	−2	0	2	0	0	(T_x, T_y)	(x, y)

Notes: (1) The D_{4h} group is a direct product of D_4 and C_i. This is indicated by the lines in the character table.

(2) The choice between which pair of C_2 axes (and operations) are labelled C'_2 and those which are labelled C''_2 is arbitrary. A redefinition will interchange B_{1g} with B_{2g} and B_{1u} with B_{2u}. Similarly the choice of the vertical planes σ_d and σ'_d is arbitrary but the σ_d planes must contain the C'_2 axes and the σ_d' planes must contain the C''_2.

(3) In this character table the strict definition of σ_d has been followed but many authors label the mirror planes containing a C'_2 axis as σ_v and those containing a C''_2 axis as σ_d.

Examples: A square prism (Figure A3.2)

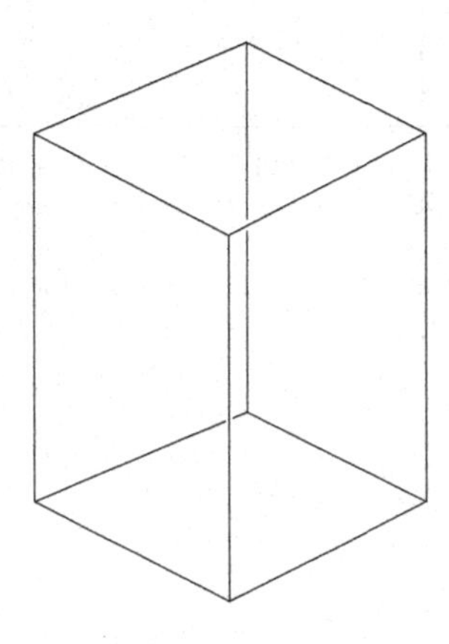

Figure A3.2 A square prism of D_{3h} symmetry.

D_{5h}	E	$2C_5$	$2C_5^2$	$5C_2$	σ_h	$2S_5$	$2S_5^2$	$5\sigma_d$		
A_1'	1	1	1	1	1	1	1	1		$x^2+y^2; z^2$
A_2'	1	1	1	−1	1	1	1	−1	R_z	
E_1'	2	2 cos 72	2 cos 144	0	2	2 cos 72	2 cos 144	0	(T_x, T_y)	(x, y)
E_2'	2	2 cos 144	2 cos 72	0	2	2 cos 144	2 cos 72	0		$\left(\frac{1}{\sqrt{2}}[x^2-y^2], xy\right)$
A_1''	1	1	1	1	−1	−1	−1	−1		
A_2''	1	1	1	−1	−1	−1	−1	1	T_z	z
E_1''	2	2 cos 72	2 cos 144	0	−2	−2 cos 72	−2 cos 144	0	(R_x, R_y)	(xz, yz)
E_2''	2	2 cos 144	2 cos 72	0	−2	−2 cos 144	−2 cos 72	0		

Notes: (1) The D_{5h} group is a direct product of D_5 and C_s. This is indicated by the lines in the character table. Irreducible representations symmetric with respect to reflection in the σ_h mirror plane are denoted by the superscript ′and antisymmetry is denoted by″.

(2) See the notes under the I_h character table.

Examples: A regular pentagonal prism (Figure A3.3) and eclipsed ferrocene (Figure A3.4)

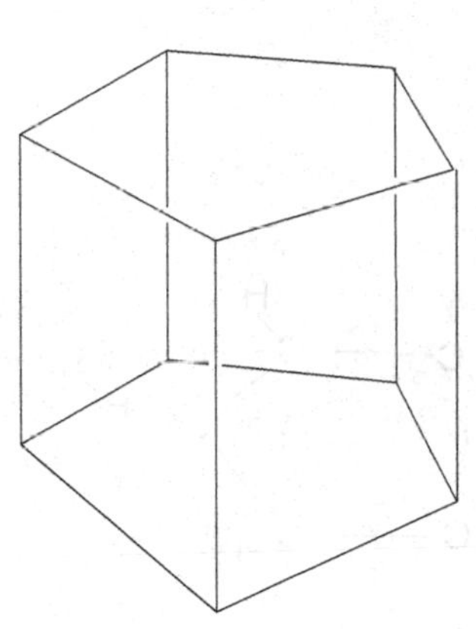

Figure A3.3 A pentagonal prism, of D_{5h} symmetry.

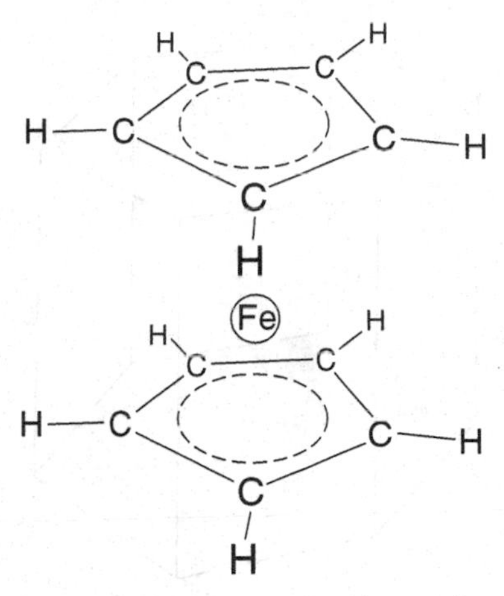

Figure A3.4 Ferrocene, $Fe(C_5H_5)_2$, in the eclipsed configuration, of D_{5h} symmetry.

D_{6h}	E	$2C_6$	$2C_3$	C_2	$3C_2'$	$3C_2''$	i	$2S_3$	$2S_6$	σ_h	$3\sigma_d$	$3\sigma_d'$		
A_{1g}	1	1	1	1	1	1	1	1	1	1	1	1		$x^2+y^2; z^2$
A_{2g}	1	1	1	1	−1	−1	1	1	1	1	−1	−1	R_z	
B_{1g}	1	−1	1	−1	1	−1	1	−1	1	−1	1	−1		
B_{2g}	1	−1	1	−1	−1	1	1	−1	1	−1	−1	1		
E_{1g}	2	1	−1	−2	0	0	2	1	−1	−2	0	0	(R_x, R_y)	(xy, yz)
E_{2g}	2	−1	−1	2	0	0	2	−1	−1	2	0	0		$\left(\frac{1}{\sqrt{2}}[x^2-y^2], xy\right)$
A_{1u}	1	1	1	1	1	1	−1	−1	−1	−1	−1	−1		
A_{2u}	1	1	1	1	−1	−1	−1	−1	−1	−1	1	1	T_z	z
B_{1u}	1	−1	1	−1	1	−1	−1	1	−1	1	−1	1		
B_{2u}	1	−1	1	−1	−1	1	−1	1	−1	1	1	−1		
E_{1u}	2	1	−1	−2	0	0	−2	−1	1	2	0	0	(T_x, T_y)	(x, y)
E_{2u}	2	−1	−1	2	0	0	−2	1	1	−2	0	0		

Notes: (1) The D_{6h} character table is a direct product of D_6 and C_i. This is indicated by the lines in the character table.

(2) The choice between which pair of C_2 axes (and operations) are labelled C_2' and those which are labelled C_2'' is arbitrary. A redefinition will interchange B_{1g} with B_{2g} and B_{1u} with B_{2u}. Similarly the choice of the vertical planes σ_d and σ_d' is arbitrary but the σ_d planes must contain the C_2 axes and the σ_d' planes must contain the C_2''.

Examples: A regular hexagonal prism (Figure A3.5) and the benzene molecule (Figure A3.6)

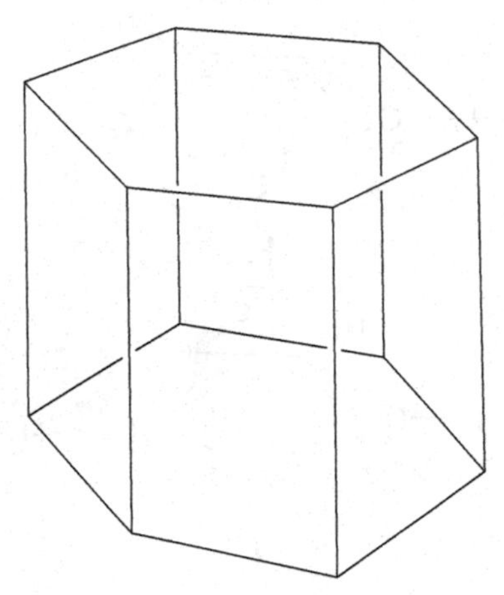

Figure A3.5 A hexagonal prism, of D_{6h} symmetry.

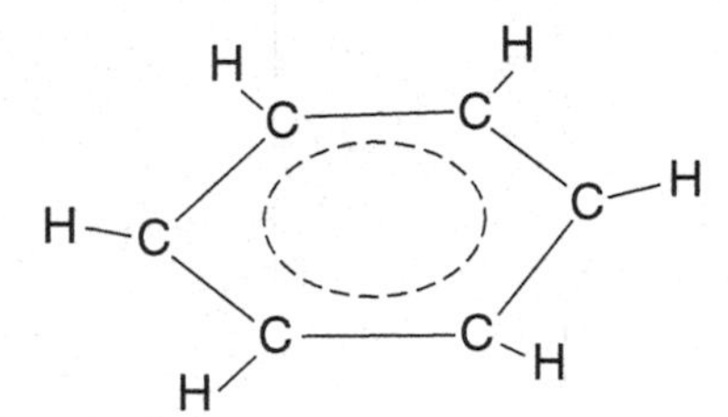

Figure A3.6 The benzene molecule, of D_{6h} symmetry.

4 THE GROUPS D_{nd}

These groups do not have the σ_h mirror plane of the D_{nh} groups. Objects with D_{nd} symmetry typically have two similar halves, staggered with respect to each other. Thus solid objects of D_{nd} symmetry are called 'antiprisms', a term which indicates the staggering.

When n is odd, the groups D_{nd} are direct products of the groups D_n and C_i (when n is even, the direct products $D_n \times C_i$ are D_{nh} groups).

D_{2d}	E	$2S_4$	C_2	$2C_2'$	$2\sigma_d$		
A_1	1	1	1	1	1		$x^2+y^2; z^2$
A_2	1	1	1	−1	−1	R_z	
B_1	1	−1	1	1	−1		x^2-y^2
B_2	1	−1	1	−1	1	T_z	$z; xy$
E	2	0	−2	0	0	$(T_x, T_y); (R_x, R_y)$	$(x, y); (xz, yz)$

Notes: (1) The x axis is taken as coincident with one C_2''.

(2) Singly degenerate irreducible representations which are symmetric with respect to an S_4 operation are indicated by an A label; antisymmetry is indicated by a B label.

(3) This group is sometimes, but increasingly rarely, called V_d,

(4) Many people find the $2C_2'$ axes and $2\sigma_d$ mirror planes difficult to locate in this group. Time spent with the examples would be time well spent.

Examples: A triangular dodecahedron (Figure A3.7) and the molecule spiropentane (Figures A3.8).

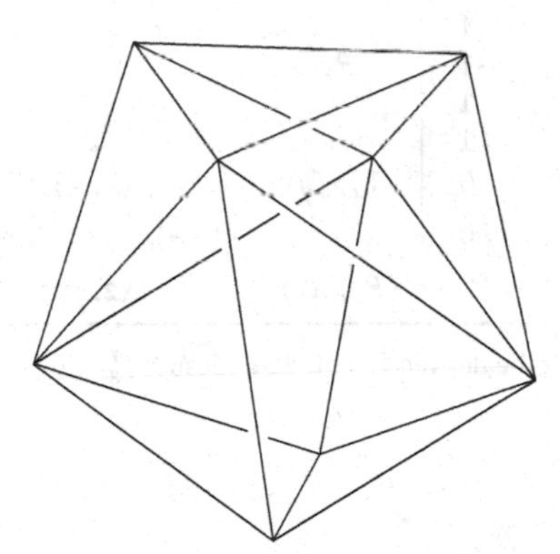

Figure A3.7 The triangular dodecahedron, of D_{2d} symmetry. Note that all of the apices of this figure lie in one of two mutually perpendicular planes.

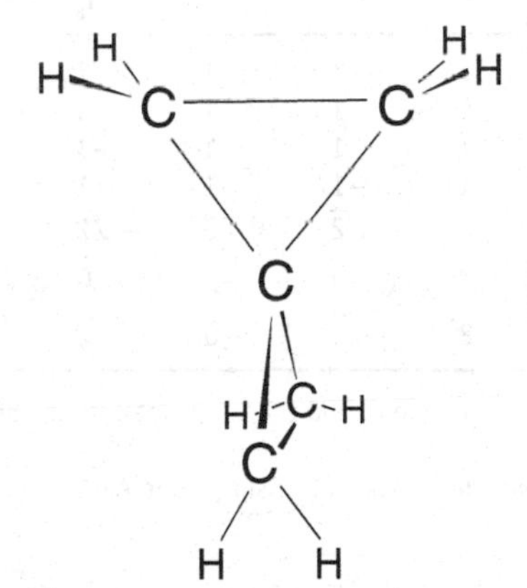

Figure A3.8 The molecular spiropentane, C_5H_8, of D_{2d} symmetry. The 'outer' carton atoms in this figure lie at four of the eight apices of the triangular dodecahedron shown in Figure A3.7.

D_{3d}	E	$3C_3$	$3C_2$	i	$2S_6$	$3\sigma_d$		
A_{1g}	1	1	1	1	1	1		$x^2+y^2; z^2$
A_{2g}	1	1	−1	1	1	−1	R_z	
E_g	2	−1	0	2	−1	0	(R_x, R_y)	$\left(\frac{1}{\sqrt{2}}[x^2-y^2], xy\right); (xz, yz)$
A_{1u}	1	1	1	−1	−1	−1		
A_{2u}	1	1	−1	−1	−1	1	T_z	z
E_u	2	−1	0	−2	1	0	(T_x, T_y)	(x, y)

Notes: (1) This group is a direct product of D_3 with C_i, indicated by the lines in the character table.
(2) The C_3 and i operations may be considered as derived from the S_6 because $S_6^2 = C_3$ and $S_6^3 = i$.
Example: The staggered ethane molecule (Figure A3.9)

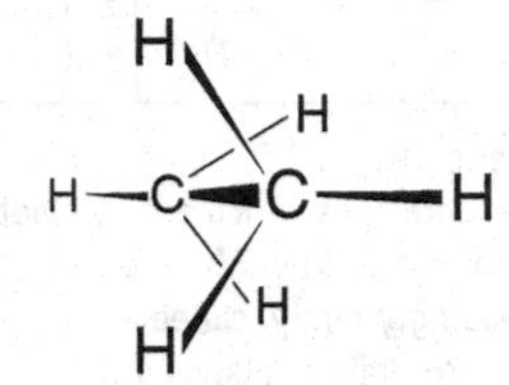

Figure A3.9 The staggered ethane molecule, of D_{3d} symmetry. This symmetry is best seen if the molecule is viewed along the C–C bond but it is difficult to draw it adequately in this orientation.

D_{4d}	E	$2S_8$	$2C_4$	$2S_8^3$	C_2	$4C_2'$	$4\sigma_d$		
A_1	1	1	1	1	1	1	1		$x^2+y^2; z^2$
A_2	1	1	1	1	1	−1	−1	R_z	
B_1	1	−1	1	−1	1	1	−1		
B_2	1	−1	1	−1	1	−1	1	T_z	z
E_1	2	$\sqrt{2}$	0	$-\sqrt{2}$	−2	0	0	(T_x, T_y)	(x, y)
E_2	2	0	−2	0	2	0	0		$\left(\frac{1}{\sqrt{2}}[x^2-y^2], xy\right)$
E_3	2	$-\sqrt{2}$	0	$\sqrt{2}$	−2	0	0	(R_y, R_x)	(xz, yz)

Note: The unique C_2 and the C_4 operations may be considered to be derived from the S_8 because $S_8^2 = C_4$; $S_8^4 = C_2$.
Example: The square antiprism (Figure A3.10)

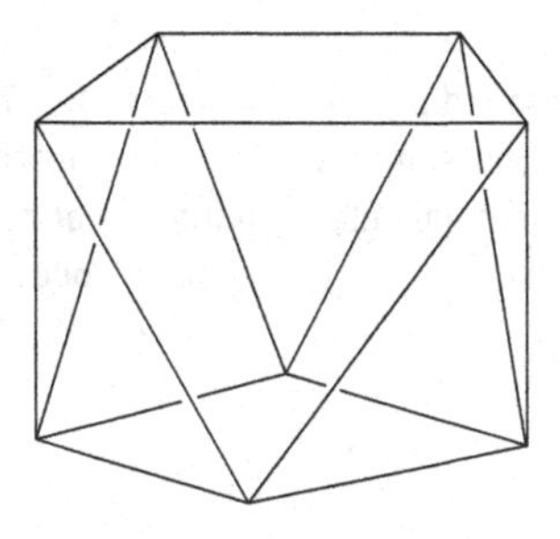

Figure A3.10 The square antiprism, of D_{4d} symmetry.

D_{5d}	E	$2C_5$	$2C_5^2$	$5C_2$	i	$2S_{10}^3$	$2S_{10}$	$5\sigma_d$		
A_{1g}	1	1	1	1	1	1	1	1		$x^2+y^2; z^2$
A_{2g}	1	1	1	−1	1	1	1	−1	R_z	
E_{1g}	2	2 cos 72	2 cos 144	0	2	2 cos 72	2 cos 144	0	(R_x, R_y)	(xz, yz)
E_{2g}	2	2 cos 144	2 cos 72	0	2	2 cos 144	2 cos 72	0		$\left(\frac{1}{\sqrt{2}}[x^2-y^2], xy\right)$
A_{1u}	1	1	1	1	−1	−1	−1	−1		
A_{2u}	1	1	1	−1	−1	−1	−1	1	T_z	z
E_{1u}	2	2 cos 72	2 cos 144	0	−2	−2 cos 72	−2 cos 144	0	(T_x, T_y)	(x, y)
E_{2u}	2	2 cos 144	2 cos 72	0	−2	−2 cos 144	−2 cos 72	0		

Notes: (1) This group is the direct product of C_5 and C_i, indicated by the lines in the character table.
(2) Before working with this group refer to the notes under the I_h character table.
(3) Many of the operations of the group may be considered to be derived from the S_{10} because $S_{10}^2 = C_5; S_{10}^4 = C_5^2; S_{10}^5 = i$.
Examples: The pentagonal antiprism (Figure A3.11), staggered ferrocene (Figure A3.12).

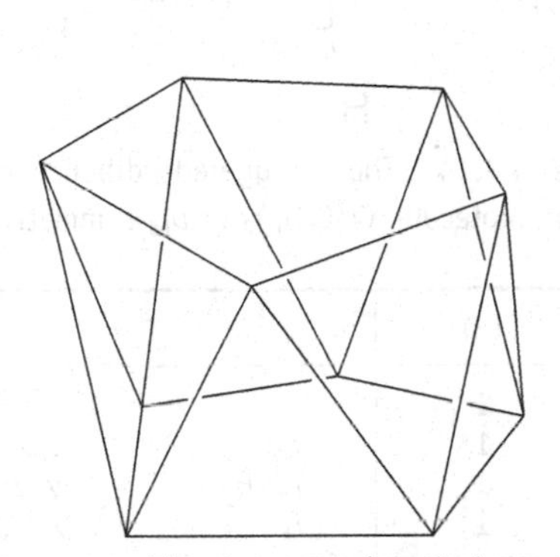

Figure A3.11 The pentagonal antiprism, of D_{5d} symmetry.

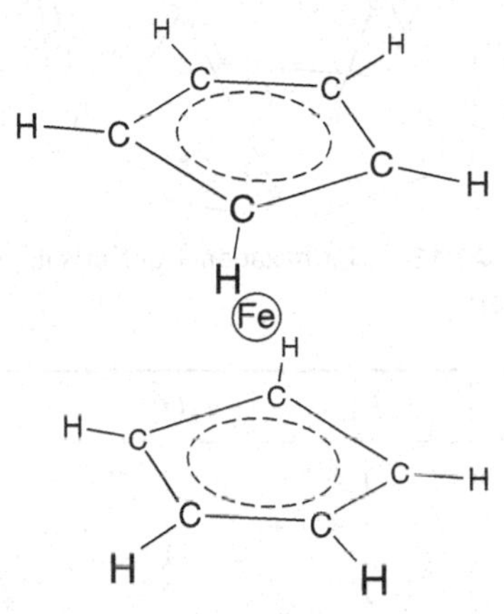

Figure A3.12 The staggered ferrocene molecule, $Fe(C_5H_5)_2$, of D_{5d} symmetry.

D_{6d}	E	$2S_{12}$	$2C_6$	$2S_4$	$2C_3$	$2S_{12}^5$	C_2	$6C_2'$	$6\sigma_d$		
A_1	1	1	1	1	1	1	1	1	1		$x^2+y^2; z^2$
A_2	1	1	1	1	1	1	1	−1	−1	R_z	
B_1	1	−1	1	−1	1	−1	1	1	−1		
B_2	1	−1	1	−1	1	−1	1	−1	1	T_z	z
E_1	2	$\sqrt{3}$	1	0	−1	$-\sqrt{3}$	−2	0	0	(T_x, T_y)	(x, y)
E_2	2	1	−1	−2	−1	1	2	0	0		
E_3	2	0	−2	0	2	0	−2	0	0		
E_4	2	−1	−1	2	1	−1	2	0	0		$\left(\frac{1}{\sqrt{2}}[x^2-y^2], xy\right)$
E_5	2	$-\sqrt{3}$	1	0	−1	$\sqrt{3}$	−2	0	0	(R_y, R_x)	(yz, zx)

Note: Many of the operations of the group may be taken to be derived from S_{12} because $S_{12}^2 = C_6; S_{12}^3 = S_4; S_{12}^4 = C_3; S_{12}^6 = C_2$.
Examples: A hexagonal antiprism (Figure A3.13), staggered dibenzene-chromium (Figure A3.14).

5 THE GROUPS D_n

These are groups of proper (i.e. pure) rotations corresponding to bodies in which there are n C_2 axes perpendicular to a principal C_n axis. To obtain solid figures of these geometries it is simplest to take a polyhedron shown for a D_{nh} or D_{nd} symmetry and to systematically introduce zig-zag edges such as used to derive Figures for the groups O and T. Molecules of D_{nh} or D_{nd} symmetries drop to D_n symmetry when the 'top' and 'bottom' parts of the molecule are given small, arbitrary, twists in opposite directions about the z axis.

D_{nh} and D_{nd} (n odd) group are direct products of D_n with either C_i or C_s. For problems in these groups it is often simplest to work in D_n symmetry and move to the full group at a later stage.

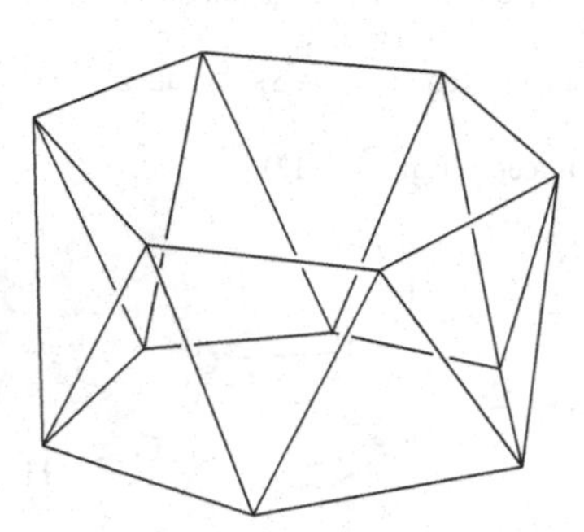

Figure A3.13 The hexagonal antiprism, of D_{6d} symmetry.

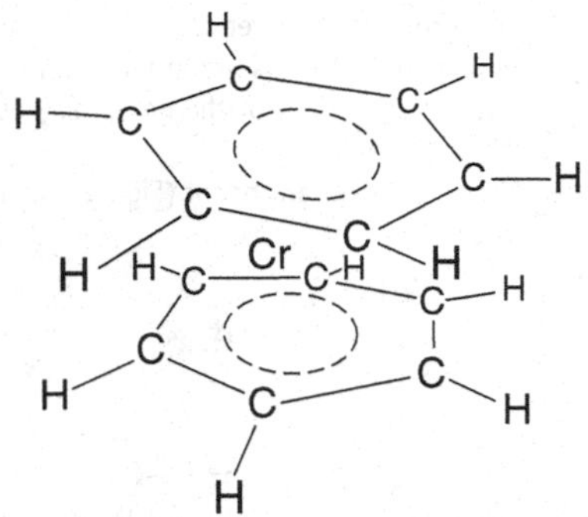

Figure A3.14 The staggered dibenzenechromium molecule, $Cr(C_6H_6)_2$ of D_{6d} symmetry.

D_2	E	$C_2(z)$	$C_2(y)$	$C_2(x)$		
A	1	1	1	1		$x^2; y^2; z^2$
B_1	1	1	−1	−1	$T_z; R_z$	$z; xy$
B_2	1	−1	1	−1	$T_y; R_y$	$y; zx$
B_3	1	−1	−1	1	$T_x; R_x$	$x : yz$

Notes: (1) Because there are three mutually perpendicular C_2 axes the choice of x, y and z is arbitrary. A relabelling of these axes will lead to an interchange of the labels B_1, B_2 and B_3.

(2) Because x^2 and y^2 transform, separately, as A, it follows that the function $x^2 - y^2$ also has A symmetry.

Example: Ethene in which the two CH_2 groups have been made non-coplanar by a counter-rotation of these two units about the C–C axis (Figure A3.15).

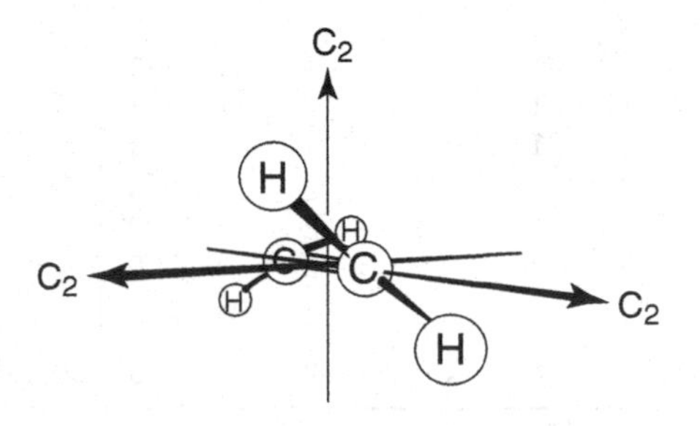

Figure A3.15 A slightly twisted ethene molecule, of D_2 symmetry.

D_3	E	$2C_3$	$3C_2$		
A_1	1	1	1		z^2, x^2+y^2
A_2	1	1	−1	T_z; R_z	z
E	2	−1	0	(T_x, T_y); (R_y, R_x)	(x, y); $\left(\frac{1}{\sqrt{2}}[x^2-y^2], xy\right)$; (zx, yz)

Example: Ethane in which the two CH_3 units have been counter-rotated about the C–C axis so that the molecule is neither eclipsed or staggered (Figure A3.16).

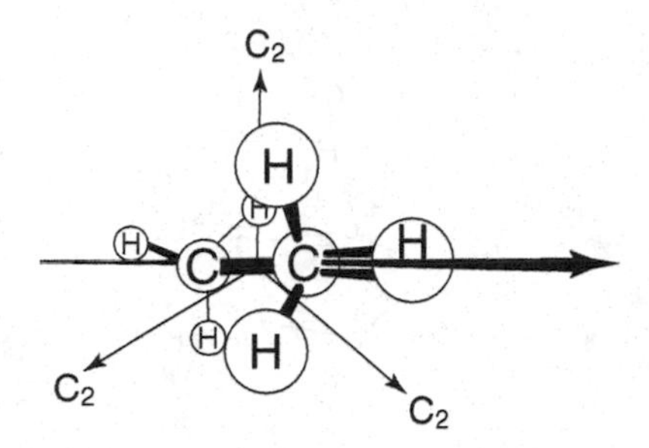

Figure A3.16 An ethane molecule which is neither eclipsed nor staggered, of D_3 symmetry.

D_4	E	$2C_4$	C_2	$2C_2'$	$2C_2''$		
A_1	1	1	1	1	1		z^2, x^2+y^2
A_2	1	1	1	−1	−1	T_z; R_z	z
B_1	1	−1	1	1	−1		x^2-y^2
B_2	1	−1	1	−1	1		xy
E	2	0	−2	0	0	(T_x, T_y); (R_x, R_y)	(x, y); (zx, yz)

Notes: (1) The x axis has been taken as coincident with one C_2'.

(2) The choice between which set of two C_2 axes is called $2C_2'$ and which is called $2C_2''$ is arbitrary. If the choice opposite to that above is taken then the labels B_1 and B_2 have to be interchanged (B_1 has a character of 1 under the C_2' operations).

Example: A twisted cube (Figure A3.17).

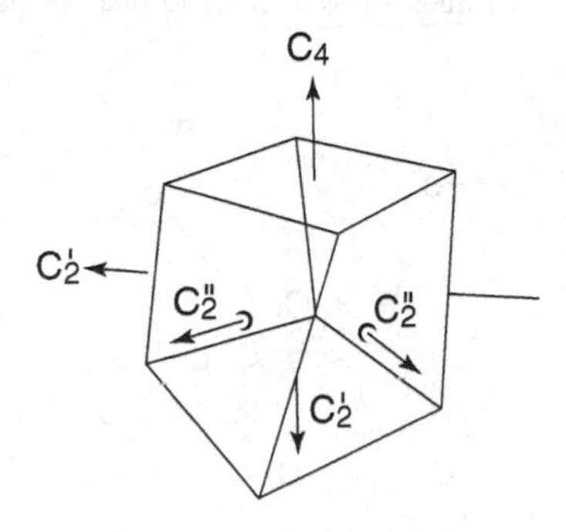

Figure A3.17 A slightly twisted cube, of D_4 symmetry.

D_5	E	$2C_5$	$2C_5^2$	$5C_2$		
A_1	1	1	1	1		$x^2+y^2; z^2$
A_2	1	1	1	−1	$T_x; R_z$	z
E_1	2	2 cos 72	2 cos 144	0	$(T_x, T_y); (R_x, R_y)$	$(x, y); (xz, yz)$
E_2	2	2 cos 144	2 cos 72	0		$\left(\frac{1}{\sqrt{2}}[x^2-y^2], xy\right)$

Note: Before working with this group refer to Note 2 under the D_{5h} character table.
Example: Ferrocene when the carbon atoms in opposite rings are neither staggered nor eclipsed (Figure A3.18).

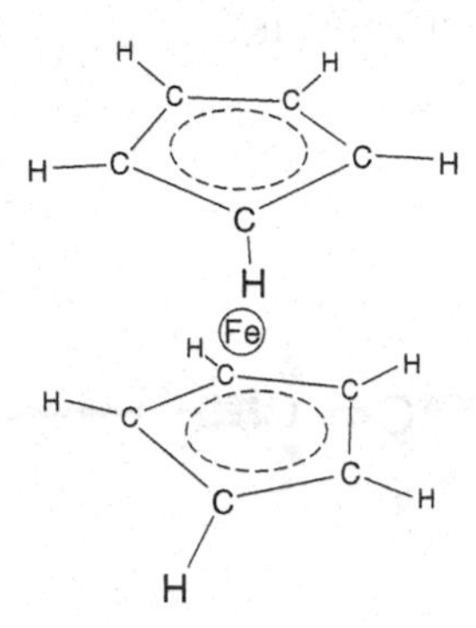

Figure A3.18 A ferrocene molecule, $Fe(C_5H_5)_2$, in which the two rings are neither staggered nor eclipsed, of D_5, symmetry.

D_6	E	$2C_6$	$2C_3$	C_2	$3C_2'$	$3C_2''$		
A_1	1	1	1	1	1	1		$x^2+y^2; z^2$
A_2	1	1	1	1	−1	−1	$T_z; R_z$	z
B_1	1	−1	1	−1	1	−1		
B_2	1	−1	1	−1	−1	1		
E_1	2	1	−1	−2	0	0	$(T_x, T_y); (R_x, R_y)$	$(x, y); (xz, yz)$
E_2	2	−1	−1	2	0	0		$\left(\frac{1}{\sqrt{2}}[x^2-y^2], xy\right)$

Notes: (1) The x axis has been taken as coincident with one C_2'.
(2) The choice between which set of three C_2 axes is called $3C_2'$ and which is called $3C_2''$ is arbitrary. If the choice opposite to that above is taken then the labels B_1 and B_2 on functions will have to be interchanged (B_1 has a character of 1 under the C_2' operations).
Example: Dibenzenechromium when the carbon atoms in opposite rings are neither staggered nor eclipsed (Figure A3.19).

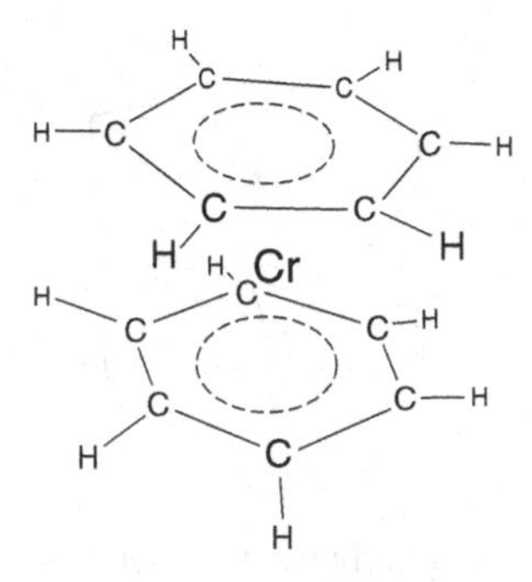

Figure A3.19 A dibenzenechromium molecule, $Cr(C_6H_6)_2$, in which the two rings are neither staggered nor eclipsed, of D_6, symmetry.

6 THE GROUPS C_{nv}

This set of groups is of considerable importance since the groups involved are of common occurrence. One problem that frequently occurs with this is that of ambiguity about the choice of x and y (z presents no problems). Thus, for C_{2v}, a change of choice interchanges the meaning of the labels B_1 and B_2. Related problems arise for all C_{nv} groups with n even.

C_{2v}	E	C_2	σ_v	σ_v'		
A_1	1	1	1	1	T_z	$z;\ z^2;\ x^2;\ y^2$
A_2	1	1	−1	−1	R_z	xy
B_1	1	−1	1	−1	$T_y;\ R_x$	$y;\ zx$
B_2	1	−1	−1	1	$T_x;\ R_y$	$x;\ yz$

Notes: (1) x is taken as lying in σ_v'.
(2) Interchange of the labels σ_v and σ_v' or (equivalently) interchange of choice of direction of x and y axes (one lies in each mirror plane) interchanges the labels B_1 and B_2 (B_1 has a character of 1 under the σ_v operation).
Example: The water molecule. See Chapters 2 and 3.

C_{3v}	E	$2C_3$	$3\sigma_v$		
A_1	1	1	1	T_z	$z;\ z^2;\ x^2+y^2$
A_2	1	1	−1	R_z	
E	2	−1	0	$(T_x, T_y);\ (R_y, R_x)$	$(x, y);\ (zx, yz);\ \left(xy, \frac{1}{\sqrt{2}}[x^2-y^2]\right)$

Example: The ammonia molecule. See Chapter 6.

C_{4v}	E	$2C_4$	C_2	$2\sigma_v$	$2\sigma_v'$		
A_1	1	1	1	1	1	T_z	$z; z^2; x^2+y^2$
A_2	1	1	1	−1	−1	R_z	
B_1	1	−1	1	1	−1		x^2-y^2
B_2	1	−1	1	−1	1		xy
E	2	0	−2	0	0	$(T_x, T_y);\ (R_y, R_x)$	$(x, y);\ (zx, yz)$

Notes: (1) x is taken as lying in one σ_v plane.
(2) Interchange of the labels σ_v and σ_v' (and the choice is arbitrary) leads to an interchange of the labels B_1 and B_2 (B_1 has a character of 1 under the σ_v operations).
Example: The BrF_5 molecule (Chapter 5).

C_{5v}	E	$2C_5$	$2C_5^2$	$5\sigma_v$		
A_1	1	1	1	1	T_z	$z;\ x^2+y^2;\ z^2$
A_2	1	1	1	−1	R_z	
E_1	2	2 cos 72	2 cos 144	0	$(T_y, T_x);\ (R_x, R_y)$	$(x, y);\ (xz, yz)$
E_2	2	2 cos 144	2 cos 72	0		$\left(\frac{1}{\sqrt{2}}[x^2-y^2], xy\right)$

Note: Before working with this group refer to Note (2) under the D_{5h} character table.
Example: η^5-cyclopentadienecarbonylnickel (Figure A3.20).

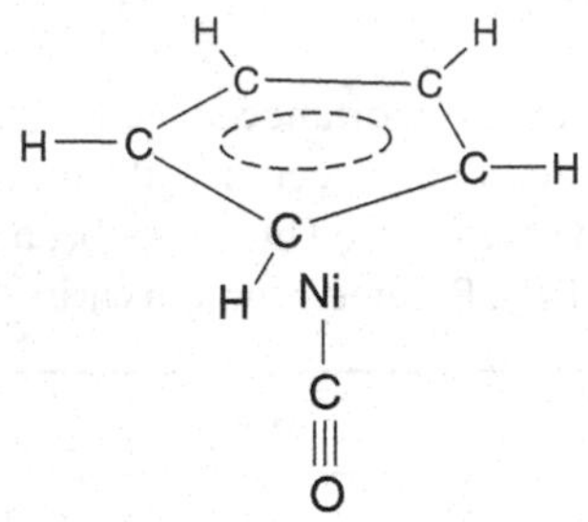

Figure A3.20 The η^5-cyclopentadiene-carbonylnickel molecule, $Ni(C_5H_5)CO$, of C_{5v} symmetry.

C_{6v}	E	$2C_6$	$2C_3$	C_2	$3\sigma_v$	$3\sigma_v'$		
A_1	1	1	1	1	1	1	T_z	$z; x^2+y^2; z^2$
A_2	1	1	1	1	−1	−1	R_z	
B_1	1	−1	1	−1	1	−1		
B_2	1	−1	1	−1	−1	1		
E_1	2	1	−1	−2	0	0	$(T_x, T_y); (R_y, R_x)$	$(x, y); (xz, yz)$
E_2	2	−1	−1	2	0	0		$\left(\frac{1}{\sqrt{2}}[x^2-y^2], xy\right)$

Note: Interchange of the labels σ_v and σ_v' (and the choice is arbitrary) leads to an interchange of the labels B_1 and B_2 on functions (B_1 has a character of 1 under the σ_v operations).
Example: The compound η^6-hexamethylbenzene, η^6-benzenechromium (Figure A3.21).

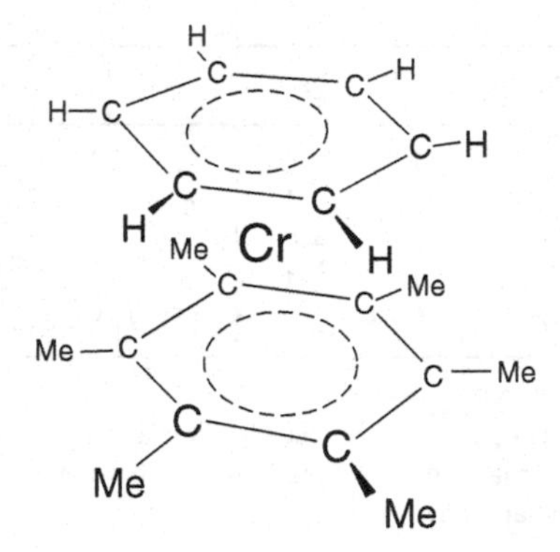

Figure A3.21 The molecule η^6-hexamethylbenzene, η^6-benzenechromium, $Cr(C_6Me_6)(C_6H_6)$ in which the two ring are eclipsed, of C_{6v} symmetry.

7 THE GROUPS C_{nh}

These groups have a derivation similar to that of the D_{nh} group – they are direct products of C_n with either C_i (n even) or C_s (n odd). The only one which has been found to be of real chemical importance is C_{2h}; however, C_{3h} is included to give an example of the n odd case.

C_{2h}	E	C_2	i	σ_h		
A_g	1	1	1	1	R_z	$x^2;\ y^2;\ z^2; xy$
B_g	1	−1	1	−1	$R_x;\ R_y$	$yz;\ zx$
A_u	1	1	−1	−1	T_z	z
B_u	1	−1	−1	1	$T_x;\ T_y$	$z;\ y$

Note: This group is a direct product of the C_2 and C_i groups, indicated in the character table by lines.

C_{3h}	E	C_3	C_3^2	σ_h	S_3	S_3^5		
A'	1	1	1	1	1	1	R_z	$x^2+y^2; z^2$
E'	$\begin{cases}1\\1\end{cases}$	ε / ε^2	ε^2 / ε	$\begin{cases}1\\1\end{cases}$	ε / ε^2	ε / ε	(T_x, T_y)	$(x, y); \left(\frac{1}{\sqrt{2}}[x^2-y^2], xy\right)$
A''	1	1	1	−1	−1	−1	T_z	z
E''	$\begin{cases}1\\1\end{cases}$	ε / ε^2	ε^2 / ε	$\begin{cases}-1\\-1\end{cases}$	$-\varepsilon$ / $-\varepsilon^2$	$-\varepsilon$ / $-\varepsilon$	(R_y, R_x)	(yz, zx)

Notes: (1) See the notes on the C_3 group for the meaning of ε and ε^2.
(2) This group is a direct product of the C_3 and C_s group, indicated in the character table by the lines.

8 THE GROUPS C_n

These are cyclic groups with character tables that look rather strange when compared with most of those encountered earlier in this Appendix. They only look strange when compared with other point groups. For many other groups – for instance, in the translation groups encountered in theories of crystal structure, but not discussed in Chapters 13 and 14 - the appearance of complex numbers is the norm. The physical meaning of exponential characters is explicitly discussed in Section 10.4.

Chapter 11 gives detailed examples of working with the C_4 group.

C_2	E	C_2		
A	1	1	$T_z;\ R_z$	$z;\ z^2;\ y^2;\ x^2;\ xy$
B	1	−1	$T_x;\ T_y;\ R_x;\ R_y$	$x; y; z; yz; xz$

C_3	E	C_3	C_3^2		
A	1	1	1	T_z; R_z	$z; x^2+y^2; z^2$
E	$\begin{cases}1\\1\end{cases}$	ε ε^2	ε^2 ε	(T_x, T_y); (R_x, R_y)	(x, y); $\left(\frac{1}{\sqrt{2}}[x^2-y^2], xy\right)$; (yz, zx)

Note: In this and two of the next three tables the notation that

$$\varepsilon = \exp\left(\frac{2\pi i}{n}\right) = \cos\left(\frac{2\pi}{n}\right) + i\sin\left(\frac{2\pi}{n}\right)$$
$$\varepsilon^2 = \exp\left(\frac{-2\pi i}{n}\right) = \cos\left(\frac{2\pi}{n}\right) + i\sin\left(\frac{2\pi}{n}\right)$$

is used, so that here (n = 3).

$$\varepsilon = \cos 120 + i \sin 120 = -\tfrac{1}{2} + i\sqrt{\tfrac{3}{2}}$$
$$\varepsilon^2 = \cos 120 - i \sin 120 = -\tfrac{1}{2} - i\sqrt{\tfrac{3}{2}}$$

Note that ε and ε^2 are complex conjugates. That is, in the case of $n = 3$, $\varepsilon^2 = \varepsilon^*$; the ε^2 notation has been used in the C_3, T, T_h and C_{3h} character tables.

C_4	E	C_4	C_2	C_4^3		
A	1	1	1	1	T_z; R_z	$z; x^2+y^2; z^2$
B	1	−1	1	−1		
E	$\begin{cases}1\\1\end{cases}$	i $-i$	−1 −1	$-i$ i	(T_x, T_y); (R_x, R_y)	(x, y); (yz, zx)

Note: It is easy to show by substitution in the equations given above that for $n = 4$, $\exp(2\pi i/n) = i$
Chapter 11 gives detailed examples of working with the C_4 group.

C_5	E	C_5	C_5^2	C_5^3	C_5^4		
A	1	1	1	1	1	T_z; R_z	$z; x^2+y^2; z^2$
E_1	$\begin{cases}1\\1\end{cases}$	ε ε^*	ε^2 ε^{2*}	ε^{2*} ε^2	ε ε	(T_x, T_y); (R_x, R_y)	(x, y); (yz, zx)
E_2	$\begin{cases}1\\1\end{cases}$	ε^2 ε^{2*}	ε^* ε	ε ε^*	ε ε^2		$\left(\frac{1}{\sqrt{2}}[x^2-y^2], xy\right)$

Note: For a definition of ε etc., see under C_3.

C_6	E	C_6	C_3	C_2	C_3^2	C_6^5		
A	1	1	1	1	1	1	T_z; R_z	$z; x^2+y^2; z^2$
B	1	−1	1	−1	1	−1		
E_1	$\begin{cases}1\\1\end{cases}$	ε ε^*	$-\varepsilon^*$ $-\varepsilon$	−1 −1	$-\varepsilon$ $-\varepsilon^*$	ε^* ε	(T_x, T_y); (R_x, R_y)	(x, y); (yz, zx)
E_2	$\begin{cases}1\\1\end{cases}$	$-\varepsilon^*$ $-\varepsilon$	$-\varepsilon$ $-\varepsilon^*$	1 1	$-\varepsilon^*$ $-\varepsilon$	$-\varepsilon$ $-\varepsilon^*$		$\left(\frac{1}{\sqrt{2}}[x^2-y^2], xy\right)$

Note: For a definition of ε, etc., see under C_3.

9 THE GROUPS S_n (n EVEN) (INCLUDES C_i)

Another set of cyclic groups denoted S_n. These only exist for n even because odd values of n do not satisfy the requirement $(S_n)^n = E$, The S_2 group is usually labelled C_i because the operations S_2 and i are identical. (this is demonstrated in Figure 9.3).

C_i	E	i		
A_g	1	1	R_z; R_x; R_y	z^2; y^2; x^2; xy; yz; zx
A_u	1	−1	T_z; T_x; T_y	x; y; z

Note: C_i often forms a direct product with another group. In this case the g and u suffixes of the C_i group are carried into the labels of the direct product group.

S_4	E	S_4	C_2	S_4^3		
A	1	1	1	1	R_z	x^2+y^2; z^2
B	1	−1	1	−1	T_z	z
E	$\begin{cases}1 & i \\ 1 & -i\end{cases}$		$\begin{matrix}-1 \\ -1\end{matrix}$	$\begin{matrix}-i \\ i\end{matrix}$	(T_x, T_y); (R_x, R_y)	$\left(\frac{1}{\sqrt{2}}[x^2-y^2], xy\right)$ (x, y); (yz, zx)

Note: See the note under the C_4 group.

10 THE GROUP C_s AND THE TRIVIAL GROUP C_1

The group C_s, like the group C_i, often participates in a direct product group. In the case of C_s it is the post-superscript primes which carry over into the labels of the irreducible representations of the product group.

C_s	E	σ		
A'	1	1	R_z; T_x; T_y	x; y; z^2; y^2; x^2; xy
A''	1	−1	T_z; R_x; R_y	z; yz; zx

C_1	E
A	1

Note: The group C_1 is trivial because it is the symmetry of an object which has no symmetry! The only symmetry operation is the identity.
In this group no bases are listed—all bases give rise to the *A* irreducible representation!

11 THE INFINITESIMAL ROTATION (LINEAR) GROUPS $C_{\infty v}$ and $D_{\infty h}$

Molecules in which all atoms lie on a common axis demand special attention because a rotation of any magnitude about this axis is a symmetry operation. The attack which proves profitable on this problem is to regard all such rotations to be (very large) multiples of an infinitesimally small rotation. That is, there is a C_∞ axis and associated operations. The character table gives the character for the operation of rotation by an arbitrary angle φ denoted C_∞^φ. Not only is there an infinite number of operations based on C_∞ there is also an infinite number of σ_v mirror planes. Fortunately, they all fall into a single class. If the linear molecule has no centre of symmetry then the appropriate group is $C_{\infty v}$. With a centre of symmetry the group is the direct product of $C_{\infty v}$ with C_i and is denoted $D_{\infty h}$. Because the groups are infinite, the usual method of reducing a reducible representation will not work. However, reduction by inspection is usually possible. Section 10.5 discusses this problem in more detail. The alternative labels for irreducible representations for $C_{\infty v}$ and for $D_{\infty h}$ antedate the system used in this book. It is the Σ, Π, Δ system which is more commonly used.

$C_{\infty v}$	E	$2C_\infty^\phi$	....	$\infty\sigma_v$		
$A_1 \equiv \Sigma^+$	1	1	...	1	T_z	$z; x^2+y^2; z^2$
$A_2 \equiv \Sigma^-$	1	1	...	−1	R_z	
$E_1 \equiv \Pi$	2	$2\cos\phi$	...	0	$(T_x, T_y); (R_x, R_y)$	$(x, y); (zx, yz)$
$E_2 \equiv \Delta$	2	$2\cos 2\phi$	...	0		$\left(\frac{1}{\sqrt{2}}[x^2-y^2], xy\right)$
$E_3 \equiv \Phi$	2	$2\cos 3\phi$	...	0		
...	.	.	...	.		

$D_{\infty h}$	E	$2C_\infty^\phi$	...	$\infty\sigma_v$	i	$2S_\infty^\phi$	...	∞C_2		
$A_{1g} \equiv \sum_g^+$	1	1	...	1	1	1	...	1		$x^2 + y^2; z^2$
$A_{2g} \equiv \sum_g^-$	1	1	...	−1	1	1	...	−1	R_z	
$E_{1g} \equiv \Pi_g$	2	$2\cos\phi$	...	0	2	$-2\cos\phi$	...	0	(R_x, R_y)	(zx, yz)
$E_{2g} \equiv \Delta_g$	2	$2\cos 2\phi$	...	0	2	$2\cos 2\phi$	...	0		$\left(\frac{1}{\sqrt{2}}[x^2 - y^2], xy\right)$
$E_{3g} \equiv \Phi_g$	2	$2\cos 3\phi$	...	0	2	$-2\cos 3\phi$	...	0		
...	.	.	...	.	.	.	...	.		
$A_{1u} \equiv \sum_u^+$	1	1	...	1	−1	−1	...	−1	T_z	z
$A_{2u} \equiv \sum_u^-$	1	1	...	−1	−1	−1	...	1		
$E_{1u} \equiv \Pi_u$	2	$2\cos\phi$	...	0	−2	$2\cos\phi$	...	0	(T_x, T_y)	(x, y)
$E_{2u} \equiv \Delta_u$	2	$2\cos 2\phi$	...	0	−2	$-2\cos 2\phi$	...	0		
$E_{3u} \equiv \Phi_u$	2	$2\cos 3\phi$	...	0	−2	$2\cos 3\phi$	...	0		
...	.	.	...	.	.	.	...	.		

Appendix 4 The fluorine group orbitals of π symmetry in SF_6

It is inevitable that in the application of group theory to chemistry some shortcuts exist – and are exploited – which circumvent tedious or difficult mathematics. The experienced worker can often astonish the inexperienced by their ability to write down the correct linear combinations for a new problem – with no apparent work. In this appendix, an attempt is made to reveal some of the tricks. Thus, although at first sight it seems an advantage to have high symmetry, this is sometimes not the case when carrying out a detailed calculation – for instance, there would be a considerable number of different interactions possible between two sets of triply degenerate orbitals in a bonding problem. In such a case it may help to pretend that the symmetry is lower than is in fact the case because the consequent reduced degeneracy forces a pairing between individual members of each set, thus reducing the number of interactions to be considered. Having paired the orbitals by this device, the low symmetry geometry can be forgotten and the correct point group used.

It is a similar trick which provides an alternative to the projection operator method of obtaining linear combinations of orbitals (Sections 5.7, 6.6, 7.3 and 8.4) and which proves to be easier to use in high symmetry cases. It uses knowledge of the correct combinations in a lower symmetry case to obtain those of a higher symmetry molecule, the lower symmetry group being a subgroup of the higher. There is no unique path in this approach – different workers might choose different low symmetry groups. For a given choice of subgroup there may be several equally valid ways of proceeding. Those experienced in the art develop a 'nose' which is based on a mixture of experience and the ability to anticipate problems that will be encountered along each alternative path. Something of this 'nose' will be evident in the next section where an attempt has been made to give the reasons for expecting a particular approach to be fruitful (or not, as the case may be).

In tackling the problem of generating the fluorine group orbitals of π symmetry in SF_6 – of O_h symmetry – a choice of lower symmetry group must first be made. It is usually sensible to choose the subgroup of the highest symmetry for which detailed results are available. In the present case this suggests that the C_{4v} subgroup of O_h be chosen because some ligand group orbitals for a molecule of this symmetry – BrF_5 – were obtained in Chapter 6. It is true that in that chapter only Br—F σ-bonding interactions were considered but

Symmetry and Structure: Readable Group Theory for Chemists Sidney F. A. Kettle

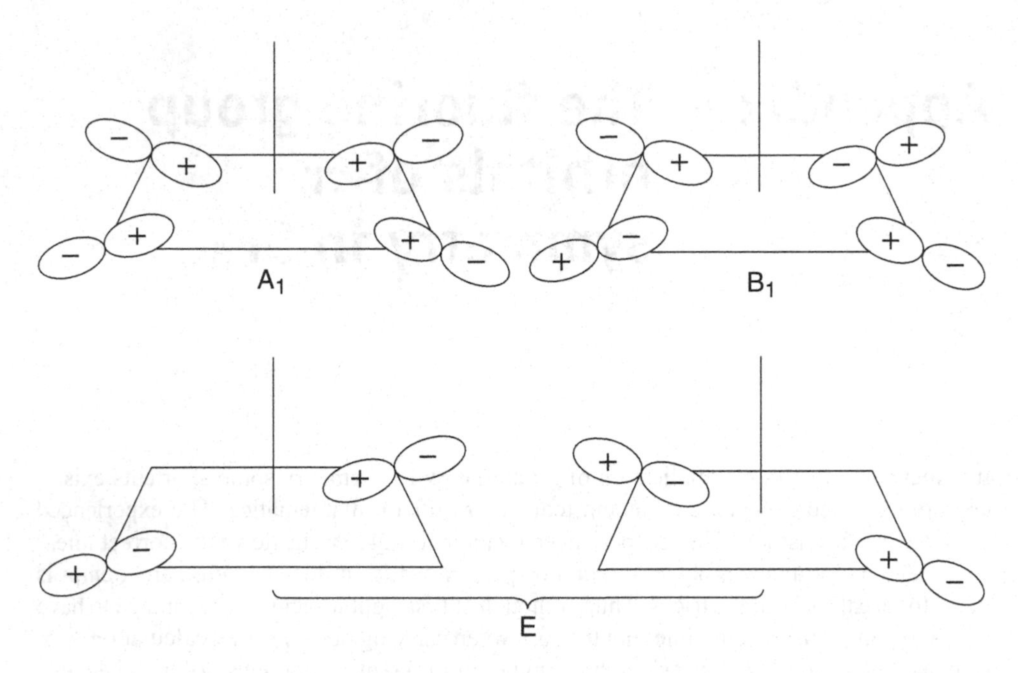

Figure A4.1 The pure p orbital representation of coplanar fluorine σ orbital symmetry-adapted combinations in BrF_5

perhaps they can be used as a base from which to obtain the π combinations. In Chapter 6 it was explicitly recognized that the fluorine σ orbitals would be mixtures of s and p atomic orbitals; for simplicity they were drawn there as pure s orbitals. In Figure A4.1 they are drawn again, but this time as pure p orbitals; the C_{4v} symmetry labels are included. Suppose the p orbitals are tilted out of the plane, as shown in Figure A4.2. The symmetry labels of Figure A4.1 remain appropriate, as do the linear combinations. This is really an indication that in BrF_5 there is no symmetry-dictated requirement that the Br—F σ-bonding orbitals have their maxima in the plane defined by the fluorine atoms. If the tilting process is now completed Figure A4.3 is obtained, which shows that the π orbital combinations have been obtained – starting from the σ! This method could be used because there is no operation in C_{4v} which interchanges – and thus compares – the 'top' with the 'bottom' of each p orbital in Figure A4.3. The trick could not have been used in the D_{4h} subgroup because the σ_h mirror plane in that group gives this comparison and so distinguishes between σ and π orbitals. Nonetheless, the combinations shown in Figure A4.3 remain correct in D_{4h} because C_{4v} is also a subgroup of D_{4h} – but the symmetry labels would have to be changed.

The next step is that of recognizing that the twelve p_π orbitals of the fluorine atoms in SF_6 can be obtained from those of the three planes of four atoms shown in Figure A4.4. Recall that the twelve π orbitals of Figure A4.4 transform as $T_{1g} + T_{1u} + T_{2g} + T_{2u}$, i.e. four different sets of triply degenerate orbitals. This triple degeneracy neatly matches the three planes and associated sets of orbitals shown in Figure A4.4. If this is exploited and the three orbitals, one from each plane, which correspond to the A_1 combination in Figure A4.3

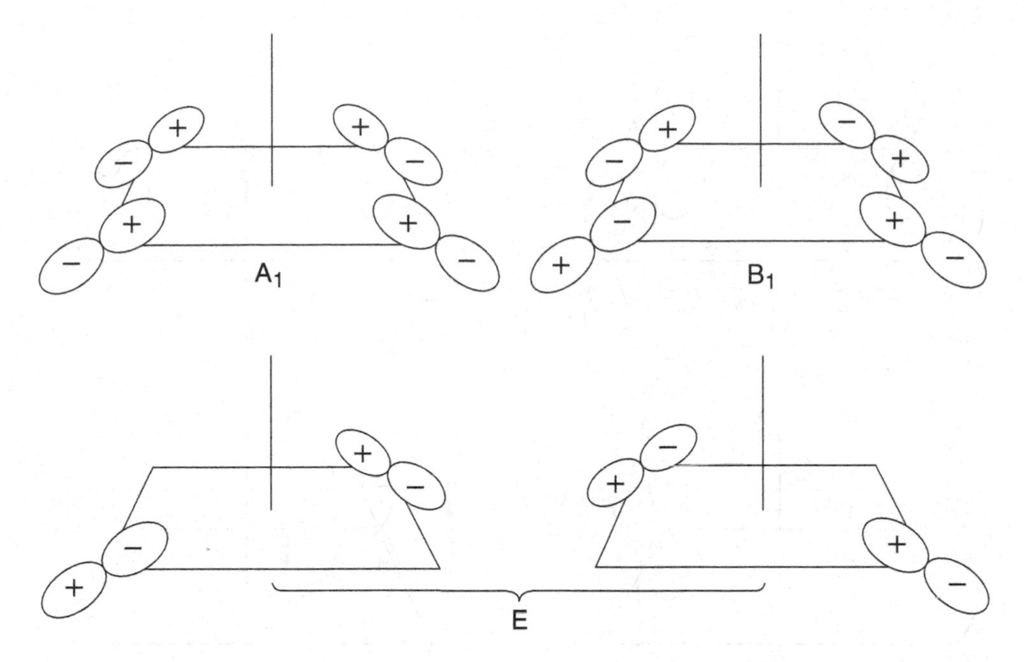

Figure A4.2 The same p orbitals as in Figure A4.1 but with each tilted out of the plane in the direction of the apical fluorine in BrF_5

are collected together:

$$\tfrac{1}{2}[p_z(B) + p_z(D) + p_z(E) + p_z(F)] \quad \leftarrow$$
$$\tfrac{1}{2}[p_y(A) + p_y(C) + p_y(E) + p_y(F)]$$
$$\tfrac{1}{2}[p_x(A) + p_x(B) + p_x(C) + p_x(D)]$$

then the T_{1u} set of ligand π orbitals in O_h is obtained, as may be checked by considering their transformations as a set. The combination shown in Figure 8.23 is indicated by an arrow. Similarly, the three combinations corresponding to the B_1 in Figure A4.3 are:

$$\tfrac{1}{2}[p_z(B) + p_z(D) - p_z(E) - p_z(F)] \quad \leftarrow$$
$$\tfrac{1}{2}[p_y(A) + p_y(C) - p_y(E) - p_y(F)]$$
$$\tfrac{1}{2}[p_x(A) + p_x(B) - p_x(C) - p_x(D)]$$

which is the T_{2u} set of ligand π orbitals in O_h, the combination shown in Figure 8.25 being arrowed.

It is at this point that anticipation cautions against plunging on and finishing the problem. There are two indications that we should pause. First, the next step would have involved three pairs of orbitals (the three sets, one from each plane, corresponding to the degenerate E set in Figure A4.3). It seems that six orbitals, apparently all degenerate, would be obtained. However, we are looking for two sets of three (T_{1g} and T_{2g}) and the two sets are not expected to be degenerate. Second, some arbitrariness has been exercised in the procedure that has been followed. In particular, when working with the planes shown in Figures A4.4

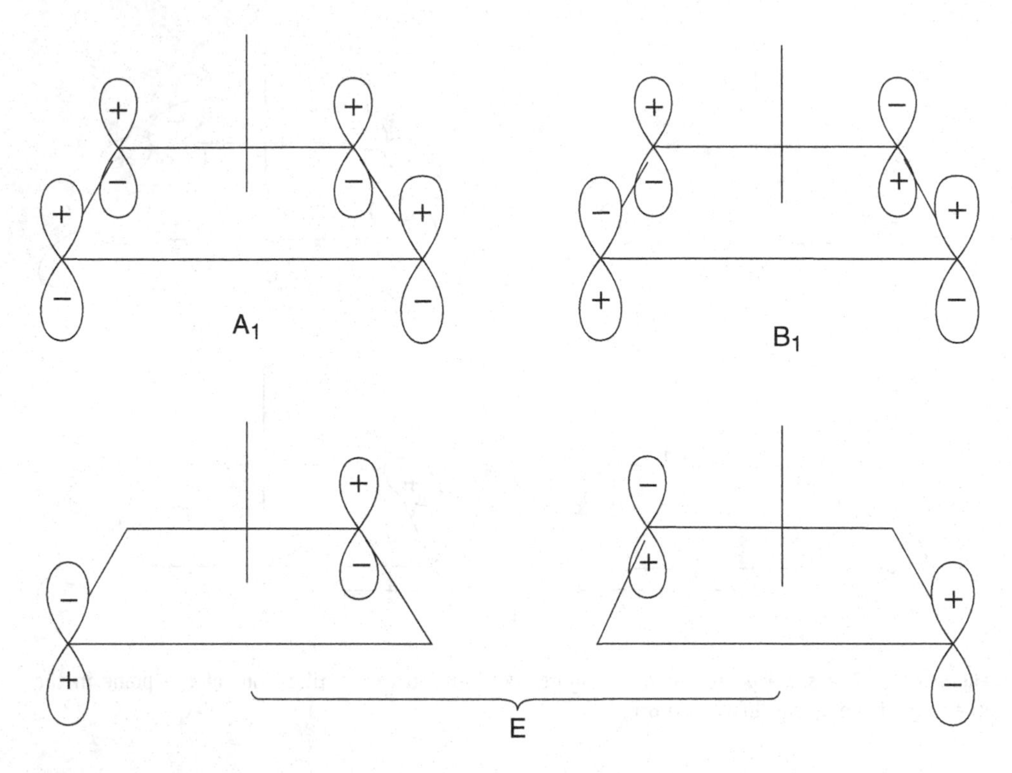

Figure A4.3 The same p orbital combinations as in Figure A4.1 but reoriented so as to be perpendicular to the original set

only p_π orbitals which have their maximum amplitude perpendicular to this plane have been considered. Equally, the choice could have been made to work with the p_π orbitals which 'lie in' the plane, as shown in Figure A4.5, although it is not immediately clear how to proceed had this alternative choice been made. Experience suggests that when six apparently degenerate orbitals are obtained, as seems to be the case here, this is because symmetry-distinct combinations have been mixed together. After all, this is always mathematically possible even if it is a step which would not be made from choice. Further, experience is that because a choice exists between 'perpendicular p_π orbitals' and 'coplanar p_π orbitals' each alternative must be expected to appear to an equal extent in the answer.

The way to extract symmetry-distinct combinations from sets in which they have been mixed together is to take suitable linear combinations of members of the mixed-up sets (a set of six orbitals in the present case). These six (un-normalized) are:

$$\psi_1 = [p_z(E) - p_z(F)] \qquad \psi_2 = [p_z(B) - p_z(D)]$$
$$\psi_3 = [p_y(A) - p_y(C)] \qquad \psi_4 = [p_y(E) - p_y(F)]$$
$$\psi_5 = [p_x(A) - p_x(C)] \qquad \psi_6 = [p_x(B) - p_x(D)]$$

Which orbitals should be combined together? It is here that the expectation of a 'coplanar' set of p_π orbitals comes to our aid. Note that the set of 'coplanar' p_π orbitals shown in

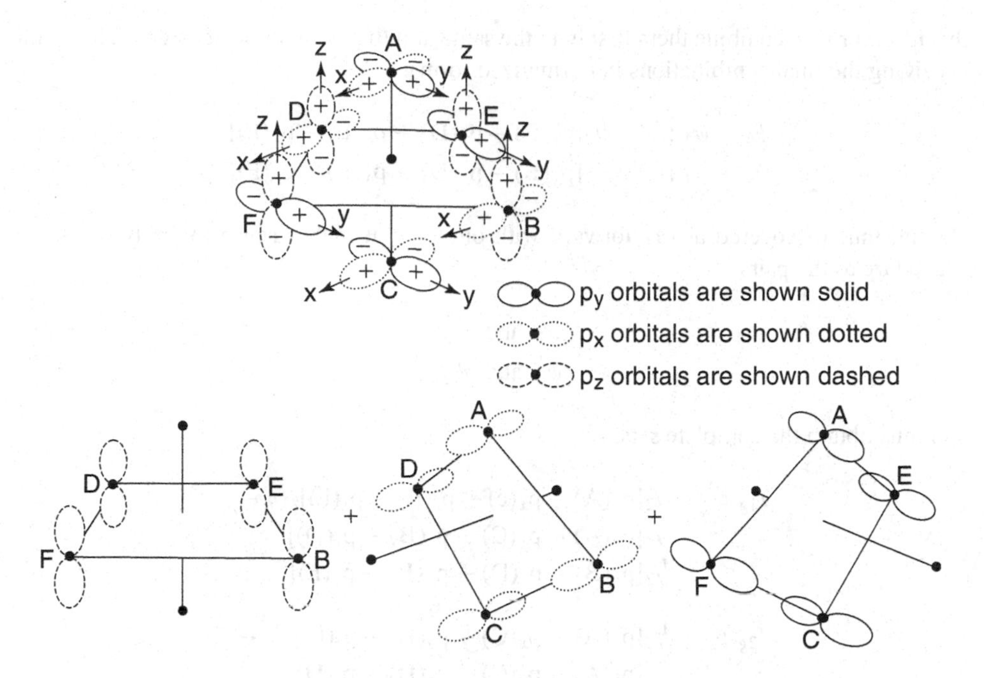

Figure A4.4 The twelve p_π orbitals of SF_6 are shown at the top (this part of the diagram is a duplicate of Figure 7.13). These twelve are the sum of the three sets of four shown at the bottom (except that in the latter we do not indicate phases). In each of the three sets the plane of four fluorine atoms is drawn, the two other fluorine atoms being represented as dots

Figure A4.5 contain contributions from

$$p_y(F), p_x(B), p_y(E) \text{ and } p_x(D)$$

that is, those orbitals contained in ψ_4 and ψ_6 in the list above. Clearly, then, we have to combine these two. Because ψ_4 and ψ_6 are symmetry-equivalent (a C_4 rotation turns ψ_4 into ψ_6) they must be expected to contribute equally to the combinations. The only way for

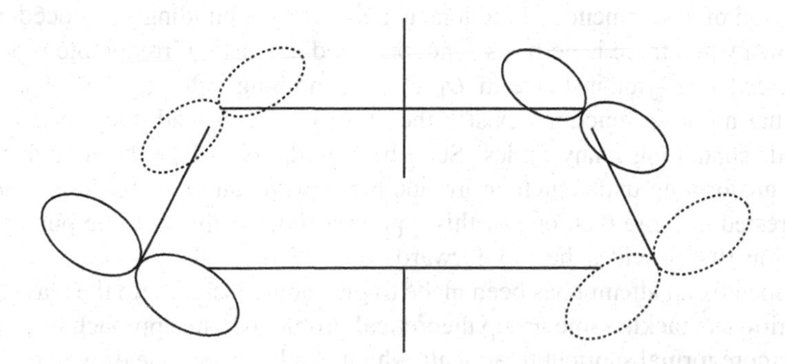

Figure A4.5 A set of four fluorine p_π orbitals which lie in the plane of the fluorine atoms (cf. Figure A4.3 and the second part of Figure A4.4, where the sets of fluorine p_π orbitals shown are all perpendicular to the plane of the four fluorine atoms)

this to occur is to combine them first with the same and then with opposite signs. The result is (giving the final combinations in normalized form):

$$\psi_6 + \psi_4: \quad \tfrac{1}{2}[\mathrm{p}_x(\mathrm{B}) - \mathrm{p}_x(\mathrm{D}) + \mathrm{p}_x(\mathrm{E}) - \mathrm{p}_x(\mathrm{F})]$$
$$\psi_6 - \psi_4: \quad \tfrac{1}{2}[\mathrm{p}_x(\mathrm{B}) - \mathrm{p}_x(\mathrm{D}) - \mathrm{p}_x(\mathrm{E}) + \mathrm{p}_x(\mathrm{F})]$$

Having thus discovered a way forward with one pair, it makes sense to apply the same procedure to the pairs:

$$\psi_1 \text{ and } \psi_5$$
$$\psi_2 \text{ and } \psi_3$$

and thus obtain the complete sets:

$$T_{1g}: \quad \tfrac{1}{2}[\mathrm{p}_x(\mathrm{A}) - \mathrm{p}_x(\mathrm{c}) + \mathrm{p}_z(\mathrm{E}) - \mathrm{p}_z(\mathrm{F})] \quad \leftarrow$$
$$\tfrac{1}{2}[\mathrm{p}_y(\mathrm{A}) - \mathrm{p}_y(\mathrm{C}) - \mathrm{p}_z(\mathrm{B}) + \mathrm{p}_z(\mathrm{D})]$$
$$\tfrac{1}{2}[\mathrm{p}_x(\mathrm{B}) - \mathrm{p}_x(\mathrm{D}) + \mathrm{p}_y(\mathrm{E}) - \mathrm{p}_y(\mathrm{F})]$$

$$T_{2g}: \quad \tfrac{1}{2}[\mathrm{p}_x(\mathrm{A}) - \mathrm{p}_x(\mathrm{C}) - \mathrm{p}_z(\mathrm{E}) + \mathrm{p}_z(\mathrm{F})] \quad \leftarrow$$
$$\tfrac{1}{2}[\mathrm{p}_y(\mathrm{A}) - \mathrm{p}_y(\mathrm{C}) + \mathrm{p}_z(\mathrm{B}) - \mathrm{p}_z(\mathrm{D})]$$
$$\tfrac{1}{2}[\mathrm{p}_x(\mathrm{B}) - \mathrm{p}_x(\mathrm{D}) - \mathrm{p}_y(\mathrm{E}) + \mathrm{p}_y(\mathrm{F})]$$

The combinations illustrated in Figure 7.22 (T_{1g}) and 7.24 (T_{2g}) are indicated by arrows.

The reader may well object that whilst a method has been given for obtaining combinations, it has not been shown that they are the ones that are required. To put it another way; how were the six combinations allocated correctly between the T_{1g} and T_{2g} sets? Formally, of course, the answer is 'by considering their transformations', but, fortunately, experience relieves us of the tedium of this step. The character table for the O_h point group given in Appendix 3 shows that T_{1g} functions have the characteristic of a rotation whilst T_{2g} functions behave like products of coordinate axes. Figure A4.6 shows how these two observations may be used as a yardstick to discriminate between T_{1g} and T_{2g} functions.

This method of assignment of functions generated by a building-up procedure appears to break down when there is no basis function listed against an irreducible representation in a character table – for instance in O_h there is nothing listed against A_{1u}. This does not mean that no basis functions exist – they always do. Instead, they tend to be rather complicated, containing many nodes. Such high nodality is usually enough to identify functions transforming under such an irreducible representation – it seldom happens that one is interested in more than one of this type at a time. If this is to be pursued in more detail, Section 10.4 provides the way forward.

In this appendix an attempt has been made to give some insight into the way that experienced practitioners tackle some group theoretical problems. The approach used is complementary to more formal shortcut treatments which can be given, one of which is described in the reprint of an article in the *Journal of Chemical Education* which follows. The case of the fluorine orbitals in SF_6 is not included in this article and the reader may find it of value to extend the treatment to include it.

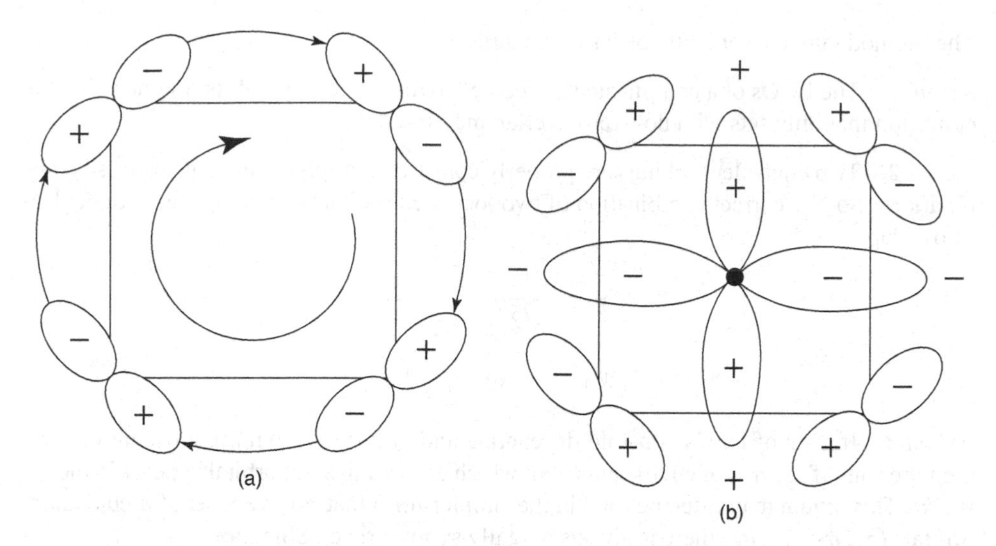

Figure A4.6 (a) A T_{1g} function compared to a rotation (central arrow). Note the arrows drawn between lobes of adjacent p_π orbitals. (b) Comparison with a typical T_{2g} function (xy, d_{xy}, etc.) – at the centre – with the nodal pattern of the corresponding T_{2g} combination of p_π orbitals

A4.1 Ligand group orbitals of complex ions[1]

Several articles which discuss ligand field theory reflect the growing interest in this refinement of simple crystal field theory (see References at end of appendix) . The reasons why covalency needs to be introduced into the latter theory are too well known to need elaboration here. Rather, we shall discuss a problem which arises in teaching the theory to undergraduate classes. As we have pointed out, the derivation of the 'correct linear combinations of ligand orbitals' in ligand field theory is a step almost invariably omitted in expositions of the subject suitable for undergraduates.[2] The reason is simple: the derivation is difficult. The derivation in the most important case – that of the octahedral complex – using a group theoretical approach has been discussed.[2] In the present article an alternative derivation is given of the form of ligand group orbitals (LGOs) which is more suitable for undergraduate tuition. In particular, no use is made of detailed group theory.

The method which we use may be termed the method of 'ascent in symmetry'. The LGOs of a complicated molecule are derived from those of simpler 'molecules' which are fragments of the complicated one, i.e. vectors appropriate to any point group are derived as linear combinations of the vectors of its subgroup, implicit use being made of the group correlation tables. The method is one of considerable power; e.g. one may obtain the LGOs for an icosahedral arrangement of equivalent σ-type orbitals in relatively simple form by this method. The standard group theoretical procedure is most unwieldy in this case, making it necessary to resort to a Schmidt orthogonalization procedure.

---

[1] Reprinted with minor alterations from an article by S.F.A. Kettle in *J. Chem. Educ.* **43** (1966), 652. Copyright 1966 by the Division of Chemical Education, American Chemical Society, and reprinted by permission of the copyright owner.

[2] S.F.A. Kettle, *J. Chem. Educ.* **43** (1996) 21.

The method may conveniently be based on three axioms:

Axiom 1 The LGOs of a complicated molecule are related to those of its fragments by the condition that only sets with non-zero overlap may interact.

Axiom 2 Two equivalent orbitals are properly considered in-phase and out-of-phase combinations. So, the correct combination of two localized orbitals σ_1 and σ_2 are, with neglect of overlap,[3]

$$\psi_s = \frac{1}{\sqrt{2}}(\sigma_1 + \sigma_2)$$

$$\psi_a = \frac{1}{\sqrt{2}}(\sigma_1 - \sigma_2)$$

Axiom 3 If a set of LGOs is d-fold degenerate and is formed from n equivalent orbitals then the sum of squares of coefficients with which each equivalent orbital appears in the set is d/n. This axiom may often be used in the simpler form that for every set of n equivalent orbitals $(\sigma_1, \sigma_2, \ldots, \sigma_n)$ there is always a totally symmetric combination:

$$\frac{1}{\sqrt{n}}(\sigma_1 + \sigma_2 + \cdots + \sigma_n)$$

We now illustrate the use of these axioms by deriving the LGO's appropriate to five different stereochemistries. In all cases we include group theoretical labels although these are not essential to the argument.

Example 1 The σ LGOs of a planar AB_3 molecule (D_{3h} symmetry). Label the σ orbitals σ_1, σ_2 and σ_3. Consider σ_1 and σ_2. From Axiom 2 the correct combinations, neglecting overlap, are:

$$\psi_s = \frac{1}{\sqrt{2}}(\sigma_1 + \sigma_2)$$

$$\psi_a = \frac{1}{\sqrt{2}}(\sigma_1 - \sigma_2)$$

By Axiom 1, of these only ψ_s can interact with σ_3 (the nodal plane implicit in ψ_a bisects σ_3). We have, then:

$$\psi_1 = \frac{1}{\sqrt{1+\lambda^2}}(\psi_s + \lambda\sigma_3)$$

and

$$\psi_2 = \frac{1}{\sqrt{1+\lambda^2}}(\lambda\psi_s - \sigma_3)$$

where the constant λ has to be determined. Now, from Axiom 3 the first of these combinations must be, in expanded form,

$$\psi_1 = \frac{1}{3}(\sigma_1 + \sigma_2 + \sigma_3)$$

[3] For consistency, a positive phase is assigned to each localized orbital.

so, by comparison of coefficients, $\lambda = 1/\sqrt{2}$ (we need only consider the positive root). It follows that

$$\psi_2 = \frac{1}{\sqrt{6}}(\sigma_1 + \sigma_2 - 2\sigma_3)$$

ψ_1 is of A_1' symmetry and ψ_a and ψ_2 together transform as E'.

Example 2 The σ LGOs of a planar AB_4 molecule (D_{4h} symmetry). Label the σ orbitals cyclically σ_1, σ_2, σ_3 and σ_4. Consider the pairs σ_1 and σ_3; σ_2 and σ_4. By Axiom 2 we have the combinations:

$$\psi_1 = \frac{1}{\sqrt{2}}(\sigma_1 + \sigma_3) \qquad \psi_2 = \frac{1}{\sqrt{2}}(\sigma_1 - \sigma_3)$$

$$\psi_3 = \frac{1}{\sqrt{2}}(\sigma_2 + \sigma_4) \qquad \psi_4 = \frac{1}{\sqrt{2}}(\sigma_2 - \sigma_4)$$

The nodal plane implicit in ψ_2 contains atoms 2 and 4. Similarly, the ψ_4 nodal plane contains atoms 1 and 3. It follows, from Axiom 1, that only ψ_1 and ψ_3 interact. From Axiom 3, one combination is

$$\psi_5 = {}^1\!/\!_2(\sigma_1 + \sigma_2 + \sigma_3 + \sigma_4), \quad \text{i.e.}\ \frac{1}{\sqrt{2}}(\psi_1 + \psi_3)$$

so the other must be

$$\psi_6 = {}^1\!/\!_2(\sigma_1 - \sigma_2 + \sigma_3 - \sigma_4), \quad \text{i.e.}\ \frac{1}{\sqrt{2}}(\psi_1 - \psi_3)$$

ψ_5 is of A_{1g} symmetry, ψ_2 and ψ_4 together transform under the E_u irreducible representation and ψ_6 is of B_{2g} symmetry.

Example 3 The σ LGOs of a tetrahedral AB_4 molecule (T_d symmetry). The derivation in this case is identical to that in Example 2. ψ_5 transforms as A_1 and ψ_2, ψ_4 and ψ_6 as T_2. In this T_2 set the Cartesian coordinates onto which ψ_2, ψ_4 and ψ_6 have a one-to-one mapping are not equivalently orientated. Two of them, those which map onto ψ_2 and ψ_4, pass through the edges of the cube corresponding to the tetrahedron, but the third passes through the mid-point of faces. The T_2 LGO set which maps onto the usual choice of Cartesian axes for the tetrahedron is:

$$\frac{1}{\sqrt{2}}(\psi_2 + \psi_4) = {}^1\!/\!_2(\sigma_1 + \sigma_2 - \sigma_3 - \sigma_4)$$

$$\frac{1}{\sqrt{2}}(\psi_2 - \psi_4) = {}^1\!/\!_2(\sigma_1 - \sigma_2 - \sigma_3 + \sigma_4)$$

and

$$\psi_6 = {}^1\!/\!_2(\sigma_1 - \sigma_2 - \sigma_3 + \sigma_4)$$

Example 4 The σ LGOs of an octahedral AB_6 molecule (O_h symmetry). We isolate four ligand σ orbitals in a plane and label them cyclically σ_1, σ_2, σ_3 and σ_4. The correct combinations for this set are given in Example 2. Above and below this plane, respectively, lie the orbitals σ_5 and σ_6.

We consider the combinations:

$$\psi_1 = {}^1\!/_2(\sigma_1 + \sigma_2 + \sigma_3 + \sigma_4)$$

$$\psi_2 = \frac{1}{\sqrt{2}}(\psi_1 - \psi_3) \qquad \psi_5 = \frac{1}{\sqrt{2}}(\psi_5 + \psi_6)$$

$$\psi_3 = \frac{1}{\sqrt{2}}(\psi_2 - \psi_4) \qquad \psi_6 = \frac{1}{\sqrt{2}}(\psi_5 - \psi_6)$$

$$\psi_4 = {}^1\!/_2(\sigma_1 - \sigma_2 + \sigma_3 - \sigma_4)$$

Axiom 1, applied by the 'nodal plane' criterion, shows that only ψ_1 and ψ_5 are non-orthogonal. The combination

$$\frac{1}{\sqrt{1+\lambda^2}}(\psi_1 + \lambda\psi_5) \text{ leads to (Axiom 3)}$$

$$\psi_7 = \frac{1}{\sqrt{6}}(\sigma_1 + \sigma_2 + \sigma_3 + \sigma_4 + \psi_5 + \psi_6) = \sqrt{\frac{2}{3}}\psi_1 + \frac{1}{\sqrt{3}}\psi_5$$

It follows, by comparison of coefficients, that $\lambda = (1/\sqrt{2})$ so that the combination:

$$\frac{1}{\sqrt{1+\lambda^2}}\,(\lambda\psi_1 - \psi_5)$$

is

$$\psi_8 = \frac{1}{\sqrt{12}}(\sigma_1 + \sigma_2 + \sigma_3 + \sigma_4 - 2\psi_5 - 2\psi_6)$$

ψ_7 is of A_{1g} symmetry, ψ_4 and ψ_8 transform as E_g and ψ_2, ψ_3 and ψ_6 as T_{1u}.

Example 5 The σ LGOs of an AB_8 Archimedean antiprismatic molecule (D_{4d} symmetry). This example again uses the results of Example 2 by considering the allowed combinations between two square planar arrangements of ligand orbitals, rotated with respect to one another by 45°. In order to use Axiom 1 the nodal planes of the two sets must be brought into coincidence. This involves the rotation of coordinate axes as discussed in Example 3. Label the ligand orbitals cyclically $\sigma_1, \ldots, \sigma_8$, those of one plane being $\sigma_1, \ldots, \sigma_4$ and those of the other $\sigma_5, \ldots, \sigma_8$. σ_5 is positioned so that viewed down the fourfold rotation axis it appears to lie between σ_1 and σ_2.
Appropriate combinations are:

$$\psi_1 = {}^1\!/_2(\sigma_1 + \sigma_2 + \sigma_3 + \sigma_4) \qquad \psi_5 = {}^1\!/_2(\sigma_5 + \sigma_6 + \sigma_7 + \sigma_8)$$

$$\psi_2 = \frac{1}{\sqrt{2}}(\psi_1 - \psi_3) \qquad \psi_6 = {}^1\!/_2(\sigma_5 - \sigma_6 - \sigma_7 + \sigma_8)$$

$$\psi_3 = \frac{1}{\sqrt{2}}(\psi_2 - \psi_4) \qquad \psi_7 = {}^1\!/_2(\sigma_5 + \sigma_6 - \sigma_7 - \sigma_8)$$

$$\psi_4 = {}^1\!/_2(\sigma_1 - \sigma_2 + \sigma_3 - \sigma_4) \qquad \psi_8 = {}^1\!/_2(\sigma_5 - \sigma_6 + \sigma_7 - \sigma_8)$$

Application of Axiom 1 shows that we must consider further combinations between the pairs:

$$\psi_1 \text{ and } \psi_5$$
$$\psi_2 \text{ and } \psi_2$$
$$\psi_3 \text{ and } \psi_7$$

The first pair gives

$$\psi_9 = \frac{1}{\sqrt{8}}(\sigma_1 + \sigma_2 + \sigma_3 + \sigma_4 + \sigma_5 + \sigma_6 + \sigma_7 + \sigma_8)$$
$$\psi_{10} = \frac{1}{\sqrt{8}}(\sigma_1 + \sigma_2 + \sigma_3 + \sigma_4 - \sigma_5 - \sigma_6 - \sigma_7 - \sigma_8)$$

while use of Axiom 3, in its more detailed form, shows that the correct combinations of ψ_2 and ψ_6 are

$$\psi_{11} = \frac{1}{\sqrt{2}}(\psi_1 - \psi_3) + \frac{1}{2\sqrt{2}}(\sigma_5 - \sigma_6 - \sigma_7 + \sigma_8)$$
$$\psi_{12} = \frac{1}{\sqrt{2}}(\psi_1 - \psi_3) - \frac{1}{2\sqrt{2}}(\sigma_5 - \sigma_6 - \sigma_7 + \sigma_8)$$

and of ψ_3 and ψ_7

$$\psi_{13} = \frac{1}{\sqrt{2}}(\psi_2 - \psi_4) + \frac{1}{2\sqrt{2}}(\sigma_5 + \sigma_6 - \sigma_7 - \sigma_8)$$

$$\psi_{14} = \frac{1}{\sqrt{2}}(\psi_2 - \psi_4) - \frac{1}{2\sqrt{2}}(\sigma_5 + \sigma_6 - \sigma_7 - \sigma_8)$$

since it is evident that ψ_{11} and ψ_{13} must be degenerate, as must also be ψ_{12} and ψ_{14}. The symmetries of these combinations are

$$\psi_4 \text{ and } \psi_5\text{: } E_2$$
$$\psi_9\text{: } A_1$$
$$\psi_{10}\text{: } B_2$$
$$\psi_{11} \text{ and } \psi_{13}\text{: } E_1$$
$$\psi_{12} \text{ and } \psi_{14}\text{: } E_3$$

Example 6 As an example of the application of the method to combinations of ligand orbitals of diatomic π symmetry we consider the LGOs of π symmetry in a tetrahedral complex. Although the standard technique can be used to obtain the correct combinations, in practice the calculation is rather difficult.

We choose axes and orientations as shown in Figure A4.7. One set of ligand π orbitals (labelled α) is 'coplanar' with the z axis. The other set (labelled β) lies in planes perpendicular to the z axis. If the x and y axes are chosen as shown some slight simplification

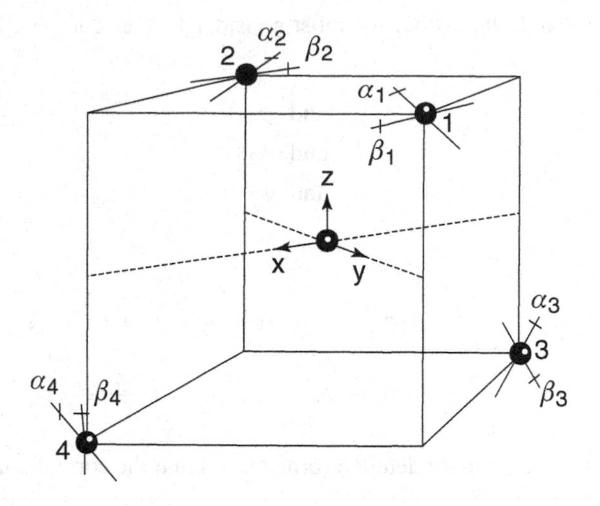

Figure A4.7 Cartesian axes and orientation of ligand π orbitals in a tetrahedral complex ion

results. Our basic combinations are

$$\psi_1 = \frac{1}{\sqrt{2}}(\alpha_1 + \alpha_2) \qquad \psi_2 = \frac{1}{\sqrt{2}}(\alpha_1 - \alpha_2)$$

$$\psi_3 = \frac{1}{\sqrt{2}}(\alpha_3 + \alpha_4) \qquad \psi_4 = \frac{1}{\sqrt{2}}(\alpha_3 - \alpha_4)$$

$$\psi_5 = \frac{1}{\sqrt{2}}(\beta_1 + \beta_2) \qquad \psi_6 = \frac{1}{\sqrt{2}}(\beta_1 - \beta_2)$$

$$\psi_7 = \frac{1}{\sqrt{2}}(\beta_3 + \beta_4) \qquad \psi_8 = \frac{1}{\sqrt{2}}(\beta_3 - \beta_4)$$

Following the usual procedure it is readily seen that

$$\psi_9 = \frac{1}{\sqrt{2}}(\psi_1 + \psi_3) = {}^1\!/_2(\alpha_1 + \alpha_2 + \alpha_3 + \alpha_4)$$

$$\psi_{10} = \frac{1}{\sqrt{2}}(\psi_1 - \psi_3) = {}^1\!/_2(\alpha_1 + \alpha_2 - \alpha_3 - \alpha_4)$$

are orthogonal to all other combinations. It follows that they must be members of degenerate sets because they contain no β component (cf. Axiom 3). Consider ψ_9. This obviously transforms like the z axis and so will be a member of a triply degenerate set of which the other components transform as x and y. It follows that $d/n = 3/8$. However, the coefficient of the α's, squared, in ψ_9 is $^1\!/_4$ so there must be an α component in the x and y transforming members. Evidently, these components are derived from $\psi_2(y)$ and $\psi_4(x)$, each of which must appear with a coefficient of $^1\!/_2$ ($^3\!/_8 - {}^1\!/_4 = {}^1\!/_8 = {}^1\!/_2[(1/\sqrt{2})]^2$). Now, ψ_6 transforms like x and ψ_8 like y so we are evidently seeking combinations of ψ_6 with ψ_4 and of ψ_8 with

ψ_2. The correct combinations are

$$\psi_{11} = \frac{\sqrt{3}}{2}\psi_6 - \frac{1}{2}\psi_4 = \frac{1}{2\sqrt{2}}[\sqrt{3}(\beta_1 - \beta_2) - \alpha_3 + \alpha_4]$$

and

$$\psi_{12} = \frac{\sqrt{3}}{2}\psi_8 - \frac{1}{2}\psi_2 = \frac{1}{2\sqrt{2}}[\sqrt{3}(\beta_3 - \beta_4) - \alpha_1 + \alpha_2]$$

where we have been careful to make sure that the phases of ψ_6 and ψ_4 mapping onto x and of ψ_8 and ψ_2 onto y are identical.

Combinations of ψ_6 and ψ_4 and of ψ_8 and ψ_2 orthogonal to ψ_{11} and ψ_{12} are

$$\psi_{13} = \frac{1}{2}\psi_6 + \frac{\sqrt{3}}{2}\psi_4 = \frac{1}{2\sqrt{2}}[\beta_1 - \beta_2 + \sqrt{3}(\alpha_3 - \alpha_4)]$$

and

$$\psi_{14} = \frac{1}{2}\psi_8 + \frac{\sqrt{3}}{2}\psi_2 = \frac{1}{2\sqrt{2}}[\beta_3 - \beta_4 + \sqrt{3}(\alpha_1 - \alpha_2)]$$

We have only to deal with ψ_5 and ψ_7, which are not orthogonal; the correct combinations are

$$\psi_{16} = ½(\beta_1 + \beta_2 + \beta_3 + \beta_4)$$
$$\psi_{17} = ½(\beta_1 + \beta_2 - \beta_3 - \beta_4)$$

The symmetries of these combinations are

$$\psi_9, \psi_{11}, \psi_{12}: T_2$$
$$\psi_{13}, \psi_{14}, \psi_{16}: T_1$$
$$\psi_{10}, \psi_{17}: E$$

References

F.A. Cotton, *J. Chem. Educ.* **41** (1964) 466.
H.B. Gray, *J. Chem. Educ.* **41** (1966) 2.
S.F.A. Kettle, *J. Chem. Educ.* **43** (1966) 21.
A.D. Liehr, *J. Chem. Educ.* **39** (1962) 135.
W. March and W.C. Fernelius, *J. Chem. Educ.* **38** (1961) 192.

Appendix 5 The Hermann–Mauguin notation

The symbolism used in the vast majority of applications of symmetry to chemistry is the Schönflies notation; it is the one used in this book. There are others; so, those reading the physics literature may well meet irreducible representations labelled Γ, ε_1, Γ_2 and so on. When dealing with the solid state, and invariably in crystallography, the notation used is the Hermann–Mauguin. In this book, the notation began to be used in Chapter 13 and, in more detail, in Chapter 14. Appendix 6 uses it extensively. This appendix largely consists of two tables which provide a connection between the Schönflies and the Hermann–Mauguin notation, although a brief descriptive introduction follows.

Whereas the concern of the group theoretician is with symmetry operations, the concern of the crystallographer is more with symmetry elements – the symmetry elements associated with a crystal structure influence both the pattern of diffracted beams in an x-ray structure determination and also which occur and which are 'missing' (systematic absences). As far as point groups are concerned, the Hermann–Mauguin notation is most simply thought of as offering an approximate answer to the question 'for a given point group which are the symmetry elements needed to uniquely define it?'. The idea here is that, as group multiplication tables show, the multiplication of symmetry operations (performing them one after the other) may lead to the generation of other symmetry operations. So, what is the smallest list of elements which uniquely specifies a particular group? Whilst the answer to this question is not provided by the Hermann–Mauguin notation, this notation certainly approximates it. Before illustrating this, a point of symbolism. In place of C_n, the notation uses n. So, C_4 is replaced by 4, C_3 by 3 and a group which contains distinct fourfold and threefold axes would be denoted 43 (remember, we are dealing with crystallographic point groups so there is no confusion with a forty-three-fold axis because one can never exist). It so happens that these two axes serve to uniquely identify the point group. This group is the one that, in Schönflies notation, is called O. Actually, in Hermann–Mauguin notation, this group is called 432, but this merely serves to illustrate the point that the notation sometimes only approximates to the minimal defining symmetry element set. Mirror planes containing a C_n axis are denoted m. So, C_{2v} becomes $2mm$ (a minimal set would be $2m$ or mm). When the mirror plane is perpendicular to a C_n axis the notation is n/m. So, the point group C_{2h} is written $2/m$, a minimal set (it is pronounced 'two over m'). A centre of symmetry is never explicitly indicated unless it is the only non-trivial symmetry element (the group C_i), when

Symmetry and Structure: Readable Group Theory for Chemists Sidney F. A. Kettle

it is written $\bar{1}$; centres of symmetry are indicated by the bar, so a threefold rotation inversion axis is denoted $\bar{3}$. Most likely to be encountered are the short Hermann–Mauguin symbols but usually longer, more explicit, ones exist. Thus, the normal, short, notation which is equivalent to O_h is *m*3*m*, the more explicit version of which is 4/*m* $\bar{3}$2/*m* (4/*m* means that perpendicular to each 4, C_4, axis is a mirror plane). In most cases the long symbols describe the symmetry characteristics of the *x*, *y* and *z* coordinate axes in this sequence; the short symbols are a contraction in which this sequence may well not appear. For cubic space groups, where all three coordinate axes are equivalent, symmetry-distinct axes are detailed (for instance, a fourfold and a threefold). A few minutes spent with this appendix (perhaps by covering up the Schönflies and working out the Schönflies equivalents of the Hermann–Mauguin) is a good way of gaining some familiarity with the Hermann–Mauguin (H–M) notation.

Point group operations

This table details the connection between individual symmetry operations (or elements) in the two systems.

Table A5.a

HERMANN-MAUGUIN	SCHÖNFLIES	Comments
1	E	
$\bar{1}$	i	A bar over a symbol in the H–M notation indicates inversion in a centre of symmetry.
2	C_2	
3	C_3	Care has to be taken to ensure that rotation is in the same sense in the two notations.
3_2	C_3^2	This is particularly important when S_4 or S_6 axes are present (see below).
4	C_4	
4_3	C_4^3	
6	C_6	
6_5	C_6^5	
m	σ	Unlike σ, the m symbol never carries suffixes.
$\bar{3}$	S_6^5	It is very important to recognise that the common practise is to use different definitions for the operations in the two notations (H–M; rotation+inversion: S; rotation+reflection) leading to apparently perverse correspondences. These can be reduced by defining rotation to be in opposite senses in the two notations (not done here).
$\bar{3}^2$	S_6	
$\bar{4}$	S_4^3	
$\bar{4}^3$	S_4	
$\bar{6}$	S_3^2	
$\bar{6}^5$	S_3	

Point groups

Although extension to other cases is straightforward, in practice, use of the Hermann–Mauguin notation is normally confined to the thirty-two crystallographic point groups. Only these are given here.

Table A5.b

HERMANN-MAUGUIN	SCHÖNFLIES	Comments
1	C_1	
$\bar{1}$	C_i	
2	C_2	
m	C_s	
$\frac{2}{m}$	C_{2h}	
222	D_2	
mm2	C_{2v}	In H–M sometimes called mm
$\frac{222}{mmm}$	D_{2h}	In H–M sometimes called mmm
4	C_4	
$\bar{4}$	S_4	
$\frac{4}{m}$	C_{4h}	
422	D_4	
4mm	C_{4v}	
$\bar{4}2m$	D_{2d}	
$\frac{444}{mmm}$	D_{4h}	In H–M sometimes called $\frac{4}{mmm}$
3	C_3	
3	S_6	
32	D_3	
3m	C_{3v}	Beware confusion with m3
$\bar{3}\frac{2}{m}$	D_{3d}	In H–M sometimes called $\bar{3}m$
6	C_6	
$\frac{3}{m}$	C_{3h}	
$\frac{6}{m}$	C_{6h}	
622	D_6	
6mm	C_{6v}	
$\frac{3}{m}m2$	D_{3h}	In H–M sometimes called $\bar{6}m2$
$\frac{622}{mmm}$	D_{6h}	In H–M sometimes called $\frac{6}{mmm}$
23	T	
$\frac{2}{m}\bar{3}$	T_h	In H–M often called m3 (beware confusion with 3m)
432	O	
$\bar{4}3m$	T_d	
$\frac{4}{m}\bar{3}\frac{4}{m}$	O_h	In H–M often called m3m

Appendix 6 Non-symmorphic relatives of the point group D_2

In Chapter 13 it was pointed out that both C_2 axes and mirror planes could be associated with one half of a primitive translation and so the interplay of these with each other increases the number of non-symmorphic space groups associated with a crystallographic point group. As a brief introduction both to this and to the detailed derivation of non-symmorphic space groups, in this appendix some of the space groups associated with the D_2 point group will be studied. This group has several advantages. In Table 13.7 it appeared without complications (under 'Orthorhombic'). Having three symmetry-distinct C_2 axes (and Cartesian axes), it will enable us to examine the interplay between different non-primitive translations. The group multiplication table of the D_2 group is small and so easily manageable. Finally, the orthorhombic system has more Bravais lattices than any other crystal system, giving an opportunity to explore all likely problem areas. However, it lacks mirror planes and so to compensate for this, as a final example, an important space group derived from the point group C_{2h} will be examined. It will also prove convenient to refer to a system containing a mirror plane within the D_2 discussion.

As a first example consider the orthorhombic crystallographic point group D_2. A diagram of axes and operations of the D_2 point group is given in Figure A6.1, in both Hermann–Mauguin and Schönflies notations (the former in projection, the latter in perspective). In both, the effect of the operations on a starting point, that labelled E, is shown. The corresponding group multiplication table is given as Table A6.1a. A feeling for this table will be essential in the development that follows. In particular, its implications for the way that the point $[x, y, z]$[1] is converted by the symmetry operations to the four other coordinate sets – and the way that the operations of D_2 convert these points into each other – need to be well understood. The reader would be well advised to stop at this stage and relate each entry in Table A6.1a to the corresponding coordinate transformations in Figure A6.1. To skip this step now may well be to invite problems later!

[1] A word of caution; the same symbols are being used to indicate an axis, as in $2(y)$, and a general point, as in x, y, z. This has been done because of the familiarity of this usage; with this word of caution no confusion should result.

Symmetry and Structure: Readable Group Theory for Chemists Sidney F. A. Kettle

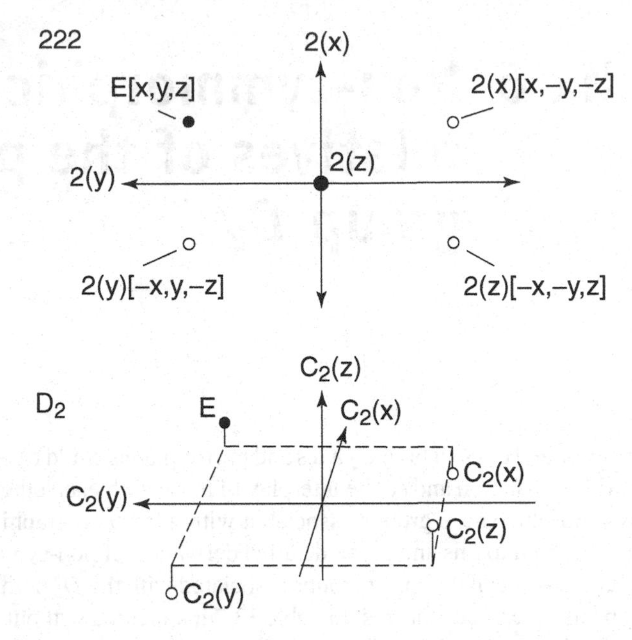

Figure A6.1 The symmetry axes of the point group D_2 (Schönflies notation), 222 (Hermann–Mauguin notation), using the symbolism appropriate to the particular notation. Whilst a perspective view has been adopted for the Schönflies diagram (lower), that given for the Hermann–Mauguin is that which will be adopted in the following figures. Note the representation of twofold rotation axes, particularly that viewed 'end-on' (that along z). In this diagram, as in the following figures, a general point in space is denoted by a solid circle and those into which it is converted by empty circles. For each, the relevant operation and coordinates are given. In the following figures it will be important to follow these carefully; it will often be helpful to compare the entries in them with those given here

Problem A6.1 Relate each entry in Table A6.1a to the corresponding coordinate transformations in Figure A6.1.

To help in understanding the symbolism Table A6.1a will now be repeated as Table A6.1b using the Hermann–Mauguin notation. This notation uses the symbol 1 for the identity but to avoid any ambiguity E will continue to be used.

Table A6.1a

D_2	E	$C_2(x)$	$C_2(y)$	$C_2(z)$
E	E	$C_2(x)$	$C_2(y)$	$C_2(z)$
$C_2(x)$	$C_2(x)$	E	$C_2(z)$	$C_2(y)$
$C_2(y)$	$C_2(y)$	$C_2(z)$	E	$C_2(x)$
$C_2(z)$	$C_2(z)$	$C_2(y)$	$C_2(x)$	E

Table A6.1b

D_2	E	2(x)	2(y)	2(z)
E	E	2(x)	2(y)	2(z)
2(x)	2(x)	E	2(z)	2(y)
2(y)	2(y)	2(z)	E	2(x)
2(z)	2(z)	2(y)	2(x)	E

Figure A6.1 and Table A6.1 refer to the symmorphic space group P222 (Primitive with three 2-fold axes), D_2^1 (the first D_2 listed). The table will now be repeated yet again but replacing some or all of the 2 rotation axes (this should be read 'twofold rotation axes') by 2_1 screw axes ('two-one screw axes'). One, two or all three of the 2 can be replaced by a 2_1 – and, because no coordinate axis has any unique properties compared with any other, they can be replaced in any order. The choices to be adopted for the three examples detailed in this appendix are given in Table A6.2.[2]

The first row in Table A6.2 corresponds to the elements in the multiplication table, Table A6.1b. To obtain a multiplication table corresponding to 'Choice 1' (P$2_1$22, D_2^2) it seems sensible to substitute $2_1(x)$ for $2(x)$ in Table A6.1b and then to ask whether this is a correct step to take. This substitution has been made in Table A6.3.

At first sight this table may appear fine, but in fact all of the entries in bold typeface present problems. All involve the $2_1(x)$ operation in some way or other. The first encountered in reading the table is the identity element resulting from the combination of two $2_1(x)$ operations. Although the entry in Table A6.3 is E, the actual outcome of combining the two $2_1(x)$ operations is a primitive translation along x. It will be necessary to return to this problem later. In the same row as this E are $2(z)$ and $2(y)$ entries, resulting from the combination of $2_1(x)$ with $2(y)$ and $2(z)$, respectively. Where has the non-primitive component of the $2_1(x)$ operation gone – there is no indication of it in the final answer? Two other entries in bold typeface are $2_1(x)$ operations which, apparently, have to result from the combination of two operations which do not have any translation component! How can this be – if it can be? The answer is that it can, indeed, be. The reason is the existence of a flexibility noted in Chapter 13 which must now be developed further.

As an aside, consider a space group derived from the C_{2h} point group in which the 2 is replaced by a 2_1. The combination of reflection in the mirror plane with inversion in the centre of symmetry of C_{2h} must now equal 2_1, rather than 2. But neither the mirror plane nor the centre of symmetry is associated with a non-primitive translation! Where can the 1 in 2_1 come from? The translation component in the combination of operations arises because this is an example of the centre of symmetry being displaced by a quarter of a primitive lattice translation (Figure A6.2). In this particular case, not surprisingly, the displacement of the centre of symmetry is along the 2_1 axis. It is important to note that the displacement takes the centre of symmetry out of the mirror plane. As a result, there is no longer a point through which all of the symmetry elements pass. We are no longer talking about a point group. Indeed, it is by no means clear that we are talking about a group at all (we can be,

[2] Although the choice of substitutions that follows has an evident logic, it is not that used in *International Tables* – in Choice 1, for instance, the 2_1 is there taken to be along z, not x.

Table A6.2

D_2	E	2(x)	2(y)	2(z)
Choice 1	E	2_1(x)	2(y)	2(z)
Choice 2	E	2_1(x)	2_1(y)	2(z)
Choice 3	E	2_1(x)	2_1(y)	2_1(z)

and this is a topic covered in Chapter 14 and again below). Is it an inevitable consequence of having non-primitive translations associated with point group operations that some of the corresponding symmetry elements no longer pass through a point? The answer is 'yes', so that the problems of the sort under discussion are common to all of the crystallographic point groups associated with non-symmorphic space groups. Can such non-coincidences resolve the problems associated with Table A6.3? Not surprisingly, the answer is 'yes'.

We now return to the D_2 non-symmorphic space groups. There are two ways in which the argument could be developed. Either the problem of the non-coincidence of symmetry elements could be treated as one requiring an answer to the question 'what is the displacement required?' or the answer to the question could simply be presented. In the latter case all would become clear in solving the problem of what is meant by 'a group' in the present context. In fact, the two approaches are linked as something of a circular argument is involved. In Figure A6.3 is given the answer for the Choice 1 (Table A6.3) space group, ($P2_122$, D_2^2). $2(y)$ and $2(z)$ are displaced relative to each other along x, although each still cuts the $2_1(x)$ axis. Also shown in Figure A6.3 are the effects of the three non-trivial symmetry operations on the point $[x, y, z]$. It was in preparation for diagrams like this that practice with Figure A6.1 was strongly recommended! In Figure A6.3 the origin has been taken as the intersection of $2_1(x)$ and $2(y)$; a primitive displacement along the x axis, a, has been indicated and set symmetrically about the origin. The $2(z)$ axis is displaced from the origin by $a/4$. This displacement could be either 'up' or 'down'; the latter has been chosen. A table which is the equivalent of Table A6.1b will now be generated but, instead of giving symmetry operations, it gives the coordinates of the points generated. This is Table A6.4. If Figure A6.1 has been thoroughly studied then the compilation of Table A6.4 should not prove unduly difficult – certainly, all the y and z entries are those appropriate for Figure A6.1. In any event, the reader should stop and check the entries in Table A6.4 (the use of Figure A6.3 is essential). There is an important point. In contrast to Table A6.3 which was obtained by simple substitution in Table A6.1b, the entries in Table A6.4 depend on whether the operations in the left-hand column operate first and are followed by those in the first row, or vice versa. Table A6.4 has been compiled with the left-hand column entries

Table A6.3

Choice 1	E	2_1(x)	2(y)	2(z)
E	E	2_1(x)	2(y)	2(z)
2_1(x)	2_1(x)	**E**	**2(z)**	**2(y)**
2(y)	2(y)	**2(z)**	E	**2_1(x)**
2(z)	2(z)	**2(y)**	**2_1(x)**	E

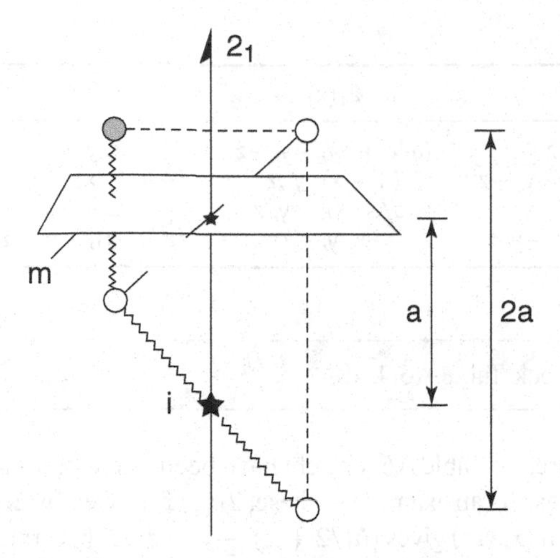

Figure A6.2 The combination of the *m* and *i* operations must be equivalent to the 2_1. Starting with the dark circle, rotating by 180° about the 2_1 axis and then translating along this axis by half a translation unit (but here denoted 2*a*) one arrives at the open circle at the bottom of the diagram. For this operation to be equivalent to *m* followed by *i* the centre of symmetry, *i* (denoted by a large star) has to be displaced from the mirror plane by one-quarter of a translation unit (here, *a*) in a direction parallel to the 2_1 axis

operating first (this makes the table easier to read). So, for instance,

$$\begin{aligned} 2_1(x)\cdot 2(y)[x,y,z] &= 2_1(x)\cdot[-x,y,-z] \\ &= \left[\frac{a}{2}+(-x),-y,-(-z)\right] \\ &= \left[\left(\frac{a}{2}-x\right),-y,z\right] \end{aligned}$$

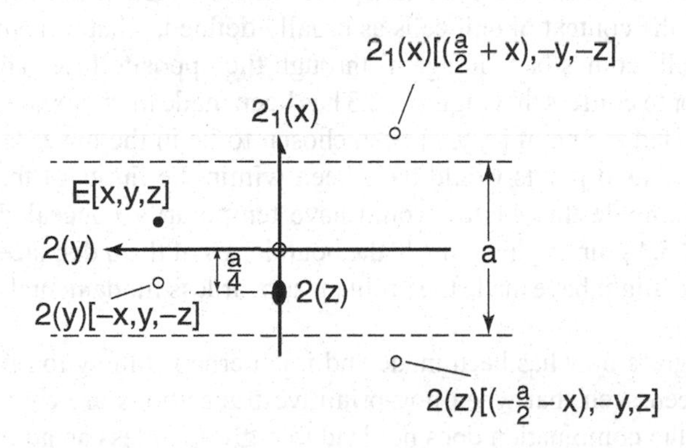

Figure A6.3 The relative arrangements of the symmetry elements of Choice 1 (those of the space group $P222_1$; note that in this label, which in the text has been treated as $P2_122$, as in the following figures, some liberty has been taken with such things as conventions for the directions of axes, always in the interests of simplifying the discussion). Note the standard convention for showing a 2_1 axis in the plane of the paper – a half-headed arrow

Table A6.4

Choice 1	E	$2_1(x)$	2(y)	2(z)
E	x, y, z	(a/2 + x), −y, −z	−x, y, −z	(−a/2−x), −y ,z
$2_1(x)$	(a/2 + x), −y, −z	(a + x), y ,z	(−a/2−x), −y, z	(−a−x), y, −z
2(y)	−x, y, −z	(a/2−x), −y, z	x, y, z	(−a/2+x), −y, −z
2(z)	(−a/2−x), −y, z	−x, y, −z	(a/2+x), −y, −z	x, y, z

Problem A6.2 Check Table A6.4.

The only problem area of Table A6.4 might have been in the bottom right-hand corner, in a square of 3×3 entries. As an example of these, $2_1(x)$ followed by $2(y)$ can be expanded as follows: $2_1(x)$ acting on (x, y, z) gives $[(a/2 + x), -y, -z]$. $2(y)$ acting on these coordinates (remember, $2(y)$ changes (x, y, z) into $(-x, y, -z)$) gives $[-(a/2 + x), -y, z]$, the same as the entry for $2(z)$, as it should. But there are problems. So, the opposite sequence, $2(y)$ followed by $2_1(x)$, gives $[(a/2 - x), -y, z]$, a different result – but D_2 is an Abelian group, and so they should be the same. The difference between the two is a, twice the non-primitive translation along the x axis, as are all differences that might be encountered in compiling Table A6.4, and indeed any other similar table compiled for a non-symmorphic space group. The differences are all members of the translation group. Because of their presence, the set of operations E, $2_1(x)$, $2(y)$ and $2(z)$ do not form a group. To form a group, a set of elements has to map onto itself when combined according to the rules of multiplication of that group – the 'closure' requirement, discussed in detail in Appendix 1. The problem is that members of another group, the translation group, are obtained and their presence prevents closure. Can the rules of multiplication of the would-be-group be modified so that this problem is avoided and the would-be-group is turned into a real group? The answer is 'yes'. The multiplication is made 'moduli primitive translations'. That is, it is made impossible to obtain primitive translations by the simple expedient of defining it to be so! Physically, and in the context of unit cells as usually defined, whatever 'moves out' of one face of the unit cell 'comes back in again' through the opposite face. This is perhaps the point for the author to confess that Figure A6.3 has been made more apparently difficult than need be the case. Had the point $[x, y, z]$ been chosen to lie in the lower left-hand quadrant then all of the generated points would have been within the limits of the a displacement indicated. To have made this choice would have temporarily concealed the fact that in compiling Table A6.4 points lying outside the boundaries of the a displacement could have been produced and might have made the problem appear less fundamental than is in fact the case.

Significant progress now has been made and it is perhaps timely to review the position reached. It has been seen that when non-primitive translations are combined with point group operations the combination does not lead to a group unless an additional restriction is placed on the group multiplication rules. This restriction only makes any sense in the context of crystals and so it is only in this context that these groups have meaning. Further, it was seen that whereas there is only one D_2 point group, it is possible to create three more groups by adding non-primitive translations – these are those listed in Table A6.3.

Table A6.3 is evidently exhaustive – one, two or three 2 axes can become 2_1, there are no other possibilities. To generate more non-symmorphic D_2 space groups we have to change to a centred lattice. All of this explains why it is reasonable to expect more non-symmorphic space groups than symmorphic. Whilst this expectation is justified, an important restriction working in the opposite direction must be noted: by invoking non-primitive translations in the definition of point – group-derived operations the corresponding primitive translation has automatically been defined. Yet, as has been seen in Figure 13.8, the definition of the non-primitive Bravais lattices of a particular crystal system invariably requires a different choice of primitive translation vectors to that appropriate to the corresponding primitive lattice (compare Figures 13.8a and 13.8b). This variation of choice is in potential conflict with the rigidity imposed by combining non-primitive translations with point group operations. In particular, for none of the so-called non-primitive Bravais lattices do the primitive translation vectors form an orthogonal (mutually perpendicular) set. For there to be an association of non-primitive translations with point group operations, perpendicular axes tend to be required (more on this later). So, the majority of non-symmorphic space groups are associated with primitive lattices. Of the 157, 113 are primitive and only 44 non-primitive, notwithstanding the fact that there are equal numbers of primitive and non-primitive Bravais lattices.

As an example of the consequences of there being two point group operations associated with non-primitive translations consider Choice 2 of Table A6.2. The space group is $P2_12_12$ (D_2^3). Following the pattern set by the previous example, in Figure A6.4 is shown a diagram which indicates how an original point $-E[x, y, z]-$ is transformed when it is operated upon by the set of operations of Choice 2. The screw axes have been taken as along x and y

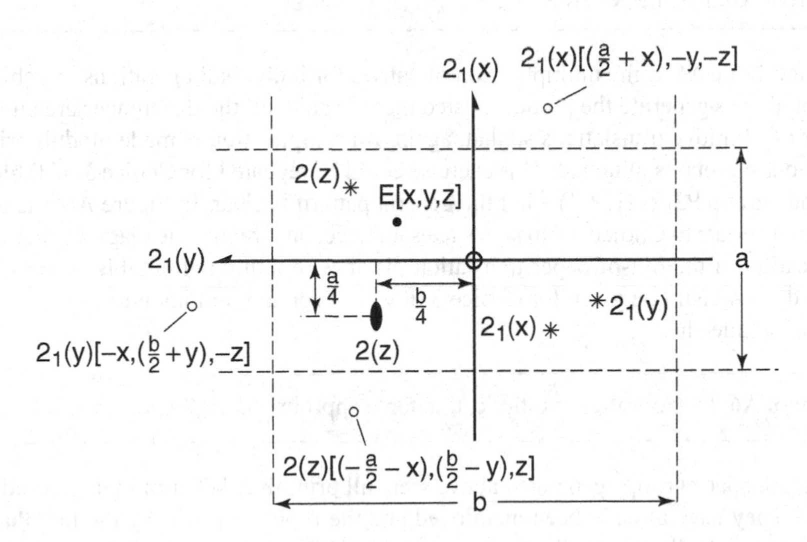

Figure A6.4 Transformations of the point $[x, y, z]$ produced by the operations $2_1(x)$, $2_1(y)$ and $2(z)$ of Choice 2 (corresponding to the space group $P22_12_1$). In this Figure, as in Figure A6.3, the transformed points are indicated by empty circles. However, in addition, the corresponding points moduli primitive translations are shown (as stars). In each case they are labelled with the same operation as that containing the primitive translations.

Table A6.5

Choice 2	E	$2_1(x)$	$2_1(y)$	$2(z)$
E	x, y, z	(a/2+x), −y, −z	−x, (b/2+y), −z	−(a/2+x), (b/2−y), z
$2_1(x)$	(a/2+x), −y, −z	(a+x), y, z	(a/2−x), −(b/2+y), z	−x, −(b/2−y), −z
$2_1(y)$	−x, (b/2+y), −z	−(a/2+x), (b/2−y), z	x, (b+y), z	(a/2+x), (b−y), −z
$2(z)$	−(a/2+x), (b/2−y), z	−(a+x), (b/2+y), −z	−(a/2−x), −y, −z	x, y, z

and their intersection has been chosen as origin. Perhaps predictably, $2(z)$ is displaced from this origin by translations along two axes, by $a/4$ and $b/4$ (in the previous example where there was just a single screw axis it was displaced by translation along a single axis). Again, as in the previous example, it could have been arranged that all transformed points fall within the bounds set by the primitive translation units shown in Figure A6.4, that is, within the area bounded by a and b. This could have been achieved by placing both the identity point, $[x, y, z]$, and $2(z)$ in the lower left-hand quadrant. A less comfortable arrangement has, in fact, been chosen in which all generated points (indicated by empty circles) fall outside the a, b bounds, safe in the knowledge that the 'moduli primitive translations' requirement means that there are equivalent points within these bounds; the latter have been indicated by stars in Figure A6.4. The actual coordinate changes associated with Choice 2 are shown in Table A6.5, which, like Table A6.4, is compiled with the operation in the left-hand column operating on the coordinates implied by the entry in the first row.

Problem A6.3 Check Table A6.5.

Just as for Table A6.4, the multiplication of entries for individual operations in Table A6.5 does not always generate the products listed there. Again, all the differences are an integer number of primitive translations so that, again, if multiplication is made moduli primitive translations a group is obtained. This exercise could be repeated for Choice 3 of Table A6.2 (the space group $P2_12_12_1$, D_2^4) – but the general pattern is clear. In Figure A6.5 is given a figure appropriate to Choice 3. No screw axes intersect and, really, the diagram should have an indication of the out-of-paper translation distance, c. Although a table akin to Tables A6.3 and A6.4 could be given for Choice 3, it would contain nothing new in principle and so is not included here.

Problem A6.4 Compile a multiplication table appropriate to Choice 3 of Table A6.2.

The three space groups generated above were all primitive. What of their centred counterparts? They have already been mentioned and the problem posed by the fact that their axes are not mutually perpendicular was recognized. The relevant Bravais lattices are the body-centred, all-face-centred and one-face-centred orthorhombic, shown in Figure 13.8b as D, E and F. We start with the latter, the one-face-centred Bravais lattice. Following the convention that z is unique, the face that is chosen to be centred is that perpendicular to this

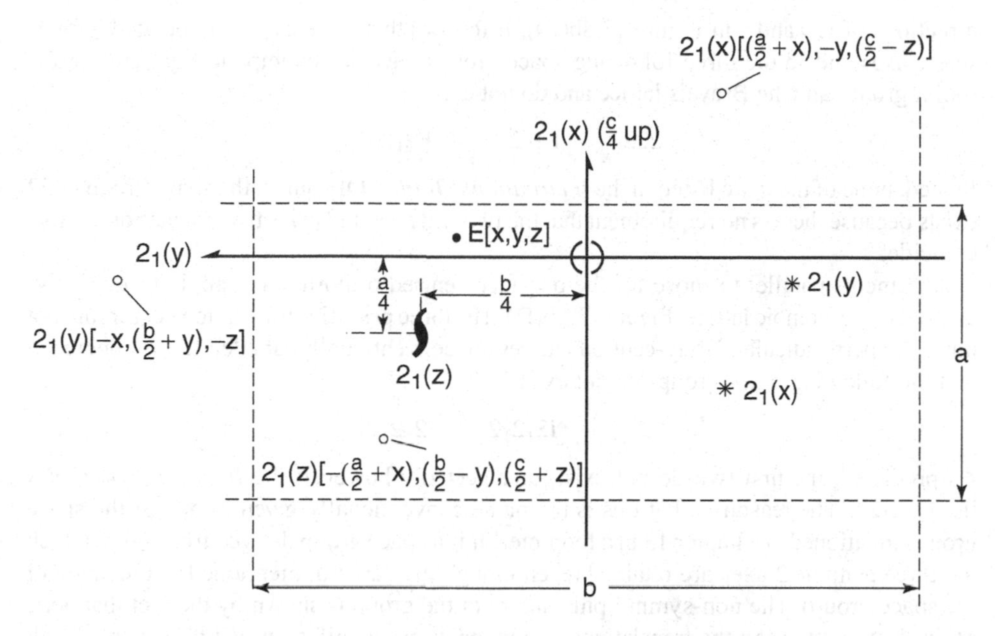

Figure A6.5 Transformations of the point $[x, y, z]$ produced by the operations $2_1(x)$, $2_1(y)$ and $2_1(z)$ of Choice 3 (corresponding to the space group $P2_12_12_1$). Note the way that a 2_1 axis is conventionally shown when viewed end on (the $2_1(z)$). In this figure, to emphasize the fact that a primitive translation does not have to be chosen to be symmetrically placed with respect to symmetry elements, one has been chosen asymmetrically placed.

axis. It is denoted C, as opposed to A and B which would characterize the other possibilities. So, the question to be answered is 'which of the following exist?':

$$C222 \qquad C222_1 \qquad C2_12_12 \qquad C2_12_12_1$$

Clearly, the first presents no problems – it was included when counting the symmorphic space groups (Table 13.7). What of the others? Here the guiding principle is that lattice and 'point group' (quotes are used to indicate that this is a convenient, but not quite correct, term) must be mutually compatible. They will be incompatible if, for example, one requires that the primitive translation vectors all be perpendicular when the other requires that one pair are not perpendicular. Now, as Figure 13.8b(F) shows, the translation vectors that define the one-face-centred orthorhombic lattice are not all mutually perpendicular (P_3 is perpendicular to P_1 and P_2 but this pair are not mutually perpendicular – if they were, a tetragonal lattice would result). So, whilst P_3 could be associated with a 2_1 axis, the other two could not. That is, a space group $C222_1$ is expected but not $C2_12_12$ or $C2_12_12_1$ – these last two require that two and three, respectively, of the primitive translation vectors defining the lattice be oriented along twofold axes of the Bravais lattice, but, as has been seen, only one is. And, indeed, there is a space group $C222_1$, D_2^5, and none of the others.

Turning now to the all-face-centred orthorhombic Bravais lattice (Figure 13.8b(E)), none of the primitive translation vectors are directed along twofold axes (these latter are in the directions of x, y and z in Figure 13.8b(E)). It follows that none can be associated with 2_1 screw axes and so the three following space groups have an incompatibility between the 'point group' and the Bravais lattice and do not exist:

$$F222_1 \qquad F2_12_12 \qquad F2_12_12_1$$

Indeed, none of them are listed in the *International Tables*. Of course, the space group F222 exists because there is no requirement that the twofold axes and primitive translation vectors coincide.

Arguments parallel to those for the one-face-centred orthorhombic apply to the body-centred orthorhombic lattice, Figure 13.8b(D). The three primitive translation vectors are not mutually perpendicular. Body-centred lattices are conventionally labelled I; it is concluded that the following space groups do not exist:

$$I222_1 \qquad I2_12_12 \qquad I2_12_12_1$$

As predicted, the first two do not exist; unexpectedly however, the *International Tables* list $I2_12_12_1$! The reason is that this is the name conventionally given to one of the space groups mentioned in Chapter 13 in a footnote.[3] It is a space group derived from D_2 in which the three genuine 2 axes are retained (even though they are not mentioned in the name of the space group). The non-symmorphic nature of the group is shown by the fact that none of the 2 axes intersect; the translational components are manifest in that the 2 axes are all displaced by ¼ translations perpendicular to the direction of the twofold axis. Each twofold axis is associated with a non-primitive translation but it is not such as to generate a screw axis (despite the label given to the group).[4] There is only one other space group of this type – the cubic group conventionally labelled $I2_13$ (a group which contains genuine 2 axes).

This appendix has shown how to obtain all of the space groups that are derived from the combination of the elements of the point group D_2 (and its derivatives) with the relevant Bravais lattices. The principles used are general and so the reader should be in a position to apply them to any such combination and thus to understand the origin of each and every space group. No doubt, there will be problems of detail which will be encountered but the broad principles are in place. The major omission is that none of the examples considered contains a glide plane. This omission is remedied in the next section, where the most commonly encountered space group, $P2_1/c$, is studied.

A6.1 The space group $P2_1/c$ (C_{2h}^5)

The first and most obvious question when one encounters the symbol $P2_1/c$ (pronounced 'pee two one over see') is 'what does it mean?'. The P and 2_1 should present no problems – a primitive lattice in some as-yet uncertain crystal system but which certainly contains a twofold rotation axis which, in this space group, has become a 2_1 screw. This group is derived from the C_{2h} point group, which means that there must be a 1:1 correspondence between the operations of C_{2h} and the point-group-derived operations of $P2_1/c$, just as exemplified earlier

[3] It can be argued that the use of the label $I2_12_12_1$ is not ideal.

[4] The diagram given in the *International Tables* shows 2_1 screw axes but these are artefacts, arising from the doubling of the primitive group (as indicated by the I symbol). The 'real' twofold rotation axes, also shown in the *International Tables*, are the important ones!

for the case of the D_2 point group. In $P2_1/c$ the C_2 (2) of C_{2h} is replaced by a 2_1 and the σ mirror plane is replaced by a glide. In the Hermann–Mauguin notation this is represented as $2/m$ (the symbol/m indicates that the mirror plane is perpendicular to the 2 axis). Clearly, a, b and c glides exist, the labels indicating half-primitive translations in the direction of the x, y and z axes respectively. How are they to be incorporated into the symbolism? The answer is that the letter a, b or c replaces the m in $2/m$ – so that one talks of $2/a$, $2/b$ and $2/c$. When incorporated into a group, the moduli primitive translation requirement has to be applied in the group multiplication. When combined with a primitive lattice these point groups derived from C_{2h} give the space groups $P2_1/a$, $P2_1/b$ and $P2_1/c$, where, following convention, the symbol is written on a single line, and not in bold type. These three are not different space groups; the only way they differ is in the choice of axis labels. Indeed, there is a fourth equivalent choice which is quite often used in crystallography – $P2_1/n$, where the 'n' indicates a translation involving half-translations along two coordinate axes. In this case it is more than a choice of axis labels which is involved, it is a change of axes (so that, for example, an axis is chosen which lies between the 'original' x and y). This may seem rather obtuse but in fact such a choice can be crystallographically convenient. Care has to be taken, however. In the symbol which corresponds to the crystallographer's 'standard' setting, $P2_1/c$, the 2_1 axis is the y axis, not the z which would be conventional in C_{2h} when used in a non-crystallographic context. One other point, in that the 'parent' point group is C_{2h}, there must be a centre of symmetry somewhere because C_{2h} contains one. In the non-symmorphic space groups $P2_1/a$, $P2_1/b$ and $P2_1/c$ this centre of symmetry is

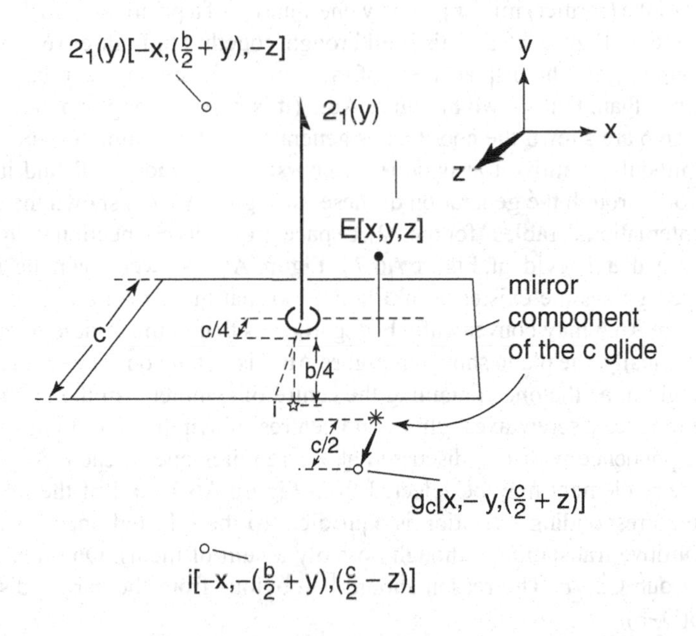

Figure A6.6 Pictorial perspective representation of the symmetry elements of the space group $P2_1/c$. The centre of symmetry is here represented by a five-point star; note its displacement from both the 2_1 and the glide plane. The glide operation is shown divided into two components; the act of reflection leads to the star; the end product of the complete operation is shown, as usual, by an empty circle

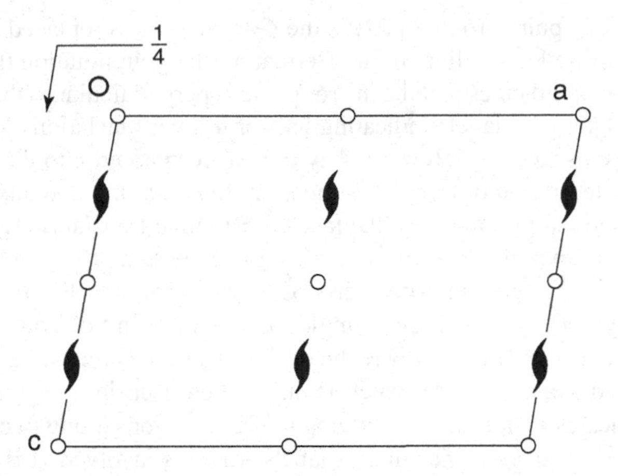

Figure A6.7 The diagram given for the space group $P2_1/c$ in *the International Tables* (viewed down the y axis). Note the way that the glide plane is $b/4$ above the plane containing the centre of symmetry is represented (top left-hand corner). Centres of symmetry are conventionally represented by circles, as here – beware confusion with a different use for this symbol in Figures A6.1–A6.6. In this diagram nine such centres of symmetry are shown but they differ only in the way that they add translations to the basic operation of inversion in any one of them. In this diagram the origin is taken as at the top left-hand corner

displaced out of the (former) mirror plane by one-quarter of a primitive translation. This has already been met in Figure A6.2. This is all brought together in Figure A6.6, which shows the space group $P2_1/c$. The displacement of the centre of symmetry is related to, but a bit more convoluted than, that shown in Figure A6.2 (it is moved along two, not one, axes).

In Figure A6.6 are shown the coordinates generated by the symmetry operations of the $2_1/c$ group (moduli primitive translations, of course). The reader will find it very useful practice to work through the generation of these. In Figure A6.7 is shown the diagram that appears in "International Tables" for the $P2_1/c$ space group; its connection with Figure A6.6 may not be immediately evident. Figure A6.7 is Figure A6.6 viewed down the y axis; unlike the latter, it recognizes the existence of a lattice, so that more than one 2_1 axis is shown (the 2_1 in Figure A6.6 may conveniently be regarded as that in the middle of the top row in Figure A6.7). Finally, the plane shown in Figure A6.7 is not the one shown in Figure A6.6; it is best thought of as the one containing the centre of symmetry in the latter. There may appear to be too many screw axes, glides and centres of symmetry in Figure A6.7 – after all, the correspondence with C_{2h} discussed above requires one of each. Any one of each sort of symmetry element may be selected from Figure A6.7 – all of the others are then such that the corresponding operation is equivalent to the selected one plus a translation (always a primitive translation, although possibly a sum of them). One final point; there is no space group $C2_1/c$. The reason should be obvious from the earlier discussion and Figure 13.8b(D–F).

Index